Computational Systems Neurobiology

Nicolas Le Novère
Editor

Computational Systems Neurobiology

Springer

Editor
Nicolas Le Novère
European Molecular Biology Laboratory
European Bioinformatics Institute
Welcome-Trust Genome Campus
Hinxton, Cambridge
United Kingdom

ISBN 978-94-017-8380-4 ISBN 978-94-007-3858-4 (eBook)
DOI 10.1007/978-94-007-3858-4
Springer Dordrecht Heidelberg London New York

Printed on acid-free paper

Springer is part of Springer Science+Business Media (www.springer.com)

Foreword

It is now well accepted that the concept "one gene-one protein-one function" is largely inadequate to understand life. The corresponding biomedical concept "one gene-one phenotype-one disease" is even more problematic. Many diseases are multifactorial, but moreover, mono- or multifactorial, the expression of almost all dysfunctions depends on the person and its environment. Systems Biology is a discipline born after second world war, based on control theory, in particular cybernetics, and systems theory. Systems Biology can be briefly summarised as the study of the emerging properties of a biological system taking into account all the necessary constituents, their relationships and their temporal variation. It relies on integration between experimental observations of different types and computer simulations of the system's dynamics. The field stayed relatively ignored during a few decades, mainly restricted to quantitative analysis of metabolic systems in 1970s and 1980s, as well as signalling pathways a bit later. The conjunction of increased computer power and high-throughput quantitative data production allowed the subject to become in a decade (1998–2008) the new paradigm in biology. A systems approach could allow significant progress in the understanding and treatment of numerous pathologies, and in particular in the field of neurological diseases. One of the consequences of the rise of Systems Biology has been a change of attitude towards mathematical models in biology (and the corresponding computer simulations). Not only are those models now used routinely, but their size has increased and they are often accompanied by an important semantic layer allowing a link to biological knowledge. Therefore it became crucial to be able to exchange and reuse those models. This need has triggered the creation of an entire domain of research and development, the "computational systems biology", comprising research groups, companies, scientific events, etc., an evolution reminiscent of the bioinformatics explosion at the beginning of 1990s.

Among the key concepts in systems biology are quantitative modelling and simulations of biological systems' behaviour, the necessity to integrate processes at different scales and the emergence at the system level of behaviours unpredictable on the basis of the behaviours of isolated components. Neurosciences occupy a unique position on the three points. Since the work of Alan Hodgkin and Andrew

Huxley half a century ago, modelling is a recognised component of research in the domain. Quantitatively describing a cellular behaviour emerging from the interaction between two different molecular components, a potassium and a sodium channel, the work of Hodgkin-Huxley can be seen as the beginning of computational systems biology. Moreover, this modelling activity is frequently pursued together with experimental work. Realistic models are developed at all scales, molecular (signalling pathways), cellular (electrophysiology, growth and differentiation), multicellular (neural networks) and tissular (cerebral electric activity). Synaptic plasticity, neuroadaptation, memory, cognition or consciousness are all emergent phenomena. Neuroscience is therefore the ideal domain where systems biology concepts can be applied.

Computational studies of neuronal systems, or computational systems neurobiology, require descriptions spanning 12 orders of magnitude in space and 24 orders of magnitude in time. Accordingly, the methods employed are extremely diverse and focus on a subset of the scale considered. When a newcomer enters the field, whether students entering Ph.D. studies or experienced researchers reorientering, the choice of reference books is overwhelming. When I started my group at the EMBL-EBI, I used to buy for each student the "yellow book", "Computational Cell Biology", by Fall et al. As the research activities of my group expanded, I progressively added other volumes, until the reading list was sufficient to populate half a shelf above each desk. Each book targeting a given audience, its content was highly technical and challenging. I dreamt of a book that I could buy for any of my students, whatever the envisioned Ph.D. topic was. This book would introduce the many different types of computational studies of neuronal systems and would aim to be an entry point for the more specific textbooks of the field, which would be chosen by each student based on their project. Since I could not find any, I decided to contact my colleagues and friends and built it. You have the result in your hands.

The first part of the book deals with molecular systems biology. Functional genomics is presented through examples of transcriptomics and proteomics studies of neurobiological interest. Quantitative modelling of biochemical systems is presented in homogeneous compartments and using spatial descriptions. The second part presents the various approaches to model single neuron physiology. The following part is naturally devoted to neuronal networks. The penultimate division is focused on the development of neurons and neuronal systems. The book closes on a series of methodological chapters. A significant freedom was granted to the authors, and the chapters are of very variable lengths and technical levels. The order is chosen so that one could start from the first chapter and read through until the end. However, each chapter is an independent entity, and you are welcome to read in the order you fancy.

Nicolas Le Novère

Contents

Chapter 1
Functional Genomics and Molecular Networks Gene Expression Regulations in Complex Diseases: Down Syndrome as a Case Study

Marie-Claude Potier and Isabelle Rivals

1.1 Introduction

Functional genomics has been or is being applied to complex diseases in the hope of finding molecular networks that are altered, as well as gene targets for treatment. The experiments were initiated as soon as tools were available, and this field of research has exploded with the commercialization of DNA microarrays and their relative affordability. Since the first development of DNA microarrays more than 20 years ago (Schena et al. 1995), the technology has improved in many aspects. Genome annotations are being updated and the probes associated to individual genes have been optimized for their selectivity and sensitivity. Although probe collections covering all the genes corresponding to various genomes are not fully optimized (Golfier et al. 2009) the data are improving and becoming consistent for powerful statistical analysis. The initial studies aiming at defining lists of differentially expressed genes have been disappointing and revealed that data analysis had to be extended using other tools than statistical tests. Many clustering methods and network analysis have been applied since. In parallel, gene ontology categorization has allowed a more functional view on the list of differentially expressed genes. Gene ontology now groups 37,078 terms with 23,050 for biological_process, 2,993 for cellular_component and 9,391 for molecular_function (www.geneontology.org 05/30/2012).

M.-C. Potier (✉)
CRICM, CNRS UMR7225, INSERM U975 and UPMC, Hôpital de la Pitié-Salpêtrière, 47 Bd de l'Hôpital, ICM, Room 5.041, 75013 Paris, France
e-mail: marie-claude.potier@upmc.fr

I. Rivals
ESA, ESPCI ParisTech, 10 rue Vauquelin, 75005 Paris, France
e-mail: isabelle.rivals@espci.fr

N. Le Novère (ed.), *Computational Systems Neurobiology*, DOI 10.1007/978-94-007-3858-4_1,

Nowadays, researchers who envision gene expression studies have always the same question in mind which is: what are the genes differentially expressed between various samples? But they want to know the answer beyond the list, meaning that they want to know what are the functions of these genes, and if they belong to a particular network or pathway. Knowing this pathway will eventually give them the key for tuning it. Of course, the first question to ask is: has it been done, published and deposited in public databases (GEO www.ncbi.nlm.nih.gov/geo/ or Arrayexpress www.ebi.ac.uk/microarray-as/ae/)? If the answer is yes, there are data available on the subject; one should then plan to produce a different set of data, keeping in mind that this new set will need to be integrated with data sets available for the ultimate meta-analysis. It is just not possible to ignore other data sets since the power of analysis will be increased along with the size of samples. Then, of course, data need to be comparable, meaning preferably performed on the same type of microarray and possibly on the same platform. If not, then microarray annotation becomes a real issue that will have to be improved in the future. Isn't it surprising that with 746,694 samples (30,299 series) in GEO the number of meta-analyses is so low? There are nowadays 95 datasets with 100–200 samples, 33 with 1,000–7,000 samples but none with more than 7,000 samples.

1.1.1 Alzheimer's Disease (AD)

Let us take the example of AD, a neurodegenerative disease which affects 25 millions of individuals worldwide, and which is becoming a real societal problem. Many gene profiling studies have been performed on AD patient samples (brain, peripheral cells) but no coherent picture of gene expression regulation in AD was obtained (Maes et al. 2007; Nagasaka et al. 2005 ; Blalock et al. 2004 ; Emilsson et al. 2006 ; Lu et al. 2004). One could argue that small sample size together with manipulating human tissues with artifacts associated to post-mortem delay have minimized the power of analysis because of high variability. Also analysis of brain samples with very heterogeneous cell composition brings another level of variability. One way around would be to analyze gene expression at the single cell level. Such analyses are still under development and will bring answers to this major problem (Bontoux et al. 2008). It might be though that the control of cellular function has both deterministic and stochastic elements: complex regulatory networks define stable states that drive individual cells, whereas stochastic fluctuations in gene expression ensure transitions and coherence at the population level (Macarthur et al. 2009). Stochastic "noise" arises from random fluctuations within the cell and is an unavoidable aspect of life at the single-cell level. Evidence is accumulating that this noise crucially influences cellular auto-regulatory circuits and can "flip" genetic switches to drive probabilistic fate decisions (Singh and Weinberger 2009). Stochastic noise in gene expression propagates through active, but not inactive, regulatory links and it was recently shown that extrinsic noise sources generate correlations even without direct regulatory links (Dunlop et al.

2008). In bacteria, it was shown that noise in expression of specific genes selects cells for competence, and experimental reduction of this noise decreases the number of competent cells (Maamar et al. 2007). This stochastic noise could have an impact on cell fate not only during development but also during disease progression. It is assumed that during development, cells acquire their fate by virtue of their lineage or their proximity to an inductive signal from another cell. However, cells can choose one or another pathway of differentiation stochastically, without apparent regard to environment or history, and this stochastic character could be critical for the maintenance of species (Losick and Desplan 2008). Although these aspects have been studied in bacteria and yeasts, it is still particularly difficult to explore in multicellular organisms and in diseases.

The experimental design applied to complex human diseases has focused on gene expression regulation in tissues or cultured cells, thus excluding the single cell resolution. Although stochastic gene expression was mentioned, it is not possible to differentiate single cell level noise from tissue complexity, cellular heterogeneity and inter-individual variability.

Recently, with the use of systems biology approaches, two studies have revealed new interesting molecular networks related to AD. The first study applied weighted gene coexpression network analysis (WGCNA) to microarray datasets analyzing brain samples (the CA1 region of the hippocampus) from AD patients and comparing to brain samples (frontal lobe) from normal elderly people (Miller et al. 2008). This analysis produced modules of co expressed genes that are functionally related with some relevant to disease progression and others conserved between AD and normal aging. In the second study, gene profiling of laser microdissected samples from the entorhinal cortex were analyzed slightly differently. Modules of highly correlated genes were constructed and among these genes regulatory cis elements were identified. New links have been identified between cardiovascular diseases, AD and diabetes (Ray et al. 2008).

Genome wide association studies (GWAS) have recently revealed the power of analyzing a very large number of samples (>1,000) (Harold et al. 2009; Lambert et al. 2009). Although getting genomic DNA samples is far much easier than getting brain samples, one would imagine that larger sample gene profiling datasets with less heterogeneous samples will improve the readout of the analysis.

1.1.2 Down Syndrome (DS)

We have been interested in another complex disease, namely Down syndrome. DS results from the presence in three copies of human chromosome 21, the smallest human autosome containing about 350 known protein-coding genes (Antonarakis et al. 2004; Epstein 1990; Lejeune et al. 1959). The mechanisms by which this aneuploidy produces the complex and variable phenotype observed in DS patients are still under discussion. The use of large scale gene expression methods such as microarrays were expected to shed light on which genes (within or outside chromosome 21) contribute to the DS phenotype as well as to the phenotypic variability. For the genes on chromosome 21, all studies have confirmed a general increase of transcription following the chromosomal imbalance, the "primary gene dosage effect". RNA samples prepared from cells or tissues of DS patients or mouse models showed a global over-expression of the three-copy genes (Ait Yahya-Graison et al. 2007; Amano et al. 2004; Dauphinot et al. 2005; FitzPatrick et al. 2002; Giannone et al. 2004; Lockstone et al. 2007; Mao et al. 2003, 2005; Potier et al. 2006; Saran et al. 2003). However, even if the mean over-expression we and others reported to be close to the expected value of 1.5, recent studies in DS cell lines have reported that about 70% of the three-copy genes were significantly below the 1.5 ratio. In these particular cell lines at least, a large proportion of the chromosome 21 transcripts were compensated for the primary gene dosage effect (Ait Yahya-Graison et al. 2007; Prandini et al. 2007).

As for non-chromosome 21 genes, results are less consistent. The aneuploidy of an entire chromosome could affect the expression of either a limited number of genes, or a large number in a more random and extensive way (Mao et al. 2005; Saran et al. 2003). Conversely classification of samples on the basis of their whole transcriptome has not been applied systematically in the published gene expression studies of DS. Rather it was unfortunately wrongly applied such as in Slonim et al. (2009). In this study they conclude to a widespread differential expression between trisomic and euploid samples based on clustering of genes differentially expressed between trisomic and euploid, excluding the chromosome 21 genes. It seems obvious that differentially expressed genes between two conditions would be able to differentiate the two conditions. Nevertheless this question regarding the regulation of gene expression for non-chromosome 21 genes is still debated, and more comprehensive studies assessing the variability among samples, tissues and development stages are needed.

We have designed several large scale gene expression studies in which we could measure the effects of trisomy 21 on a large number of samples in tissues or cells that are affected in DS (Dauphinot et al. 2005; Laffaire et al. 2009; Moldrich et al. 2009). All were performed with the Ts1Cje mouse model of DS which is a segmental trisomy of mouse chromosome 16 (MMU16) with many genes orthologous to human chromosome 21 (HSA21) present in three copies (about 95). This mouse model has the advantage of being available as large colonies of mice on B6C3SnF1/Orl mixed genetic background and rapidly screened (Sago et al. 2000). Experiments were designed in order to correlate gene expression changes with the phenotype observed. Two data sets focused on cerebellum since adult Ts1Cje mice show a reduction in cerebellar volume that parallel the observations in DS patients and in another mouse model of DS (Ts65Dn mice) (Baxter et al. 2000; Olson et al. 2004). The reduced size of the cerebellum and the reduced cerebellar granule cell number in Ts65Dn adults originate around birth because of a defect in granule cell precursor proliferation (Roper et al. 2006). In our studies, three early postnatal time points that are crucial for cerebellar development were investigated which could provide a read-out of genes involved in cerebellar hypoplasia in DS. These three time points correspond to birth (P0) and postnatal days 15 (P15) and

30 (P30). During the P0-P10 time period granule cells proliferate and migrate from the external to the internal granule cell layer and Purkinje cells start differentiating and growing their highly dense dendritic tree. We quantified the proliferation of granule cell precursors on fixed cerebellum slices of Ts1Cje and euploid mice at P0, P3 and P7 using immunohistochemistry and histology. A significant 30% decrease of their mitotic index was observed at P0 but not at P3 and P7, in agreement with the results obtained in Ts65Dn mice (Roper et al. 2006). Finally and in order to find gene expression variations in cerebellar regions rich in granule cell precursors, external granule cell layers of newborn Ts1Cje and euploid mice were dissected and analyzed on microarrays.

We also integrated data sets that contained a number of samples that was sufficient for statistical analysis ($n \geq 4$). These included the studies of Mao et al. and Saran et al. from 2003 (Mao et al. 2003; Saran et al. 2003). The first dataset contains gene expression profiles of human fetal cortex and cultured astrocytes from four Down syndrome cases and four controls. The second study produced gene expression profiles of the adult cerebellum from the Down syndrome mouse model Ts65Dn.

We included in the meta analysis the data set from Amano et al. (2004) from whole brain of newborn Ts1Cje mice (Amano et al. 2004), the one from 2007 of Lockstone et al. (2007) and Pevsner et al. (unpublished GEO GSE9762) on adult cortex and cultured fibroblasts respectively, from DS patients and controls. Finally, we failed to analyze the data set from Slonim et al. (2009) on uncultured amniotic fluid supernatants from DS and euploid fetuses (Slonim et al. 2009). Indeed, from all the samples published, less than 1,000 genes were expressed in all experiments, which were not representative enough for the analysis to be meaningful.

1.2 Elements of Microarray Statistical Analysis

The aim of this section is not to propose an exhaustive panorama of the existing methods for the analysis of microarray data, but rather to give the necessary and sufficient technical elements needed in order to understand and to reproduce the statistical treatments that we or the authors we cite have applied to the various data sets surveyed in this chapter.

1.2.1 Data Normalization

In addition to the variability of interest that is due to the difference between diseased (here DS) and normal tissue, observed expression levels are also subject to the variability introduced during sample preparation, the manufacture and the processing of the arrays (labeling, hybridization and scan). Even if some of this unwanted variability can be controlled using appropriate experimental design and procedures, for example by having all experiments performed at a single time point by a single operator, some of it can not be controlled, but still needs to be corrected.

The most famous of these sources is perhaps the dye bias for cDNA microarray experiments, where the efficiency, heat and light sensitivities differ for Cy3 and Cy5, resulting in a systematically lower signal for Cy3. For cDNA microarrays, the normalization procedure proposed in (Dudoit et al. 2002) was shown to be efficient. It is based on Cleveland's robust locally weighted regression for smoothing scatterplots (Cleveland 1979), and consists in fitting a *lowess curve* to the MA plot of log intensities[1] of the red and green labels and considering the residuals as the normalized log ratios.

This approach is not directly applicable to single color arrays, such as the Affymetrix or Illumina arrays considered in this chapter. However, contrarily to the current perception that the lowess normalization is only suited for normalizing two single color arrays at a time, Sasik et al. (2004) showed that lowess can indeed be applied across $n > 2$ arrays, assuming that most genes expressions do not change notably across the n experiments.

In practice, multiple lowess proves quite similar to *quantile normalization*, which is a much lighter procedure. The principle of quantile normalization is to make the distribution of the probe intensities equal to a reference distribution for each of the n arrays. This reference distribution is the mean distribution of the n arrays, computed by sorting all p probe intensities of each array in increasing order, and computing the ith reference intensity value as the mean of the ith intensity values of the n arrays. Boldstad et al. showed the efficiency of the method, which is commonly used for the normalization of Affymetrix data (Bolstad et al. 2003).

Let us illustrate this efficiency with an example exhibiting a known undesirable effect. Gene expression was measured twice on cell lines from 12 DS patients at a 2 month interval on Illumina chips with 48,701 probes, the labeling being the same for the two hybridizations. Figure 1.1a shows the raw intensity values for the 24 arrays, those of the first hybridization in black, those of the second in grey: the two groups differ visibly. One also notices differences among the first and second hybridization, the arrays being located on two different Illumina chips (there were up to six arrays on the considered Illumina chips). Figure 1.1b shows the mean distribution used for quantile normalization.

In order to demonstrate the efficiency of quantile normalization, we performed a PCA (see next section for further technical details about PCA) of the raw and of the quantile normalized data. Both are shown in Fig. 1.2. Whereas the arrays are grouped according to the hybridization when considering the raw data (Fig. 1.2a), they are clearly grouped two by two when using the normalized data (Fig. 1.2b), i.e. two arrays corresponding to the same tissue are now very close. Furthermore, the markers used for the arrays correspond to the chip they belong to. With the raw data, a chip effect can be noted (for arrays 1, 2 and 3 for example), which lessens considerably after normalization.

[1]Transforming expression data to a log scale (any base) reduces the asymmetry of the distribution of the intensities and homogenizes their variance. Here, probe intensities are systematically $\log_2$ values.

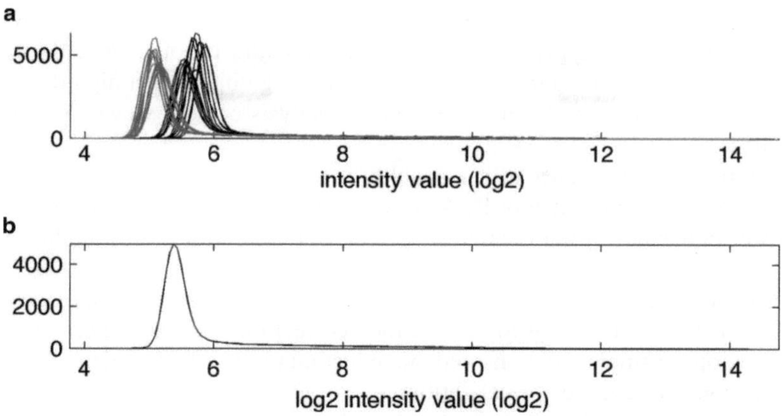

Fig. 1.1 Distributions of the probe intensities for the 12 DS patients in *black* for the first hybridization, and in *grey* for the second one. (**a**) Raw data; (**b**) Mean distribution

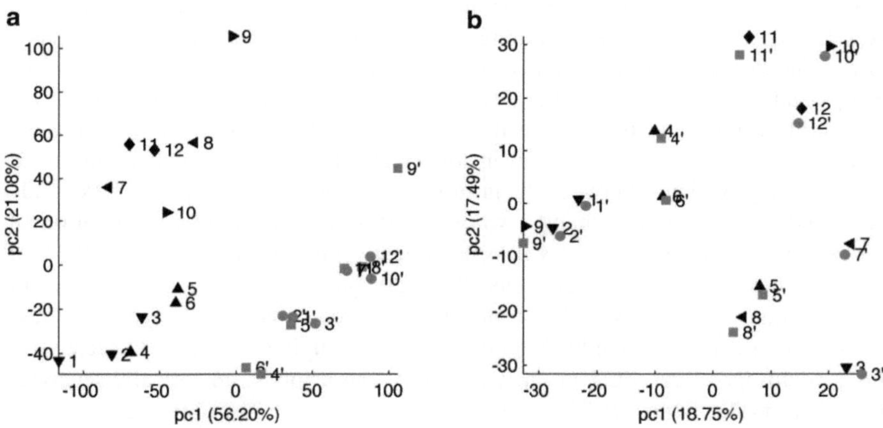

Fig. 1.2 PCA of the 24 arrays along the first two principal axes, each sample being originally represented by the intensities of 48,701 probes. The arrays that where first hybridized are shown in *black*, the second in *grey*. Identical markers denote arrays located on the same Illumina chip. (**a**) Raw data (48,701 transcripts); (**b**) Normalized data (48,701 transcripts)

This illustration using PCA leads us to the second part of the analysis, that of data visualization prior to differential analysis.

1.2.2 Dimensionality Reduction and Data Visualization

The result of a microarray experiment involving n arrays with p probes presents itself as a $n \times p$ matrix of – now normalized – intensities, which can be viewed as the representation of n tissues by the intensities of their p genes or probes (typically hundreds or thousands), or conversely, as the representation of the p probes by their expression in n tissues (typically tens or even less). In this chapter, we will focus on the first view, which raises the problem of visualizing objects in a high dimension space, see McLachlan et al. for an exhaustive analysis of both views (McLachlan et al. 2004).

A common way to reduce dimensionality is to carry out a principal component analysis (PCA): the principle of PCA is precisely to project multidimensional data to a lower dimension space retaining as much as possible of the variability of the data.

A first purpose of such a PCA prior to differential analysis is to detect outliers and possible biases, as well as to validate their correction by a proper normalization: in the previous example, PCA showed the reduction of the effect of having different hybridizations by quantile normalization.

A second goal may be to exhibit groups of tissues, especially according to the known differences between them, such as normal and DS tissues. In this context, we must insist that PCA is an unsupervised procedure, whose only property is that the projection in the d-dimension space generated by the d first principal axes has the highest variance among all possible projections in a d-dimension space. The direction of maximum variance being composed of variance *within* the groups and variance *between* the groups, the first principal components need not necessarily reflect the direction in the probe space that is best for revealing the group structure of the tissues. However, conversely, if PCA indeed reveals clusters, it implies a large variance between the groups, i.e. the presence of many differentially expressed probes. In this chapter, whose main object is the characterization of DS versus normal tissues, we will systematically present three different PCA of the data: the PCA on all transcripts, the PCA on the HSA21 chromosome (or the equivalent part of MMU16 chromosome in three copies in the case of mouse models) transcripts, and the PCA on the remaining euploid transcripts. Because of the gene dosage effect, PCA on the three-copy transcripts systematically separates normal from DS tissues. If PCA without the three-copy transcripts does, it means that the whole transcriptome is affected by DS. This might be a useful and complementary information to differential analysis, especially in the case of less powerful experiments (i.e. with too few samples) where only a few genes can be determined as significantly differentially expressed. We could have completed the PCA with a cluster analysis, however for all the data sets presented in the next section, hierarchical clustering never exhibits two separate clusters of DS and euploid samples when PCA does not (while the opposite case often occurs).

Now, a few technical details need to be clarified. The lower dimensional space used for the PCA projection is the space generated by the eigenvectors of the feature (probe) correlation matrix corresponding to its largest eigenvalues, called principal axes, see for example Johnson and Wichern (2002). In many applications, it happens that some features have completely different scalings. For example, one of the features may have been measured in meters and another one, by design or accident, in micrometers. Since eigenvalues are scale dependent, it might be appropriate in such cases to rescale all features to the same scale, which amounts to use the

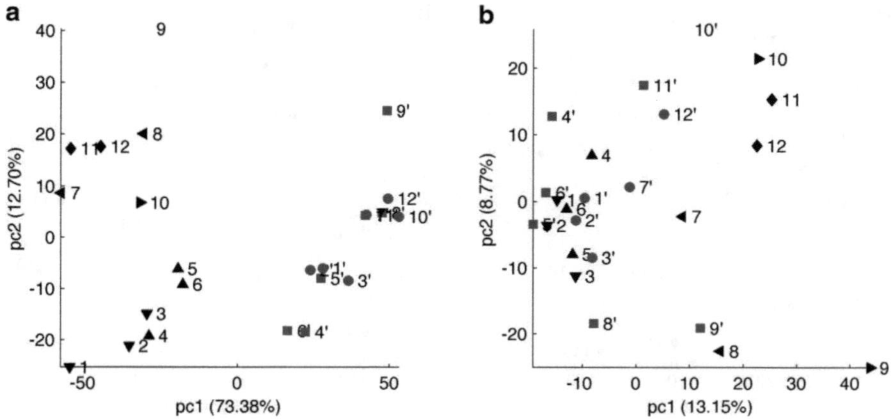

Fig. 1.3 Same as in Fig. 1.2, except that the PCA is performed on the correlation matrix (i.e. with rescaled probe intensities). (**a**) Raw data (48,701 transcripts); (**b**) Normalized data (48,701 transcripts)

correlation matrix of the features, instead of their covariance matrix. In the case of gene expression, rescaling leads to give low- or unexpressed genes (the variance of which corresponds to noise) the same importance as highly expressed genes, which is indeed not desirable.

To illustrate this, Fig. 1.3 shows the two PCA of the 12 DS tissues hybridized twice, this time with rescaled intensities. On the raw data, the main variability being due to the different hybridizations, the projection is quite similar as when performed on the un-rescaled data. But on the normalized data, where this effect has been removed, we see that we have lost the close neighborhood of the couples of arrays corresponding to the same tissue. Thus, all PCA presented in this chapter are performed on the normalized, un-rescaled probe intensities. On PCA figures, the percentage indicated in parentheses in a principal axis label corresponds to the proportion of the variance explained by this axis.

Finally, whatever the platform, intensity values are usually provided with "calls" (present, absent, marginal) and/or detection *p*-values. The PCA shown in the next section have been performed on the transcripts considered present or with expression *p*-values lower than 5% or 1% for all *n* arrays (when the *p*–values were not available, we chose a cutoff on the probe intensity so as to obtain the same proportion of expressed transcripts). For our example, 10,626 transcripts are considered present on all arrays with a threshold of 5% on the detection *p*-value, and the PCA on these 10,626 transcripts is shown on Fig. 1.4. The benefit of removing the non-expressed genes is especially noticeable on the raw data, where the couples are now visible (though still much less than on the normalized data).

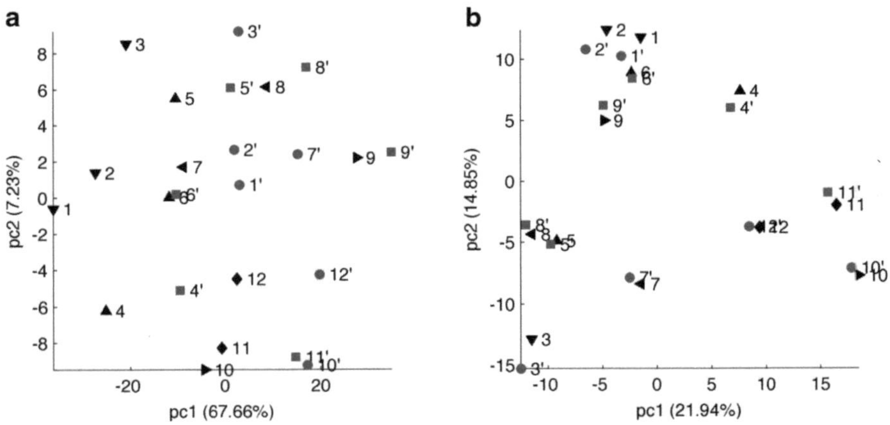

Fig. 1.4 Same as in Fig. 1.2, but only for the 10,626 transcripts considered present on all 24 arrays. (**a**) Raw data (10,626 transcripts); (**b**) Normalized data (10,626 transcripts)

1.2.3 Differential Analysis at the Gene Level

The purpose of differential analysis at the gene (or transcript) level is to identify genes whose expression level differs from a condition to another, based on statistical hypothesis testing. Almost all experiments analyzed in this chapter involve two groups of tissues, normal and DS tissues, usually unpaired. Thus, the traditional *t*-test is relevant for our purpose, which reformulates the question of differential expression of gene *i* in terms of a null hypothesis H_{0i} "there is no difference of mean expression for the transcript *i* between the normal and the DS tissues". Student's *t*-test is indeed the test that was used in almost all reviewed papers, and that we used for the experiments for which no analysis was published. Once the *t*-statistic is computed, the classical decision rule to accept or reject H_{0i} consisting in controlling the type I error probability can be applied for declaring each gene differentially expressed (DE) or not.

However, the specificity of microarray differential analysis lies in the large number of tests to be performed: as many as probes on the array, or at least, as expressed transcripts. The question of differential expression must hence be restated as a multiple testing problem. The first attempts to solve this problem aimed at controlling the Family Wise Error Rate (FWER), that is the probability to have at least one false positive, and the procedures of Bonferroni and Sidak are the most widely used to this end. An alternative approach has been proposed in Benjamini and Hochberg, based on the principle that the designer of a microarray experiment is ready to accept some type I errors, provided that their number is small as compared to the total number of rejected hypotheses (i.e. of genes decided DE) (Benjamini and Hochberg 1995). This approach aims at controlling the False Discovery Rate (FDR), i.e. the expected proportion of false positives among the total number of positives.

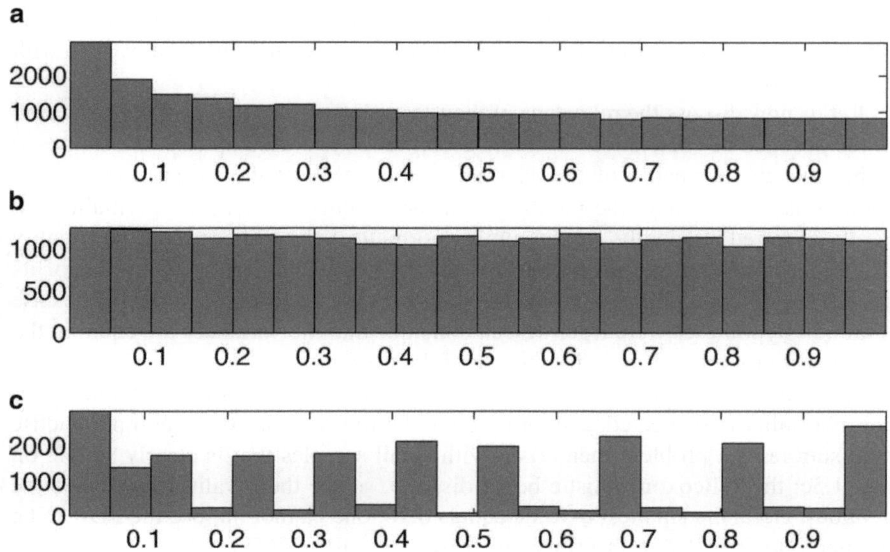

Fig. 1.5 Histograms of the *p*-values of three statistical hypothesis tests: (**a**) Student's test of equality of the means, (**b**) Fisher's test of equality of the variances, (**c**) Wilcoxon's non parametric rank sum test

Storey and Tibshirani proposed to define an equivalent of the *p*-value for the control of the FDR, the *q*-value (Storey and Tibshirani 2003). If genes with *q*-values smaller than 5% are decided DE, then there is a FDR of 5% among the DE genes. In practice, the *q*-values can be computed from the *p*-values, and are often called "adjusted *p*-values". Most papers reviewed here use the *q*-values corresponding to Benjamini and Hochberg's rule to control the FDR, possibly with an estimation of the number m_0 of true null hypotheses H_{0i} (i.e. the number of not DE genes), see Storey and Tibshirani (2003).

Let us take the example of Pevsner's data available on GEO without published analysis. We analyze the expression of human skin fibroblasts, from five normal and five DS individuals, as measured by Affymetrix arrays involving 54,675 probes (we use the calls and normalized intensity values calculated by the MAS 5 or GCOS software as available on GEO).

We performed the *t*-test for the transcripts which were considered present at least three times in both conditions, i.e. for 22,606 transcripts. The histogram of the corresponding *p*-values is shown in Fig. 1.5a. Their distribution is far from being uniform, which means that many genes are differentially expressed. As a matter of fact, when controlling only an individual type I error risk of 5% using the *p*-values, 2,938 transcripts are decided DE.

The number of true null hypotheses m_0 is roughly given the number of *p*-values in the flat part of the histogram (the ones which would correspond to the uniform distribution). It can be estimated at 17,108 according to Storey and Tibshirani (2003)

(with the tuning parameter $\lambda = 0.5$). Using this estimate for the computation of the q-values, only 11 transcripts are decided DE when imposing a FDR of 5% (76 with a FDR of 10%).

Let us now discuss the relevance of the t-tests. For a t-test to be valid, in addition to the absence of correlation of the measurements, two assumptions are supposed to be true: the normality of the data, and the equality of the variance in the two conditions. If, like here, the number of measurements is small, the normality can hardly be tested efficiently. But (assuming normality), the Fisher test of the equality of the variances can indeed be performed. For our example, the histogram of its p-values is shown on Fig. 1.5b. The distribution being uniform, as would be the case if all null hypotheses were true, we can conclude that the variances are equal in the two conditions, and this justifies the use of the t-test.

If the previous Fisher test establishes that many variances are different, or if non-normality is suspected, a solution could be to use Wilcoxon's non parametric rank sum test. A problem then arises with small samples that is clearly visible on Fig. 1.5c: the Wilcoxon statistic being discrete, so are the p-values and hence the q-values. Here, the smallest q-value equals 0.26, one cannot impose the FDR to be smaller than 26% (627 transcripts are DE with a FDR of 26%).

Thus, in the situations where the assumptions that the data is normal and/or that the variances are unequal are really unsuitable, the best alternative is to estimate the empirical distribution of the t-statistics using permutation methods such as bootstrap or permutations, see Westfall and Young (1992). A particularly popular and efficient permutation method is the Significance Analysis of Microarrays (SAM) proposed by Tusher et al. (2001).

1.2.4 Differential Analysis at the Gene Set Level

In order to take full advantage of the differential analysis at the gene level, which merely provides an unstructured list of DE genes, an integration at a higher level is necessary. Thus, the identification of predefined sets of biologically related genes enriched or depleted with DE genes has become a routine part of the analysis and of the interpretation of microarray data.

Gene sets can be built on several criteria. These criteria can be based on the available annotation sources such as GO, the Gene Ontology project, KEGG, the Kyoto Encyclopedia of Genes and Genomes, or GenMAPP for example. In the case of DS studies, other gene sets of interest are the HSA21 genes, or even genes belonging to the specific bands of HSA21, as analyzed in Slonim et al. (2009).

The first and most common approaches used to identify gene sets enriched or depleted in DE genes are based on the two-by-two contingency table obtained by classifying the genes into "being DE or not DE" on one hand, and "belonging to the gene set S of interest or not" on the other hand. The statistical significance of the overlap between being DE and belonging to S can be established more or less equivalently using the hypergeometric test, Fisher's exact test or chi-square

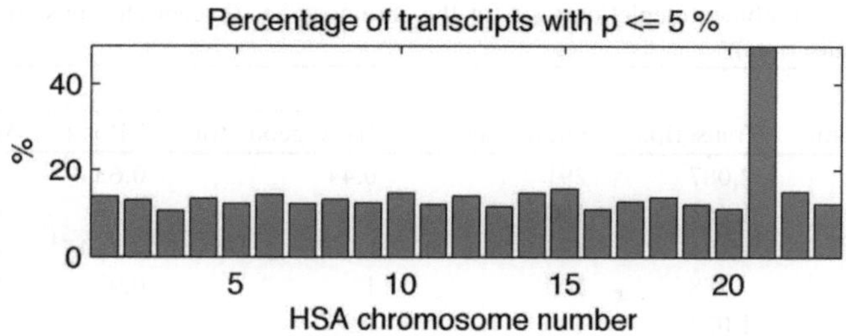

Fig. 1.6 Percentage of DE transcripts for each chromosome (the 23th is the X)

tests, as proposed by many GO processing tools, see Rivals et al. (2007) for a review. Though these approaches are quite efficient, their limitation is to require a preliminary categorization into DE and not DE genes, i.e. they necessitate the choice of a significance cutoff (be it in the form an individual type I error risk, or of a FDR), which is always arbitrary.

More recently, several methods have been proposed that avoid categorizing the genes into DE and not DE, by simply using the t-statistics or the associated p-values. For example, Sartor et al. (2009) propose a very intuitive logistic regression approach, *LRpath*. Given a gene set S of interest, a target variable y is defined as having value 1 for the genes in S, and value 0 for the others. The –log(p-value) is used as explanatory variable x, and y is modeled by a logistic function of x, $1/(1+\exp(-(ax+b)))$. If the slope a is found significant according to a classic Wald test, the subset is decided significantly enriched ($a>0$) or depleted ($a<0$) in DE genes.

Let us illustrate the enrichment/depletion analysis using the hypergeometric test and LRpath on the example of Pevsner's data, simply defining the gene subsets of interest according to the chromosomes they belong to. For the hypergeometric test, we define a threshold of 5% for the p-values, i.e. genes with $p \leq 5\%$ are considered DE. The percentage of DE transcripts for each chromosome is shown on Fig. 1.6: with 49% of DE transcripts, chromosome 21 appears clearly enriched. But is chromosome 3 significantly depleted with 10.6%?

The p-values of the hypergeometric test and of LRpath are shown in Table 1.1. It is interesting to note that they often disagree (Spearman's $\rho = 0.34$), hence the interest for the recent approaches avoiding categorization into DE/not DE.

On the other hand, we have computed Wilcoxon's rank sum statistic for the p-values (one group being the set S corresponding to one chromosome, the second all the other transcripts), which is in good agreement with LRpath (Spearman's $\rho = 0.80$): we see that this simple test is a good indicator for enrichment/depletion.

For the chromosomes for which the three tests agree (chromosomes 3, 1 and 21), Fig. 1.7 shows the results of the logistic regression for LRpath. Chromosomes 3 and 15 are depleted in DE genes, whereas chromosome 21 is enriched, as expected.

Gene set enrichment analysis (GSEA), in the version proposed by Subramanian et al. (2005), is used for example in Slonim et al. (2009) in order to detect

Table 1.1 Enrichment/depletion tests for the chromosomes (except chromosome Y with too few transcripts)

			p		
Chromosome	Transcripts	DE transcripts	Hypergeometric	LRpath	Wilcoxon
1	2,087	291	0.44	0.63	0.058
2	1,461	190	0.7	0.36	0.36
3	1,184	125	0.003	$7.2 10^{-4}$	0.0054
4	788	105	1	0.031	0.015
5	1,054	130	0.33	0.32	0.12
6	1,079	155	0.35	0.18	0.2
7	954	117	0.32	0.42	0.87
8	712	94	0.94	0.56	0.3
9	823	102	0.43	0.99	0.65
10	839	123	0.29	0.14	0.18
11	1,103	134	0.23	0.28	0.92
12	1,095	154	0.52	0.072	0.061
13	430	50	0.31	0.59	0.5
14	700	103	0.32	0.042	0.073
15	675	106	0.085	0.92	0.78
16	839	92	0.037	0.0045	0.035
17	1,117	143	0.6	0.44	0.34
18	329	45	0.92	0.67	0.35
19	1,109	135	0.24	0.029	0.058
20	561	62	0.11	0.2	0.13
21	219	107	$7.9e{-}11$	0	0
22	427	64	0.36	0.93	0.86
X	665	82	0.46	0.037	0.038

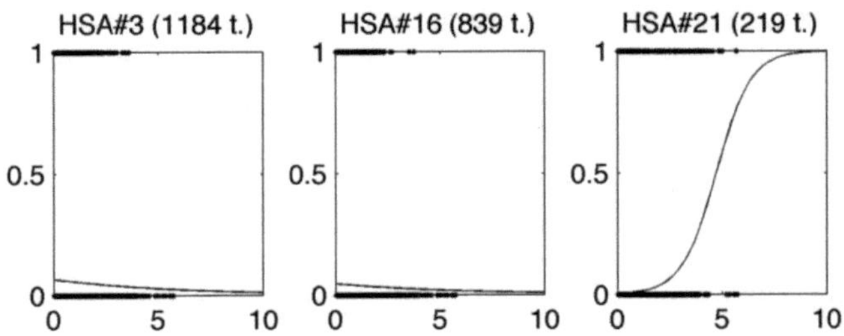

Fig. 1.7 Logistic regression on $-\log_{10}(p)$ for the chromosomes significantly enriched or depleted in DE transcripts

enriched bands on chromosome 21. Like the Wilcoxon test, GSEA uses the complete distribution of the *p*-values, divides the genes into the set *S* of interest and the rest, and ranks them according to the *p*-value. But the enrichment score is computed by walking down the list, increasing a running sum statistic when a gene in *S* is encountered, and decreasing it when the gene is outside *S*; the enrichment score is the maximum deviation from zero encountered during the walk, and its significance is evaluated by estimating the null distribution through permutations (i.e. the correlation structure of the gene expression is taken into account, what the simple Wilcoxon test does not).

Finally, let us mention ProbeCD, the method proposed by Vencio and Schmulevitch (2007). ProbCD not only presents the advantage of not requiring the choice of a significance cutoff, but it is also able to take the uncertainty in the gene annotation into account.

1.3 Results

Table 1.2 summarizes the data sets considered in this study. As detailed in the previous section, PCA was applied to the normalized datasets, on the transcripts expressed across all arrays. Three different PCA were systematically performed: one with all expressed transcripts, another with the expressed three-copy transcripts only (those of HSA21 or of the triplicated part of MMU16), and the last with the remaining euploid transcripts. The ultimate goal of this analysis was to visualize whether samples would be grouped according to their genotype (DS or control) and in which conditions (with all genes, with triplicated genes only and/or with the euploid genes only). Figures 1.8, 1.9, 1.10, 1.11, 1.12, 1.13 and 1.14 show the results from these PCA applied to the data sets. On all of them, DS samples are shown in black, control samples in white.

From the analysis including the three copy genes only, samples from DS models are very clearly separated from samples from euploid controls. This is due to the global overexpression of the three copy genes that has been largely described previously. Indeed, in DS, three-copy genes are globally over-expressed by a mean factor of 1.5. However at the single gene resolution, this 1.5 overexpression does not strictly apply and several comprehensive studies have shown that compensation and amplification mechanisms do exist. Compensated three-copy genes will not be over-expressed while amplified three-copy genes will be over-expressed by a factor significantly higher than 1.5 (Ait Yahya-Graison et al. 2007; Prandini et al. 2007).

When comparing the PCA performed on all expressed genes and applied to the various sets of data, the results are quite different. With three sets of data (Figs. 1.8, 1.9, 1.10), samples from DS models are separated from samples from euploid controls, although comparatively less than when the analysis is applied to three copy genes only. With one set of data (Fig. 1.11), the separation is present in a lesser extent. Finally, with the last three sets of data all samples are mixed and no separation is clearly depicted (Figs. 1.12, 1.13 , 1.14).

For the datasets with a clear separation, we tested the influence of the three-copy genes. We removed them and run the PCA on all expressed genes except the three-copy genes. The right panels of Figs. 1.8, 1.9 and 1.10 show the same projections than the right panels, thus suggesting that the categorization into normal and DS samples is not due to the overexpression of the three-copy genes only but rather to a modification of the whole transcriptome.

Table 1.2 List of datasets used in this study

Authors	Title	Journal	Sample type	Number of DS samples	Number of control samples	Microarray type	Data location
Mao et al.	Global up-regulation of chromosome 21 gene expression in the developing Down syndrome brain	Genomics, 2003	Human fetal cortex	4	4	Affymetrix U133A	Science direct
Saran et al.	Global disruption of the cerebellar transcriptome in a Down syndrome mouse model	Hum Mol Genet, 2003	Cerebellum from adult Ts65Dn mice	6	6	Affymetrix U74Av2	Private communication
Amano et al.	Dosage-dependent over-expression of genes in the trisomic region of Ts1Cje mouse model of Down syndrome	Hum Mol Genet, 2003	Whole brain of new born Ts1Cje mice	6	6	Affymetrix U133A	GEO
Dauphinot et al.	Cerebellum transcriptome during postnatal development of the Ts1Cje mice, a model for Down syndrome	Hum Mol Genet, 2005	Cerebellum from P0, P15 and P30 Ts1Cje mice	3 × 2	3 × 2	Affymetrix U74Av2	GEO
Lockstone et al.	Gene expression profiling in the adult Down syndrome brain	Genomics, 2007	Human adult prefrontal and dorsolateral cortex	7	8	Affymetrix U133A	GEO
Pevsner et al.	Unpublished	–	Human cultured fibroblasts	5	5	Affymetrix U133 plus 2.0	GEO
Laffaire et al.	Gene expression signature of cerebellar hypoplasia in a mouse model of Down syndrome	BMC Genomics 2009	Cerebellum external granular layer	9	9	Illumina mouse-6 v1.1	GEO
Slonim et al.	Functional genomic analysis of amniotic fluid cell-free mRNA suggests that oxidative stress is significant in Down syndrome fetuses	Proc Natl Acad Sci, 2009	Human uncultured amniotic fluid	7	7	Affymetrix U133 plus 2.0	GEO

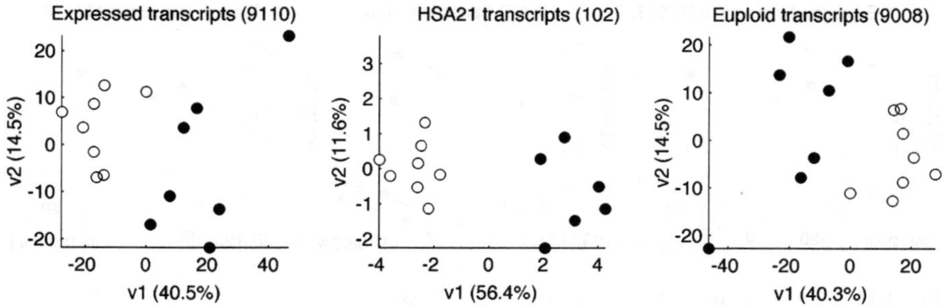

Fig. 1.8 PCA of the data described in Lockstone et al. (2007)

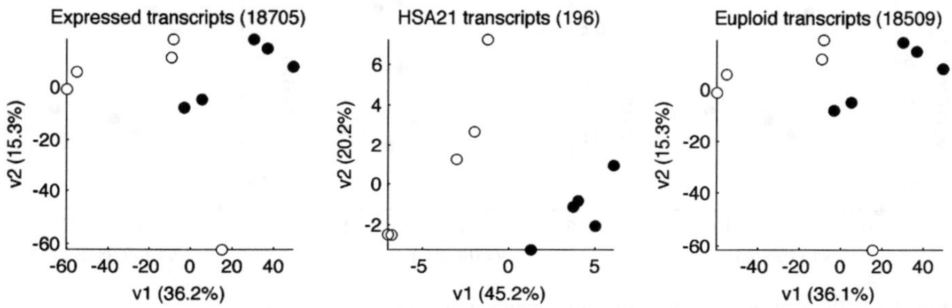

Fig. 1.9 PCA of the data described in Pevsner (GEO GSE9762)

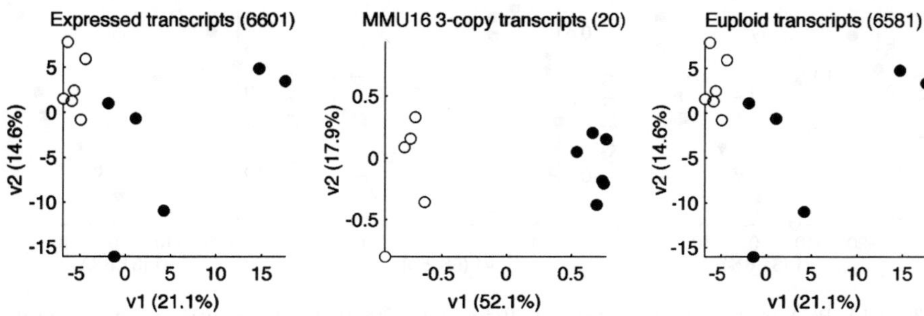

Fig. 1.10 PCA of the data described in Saran et al. (2003)

We tried to analyze the reasons why datasets would behave differently towards PCA. One obvious reason would be that there is a factor which is stronger than the genotype (DS or control) that drives the separation of samples. This is the case for samples that include different time points during development in the same analysis (Figs. 1.13, 1.14). On Fig. 1.13, samples segregate with the litter. In this particular analysis the external granular layer of the cerebellum was dissected at birth (P0) from the Ts1Cje mice. What is called P0 can in fact be between birth and P1 depending on the time of birth during the day or during the night. According to

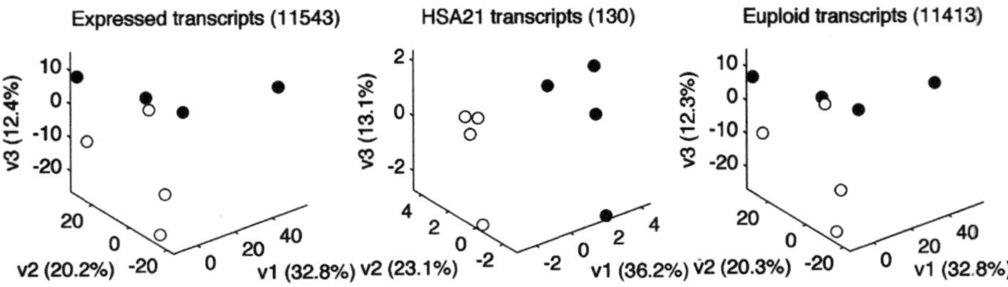

Fig. 1.11 PCA of the data described in Mao et al. (2003)

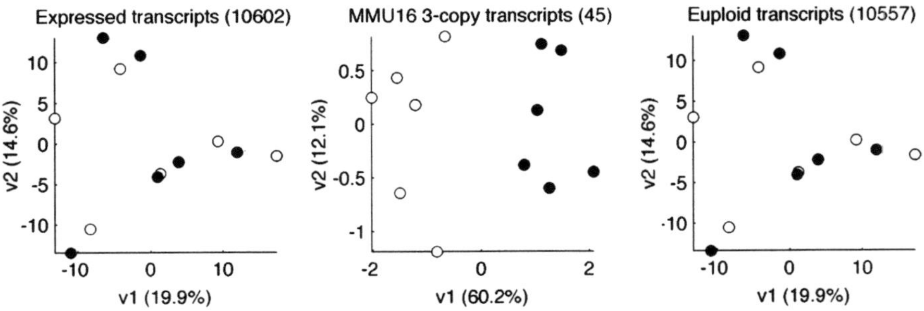

Fig. 1.12 PCA of the data described in Amano et al. (2004)

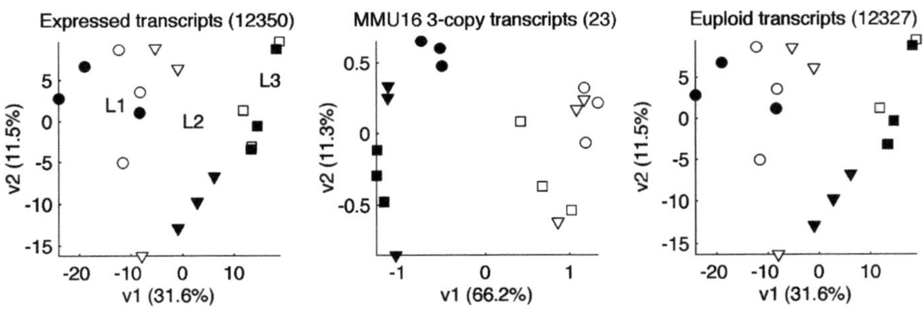

Fig. 1.13 PCA of the data described in Laffaire et al. (2009). The three markers correspond to three different litters

the PCA, samples were separated according to the litter, indicating that the up to 12–24 h can impact seriously on the transcriptome of this particular cell type. On Fig. 1.14, it is clear that the impact of development on gene expression is much bigger than the impact of trisomy 21, as was discussed previously (Dauphinot et al. 2005).

In the case of the data set from Amano et al., again whole brains were obtained at birth with possibly an up to 24 h difference between litters and even between pups. It is known that the embryos from a litter are not totally equivalent in term of development depending on their position in the uterus.

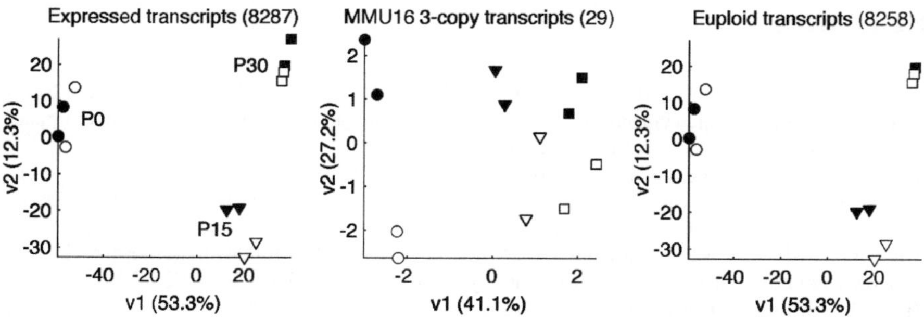

Fig. 1.14 PCA of the data described in Dauphinot et al. (2005). The three markers correspond to three different developmental stages (P0, P15, P30)

1.4 Conclusion

Functional genomics needs to be applied to complex diseases. In the case of Down syndrome, we have used gene expression profiling in various human samples or in mouse models and shown that, when we selected the three-copy genes for the analysis, samples were separated according to their genotype (DS or euploid) in all data sets. This is due to the global over-expression of the three-copy genes in DS or in mouse models. When using all expressed genes, samples were separated according to their genotype only in some datasets. This suggests that, in the datasets with no separation, there is a factor other than trisomy that strongly impacts on the transcriptome. We have shown that this factor can be the postnatal development of the cerebellum.

It now remains to be shown whether, beside the global over-expression of the three-copy genes, there will be a common set of genes that is modified in all samples analyzed. We and others have tried to search for this group of genes without any frank success. To get a more precise answer, very large sets of data will need to be generated, or alternatively, gene profiling should be obtained from single cells either trisomic or euploid. At the present time, gene expression profiles are obtained from samples that are too variable (different tissues or cells, different time points during development, different individuals with too many inter-individual variations and not enough samples). If the common set of dysregulated genes does not exist, it suggests that the most important trend is the overexpression of the three-copy genes themselves that secondarily impacts on the whole transcriptome in a "stochastic" way.

Acknowledgements The authors wish to thank The European program AnEUploidy and the Fondation Jérôme Lejeune for their financial support.

References

Ait Yahya-Graison E, Aubert J, Dauphinot L, Rivals I, Prieur M, Golfier G et al. (2007) Classification of human chromosome 21 gene-expression variations in Down syndrome: impact on disease phenotypes. Am J Hum Genet 81(3):475–491

Amano K, Sago H, Uchikawa C, Suzuki T, Kotliarova SE, Nukina N et al. (2004) Dosage-dependent over-expression of genes in the trisomic region of Ts1Cje mouse model for Down syndrome. Hum Mol Genet 13(13):1333–1340

Antonarakis SE, Lyle R, Dermitzakis ET, Reymond A, Deutsch S (2004) Chromosome 21 and Down syndrome: from genomics to pathophysiology. Nat Rev Genet 5(10):725–738

Baxter LL, Moran TH, Richtsmeier JT, Troncoso J, Reeves RH (2000) Discovery and genetic localization of Down syndrome cerebellar phenotypes using the Ts65Dn mouse. Hum Mol Genet 9(2):195–202

Benjamini Y, Hochberg Y (1995) Controlling the false discovery rate: a practical and powerful approach to multiple testing. J R Stat Soc B 57(1):289–300

Blalock EM, Geddes JW, Chen KC, Porter NM, Markesbery WR, Landfield PW (2004) Incipient Alzheimer's disease: microarray correlation analyses reveal major transcriptional and tumor suppressor responses. Proc Natl Acad Sci U S A 101(7):2173–2178

Bolstad BM, Irizarry RA, Astrand M, Speed TP (2003) A comparison of normalization methods for high density oligonucleotide array data based on variance and bias. Bioinformatics 19(2):185–193

Bontoux N, Dauphinot L, Vitalis T, Studer V, Chen Y, Rossier J et al. (2008) Integrating whole transcriptome assays on a lab-on-a-chip for single cell gene profiling. Lab Chip 8(3):443–450

Cleveland WS (1979) Robust locally weighted regression and smoothing scatterplots. J Am Stat Assoc 74:829–836

Dauphinot L, Lyle R, Rivals I, Dang MT, Moldrich RX, Golfier G et al. (2005) The cerebellar transcriptome during postnatal development of the Ts1Cje mouse, a segmental trisomy model for Down syndrome. Hum Mol Genet 14(3):373–384

Dudoit S, Yang YH, Callow MJ, Speed T (2002) Statistical methods for identifying genes with differential expression in replicated cDNA microarray experiments. Stat Sin 12:111–139

Dunlop MJ, Cox RS 3rd, Levine JH, Murray RM, Elowitz MB (2008) Regulatory activity revealed by dynamic correlations in gene expression noise. Nat Genet 40(12):1493–1498

Emilsson L, Saetre P, Jazin E (2006) Alzheimer's disease: mRNA expression profiles of multiple patients show alterations of genes involved with calcium signaling. Neurobiol Dis 21(3): 618–625

Epstein CJ (1990) The consequences of chromosome imbalance. Am J Med Genet 7(Suppl):31–37

FitzPatrick DR, Ramsay J, McGill NI, Shade M, Carothers AD, Hastie ND (2002) Transcriptome analysis of human autosomal trisomy. Hum Mol Genet 11(26):3249–3256

Giannone S, Strippoli P, Vitale L, Casadei R, Canaider S, Lenzi L et al. (2004) Gene expression profile analysis in human T lymphocytes from patients with Down syndrome. Ann Hum Genet 68(Pt 6):546–554

Golfier G, Lemoine S, van Miltenberg A, Bendjoudi A, Rossier J, Le Crom S et al. (2009) Selection of oligonucleotides for whole-genome microarrays with semi-automatic update. Bioinformatics 25(1):128–129

Harold D, Abraham R, Hollingworth P, Sims R, Gerrish A, Hamshere ML et al. (2009) Genome-wide association study identifies variants at CLU and PICALM associated with Alzheimer's disease. Nat Genet 41(10):1088–1093

Johnson RA, Wichern DW (2002) Applied multivariate statistical analysis. Prentice-Hall, Upper Saddle River

Laffaire J, Rivals I, Dauphinot L, Pasteau F, Wehrle R, Larrat B et al. (2009) Gene expression signature of cerebellar hypoplasia in a mouse model of Down syndrome during postnatal development. BMC Genom 10:138

Lambert JC, Heath S, Even G, Campion D, Sleegers K, Hiltunen M et al. (2009) Genome-wide association study identifies variants at CLU and CR1 associated with Alzheimer's disease. Nat Genet 41(10):1094–1099

Lejeune J, Gautier M, Turpin R (1959) Study of somatic chromosomes from 9 mongoloid children. C R Hebd Seances Acad Sci 248(11):1721–1722

Lockstone HE, Harris LW, Swatton JE, Wayland MT, Holland AJ, Bahn S (2007) Gene expression profiling in the adult Down syndrome brain. Genomics 90(6):647–660

Losick R, Desplan C (2008) Stochasticity and cell fate. Science 320(5872):65–68

Lu T, Pan Y, Kao SY, Li C, Kohane I, Chan J et al. (2004) Gene regulation and DNA damage in the ageing human brain. Nature 429(6994):883–891

Maamar H, Raj A, Dubnau D (2007) Noise in gene expression determines cell fate in *Bacillus subtilis*. Science 317(5837):526–529

Macarthur BD, Ma'ayan A, Lemischka IR (2009) Systems biology of stem cell fate and cellular reprogramming. Nat Rev Mol Cell Biol 10(10):672–681

Maes OC, Xu S, Yu B, Chertkow HM, Wang E, Schipper HM (2007) Transcriptional profiling of Alzheimer blood mononuclear cells by microarray. Neurobiol Aging 28(12):1795–1809

Mao R, Zielke CL, Zielke HR, Pevsner J (2003) Global up-regulation of chromosome 21 gene expression in the developing Down syndrome brain. Genomics 81(5):457–467

Mao R, Wang X, Spitznagel EL Jr, Frelin LP, Ting JC, Ding H et al. (2005) Primary and secondary transcriptional effects in the developing human Down syndrome brain and heart. Genome Biol 6(13):R107

McLachlan GJ, Do K-A, Ambroise C (2004) Analyzing microarray gene expression data. Wiley, New York

Miller JA, Oldham MC, Geschwind DH (2008) A systems level analysis of transcriptional changes in Alzheimer's disease and normal aging. J Neurosci 28(6):1410–1420

Moldrich RX, Dauphinot L, Laffaire J, Vitalis T, Herault Y, Beart PM et al. (2009) Proliferation deficits and gene expression dysregulation in Down's syndrome (Ts1Cje) neural progenitor cells cultured from neurospheres. J Neurosci Res 87(14):3143–3152

Nagasaka Y, Dillner K, Ebise H, Teramoto R, Nakagawa H, Lilius L et al. (2005) A unique gene expression signature discriminates familial Alzheimer's disease mutation carriers from their wild-type siblings. Proc Natl Acad Sci U S A 102(41):14854–14859

Olson LE, Roper RJ, Baxter LL, Carlson EJ, Epstein CJ, Reeves RH (2004) Down syndrome mouse models Ts65Dn, Ts1Cje, and Ms1Cje/Ts65Dn exhibit variable severity of cerebellar phenotypes. Dev Dyn 230(3):581–589

Potier MC, Rivals I, Mercier G, Ettwiller L, Moldrich RX, Laffaire J et al. (2006) Transcriptional disruptions in Down syndrome: a case study in the Ts1Cje mouse cerebellum during post-natal development. J Neurochem 97(Suppl 1):104–109

Prandini P, Deutsch S, Lyle R, Gagnebin M, Delucinge Vivier C, Delorenzi M et al. (2007) Natural gene-expression variation in Down syndrome modulates the outcome of gene-dosage imbalance. Am J Hum Genet 81(2):252–263

Ray M, Ruan J, Zhang W (2008) Variations in the transcriptome of Alzheimer's disease reveal molecular networks involved in cardiovascular diseases. Genome Biol 9(10):R148

Rivals I, Personnaz L, Taing L, Potier MC (2007) Enrichment or depletion of a GO category within a class of genes: which test? Bioinformatics 23(4):401–407

Roper RJ, Baxter LL, Saran NG, Klinedinst DK, Beachy PA, Reeves RH (2006) Defective cerebellar response to mitogenic Hedgehog signaling in Down [corrected] syndrome mice. Proc Natl Acad Sci U S A 103(5):1452–1456

Sago H, Carlson EJ, Smith DJ, Rubin EM, Crnic LS, Huang TT et al. (2000) Genetic dissection of region associated with behavioral abnormalities in mouse models for Down syndrome. Pediatr Res 48(5):606–613

Saran NG, Pletcher MT, Natale JE, Cheng Y, Reeves RH (2003) Global disruption of the cerebellar transcriptome in a Down syndrome mouse model. Hum Mol Genet 12(16): 2013–2019

Sartor MA, Leikauf GD, Medvedovic M (2009) LRpath: a logistic regression approach for identifying enriched biological groups in gene expression data. Bioinformatics 25(2):211–217

Sasik R, Woelk CH, Corbeil J (2004) Microarray truths and consequences. J Mol Endocrinol 33:1–9

Schena M, Shalon D, Davis RW, Brown PO (1995) Quantitative monitoring of gene expression patterns with a complementary DNA microarray. Science 270(5235):467–470

Singh A, Weinberger LS (2009) Stochastic gene expression as a molecular switch for viral latency. Curr Opin Microbiol 4:460–466. http://www.doodle.com/yix62vkik5gks8v612

Slonim DK, Koide K, Johnson KL, Tantravahi U, Cowan JM, Jarrah Z et al. (2009) Functional genomic analysis of amniotic fluid cell-free mRNA suggests that oxidative stress is significant in Down syndrome fetuses. Proc Natl Acad Sci U S A 106(23):9425–9429

Storey JD, Tibshirani R (2003) Statistical significance for genomewide studies. Proc Natl Acad Sci U S A 100(16):9440–9445

Subramanian A, Tamayo P, Mootha VK, Mukherjee S, Ebert BL, Gillette MA et al. (2005) Gene set enrichment analysis: a knowledge-based approach for interpreting genome-wide expression profiles. Proc Natl Acad Sci U S A 102(43):15545–15550

Tusher VG, Tibshirani R, Chu G (2001) Significance analysis of microarrays applied to the ionizing radiation response. Proc Natl Acad Sci U S A 98(9):5116–5121

Vencio RZ, Shmulevich I (2007) ProbCD: enrichment analysis accounting for categorization uncertainty. BMC Bioinforma 8:383

Westfall PH, Young SS (1992) Resampling-based multiple testing. Wiley, New York

Chapter 2
Reconstructing Models from Proteomics Data

Lysimachos Zografos, Andrew J. Pocklington, and J. Douglas Armstrong

Abstract The synaptic proteome is a highly complex and dynamic structure composed of more than 2,000 distinct proteins. The constant improvement of synaptic fraction preparation, protein complex isolation and mass spectrometry identification methods has led to a great accumulation of synaptic proteomics data. In order to gain a better insight of how the synaptic proteome is organised in molecular complexes identified from the biochemical analysis of neural tissues we have developed and combined a series of methods for reconstructing and analysing protein interaction network models from synaptic proteomics data. These methods cover every aspect of the reconstruction, ranging from how to annotate the proteins and acquire the protein interaction data to how to interpret and analyse the resulting models. This chapter gives a detailed overview of these methods as well as example applications to case study proteomics datasets.

2.1 Introduction

2.1.1 The Synapse and the Postsynaptic Density

Information in the nervous system is encoded in patterns of action potentials – electrical pulses generated in neurons, and transmitted from one to another at specialised junctions known as synapses. At chemical synapses, the most abundant type in the nervous system, action potentials propagating through the presynaptic

---

L. Zografos (✉) • J.D. Armstrong
School of Informatics, University of Edinburgh, Edinburgh, UK
e-mail: l.zografos@ed.ac.uk; jda@inf.ed.ac.uk

A.J. Pocklington
School of Medicine, Cardiff University, Cardiff, UK
e-mail: PocklingtonAJ@cardiff.ac.uk

N. Le Novère (ed.), *Computational Systems Neurobiology*,
DOI 10.1007/978-94-007-3858-4_2,

neuron are converted into release of a neurotransmitter, such as glutamate. This diffuses across the synaptic cleft and binds to receptors on the postsynaptic cell, resulting in transient local depolarization of the cell membrane. When the postsynaptic neuron becomes sufficiently depolarised, due to input from one or more synapses, a new action potential is generated. Synaptic input is also processed by the postsynaptic signalling machinery, which is closely linked to the intracellular side of the post-synaptic membrane in a structure known as the postsynaptic density (PSD). The PSD is a complex, dynamic structure composed of ~2,000 distinct proteins (Bai and Witzmann 2007; Choudhary and Grant 2004; Li et al. 2004; Collins et al. 2006; Emes et al. 2008; Li and Jimenez 2008; Trinidad et al. 2008; Fernández et al. 2009; Croning et al. 2009), of which ~100 are thought to be present at an individual synapse (Sheng and Hoogenraad 2007; Selimi et al. 2009). Physical interactions organise these proteins into signalling pathways that coordinate changes in synaptic strength (the amount of depolarisation caused by activation of the synapse) in response to patterns of neuronal activity. These changes in synaptic strength, known as synaptic plasticity, alter the flow of activity in neuronal networks and are widely thought to form the basis of behavioural learning and memory.

2.1.2 Mental Disease and the Synapse

Given the importance of synaptic signalling to normal brain function and development, it is natural to expect that mutations affecting synapse proteins may contribute to human psychiatric disorders. Indeed, functional genetic studies have shown that disruption of PSD proteins linked to glutamate receptor signalling alters cognitive function in rodents (Migaud et al. 1998; Husi et al. 2000; Grant 2003), while drugs acting at synapses via antagonism of the glutamatergic NMDA receptors have long been known to result in a schizophrenia-like psychosis with cognitive disturbance. However it is only comparatively recently that clear evidence has started to appear for a specifically synaptic involvement in complex psychiatric disorders such as autism (Jamain et al. 2003; Moessner et al. 2007; Berkel et al. 2010; Pinto et al. 2010; Hamdan et al. 2011) and schizophrenia (Kirov et al. 2009b). The earliest genetic studies, focusing on candidate genes, were based on small samples with only sufficient power to reliably detect disease-relevant mutations of relatively large effect. As a result, most reported genes failed to replicate in subsequent studies, and there was little consensus on which genes were the most strongly supported (for a review of schizophrenia studies see Harrison and Weinberger 2005). To support the equivocal genetic data, comparisons were also made of gene expression and protein abundance between affected and unaffected individuals, some identifying differences in synaptic proteins. These studies were also of limited impact due to small sample sizes and problems in interpretation, it being unclear if the changes identified were primary causes of disease or secondary effects due to compensatory mechanisms, medication, etc. When genome-wide association studies (GWAS) of common single nucleotide polymorphisms (SNPs) started to be performed, it became clear that conditions such as schizophrenia and bipolar disorder are highly polygenic, with potentially thousands of SNPs of small effect contributing to susceptibility (Purcell et al. 2009). With the exception of late-onset Alzheimer's disease, where evidence points to disruption of cholesterol and immune pathways (Jones 2011), the SNPs that have so far reached genome-wide levels of significance have not yet converged on a clear set of disease-relevant processes. Arguably the most productive area of research to date has been the study of rare structural variants, with early studies identifying a translocation of DISC1 (Millar et al. 2000), and a microdeletion causing Velocardiofacial syndrome as conferring increased risk of schizophrenia (for a recent review see Karayiorgou et al. 2010). Genome-wide studies of copy number variants (CNVs), in which extended genomic sequences are duplicated or deleted, have discovered that large, rare CNVs contribute to both autism and schizophrenia (Redon et al. 2006; Walsh et al. 2008; Stone et al. 2008). Many CNVs disrupt multiple genes, making identification of the underlying risk factors difficult. Where it has been possible to link a CNV to disruption of a single gene, strong evidence for involvement of the trans-synaptic machinery has been found, with the identification of rare mutations in NRXN1 (Kim et al. 2008), NLGN3 (Jamain et al. 2003), NLGN4X (Jamain et al. 2003), SHANK2 (Berkel et al. 2010; Pinto et al. 2010), SHANK3 (Moessner et al. 2007), SYNGAP1 (Hamdan et al. 2011; Pinto et al. 2010) and DLGAP2 (Pinto et al. 2010) in autism, and NRXN1 (Kirov et al. 2009a) in schizophrenia. Interestingly, almost all of these genes regulate synapse structural organisation: the presynaptic neurexins (NRXN) and their postsynaptic binding partners the neuroligins (NLGN) are cell-adhesion molecules with a key role in synapse development and differentiation (Craig and Kang 2007); while SHANK2, SHANK3 and DLGAP2 are PSD scaffolding molecules organising postsynaptic signal transduction pathways (SYNGAP1 being one important component of such pathways). Psychiatric genetics is now beginning to explore whole exome/genome sequencing, which will allow the identification of rare SNPs and small insertions and deletions. Understanding the functional organisation of the synapse should help elucidate the mechanisms by which these novel mutations contribute to disease and isolate the specific aspects of neuronal function being disrupted in different disorders and cross-disorder phenotypes.

2.1.3 Proteomics of the Synapse

The characterisation of synaptic complexes has generally been performed by mass spectrometry (MS) applied to a protein sample: enzymes cut the proteins into fragments which are then ionized, fired through an electromagnetic field and their mass to charge ratio measured by a detector. The abundance of individual peptides is calculated from the resulting spectrum, and clusters of peptides corresponding to individual proteins (or sets of closely related proteins) identified. Care must be taken that ambigiously identified sets of proteins do not overly influence the results of any subsequent analysis, e.g. by removal, or creation of a single composite 'protein' which reflects the functional properties of the set as a whole. The ability to reliably detect a protein will depend on its abundance, the number of characteristic peptide fragments it is cleaved into, and how well these peptides ionise and 'fly' within the machine. With improving technology it has become possible to identify low abundance proteins, an inevitable side-effect of which is the increased identification of trace contaminants. This problem may be reduced by improvements in isolation techniques, more extensive validation of identified proteins and the removal of known common contaminants from results. The most prevalent MS method is liquid chromatography-mass spectrometry (LC-MS), although matrix-assisted laser desorption/ionization reflector time-of-flight mass spectrometry (MALDI-TOF-MS) has also been used, primarily in early publications.

A broad overview of synapse composition can be obtained by extracting synaptic fractions from homogenised brain tissue through synaptosome isolation methods such as the one described by Carlin et al. (1980) and Wu et al. (1986). In order to isolate specific components of the synaptic machinery, affinity or immunoprecipitation methods can be used, in which a "bait" protein is immobilised on resin via interaction with an antibody against an epitope or a genetically engineered tag; contaminants are removed with repeated washes; then the complex of "prey" proteins binding to the bait (both directly and through interactions with other proteins) are eluted and characterised. A similar approach is the use of resin with a bound synthetic peptide acting as an artificial protein interaction domain. In analysing the composition of such complexes, it must be remembered that affinity and immunoprecipitation based methods are susceptible to biases ranging from non-specificity of the affinity reagent to potential inability of a genetically tagged protein to be post-translationaly modified in order to interact with some of its partners. Additional problems may arise if the immunoprecipitation epitope or affinity tag overlap interaction domains required by prey proteins or from the presence of promiscuous non-specific interactors. The transgenic Tandem Affinity Purification (TAP) method (Puig 2001) can tackle some of these issues by using two consecutive purifications with two different affinity tags.

There have been several global studies of synapse composition, predominantly in mouse or rat. One of the earliest high throughput studies identified proteins in a series of 26 prominent multi-protein bands from synaptosome preparations using MALDI-TOF-MS (Walikonis et al. 2000). Other proteomic studies focusing on the PSD followed, including Jordan et al. (2004), Peng et al. (2004), Yoshimura et al. (2004) and Collins et al. (2006). Although isolation and identification methods have improved over the years, the latter studies revealed protein sets with an overlap in the area of 50%. An attempt to define a consensus PSD was made by Collins et al. (2006). Utilizing 1D gel electrophoresis of synaptosome protein extracts and LC-MS, 698 proteins were identified in the mouse postsynaptic terminal, of which 620 had previously been found in PSD preparations. These were combined with data from other studies (Walikonis et al. 2000; Jordan et al. 2004; Peng et al. 2004; Yoshimura et al. 2004) to produce a list of 1,126 postsynaptic proteins, of which 446 were found in two or more studies. More recent additions to these lists come from Hahn et al. (2009) and Bayés et al. (2010) who studied the human

PSD, Trinidad et al. (2008) and Coba et al. (2009), with the latter two focusing on phopshorylation and signaling associated proteins. With these constant additions the total number of proteins in the PSD has risen to more than 2,000 (Bayés and Grant 2009). Studies of the presynaptic machinery have also been made, with synaptic vesicles being characterised by Morciano et al. (2005), Burre et al. (2006), Takamori et al. (2006) and Grønborg et al. (2010) while the presynaptic active zone was studied by Morciano et al. (2009). Variations in synapse composition are likely to be of fundamental importance in shaping the transmission and processing of information between different cell-types. Recent studies have sought to characterise the synaptic machinery present at a single synapse type (Selimi et al. 2009) and to compare the composition of synaptic vesicles at GABA and glutamate synapses (Grønborg et al. 2010).

The first studies to isolate protein complexes from within the PSD focused on the NMDA receptor (NMDAR), which is coupled to signalling pathways via MAGUK-family (and other) scaffold proteins and plays a major role in the induction of synaptic plasticity. Initially 100 proteins were identified in isolates using an antibody to the NR1 subunit of the receptor then 170 by peptide-affinity purification using a MAGUK-binding peptide from the C-terminus of the NR2B subunit (see Husi and Grant 2001). The combined set of 186 proteins, referred to as NMDA Receptor Complex/MAGUK-Associated Signalling Complex (NRC/MASC) has been the subject of a number of subsequent analyses and will appear as an example in following sections. Other complexes isolated using similar approaches include the glutamatergic mGluR5 (Farr et al. 2004), serotonergic 5HT-2C (Bécamel et al. 2002), and nicotinic alpha-7 receptor complexes (Paulo et al. 2009). More recently the transgenic TAP technique was applied by Fernández et al. (2009) to characterize 118 proteins in complexes containing PSD-95, one of the most abundant postsynaptic scaffold proteins at excitatory synapses (Nourry et al. 2003; Peng et al. 2004).

2.1.4 *Organisation and Analysis of the Synapse Proteome*

Physical interactions are central to synapse functional organisation, shaping the synaptic machinery into localised signalling microdomains. Postsynaptically, these domains are centred upon cell-surface channels and receptors. The structural core of the PSD is composed of cytoskeletal, scaffold and adaptor proteins that connect and organise the various receptors, channels, cell adhesion and membrane bound proteins. The scaffold also regulates interaction of the receptors and membrane bound proteins with various signaling molecules and enzymes including kinases, phosphatases, proteases and G-protein signalling molecules. The proteomic studies outlined above give multiple, sometimes contrasting views of this machinery. For example, PSD-95 has been shown to form a lattice-like structure directly under the postsynaptic membrane (Kim and Sheng 2004; Sheng and Hoogenraad 2007). This is reflected in the composition of the PSD-95 complex, which contains diverse channels and receptors and upstream signalling proteins (Fernández et al. 2009). This 'horizontal' section through the postsynaptic apparatus is complemented by the 'vertical' sections provided by isolation of receptor complexes such as NRC/MASC, which retrieve a greater depth of downstream signal transduction molecules recruited by the receptor.

Ultimately, biophysically realistic models of the synapse must take this spatial organisation into account, along with the stoichiometry of synapse proteins and the rate constants determining their interaction. While modelling environments have been developed (Franks and Sejnowski 2002), much of the detailed information they require has yet to be gathered. Given the importance of interactions in shaping the synaptic machinery, much can be learned from the analysis of protein–protein interaction networks (PPINs). These can be used to create static models onto which detailed annotations concerning the functions and phenotypes of synapse proteins can be mapped and then analysed using a combination of visaulisation, graph theory and statistics methods. The following sections provide a step-by-step overview of network reconstruction and analysis, using examples from our own work on the NRC/MASC and PSD-95 associated proteins complexes (Collins et al. 2006; Husi and Grant 2001; Pocklington et al. 2006; Fernández et al. 2009).

2.1.5 The Modelling Pipeline

The process of network reconstruction and analysis is summarised in Fig. 2.1. To analyse the PSD we designed a custom workflow and applied it to multiple proteomics datasets. There are various commercial tools available that can perform similar tasks such as Ingenuity's IPA (http://www.ingenuity.com/) and GeneGo's MetaCore (http://www.genego.com/) which incorporate custom databases of annotations and interactions. While these tools can be used to quickly construct annotated PPINs, care must be taken that the data is of the correct type (e.g. excluding genetic interactions when reconstructing a protein complex) and quality (e.g. is computational annotation acceptable?) for your purpose. They typically provide a fixed workflow which offers some alternative options and a good GUI, but are generally not very customisable. Before spending large sums of money on such software, it is worth making sure that it is sufficiently flexible. Depending on your needs this flexibility, usually offered by custom, rather than commercial tools, may include the ability to: check data provenance (e.g. via linked PubMed ids, virtually indispensable); filter the data based on your quality requirements, both manually and through simple rule-based filtering; incorporate qualitative/quantitative data of your own (e.g. task-specific annotations, expression data); combine annotations with each other or with quantitative data to generate new annotations (e.g. all channels and receptors with high expression in hippocampus); and give sufficient control over statistical testing (the ability to define an appropriate reference set is vital when performing enrichment analyses). It is also worth noting that the information incorporated in such tools can be biased towards particular areas of research,

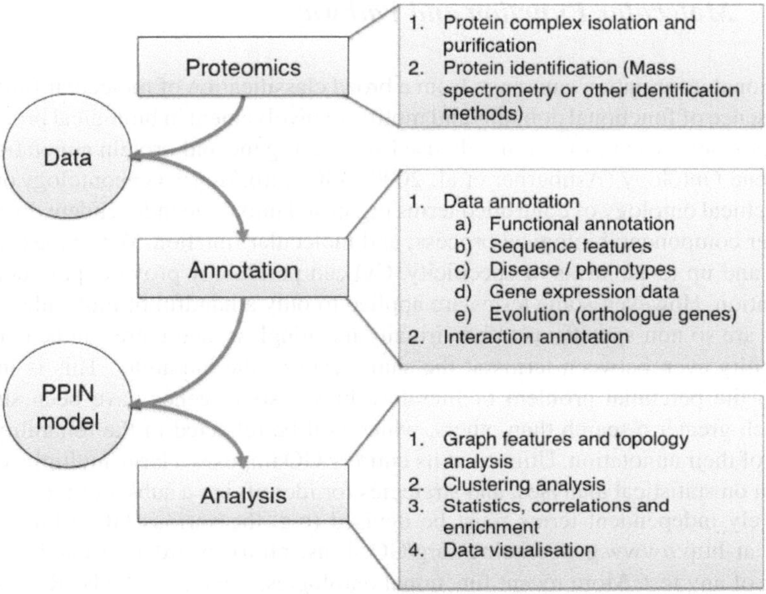

Fig. 2.1 The modelling pipeline applied to PSD proteomics data

which may not overlap well with your own. We have found that while commercial software might be fine for a quick first-pass analysis (if one can afford the licenses), having control over all components of the workflow allows for efficient, agile and potentially more insightful research. The following sections provide a step-by-step overview of the modelling pipeline, using examples from our own work (Husi and Grant 2001; Collins et al. 2006; Pocklington et al. 2006; Fernández et al. 2009).

2.2 Data Annotation

The first step when faced with a set of proteins is to collate information concerning their biological roles. Relevant features include known functional characteristics, phenotypes and disease associations, patterns of expression and evolution. Analysing these annotations can often yield insight into a complex even before constructing a network model. Databases dedicated to synapse-related genes that draw upon these resources include G2CDB (Croning et al. 2009) (http://www.genes2cognition.org/db/) and SynDb (Zhang et al. 2007) (http://syndb.cbi.pku.edu.cn/). Techniques for manual literature curation, suitable for small complexes or obtaining more comprehensive coverage for specific annotations of interest, will be discussed in Sect. 2.4.

2.2.1 Molecular Function and Pathways

Functional information can range from a broad classification of molecular function, to presence of functional domains and motifs, to involvement in biological processes and pathways. The most commonly used source of gene and protein annotations is the Gene Ontology (Ashburner et al. 2000) (GO, http://www.geneontology.org), a hierarchical ontology of controlled terms organised into three independent domains: cellular component; biological process; and molecular function. With thousands of terms and up to 12 levels of specificity, GO can potentially provide quite detailed annotation. However, some terms are applied to only a handful of molecules while others are so non-specific as to be virtually meaningless, and there can be extreme variability even between terms at the same level of the hierarchy. This is in part due to the perennial problem of literature bias – some genes have been subject to much greater research than others, which will be reflected in the reliability and depth of their annotation. Utilised in its entirety GO imposes a large multiple testing burden on statistical analyses, and strategies for identifying a subset of meaningful, relatively independent terms must be devised (e.g. the various GO Slim subsets found at http://www.geneontology.org/GO.slims.shtml) in order to maximise the power of any test. More recent functional ontologies, such as PANTHER (Thomas et al. 2003; Mi et al. 2006) (http://www.pantherdb.org/), have tried to deal with some of the above issues. While PANTHER terms fully map to a subset of GO (and are backed up by the same types of evidence) they are much better balanced, being only three levels deep with more evenly sized gene sets at each level. The PANTHER ontology also attempts to predict the functions of uncharacterized genes based on their evolutionary relationship to those with experimentally determined functions.

Protein domains – peptide sequences encoding structured, functional units – provide another important source of information (especially where gene-level data is poor). Sources of domain classifications include InterPro (Hunter et al. 2009) and PFAM (Finn et al. 2010). The Uniprot database (Magrane and Consortium 2011) annotates the amino acid sequence of each protein for known domains and other functional features such as binding and phosphoylation sites. These more detailed annotations will become increasingly useful as researchers investigate the role of alternative splicing, posttranslational modification and genetic variation on the functional properties of complexes. Tools such as ELM (Gould et al. 2010) can also be used to identify short linear peptide motifs regulating subcellular targeting, physical interactions, phosphorylation and other processes. As these are short and often highly degenerate they can easily occur by chance and many predicted sites will not be functional. Similar caveats apply to some gene regulatory elements, such as the transcription factor binding motifs found in TRANSFAC (Matys et al. 2003) and JASPAR (Sandelin et al. 2004), which may also be of interest.

Information on involvement in biological processes and 'pathways' can be obtained from GO, PANTHER, KEGG (Kanehisa et al. 2010) and Reactome (Matthews et al. 2009). However, with the exception of well studied metabolic processes (and even here novel observations are still being made), most functional pathways are still poorly defined. To use an example from our own experience, where single gene studies have painstakingly sketched out the relationships between major components of synaptic signalling pathways, high-throughput assays reveal a much greater range of molecules and interactions (Coba et al. 2008). We think it is fair to say that the full molecular complexity of many biological processes has yet to be uncovered, and is not captured by pathway definitions currently found in databases.

For compactness some annotation files will only contain the lowest level terms for each gene, in which case it will be necessary to download the full ontology and assign all parent terms. None of the annotation resources mentioned above capture the entire literature, and the overlap between similar terms in different ontologies can be surprisingly low in some cases. Although all annotations can be improved by manual curation, this is a time-consumming process best reserved for highly focussed studies, such as refining the results of analyses based on one of the comprehensive ontologies. Evidence codes summarising the type of information linking an annotation to a gene (e.g. as supplied by GO) can be very useful for simple filtering without recourse to manual checking of references. Since each gene can have multiple functional annotations, these may capture pleiotropic effects in diverse cell-types that in some cases can be misleading (Inlow and Restifo 2004). On the other hand, this can be hard to disentangle from genes which truly have multiple functions – some of which were first noted in one cell-type, some in another – and whose disruption may have a more widespread effect on a complex than disruption of a highly specialised, single-function gene. The question of literature bias must always be kept in mind, especially when analysing the overlap between annotations. The fact that a gene has been extensively studied in one context (e.g. synaptic signalling) may make it more likely to have been studied in another (e.g. as a candidate gene for schizophrenia), making annotations based on these studies non-independent.

When annotating the NRC/MASC and PSD-95 associated proteins complexes we drew heavily upon manual literature curation. To save time we only gave a single functional classification to each molecule (which was rather problematic for some), although this was supplemented by protein domain annotation. Curation and checking were admittedly rather laborious (not to say mind-numbingly tedious at times), but it did allows us to get a much better feel for the complex as a whole. As we started to work with much larger datasets manual annotation became more problematic, and we have since tended to use annotation databases (GO, PANTHER, InterPro, OMIM among others) more extensively. However, when asking a highly specific question where data quality is paramount, we still consider focused literature curation (perhaps drawing upon pre-compiled annotations to get a head start) to be the best option.

2.2.2 Diseases and Phenotypes

Data covering involvement of genes in human mendelian disorders is collated in the Online Mendelian Inheritance in Man (OMIM) database (McKusick and Amberger 1994; McKusick 2007) (http://www.ncbi.nlm.nih.gov/omim/). OMIM also covers complex disorders, but these are better dealt with by other resources. There are now a number of sites, e.g. Alzgene (Bertram et al. 2007) (http://www.alzgene.org), PDgene (Yu et al. 2008c) (http://www.pdgene.org) and SZgene (Allen et al. 2008) (http://www.szgene.org) , that collate the results of genetic association studies for a particular disorder, perform meta-analyses and provide ranked lists of associated genes or loci, making them a useful gateway into the field for non-experts. While it is possible to use the top hits from these or other lists as a disease annotation, this does throw away a lot of information and can introduce the problem of literature bias. With many genome-wide studies of SNPs and CNVs now available through dbGAP (Wooten and Huggins 2011) (http://www.ncbi.nlm.nih.gov/gap) it makes much more sense to use individual studies in their entirety, with access to multiple datasets allowing replication of results. It can take a while for access to genetic data to be approved, so it is best to identify the resources you will need and apply well in advance.

Genetic and pharmacological manipulations of model organisms have uncovered developmental, physiological and behavioural roles for many genes. A substantial amount of this phenotypic data is available from organism-specific databases such as MGI (Blake et al. 2002; Eppig et al. 2005, 2007; Bult et al. 2008) (Mouse Genome Informatics, http://www.informatics.jax.org/), RGD (Shimoyama et al. 2011) (http://rgd.mcw.edu), Flybase (Gelbart et al. 1997) (http://www.flybase.org) and Wormbase (Stein et al. 2001) (http://www.wormbase.org). MGI and RGD both use the Mammalian Phenotype ontology (Smith and Eppig 2009), allowing them to be easily combined if necessary. As with virtually all resources, the databases listed above do not encapsulate the entire literature and can always be supplemented by text mining. As noted earlier, the representation of genes in the literature may be biased. Depending on the source, phenotype annotations can be based on a diverse array of evidence, and the ability to filter out certain types of study and refer back to the original literature are essential for ensuring data quality. We would strongly recommend separating single gene data from multi-gene manipulations, and suggest that the relevance of transgenic studies be carefully considered. Some additional resources have appeared in the literature (e.g. PhenomicDB Kahraman et al. 2005; Groth et al. 2007, http://www.phenomicdb.de/) that integrate data from multiple sources (including the databases listed above). These may also be of use, although care must still be taken to ensure data quality.

When collating NRC/MASC genes linked to human disease, we did not have access to GWAS data (the earliest studies had yet to be published). Instead we employed literature curation to identify cases where disease association had been directly linked to a genetic variant. The results of analysing these annotations were highly speculative, as we would be the first to admit, and have yet to be confirmed in a more robust dataset. Rodent phenotypes (synaptic plasticity and behaviour) were also manually curated, being derived from single-gene studies, and we would still perform their curation this way today.

2.2.3 Gene and Protein Expression

Complexes are typically isolated from either whole brain preparations, or in some cases a particular anatomical region. Analysis of expression data can indicate the ways in which complex composition and function may vary across brain regions, cell-types and developmental stages. Expression can be measured in multiple ways. Western blots can detect the presence of a protein in a tissue, while immunohistochemistry allows its localisation to be determined as well. These methods are semi-quantitative at best, as is in-situ hybridisation which highlights mRNA localisation in a tissue. Microarrays and RNA sequencing (RNAseq) both measure RNA abundance in a quantitative manner. In addition to producing more detailed information, exon arrays give more accurate gene-level expression measurements compared to older microarray chips, with the emerging RNAseq technology providing the cleanest data to date. When drawing upon multiple types of expression data it must be kept in mind that neurons can span multiple anatomical regions, with their cell-body (and most of the RNA) in one and axons and dendrites (and many proteins) extending into others. Useful resources include the MGI gene expression database (Ringwald et al. 1997, 2001; Smith et al. 2007; Finger et al. 2011), the Brain Gene Expression Map (BGEM) (Magdaleno et al. 2006) and GENSAT (Heintz 2004) (http://www.gensat.org/index.html) for mouse, and the Allen Brain Atlas (Jones et al. 2009) (http://www.brain-map.org/) for both mouse and human. Individual high quality datasets (Doyle et al. 2008) can also be identified in the literature. Many of these can be downloaded from the large public repositories ArrayExpress (Parkinson et al. 2009) and the Gene Expression Omnibus (Barrett et al. 2009, 2011) (GEO, http://www.ncbi.nlm.nih.gov/geo/), both of which can be searched directly.

When investigating NRC/MASC variation between brain regions (Emes et al. 2008), expression data was experimentally collected using western blotting, immunohistochemistry or in-situ hybridisation on any genes/proteins for which the necessary reagents were available, and also extracted from an early microarray study comparing mouse brain regions (Zapala et al. 2005). Each of these datasets was analysed separately and the results evaluated for consistency. To make them comparable, expression levels for each type of data were converted to a rough semi-quantitative scale ('none', 'low', 'medium' and 'high') prior to analysis. Microarray expression levels were also analysed in a quantitative fashion, to ensure our results were not an artefact of the discrete scale we had to impose on the other datasets. The generation of this data was a mammoth task, one that is unlikely to be repeated now that transcriptome-wide expression studies are commonplace.

2.2.4 Evolution and Orthology

Nervous systems and the behaviours they support vary tremendously in complexity. By identifying and comparing orthologous genes we can start to investigate the functionality out of which a complex may have evolved and the way its various

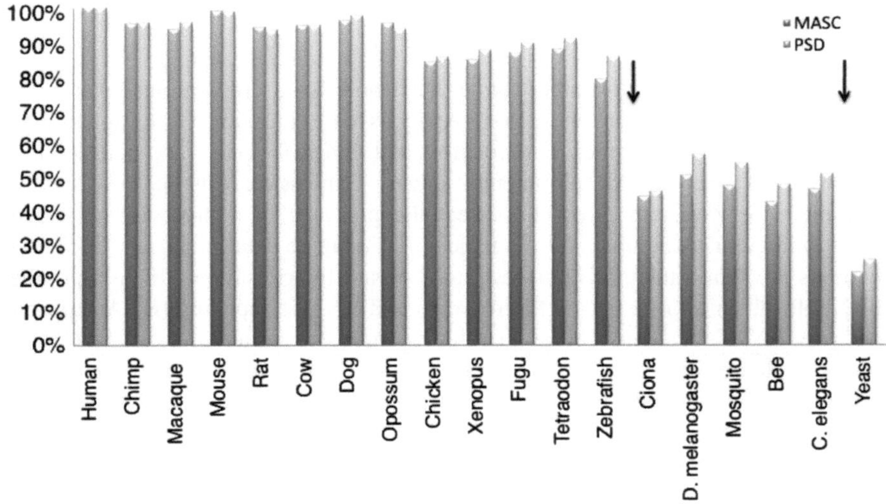

Fig. 2.2 Comparison of PSD and MASC orthologues across species. The occurrences of human PSD and MASC orthologues found in each of the 19 species are shown as a percentage of those found in human. *Black arrows* mark major events of gene family expansions (Data from Emes et al. 2008)

components have developed. This may in turn shed light on the relative importance of particular complexes or classes of molecules in the behavioural complexity of different species. More pragmatically, identifying orthologous genes allows annotations to be transferred from one species to another e.g. when wanting to investigate the relevance to human disorders of complexes characterised in rodents. The Ensembl Compara database (Vilella et al. 2009, http://www.ensembl.org/) concentrates upon vertebrates, but also contains a number of invertebrate and unicellular species comonly used as model organisms. The Inparanoid database (Berglund et al. 2008; Ostlund et al. 2010) covers a more diverse range of organisms. Information retrieved from these databases will typically contain a large number of many-to-one and many-to-many mappings. These will need to be resolved, identifying the most closely related cross-species pair. If this sounds like a lot of work (which it can be), MGI has pairwise one-to-one mappings betweem mouse, human and rat that can be downloaded.

In an attempt to investigate the evolutionary origins of the PSD and its architecture a comparative genomics approach was proposed and applied by Emes et al. (2008). In their work one-to-one mappings of orthologues of genes encoding for proteins in the PSD were retrieved for 19 species (Fig. 2.2). The species studied comprised a wide range of animals with nervous systems of differing anatomical complexity, including invertebrates, non-mammalian vertebrates, mammals and also an out-group that does not possess a nervous system (*S. cerevisiae*). The authors observed that approximately 23% of all mammalian synapse proteins were detected in yeast (21.2% MASC, 25.0% PSD) and 45% were detected in invertebrates (46.2% MASC, 44.8% PSD). Therefore, a substantial proportion of genes encoding MASC and PSD orthologues precede the origins of the nervous system, with apparent stepwise expansions following the divergence of metazoans from eukaryotes and vertebrates from invertebrates. Further investigation suggested that most functional types of synaptic proteins were present in early metazoans and that the proto-synapse constructed from this core functionality has been elaborated on during the evolution of invertebrates and vertebrates. Other studies confirmed that many families of synapse and cell signalling genes are present in the phylum Porifera (sponges), supporting the hypothesis that core synaptic signalling components were present at the base of animal kingdom (Yasuyama et al. 2002; Ruiz-Cañada et al. 2002; te Velthuis et al. 2007). A more careful examination of the appearance of individual functional families of PSD genes shows that this expansion appears to have primarily involved gene family expansion and diversification among upstream signalling and structural components (receptors, scaffolding proteins, and cytoskeletal, adhesion and signal transduction molecules). The main gene expansions can be seen in Fig. 2.2 as "jumps" in percentage of human PSD genes marked by the arrows.

2.2.5 Notes on Data Annotation

Various issues can arise during the annotation process. We have found that a typical problem in data annotation is keeping the protein and gene identifiers up-to-date. This type of analysis is done at a later stage, after the proteomics results have been obtained and it is very usual for some of the database IDs of the proteins or associated genes to have changed or become "stale" resulting to a chain of misannotation events. A solution to this can be automated by running sanity check scripts on the dataset frequently and manually checking any inconsistencies. Also choosing a database which uses identifiers that do not become "stale" often is suggested. We have found that MGI identifiers rarely become "stale".

Other issues that we have encountered involved the use of annotation resources such as GO and PANTHER. The reader has to keep in mind that these annotations are based on various types of evidence which are denoted by an attached evidence code (http://www.geneontology.org/GO.evidence.shtml). We suggest the reader chooses a set of standards as to which evidence codes they deem acceptable and apply the systematically throughout the annotation procedure. The same standards should be applied in cases of ambiguous annotation or similar issues that can only be resolved by a human annotator.

Also, sometimes there are cases where an annotation will not fit the ontology format of GO or PANTHER since the user will want to have more control over parameters. An example of that would be differential expression of a gene in different cell types, where the user would want to control things like the threshold of the ratio of gene expression in different brain areas for example. These types of annotations have to be accommodated manually.

The process of annotating is lengthy. Rigorous systematic annotation can be augmented and assisted by partial automation. Services like Ensembl's Biomart (http://www.ensembl.org/biomart/) can simplify and streamline the retrieval of such information. However we suggest, specially in cases of smaller datasets, that everything is manually checked. This manual curation has to be done using a set of rules as to what is accepted and what not and these rules have to be followed for all the annotation curation so everything under scrutiny is evaluated meaningfully. As an extension of that we should note that reusing annotations from old data can save a lot of time and effort, so good archiving and regular updating of central data repository is imperative.

As a final point, the reader should note that cases of imbalances in literature affects most annotations. These imbalances range from bias towards the study of specific genes and their products, or specific contexts of the latter. Sources of imbalance, however, could be even more subtle and could for example have to do with biases in the experimental methods uses to study specific gene products.

2.2.6 *Statistical Significance*

2.2.6.1 Annotation Significance

Once the constituent protein parts of a complex have been annotated with specific attributes regarding for example their functional classification or involvement in a certain phenotype, questions arise regarding the relation of these attributes. For instance we can ask questions like is a functional family A significantly associated more with module K? Here the notion of significance represents a number of co-occurrences that is higher than expected at random. The simplest approach is to use Fisher's exact test (Fisher 1922), either one- or two-sided depending on whether the test is specifically for enrichment, or fro both enrichment and depletion.

2.2.6.2 Considerations

When implementing this method the reader has to take a few things into account. The first is that sets of annotation variables are seldom independent and they range from mutually exclusive (e.g. the chromosomal location of the gene a protein is associated with) to redundant and overlapping. Another issue in the use of the method is our partial knowledge of the datasets. For example if a protein is not known to be associated with a disease mechanism, that could be either, a true negative or the result of a bias of the experimental observations (false negative). For these reasons this method should be used carefully and its results should be trusted if appearing consistently and repeatedly.

2.2.6.3 Multiple Testing

Multiple testing is a general statistical concept of considering multiple statistical inferences simultaneously. In a more specific context, when dealing with annotation data, one can use multiple replicates of randomized data point sets in order to assess the a null hypothesis, e.g. of an annotation having a high count due to chance. An example of that is that if an annotation appears k times in a protein list of N proteins, one can randomly sample multiple N sized samples the proteome in hand and see how many of the protein possess that annotation.

Another important application of multiple testing, summarised by Noble (2009) is correcting p-values obtained from tests like the one mentioned above – particularly in the cases of larger datasets. The most popular multiple testing p-value correction methods are Bonferoni and false discovery rate (FDR) estimation (Benjamini and Hochberg 1995). Application of the Bonferoni method means that a p-value p is accepted if $p < a/n$, where a is the confidence threshold and n the number of separate tests. This approach can sometimes be too strict so the FDR estimation method or the Benjamini–Hochberg variation of the FDR procedure can be used as an alternative. In the former the FDR is computed using the empirical distribution of the null hypothesis while the latter uses the p-values (see also Benjamini and Yekutieli 2001).

2.2.6.4 Software Alternatives

A number of tools have been developed to perform gene set enrichment analyses including DAVID (Dennis et al. 2003; Hosack et al. 2003), FuncAssociate (Berriz et al. 2003), MAPPFinder (Doniger et al. 2003), GoMiner (Zeeberg et al. 2003), GoSurfer (Zhong et al. 2004), FatiGO (Al-Shahrour et al. 2004) and BINGO (Maere et al. 2005). These tools used similar variants of Fisher's exact test or the Hypergeometric test and the Z-statistic. The Hypergeometric test is identical to the corresponding one-tailed version of Fisher's exact test. Additionally other alternatives for computing annotation significance are available including Barnard's test (Barnard 1945), Chi-square tests (e.g. Vêncio and Shmulevich 2007; Prifti et al. 2008) and Bayesian methods (e.g. Antonov et al. 2008). At the time we chose to develop our own approach in a bottom-up manner and also in order to be able to easily incorporate in our workflows. However, currently there are many tools that implement all the above methods. Of these we highlight GSEA (Subramanian et al. 2005) and a later updated version of DAVID (Huang et al. 2009). Computations in the former are based on a variation of the Kolmogorov-Smirnov test while the latter uses a Fisher's exacts test in combination with multiple testing p-value correction (Benjamini, Bonferoni and false discovery rate methods). A large and constantly updated list of these tools can be found on the gene ontology website (http://www.geneontology.org/GO.tools.shtml).

Table 2.1 Ten most common protein domains in NRC/MASC proteins

Domain	n-fold enrichment compared to genome
IQ calmodulin-binding region	12.1
PDZ/DHR/GLGF	7.3
Serine/threonine-protein kinase domain	6
C2 calcium-dependent membrane targeting	5.9
Src homology-3 domain	5.3
Pleckstrin homology	4.7
Pleckstrin homology-type	4.5
Small GTP-binding protein	3.2
Protein kinase, catalytic domain	3.1
Calcium-binding EF-hand	2.9

Table 2.2 Ten most common protein domains in the PSD-95 associated proteins complex

Domain	n-fold enrichment compared to genome
Ionotropic glutamate receptor	77.7
NMDA receptor	77.7
Glutamate receptor-related	77.7
Extracellular ligand-binding receptor	28.1
Serine/threonine-protein kinase domain	6.4
Pleckstrin homology	4.3
Serine/threonine-protein kinase, active site	4.3
Src homology-3 domain	4.1
Protein kinase, ATP binding site	3.1
Protein kinase, catalytic domain	2.9

2.2.7 Analysis of Synapse Complexes

For both the NRC/MASC and the PSD-95 associated proteins complexes membrane spanning channels, receptors and adhesion proteins, together with their associated signal transduction machinery, including adaptors and enzymes, account for the majority of the proteins, revealing the central role of those families for the functionality of the PSD molecular machine. The protein domains most commonly found in NRC/MASC and the PSD-95 associated proteins complex (Tables 2.1 and 2.2) were highly enriched (3–77-fold) when compared to their occurrence frequency in the genome as a whole. In both datasets, these top ten domains represent key functionality associated with synaptic signalling: calcium binding (calcium-binding EF hand, C2, IQ calmodulin-binding region), G-protein-coupled signal transduction (small GTP-binding protein domain, extracellular ligand-binding receptor), phosphorylation (serine/threonine protein kinase, protein kinase, catalytic domain), scaffolding (SH3, PDZ/DHR/GLGF), membrane localisation (Pleckstrin homology type, Pleckstrin type, C2) and neurotransmitter related signaling (ionotropic glutamate receptor, NMDA receptor, glutamate receptor-related). Based on this kind of annotation one can also compare different bait proteins and proteomics protocols. For example in this case if one compare the family and subfamily distributions for the two complexes, some of the basic differences can be highlighted. In the case of family distribution (Fig. 2.3) one can see what are the main types of proteins found in these two PSD sub-complexes. Big differences are observable in the Enzymes and Receptors/Channels/Transporters families. If we focus on the Receptors/Channels/Transporters family in Fig. 2.4 one can observe cases like the Voltage-gated K^+ channels and the Inward rectifying K^+ channels as extreme examples that are absent from the MASC dataset. Differences like these can be attributed to the nature of the bait proteins and their interactors. When one compares the two complexes there is an overlap of 48 proteins. The differences between the complexes highlight the different sets of protein isolated by the different complex purification methods. For example an important set of proteins that was recovered using the TAP method consisted of the AMPA receptors and K^+ channels, which are very significant factor of synaptic signal transduction.

2.3 Protein Interactions

Once a list of constituent parts of a protein complex is obtained and annotated the next step is gathering the connectivity information in order to produce a PPIN model. Connectivity in the case of PPINs comes from binary interaction information. The following section describes the ways that interaction data may be obtained. There is a wide variety of data resources for protein–protein interactions ranging from single interaction studies to high throughput whole interactome studies.

2.3.1 Experimental Data Resources

The constant improvement of protein complex affinity purification, mass spectrometry identification and other high throughput methods like Yeast Two-Hybrid (Y2H) screening (Young 1998) and the mammalian protein interaction oriented LUMIER method (Barrios-Rodiles et al. 2005) have resulted in a great accumulation of protein–protein interaction data. Beyond the volume of research done on smaller complexes and interactions in a low throughput manner (e.g. affinity purification methods), as of today there are also a number of whole interactomes available, including organisms like yeast (Uetz et al. 2000; Schwikowski et al. 2000; Ho et al. 2002; Gavin et al. 2002, 2006; Krogan et al. 2006; Yu et al. 2008a), *C. elegans* (Walhout et al. 2000, 2002), *D. melanogaster* (Stuart et al. 2007), *H. pylori* (Rain et al. 2001) and human (Bouwmeester et al. 2004; Barrios-Rodiles et al. 2005; Rual et al. 2005; Lim et al. 2006; Ewing et al. 2007). Several issues have been

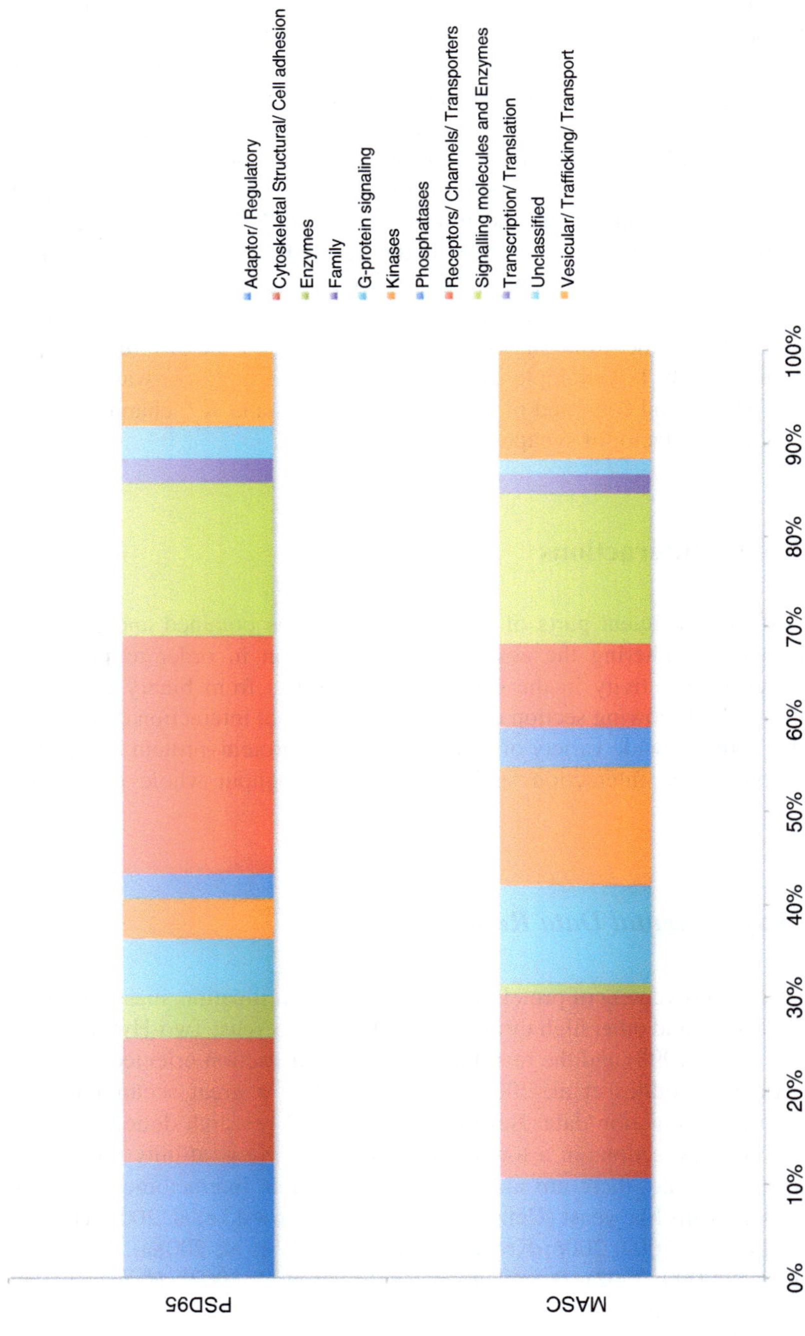

Fig. 2.3 Protein family distribution in the NRC/MASC and PSD-95 associated proteins complexes

Fig. 2.4 Subfamily distribution for the Receptors/Channels/Transporters family in the NRC/MASC and PSD-95 associated proteins complexes

raised over the years with regard to Y2H screening, which at the moment is the most high throughput of the experimental methods. The critique has focused on the high rate of false positives, analysed by Vidalain et al. (2004) as biological and technical false-positives and has caused dispute over the use of Y2H screening data in manually curated interaction models unless there is other supporting evidence. The first category includes interactions that occur in yeast cells, but do not occur in vivo in the organism of study, because there is no way to simulate differential gene expression and protein localisation. The only way to eliminate these is by obtaining this type of information regarding the studied proteins, which is not always available. The second category of technical false positives includes protein interactions that are identified in Y2H screens due to technical limitations of the system. Various approaches and frameworks have been proposed to minimise the false positives. These involve changes in the method itself (Vidalain et al. 2004), the use of combined results obtained by other methods (Mering et al. 2002) or the use of statistical methods in combination with functional annotation in order to estimate the quality parameters of a Y2H screening experiment (Venkatesan et al. 2009).

2.3.2 Databases

Database technology allows us to take advantage of the accumulation of data coming from low and high throughput methods by organising it in databases. Because of the various different experimental approaches and data sources there is a lot of variety in the available data as well. This variety stems from the methods, different species, types of interactions (binding, phopshorylation etc), dataset quality, data point confidence and type of experiments.

Although several protein interaction databases are publicly available (see http://ppi.fli-leibniz.de/jcb_ppi_databases.html), we will focus on databases that include mammalian data. Databases have two major focuses, either being central protein complex repositories or curated databases of protein interactions focused on a specific set of organisms or type of interaction. An example of the former is IntAct (Hermjakob et al. 2004) (http://www.ebi.ac.uk/intact/) which is the one of the central repositories for protein interactions. It is managed by the European Bioinformatics Institute (EBI) and contains a mixture of literature curated entries and data submissions. The other category of databases include entries that usually come from manual or semi-automated curation of the literature or collections of high throughput interaction screening experiments. Examples of such databases are can be seen on Table 2.3.

UniHi is a very comprehensive database of the computational and experimental based human protein interaction networks and has been extensively used in our workflows. It is based on merging different whole interactome maps from different data sources. These include interactions from BioGrid, IntAct and DIP among others as well as Y2H screening data.

Table 2.3 Protein interaction databases used for the reconstruction of PSD PPIs

Database	URL	Reference
BioGRID	http://www.thebiogrid.org/	(Stark et al. 2006)
DIP	http://dip.doe-mbi.ucla.edu/dip/	(Salwinski et al. 2004)
HOMOMINT	http://mint.bio.uniroma2.it/HomoMINT/	(Persico et al. 2005)
HPRD	http://www.hprd.org/	(Peri et al. 2003)
MINT	http://mint.bio.uniroma2.it/mint/	(Ceol et al. 2010; Cesareni et al. 2008)
UniHi	http://www.mdc-berlin.de/unihi	(Chatr-aryamontri et al. 2007, 2008)

Parsing data from these resources can usually be done automatically – when the data is available as for download or bulk searches. This is done by mapping gene IDs to the database's internal IDs and then retrieving all the relevant interactions. In some cases, where it is possible, setting confidence cut-offs is very useful since many of the interactions have a low confidence score.

2.3.3 Homology Data

When two pairs of orthologue proteins from different species interact they are referred to as interlog pairs. Interlog prediction is a good way of inferring interactions but is also a thorny subject when only using the sequence homology as a criterion of similarity between interacting pairs. There are cases of very big length and sequence differences between orthologues in distant lineages, for example the NR2 subunit of the NMDA receptor in mice and flies (Ryan and Grant 2009) and in these cases not all interlog interactions might take place. Interlog data should be used carefully and ideally filtered using a confidence score based on homology but also correlation of gene expression or functional annotations. An example of this approach has been implemented in DroID (Yu et al. 2008b), a database of interactions for *D. melanogester.*

2.3.4 Other Ways of Acquiring Interaction Data

While protein interaction databases like the ones mentioned in the previous paragraphs rapidly provide a useful set of candidate interactions their coverage is far from complete. A large body of molecular interaction data is effectively buried within a corpus of hundreds of thousands of scientific papers in the existing literature. The volume of this corpus along with other issues (e.g. ambiguous terms and non-machine readable formats) makes the application of text-mining methods for information extraction imperative (Rzhetsky et al. 2008). This will be thoroughly described in more detail in Sect. 2.4.3.

2.4 Mining Data from the Literature

2.4.1 Text Mining

Text mining is the derivation of information from a corpus of text with the application of natural language processing (NLP) tools and computational search and classification methods. The use of text mining is imperative for the augmentation of annotations and identification of interactions between proteins when dealing with large complexes. Although a discipline by itself, we will briefly present the principles we followed in order to perform text mining for the reconstruction of PSD complexes in the following paragraphs.

2.4.1.1 Corpus Compilation and Indexing

For biomedical text mining applications, local mirrors of PubMed are the standard type of corpus. These include all the information and meta-information, e.g. MeSH terms of the article along with the relevant abstract. It has also been empirically found that figure legends provide an excellent additional corpus as well, although correctly extracting them from the papers still poses a challenge. Regardless of the goal and the methods used for text mining, the corpus in hand has to be indexed in order to optimise the search of terms. This can be achieved with freely available software like Lucene (http://lucene.apache.org/). There are also some optional steps in order to maximise the efficiency of text mining. These include the processing of the corpus with NLP tools for the purposes of: tokenisation and sentence detection, part of speech (POS) tagging and abbreviation detection. It also includes the step of named entity recognition (NER). In NER parts of the text referring to biological entities are tagged using a classifier trained for the biomedical domain (Alex et al. 2007), identifying terms the represent protein names or interaction terms. Currently there are freely available tools like the ones offered by NaCTeM (http://www.nactem.ac.uk) which can be used for all the NLP steps. This collection of tools could perform most of the text tokenisation and POS tagging. The aforementioned, used in combination with tools like Biothesaurus (Liu et al. 2006) or Biotagger-GM (Torii et al. 2009) for the NER can potentially provide higher efficiency.

2.4.1.2 Queries and Query Expansion

After analysing the corpus queries are submitted to retrieve abstracts containing information on the gene or gene product of choice. Obtaining the right keywords to retrieve "relevant" abstracts is another challenge. This is mainly because of the number of potential synonyms every gene or protein might have, as well as their potential spelling variations. For some common terms or initials lists of variations can be manually compiled and combined with a list of synonyms for each protein and its associated gene name. Acquiring the gene and protein synonyms can be done by mining public database entries' "name", "synonyms" and "gene name" fields or using some specialised service like BioMinT/GPSDB (Pillet et al. 2005) (http://biomint.pharmadm.com/). Query expansion is a key step in compiling the list of synonyms. The simplest form of query expansion would be to try all possible combinations of spelling taking special characters like spaces and hyphens as well as spelling variations generated by converting trailing numbers to Roman numerals or vice versa into account. Although other than pre-compiled thesauruses which can be very useful (Sasaki et al. 2010) there is no stand-alone solution for this problem. We have found that collecting all available synonyms for a gene and protein entity and applying rules like the one mentioned above will generate rich enough lists. An alternative to this is using the EFetch utility, as mentioned above, which uses PubMed's built in query expansion system although that was recently found to reduce precision in some cases (Schuemie et al. 2010). More recent innovations addressing this problem are based on the query itself, expanding it based on the biological context of the gene or protein, like for example QuExT (Matos et al. 2010).

2.4.1.3 Overview of the Text Mining Process

Once the corpus is prepared and the list of queried terms is compiled it is easy to programmatically automate a process of submitting queries with the keywords of interest to the corpus. If attempting to mine for protein annotations these queries will include all the synonyms of a specific gene or gene product (after query expansion) in all combinations with all the annotation terms of interest (e.g. ("gene name" or "synonym 1") and "disease"). If mining for protein interactions these queries will include all combinations of synonyms for a given pair of potentially interacting proteins. This will generate lists of results or "hits" which can then be prioritsed and curated.

2.4.2 Mining the Literature for Annotations of PSD Proteins

Although annotation retrieval can be more difficult than protein interaction retrieval due to the more evasive nature of the types of annotations, annotating for specific features can be easier once the keywords are defined. In the case of the PSD annotation we initially used a combination of Lucene, Rainbow (Andrew McCallum, http://www.cs.cmu.edu/~mccallum/bow/) and Weka (Hall et al. 2009) for text classification. At a later stage and in order to look for newer data the EFetch utility (http://eutils.ncbi.nlm.nih.gov/entrez/eutils) provided by PubMed was used. EFetch utilises PubMed's search function but can be used in programmatic workflows. The abstracts retrieved from this process should always be manually checked.

2.4.3 *Mining the Literature for Protein Interactions of PSD Proteins*

2.4.3.1 Interaction Retrieval

We have used two text-mining solutions for the reconstruction of the PSD complexes. These solutions were the TXM pipeline (Alex et al. 2008), as well as a simpler solution developed in house using Lucene, query expansion and a Weka based text classifier. Application of TXM proved to be useful since it included a Relation Extraction (RE) analysis feature. This analysis utilises information within the structure of the text in order to compute a confidence for each hit, based among others on the proximity of the references of the two potentially interacting entities in the text or the presence of interaction associated terms.

2.4.3.2 Interaction Curation and Curation Standards

Text-mining tools and methods keep getting more accurate, however the results always contain false positive hits and have to be quality controlled manually. Furthermore, it is good practice to recheck the evidence supporting interactions retrieved from databases using the same criteria as in the text mining result check. This curation is a form of manual quality control, which is performed by reading the abstract of the paper linked as supporting evidence and verifying that interaction is between the proteins and the methods used for verifying it are reliable. This manual checking of individual papers, although in 90% of the cases just the abstract provides the related information, is the single major bottleneck of the modelling pipeline. Given the size of the text-mining task and volume of resulting hits, combined with the data retrieved from databases, the procedure of quality control needs to be stream-lined.

Stream-lining the curation process can be done using computer assistance. Since curation has to be manual the assistance will be in managing the curated data and allowing multiple experts to collaborate. This can be easily managed by a server side application that includes a web-based curation interface, where curators can log in and a database manipulation back-end to store and manage the curated text-mining results. Any such application should be designed to provide an easy way to perform the quality control. After noticing that there is a lack of freely available software to assist with this process the authors of this chapter have implement a free software solution which is available upon request.

When multiple curators collaborate, curation standards are imperative. These standards dictate if a potential physical interaction is accepted as a true positive or not and should be followed as an intact set of instructions throughout the data curation. The common curation standard used for the PSD complexes includes the following rules:

- Clear mention of physical interaction in the abstract or full text (any evidence except "prior experimental knowledge", unless backed up by experimental evidence).
- Do not accept as true positive if the only supporting data only comes from colocalisation, protein complex pulldown, interlog pairs from distant lineages or Y2H with no other supporting evidence. Regarding the last point we refer the reader to the relevant paragraphs of Sect. 2.3.1. We advice that if Y2H data is used then it should be coming from datasets that have been thoroughly reviewed and shown to consider and tackle the caveat of false positives successfully.

2.4.3.3 Interaction Mining for the PSD Complexes

The manual curation procedure for the NRC/MASC and PSD-95 associated proteins complexes lasted for 2 weeks or so per complex and required going through all the abstracts and papers, where needed, provided as evidence in protein interaction databases entries (UniHi, BioGRID, MINT, IntAct) plus all the set of hits produced by the text mining. Out of the 186 proteins in NRC/MASC 248 interactions were found between 105 proteins. In the PSD-95 associated proteins complex, we found 163 interactions between 50, out of which 40 form a major connected component (MCC). The volume of evidence one needs to curate is in the area of 3,000 or 4,000 papers (for a high confidence cut off). This might sound like a daunting task, but in most of the cases the abstract alone is enough to verify an interaction as a true positive.

2.5 Model Analysis

The premise of network biology is that we can integrate the data in a network model (e.g. a protein–protein interaction network) and abstract this network to a graph. The term graph is used as the mathematical equivalent of the network. Application of the modelling pipeline up to this point would result to a "coloured" graph. The term coloured in graph theory refers to specific attributes (colours) of the nodes and edges. Attributes can be of any type (i.e., gene name, family, subfamily, disease correlation). Graphs are useful data abstractions because they can be manipulated by a wide variety of algorithms with which, combined with visualisation, we can ask questions about the topology, architecture and most importantly latent structures or patterns in the network that we could not be seen otherwise.

2.5.1 Biological Networks and Basic Graph Theory Terms

Before we discuss any specific algorithms or workflows its critical to mention some principles and central concepts of graph theory that are the basis of any analysis performed to protein interaction networks. The foremost important is the definition of a graph. A graph is an abstract representation of a set of objects, called nodes or veritces, where some pairs of the objects are connected by links, called edges. These links could have or not have a specific direction, resulting to directed and undirected graphs respectively (there are "mixed" graphs as well).

The formal definition of a graph G is: $G = \{N, E\}$, where N is a set of k nodes, $N = \{n_1, n_k\}$ and E is a set of l edges, $E = \{e_1, e_l\}$. Each edge e is defined as a relation of incidence that connects two nodes from N. The graphs resulting from the PPINs usually (1) finite: i.e. have a finite set of nodes and edges, (2) unweighted, i.e. no special value is associated with each edge, for example an association coefficient (not easy to obtain for all edges of networks of this size) representing the strength of the interaction and (3) undirected, since binding is an bidirectional process. Note that the last is not true when modification (e.g. phopshorylation) interactions are included.

Graphs can be used for computations but first they have to be represented in a manipulatable form. This form is the adjacency matrix, A. A is a symmetric matrix defined as:

$$A_{ij} \begin{cases} = 1 & \textit{if there is an edge between } n_i \textit{ and } n_j \\ = 0 & \textit{otherwise} \end{cases}$$

So for example for the G graph in Fig. 2.5 (top left panel) would be:

	A	B	C	D	E
A	0	1	0	1	0
B	1	0	1	1	1
C	0	1	0	0	0
D	1	1	0	0	0
E	0	1	0	0	0

There are a number of ways to describe and summarise a network, its topology and overall architecture. First we will look at the methods that describe the global topology of a graph as well as some basic properties of the nodes. Graph metrics of this type that are used for biological networks are:

- Degree: the degree of node i, k_i s the number of edges connected to it. (Fig. 2.5 top right panel).
- Distance: the distance d_{ij} between nodes i and j is the shortest path (counted in edges) between length them (Fig. 2.5 bottom left panel).
- Diameter: the network's diameter is the greatest path length distance between any two nodes in the network (Fig. 2.5 bottom right panel). Formally $D = max\{d_{ij} | i, j \in N\}$.

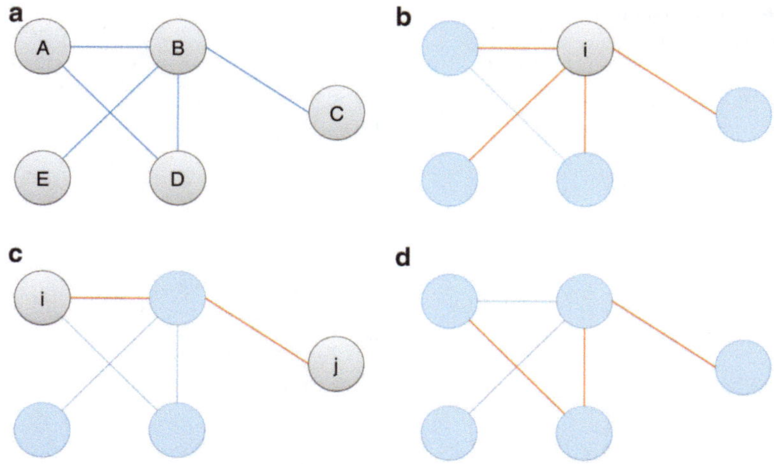

Fig. 2.5 Illustration of central graph concepts: graph as represented by an adjacency matrix (*top left*), degree of a node (*top right*), distance (*bottom left*) and network diameter (*bottom right*)

- Clustering coefficient: this measure is calculated for each node and shows the degree to which the neighbors of a particular node are connected to each other. It is defined as $C_i = \frac{2e_i}{(k(k_i-1)}$, where e_i are the number of edges between the k_i nodes that connect to node i.

Distance in networks is measured by the path length, which tells us how many edges we need to cross in order to travel between any two nodes. As there are many alternative paths between two nodes, we chose the shortest path, i.e the path with the smallest number of links between the selected nodes. Formally the shortest path problem is the problem of finding a path between two nodes such that the sum of the weights of its constituent edges is minimized. In unweighted graphs like most protein interaction networks all edge weights are 1. The shortest path problem is subdivided into four categories: (a) the single source (all shortest paths from a given node), (b) the single destination (all shortest paths to a given node), (c) the single pair and (d) all-pairs shortest path problems. There are different algorithmic approaches that can solve some of the categories. Some of these algorithms are: Dijkstra's algorithm (Dijkstra 1959), the Bellman–Ford algorithm (Bellman 1958) and the A* search algorithm (Hart et al. 1968) among others.

The concept of shortest paths allows us to introduce the notion of betweeness in graphs. Betweeness is a measurable property of nodes and edges. More specifically there are two types of betweenness in a graph:

- Node betweeness, which for a node l is defined as:

$$b_l^n = \sum_{ij} p_{ij}(l)/p_{ij} \tag{2.1}$$

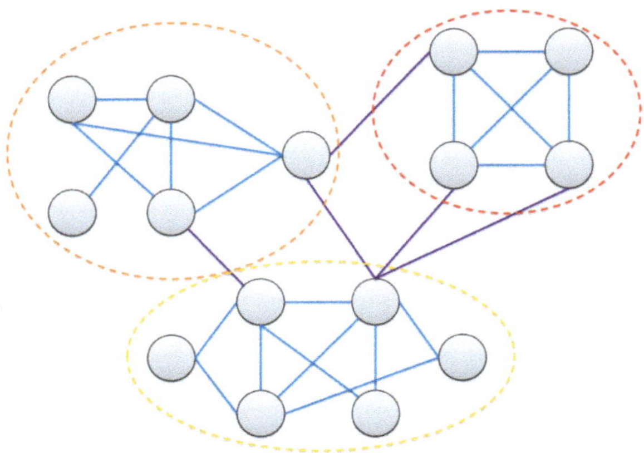

Fig. 2.6 An illustration of community structure

where $p_{ij}(l)$ is the number of shortest paths between nodes i and j that go through node l and p_{ij} is the total number of shortest paths between nodes i and j.

- Edge betweeness, which for a node l is defined as:

$$b_k^e = \sum_{ij} p_{ij}(k)/p_{ij} \tag{2.2}$$

where $p_{ij}(k)$ is the number of shortest paths between nodes i and j that include edge k and p_{ij} is the total number of shortest paths between nodes i and j.

2.5.2 Community Structure in Protein Interaction Networks

2.5.2.1 Community Structure

Studies on all types of networks have shown that one of the properties that seems to be universal in complex networks is community structure (Ravasz et al. 2002; Guimerà and Amaral 2005; Lagomarsino et al. 2007). Community structure is the segregation of nodes into groups, sometimes called clusters or communities. The characteristic feature of network clusters is that the nodes form densely connected groups or sub-graphs with sparse connections between them. An example of that can be see in Fig. 2.6. Notice how the three communities in the orange, yellow and red circles have more intra-connections rather than inter-connections between them.

2.5.2.2 Clustering Algorithms

The identification of clusters or "clustering" within networks is a well known problem. Since it is a more general data analysis problem numerous solutions have been proposed over the years. Clustering in biological networks is similar to the problem of graph partitioning in computer science and hierarchical clustering in the social sciences. In this section we will focus on clustering solution presented in the domain and context of Systems Biology. However these approaches almost fully overlap with approaches used for the network sciences. Due to the very high number of algorithms available we refer the reader to an excellent recent review by Wang et al. (2010) and in this section we will present some of the widely used algorithms based on the classification used to present them in the aforementioned review. The algorithms can be split into two major categories: graph-based and combination based. The first category encompasses algorithms that are based solely on the structure of the graph and act independently of the annotation of the nodes and the second category encompasses algorithms that use such information. We should also mention that, depending on their definition of a cluster, these algorithms can either identify overlapping or non-overlapping clusters. Overlapping clusters can potentially represent multiple complexes that a protein can belong to as a result of transient associations, differential expression or presence of all the components.

Furthermore we should add that clustering, independently of which method is being used, has to performed with two problems taken into account. These are (1) the presence of false positive and negative interaction data (2) the fact that we, in reality, do not know the number of clusters an algorithm should produce. Although these cannot be eradicated the former can be partially minimized with careful data curation. Regarding the second, it has to be taken into account that the results of clustering algorithms represent a mathematical computation or optimisation and do not necessarily accurately reflect biological reality since an algorithm will always generate an output.

Graph-Based Algorithms

These algorithms are based on local search around dense sub-graphs, hierarchical clustering or parameter optimisation.

Local search algorithms: In this category the clusters are define as densely connected sub-graphs of the main network. The density of a sub-graph is defined as $d = 2n_{edges}/(n_{nodes} - 1)n_{nodes}$ (Spirin and Mirny 2003) and reaches its maximum of 1 in a sub-graph where every two nodes are connected by an edge. In this case the sub-graph is called a clique (Erdös and Szckeres 1987). Enumerating cliques in a graph is an NP-complete problem, however protein networks make the enumeration less difficult because of their sparseness. Various solutions have been proposed implementing clique based methods including supermagnetic clustering (SPC) with Monte Carlo (MC) optimisation (Spirin and Mirny 2003) and a quasi clique based method (Bu et al. 2003) among others. A widely used algorithm of this type is the molecular complex detection (MCODE) algorithm (Bader and Hogue 2003). Although these algorithms tackle the issue of missing edges – or unknown interactions – in proteins interaction networks they suffer from issues related to the topology of the graphs as illustrated by Altaf-UI-Amin et al. (2006).

Hierarchical clustering algorithms: Hierarchical clustering algorithms, widely used in other disciplines, can be either agglomerative or divisive, depending on whether it adds or removes edges to or from the network. For a network of n nodes, in agglomerative methods, node similarity is computed with some method, either distance or some other similarity metric and an edge is then added to a network of n nodes and no edges. The complete process is mapped to a dendrogram, which when cut at a certain level, can give specific community configurations. In divisive methods on the other hand, we start from a full network and remove edges between the least similar nodes. One type of divisive algorithms are clustering coefficient based algorithms. Clustering coefficient represents a more local view of the network centered around a node and has been used as the basis for algorithms proposed by Radicchi et al. (2004) and Li et al. (2008). Newman and Girvan (2004) proposed a betweeness based divisive hierarchical clustering algorithm. We have chosen this for our approach because it is a simple and elegant algorithm that allows not only to group proteins in clusters but to tease out hierarchical structures via the use of the dendrogram. Also this is a non-heuristic algorithm that runs within tractable time for the dataset sizes in hand and when tested it produced biologically meaningful results for our models . The metric used in this case is edge betweenness. Edges with high betweenness tend to be edges where the flow of information converges. This is because according to Eq. 2.2, more shortest paths pass from these edges, thus they are parts of the path of least resistance. An interesting property of these paths is that they usually are tend to connect the segregated clusters of the network, as more thoroughly discussed in Sect. 2.5.3. The authors propose three different flavours of betweeness metrics for the implementation of the algorithm: based on shortest paths, random walk betweeness and current flow betweeness. The steps of the Newman and Girvan algorithm are:

1. Compute betweeness score for all edges in the network
2. Find the edge of highest betweeness and remove it
3. Recomputes betweeness score for all edges in the network
4. Repeat from 2

Up to this point application of the algorithm would eventually start breaking up the network in sub-networks. If one a priori knew how many communities were in a network the algorithm would be stopped when reaching that number. However, this is not the case in most practical applications. For that reason the authors also define a modularity measure or modularity quantity, Q. If one considers a particular division of a network into k communities, we can define a k x k, e matrix whose elements e_{ij} are the fractions of edges all edges in the network that connect nodes in k_i with nodes in k_j, considering all edges in the original network including the ones removed so far. The trace of the aforementioned matrix, $Tr\,e = \sum_i e_{ii}$ would, in practice give the fraction of edges in the network that connect nodes from the same community. However the trace is not a good indicator because there are cases where $Tr\,e = 1$, without that being the best configuration, e.g. if all nodes were in the same community. For that reason the authors define the row or column sums $a_i = \sum_j e_{ij}$, which represent the fraction of edges that connect to nodes in community i. In a network where all edges connect nodes without regard to which community they belong to, that would mean $e_{ij} = a_i a_j$. Thus Q is defined as:

$$Q = \sum_i (e_{ij} - a_i^2) = Tr\,e - ||e||^2$$

where $||e||$ is the sum of elements of e.

In practice Q measures the fraction of edges between nodes of the same community over the edges between nodes of different communities. Q is monitored as the algorithm progresses and once all edges have been removed we can trace back to the configuration that resulted to the maximum value of Q, which also is the optimal community structure. By definition Q is found to be between 0 and 1, with low values reflecting configurations that are no better than random. Empirically, in biological networks the value of Q lies between 0.3 and 0.7

The Newman and Girvan algorithm can be easily implemented in most programming using libraries that perform the basic graph computations, like the Graph C++ library from Boost (http://www.boost.org/) and its port for Matlab (Mathworks Inc, http://www.mathworks.com/) MatlabBGL (http://www.stanford.edu/~dgleich/programs/matlab_bgl/). It should be noted that this algorithm has heavy demands on computational resources, running in $O(e^2 n)$ time on an arbitrary network with e edges and n nodes, or $O(n^3)$ on a sparse network, where $n \sim e$. This restricts the algorithm to networks of a few thousand nodes. For that reason there has been a later modification (Clauset et al. 2004) based on more sophisticated data structures.

Along similar lines, authors have proposed other divisive algorithms as well with variations on the distance measures. These include HCS (Przulj et al. 2004) which is based on the minimum cut heuristic rule for grouping nodes, i.e. a configuration that separates two groups of nodes with the minimum number of edges between them (Hartuv and Shamir 2000) and UVCLUSTER (Arnau et al. 2005) which is based on shortest paths instead of edge betweeness.

Parameter optiomisation algorithms: From a machine learning perspective parameter optiomisation is based on the definition of a cost function which is then minimized searching through different cluster configurations. An example of such cost function is modularity Q discussed earlier. In these algorithms different configurations are randomly sampled and the cost function is optioned by selection. Newman (2004) proposed a greedy optimization of the modularity (Q) quantity and exploitation of shortcuts in the aforementioned optimization problem. Different approaches to the optiomisation of Q have been proposed by various authors including external optiomisation and spectral approach (Brandes et al. 2006; Newman and

Girvan 2004; Ruan and Zhang 2007). Markov Cluster Algorithm (MCL) (Enright et al. 2002) is also a widely used parameter optimisation algorithm which uses random walks on the network and then computing all the transition probabilities between nodes in the network. It is also worth mentioning that in the case of directed networks other measures can be used like, for example, the paths of least resistance in the information flow between two nodes (Hwang et al. 2006, 2008).

Combination Based Algorithms

Unlike graph-based algorithms, combination based algorithms are not solely based on the graph that the protein interaction network represents but also the properties of its nodes. By taking the latter into account these algorithms reduce the effects of false positive or false negative interactions. Properties can include genomic data (Jiang and Keating 2005; Zhang et al. 2008), structural features of the proteins (Dittrich et al 2008), gene co-expression data (Jansen et al. 2002; Hanisch et al. 2002; Ideker et al. 2002; Segal et al. 2003; Cho et al. 2006; Cline et al. 2007; Maraziotis et al. 2007; Lu et al. 2006; Ulitsky and Shamir 2009; Jung et al. 2008) and ontology annotations (Lubovac et al. 2006; Ulitsky and Shamir 2009). These properties are integrated into frameworks like AVID (Jiang and Keating 2005), PSIMAP (Park et al. 2005) and MATISSE (Ulitsky and Shamir 2009) which usually utilise graph-based algorithms in light of the node properties data. Although these methods are evolving constantly the fact that all the availability of data of all properties examined is not always guaranteed and that has to be taken into account.

Within the class of combination based algorithms there is the variation of ensemble frameworks which use combinations of clustering methods and integrate their results into a common consensus. This type of approach was first proposed by Asur et al. (2007), followed by Greene et al. (2008) and Simpson et al. (2010). Although still in development as an approach ensemble clustering shows promise, if the choice of parameters, like which basic clustering methods to use and how to build the consensus, is done with care. Another consideration is to be taken into account is how well these algorithms take that into account pleiotropic function of proteins.

2.5.3 *Network Topology Features*

A widely reported feature of many networks, including biological, are their "scale-free architectures" (Barabási and Albert 1999; Albert et al. 2000; Jeong et al. 2001; Barabási et al. 2003). The scale-free architecture of biological networks implies that the great majority of nodes only have a few edges connecting them to other nodes. On the contrary there are a few nodes in the network that have many edges connection them to other nodes (Barabási and Albert 1999). The scale-free property of the architecture can formally described by the degree distribution of nodes which approximates a power law, $P(k) \sim k^{-\gamma}$, where k the node degree. The "scale-free"

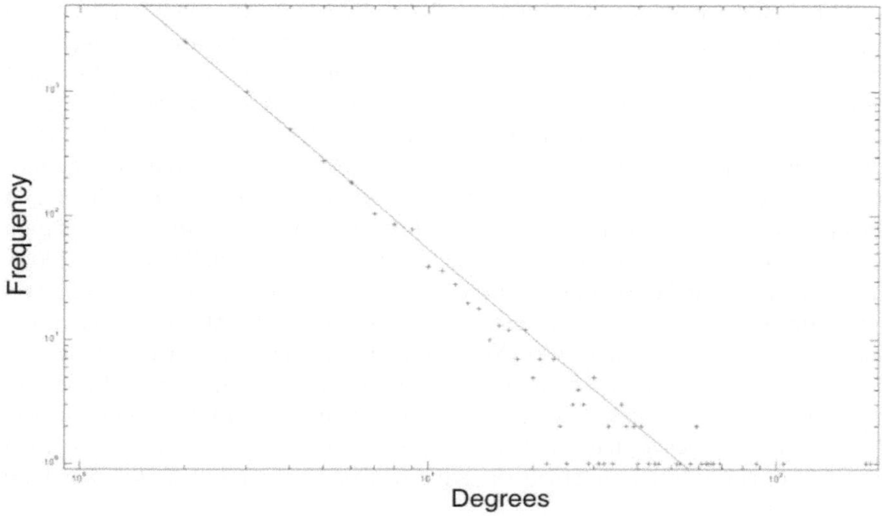

Fig. 2.7 Degree distribution in scale-free network. The network has 5,000 nodes and was generated using the Barabasi–Albert (Barabási and Albert 1999) model of preferential attachment (2 nodes per step). Notice how the probability of a node connecting with many nodes decreases according to the $k^{-\gamma}$ power law

term comes from the dependence of the $\frac{P(ak)}{P(k)}$ ratio only from a. This is illustrated in Fig. 2.7. Power law curve fitting has been addressed many times in the literature with different methods, including least squares fitting as the most popular approach. The choice of method however is crucial since substantial inaccuracies might arise (Clauset et al. 2007). Also, we have to note two important points. The first is that the intuitive assumption that a scale-free network's sub-networks are scale-free does not always hold (Stumpf et al. 2005). The second point applies to datasets that give rise to scale-free networks. In this cases the reader must be aware that the scale-free architecture might be an artifact caused by regularities and biases in the selection of the dataset (Han et al. 2005) and not reflect any biological importance. An observable consequence of the scale-free structure in biological networks is that only a few nodes have many interactions and this can attribute to robustness against random mutations (Albert et al. 2000). Except robustness again random mutations the scale-free architecture implies, by definition, that there are nodes in the network that have more connections than others. When this is interpreted in the context of community structure of networks the notion of hub nodes emerges. Hub nodes are nodes that interact with many partners. For that reason and in contrast with non-hub nodes, hub nodes are extremely sensitive to the targeted mutation (Jeong et al. 2001). Hub nodes are points of convergence and in some cases connect different functional modules of the network that appear in the form of communities. There are two types of hubs in biological networks: party hubs, where most of the interactions are simultaneous, and date hubs, where different interactions take place at different times (Han et al. 2004).

Although the high degree of nodes in biological networks could imply importance of a specific node, betweenness is often used as a measure as well. By definition, nodes and edges of high betweenness accumulate the majority of shortest paths passing through them. Therefore, seen by a "path of least resistance" principle these nodes become the central points controlling the direction of information passing in the network. Newman and Girvan argue that high betweenness implies nodes or edges that connect modules in the network (Newman and Girvan 2004) and thus possibly promote their crosstalk. Additionally it has found that clustering on betweenness results to clusters with similar functional annotation (Dunn et al. 2005). However, although these claims might seem intuitive, there has been a lack of direct supporting evidence. Yu et al. presented some evidence by bioinformatic analysis of yeast protein interaction networks (Yu et al. 2007). In their work, Yu et al. defined high betweenness nodes as "bottleneck" nodes and dissected the types of nodes to hub bottleneck, non-hub bottleneck, hub non-bottleneck, non-hub non-bottleneck. They showed evidence that non-hub bottleneck nodes tend to be essential when involved in non-transient interactions, are rarely parts of large complexes and are joints for crosstalk.

Taking all this into account as well as some contradicting evidence in the literature, like the findings of Goh et al. (Kahng et al. 2003), who showed a correlation between betweenness and degree in social networks, the reader has to be cautious about the use of these measures for such predictions. Yu et al. (2007) also argue that degree might be a better predictor for protein interaction networks specifically. In any case even partial corroboration of predictions by previous biological knowledge is advised.

2.5.4 *Patterns in Networks: Motifs and Graphlets*

2.5.4.1 Network Motifs

Introduced by Uri Alon's group (Milo et al. 2002; Shen-Orr et al. 2002; Alon 2007) in 2002, network motifs are significantly enriched reoccurring connectivity patterns of sub-graphs that are found in networks and have become a concept widely applied in the analysis of the architecture of all types of, including biological, networks. An example of a network motif can be seen in Fig. 2.8. These small circuits are the building blocks of networks and their functional significance is reinforced by the high degree of evolutionary conservation in the yeast interactome (Wuchty et al. 2003; Meshi et al. 2007) and the convergent evolution patters observed in transcription regulation networks (Hinman et al. 2003; Conant and Wagner 2003; Hinman and Davidson 2007).

The premise of the proposed motif detection algorithm by Milo et al. (2002) is based on a simple, but yet very computationally intense graph theory analysis. Initially the network is scanned and all sub-graphs of size n, with $n = \{1, 2, \ldots, n_{max}\}$ are enumerated. In practice this is very hard for $n_{max} > 7$. Any given observed network might contain any number of n-sized sub-graphs, so in order

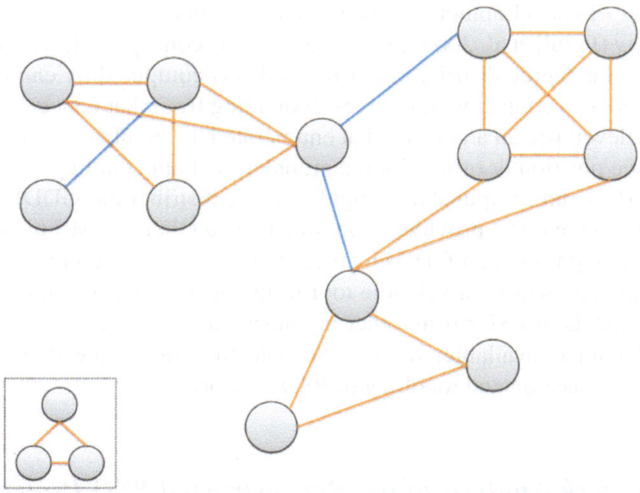

Fig. 2.8 An example of multiple occurrence of a simple triangular motif (*in box*) in a network

to focus on the significant ones, the authors validate significance by multiple testing. This validation is done, by creating an ensemble of randomised versions of the original observed network and re-enumerating the occurrences of each different type of n-sized sub-graph found in the observed network. These randomised versions have the same number of nodes and edges, the same degree for each node and the same number of $(n-1)$-sized motif appearances but different configuration of edges. This way the randomised networks have the same node characteristics and also ensure that there is no high significance assigned to n-sized motif patterns just because their $(n-1)$-sized sub-patterns appear more frequently. Network motifs are those patterns for which the probability P of appearing in the randomised networks an equal or greater number of times than in the real network is lower than a cutoff value p-value α, which indicates the significance of the motif. There are several implementations of motif detection algorithms, based on the original proposed in (Milo et al. 2002). Here we will mention MFINDER (Kashtan et al. 2004), a command line tool easily integrated in workflows and FANMOD (Wernicke and Rasche 2006), which offers a graphical user interface (GUI) application with more versatile output capabilities.

2.5.4.2 Graphlets

Graphlets are based on a concept almost identical to motifs, initially proposed by Przulj in 2006 (Przulj et al. 2006; Przulj 2007). Graphlets are small connected non-isomorphic sub-graphs of the network. The difference in definition with motifs is that they do not have to be significantly enriched. However the main difference is in the manner of use. Graphlets can be used as systematic measures of a network's local topology (Przulj et al. 2006) by generalising the concept of degree distribution. In particular, the degree distribution of the nodes is computed by enumerating the number of nodes in contact with k edges. Extending from that concept, Milenković et al. (2008) developed an algorithm that enumerated the number of nodes in contact with a specific position or orbit within a graphlet, without counting automorphic or symmetric orbits and computed the graphlet degree distribution (GDD). Computing and normalising the GDD provides a measure that can then be used to compare two networks by comparing their GDDs by means of distribution comparison.

Using graphlets will be a valuable tool in the near future, when there are high confidence models of PSDs from other organisms or different brain regions. Once this type of data is available, we will be able to address questions rearing the differential organisation and topology of PSD networks.

2.5.5 *Network Analysis of the Reconstructed PSD Protein Complexes Network Models*

2.5.5.1 Network Topology

Examining the shortest paths in both networks shows how the modules and pathways involved interact. The average number of interactions separating any pair of proteins is very low, with the average shortest path (ASP) length between any two nodes being 3.3 for NRC/MASC and 2.25 for the PSD-95 associated proteins interaction network. These short average distances between proteins nodes in both networks imply a crosstalk between the signal transduction pathways. Also one can notice how specific proteins have central positions in the architecture of the network and how these positions reflect their functionality. PSD-95 for example, which was also used as the bait for the isolation of the PSD-95 associated proteins complex, has an ASP to all other proteins of 1.3 edges showing how central that protein is for the organisation of these complexes – since as a central member of the membrane-associated guanylate kinase (MAGUK) family, it operates as a scaffolding and is involved in anchoring many cental PSD molecules such as NMDA and AMPA receptors and ion channels.

2.5.5.2 Modular Structure

In the NRC/MASC network, the connected component was found to possess a clearly modular structure (Q = 0.56) (Fig. 2.10), with ~75% of its proteins contained in the five largest clusters. Cluster 1 contains all ionotropic glutamate receptor proteins ($p < 10^{-3}$) and a large number of scaffolding molecules. In total, ~50% of its proteins are essential to normal synaptic plasticity ($p < 10^{-2}$) and ~40% are implicated in schizophrenia ($p < 10^{-2}$). Cluster 2 appears specialised for metabotropic/G-protein coupled signalling ($p < 10^{-2}$ for both) with half of its proteins having known behavioural phenotypes ($p<10^{-2}$). Cluster 3, the largest, is strongly connected to clusters 1 and 2. Its size and centrality within the network suggest that it assimilates signals from various sources and co-ordinates common effector mechanisms. The cluster also contains a concentration of tyrosine kinases ($p<10^{-3}$) which are a point of convergence for multiple signalling pathways regulating NMDA receptor activity (Salter and Kalia 2004). This could suggest that cluster 3 integrates the ionotropic and metabotropic signals of clusters 1 and 2. Cluster 4 encapsulates the well-studied MAPK–ERK signalling pathway. ERK activation has been linked to transcription, protein synthesis, regulation of AMPA receptors and structural plasticity (Thomas and Huganir 2004). Cluster 5 is another MAPK pathway (Ser/Thr kinases, $p<10^{-2}$). Clusters 6, 7, 11 and 13 mediate interactions with the cytoskeleton. One other interesting feature of the modular structure of the NRC/MASC network is that clusters can be grouped based on their functionality annotation to two components, the upstream and the downstream, with three major cluster groups: input, information processing and output or effector clusters. Input clusters are part of the upstream component and relay the extracellular message of neurotransmitter release. The information processing clusters are also part of the upstream component and propagate that signal and take molecular decisions that affect similar future propagations. Finally the output clusters which form the downstream component have the role of effectors which produce the desired changes according to the aforementioned decisions. This meta-structure is highlighted in Fig. 2.10.

The PSD-95 associated proteins complex network (Fig. 2.11) also has a clear modular structure ($Q = 0.37$). One can note the location and proximity of the receptors and channels responsible for the postsynaptic depolarization and subsequent action potential generation. All NMDA, AMPA and kainate glutamate receptors were restricted to Cla and Clb and the voltage-dependent K^+ channels were found in Cla and Clc (entirely comprised of K^+ channels). These channels are known to couple to plasticity mechanisms (Kim et al. 2007). It must be also noted that Cla contains important signalling enzymes involved in plasticity, including CAMKII (Frankland et al. 2001) and SYNGAP (Komiyama et al. 2002). It therefore seems that Cla, Clb and Clc are enriched with membrane proteins responsible for the electrical properties of the postsynaptic terminal. Out of 28 schizophrenia-implicated proteins, 20 were mapped into the network model (see also Fig. 2.12). It is worth mentioning that comparing the NRC/MASC and PSD-95 associated proteins complex networks revealed an overlap between the two datasets that was withing the major connected components was centered around Clusters 1 and Cla respectively (10/16 proteins, $p<10^{-3}$).

2.5.5.3 Effect of Topology on Function

An important aspect of the networks topology emerges when we overlay electrophysiology data on the network. In contrast with annotation correlations which

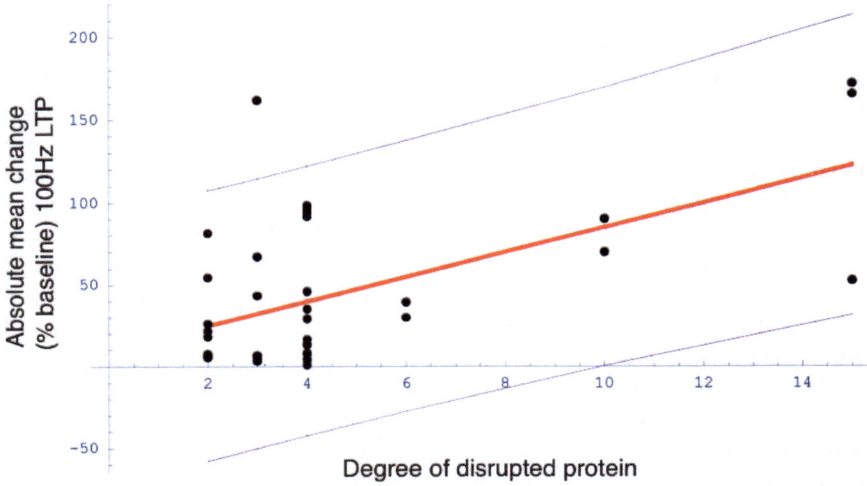

Fig. 2.9 Absolute change (% baseline) in 100 Hz LTP is shown as a function of the degree of the protein disrupted. A significant linear fit was also evident for the full data set of 36 points (Data from Pocklington et al. 2006)

have are the more predictive side of these models, we can also demonstrate the effect on of a node's position and importance in the network on physiology. Quantitative data on the perturbation of LTP/LTD caused by disruption of individual proteins by mutation were available for a subset of MASC proteins with synaptic plasticity phenotypes. Despite the inherent variability of this data due to differences in experimental protocols, protein degree and quantitative perturbation of LTP on disruption were found to be strongly correlated (linear fit: $p<10^{-3}$, Pearson correlation 0.85). Figure 2.9 illustrates this concept by showing the absolute change (% baseline) in 100 Hz LTP as a function of the degree of the protein disrupted. What is visible in this figure is that the higher the degree of a node the higher is the effect of its disruption to the function of the network as a whole.

2.6 Visualisation

2.6.1 Visualisation Software

In the data intense environment of modern science, data visualisation is an independent branch by itself. It is of great importance to Systems Biology and data driven science in general. Since the models discussed in this work are descriptive and integrate annotations from various sources , this means that they carry a lot of information. Good and informative visualisation can convey abstract, complex information in intuitive ways, taking advantage of the human eye's bandwidth and allow us to understand large amounts of information at once.

There are many software solutions for data visualisation. These vary from general graph visualisation tools that implement graph drawing algorithms like GraphViz (http://www.graphviz.org/) to more biologically oriented software, on which we will focus. The former can be applied as parts of workflows since they can be integrated in many scripting languages. The latter allow interaction with the model via a graphical user interface and a set of different functionalities such as a selection of drawing algorithms, analysis plug-ins and visualisation schemes and options. Although interactive interface design allows some exploratory analysis to the models the choice of tool is mostly a matter of preference (note: tool performance varies with the dataset size and the type of information available).

Cytoscape (Shannon et al. 2003) is an open source platform for visualizing molecular interaction networks and biological pathways and integrating these networks with annotations, gene expression profiles and other state data. Features that make Cytoscape a useful tool are the support of many different formats and standards for input and output (e.g. plain text, csv, spreadsheets, gml, xgmml) as well as an application programming interface (API) which allows easy integration in all workflows, an intuitive graphical interface, support of database web services for data import, efficient 2D visualisation with most visual parameters (graphics, colours, sizes, transparency, etc.) customisable according to attributes, a variety of layout algorithms and enhanced navigation. Its plug-in architecture is probably the strongest feature of Cytoscape, because it allows the community to design and implement freely available add-ons. These plug-ins can perform functions such as community search or annotation and streamline the network analysis process. BioLayout3D (Theocharidis et al. 2009) is the continuation of the BioLayout (Enright and Ouzounis 2001) project, which started as an implementation of a graph layout algorithm. BioLayout3D offers rendering in 3D, which is useful for navigating in and exploring larger networks. Although BioLaoyout3D offers some network analysis features, due to its non plug-in based architecture, these capabilities are more limited. Beyond just visualising PPINs the concept of connections (edges or interactions) between agents (nodes or proteins) has been expanded by Köhler et al. (2006) into a wider concept of relations between entities, where an entities can be proteins and the relation an interaction or entities can be an annotation and a protein and the relation an association. This more complicated knowledge model is implemented in Ondex (http://www.ondex.org), a software package that allows data from diverse biological data sets to be linked, integrated and visualised through graph analysis techniques.

2.6.2 *Visualising Networks*

There can be many different approaches in visualising a protein interaction network. The typical approach is to draw the network in some informative manner, e.g. the nodes separated in communities or by subcellular location and highlight nodes with specific features, e.g. of a specific classification or associated with some disease.

Sometimes, when the datasets are too big its more informative to visualise more collapsed versions of a network. Minimizing the clutter of visualisation by making the infomation more compact without reducing it can give a quick overview of a dataset or even provide a collapsed representation that can tease information out of the model. That could be achieved in the form of a meta-network. Meta-networks are the same models visualised with the methods mentioned earlier, but in a collapsed form. An example of such collapsed form can be obtained for example by grouping the nodes based on a common property (e.g. family) and assigning a meta-edge if nodes with that property interact in the network. Information is visualised in order to make a structured collection of data shorter and concise. In the same spirit visualising networks using meta-networks is a way of summarising the existent model into something even more concise. This higher-level view might allow observations that could not be made otherwise due to limitations in visualising large datasets.

A more sophisticated concept of visualisation which stands between collapsed representation, network layout algorithm and clustering is power graph (Royer et al. 2008). Power graph analysis is based on the concept of power node and power edge representation. A power node represents a fully connected clique, i.e. a group of nodes that are fully connected and this representation can be used to generate graph primitives like the one of a star motif or a biclique. A power edge represents the case of two power nodes of any size, where all the respective member nodes are connected. Royer et al. presented an algorithm that scans the graph for power nodes and generates a layout using power edges. Power graphs can compress up to 85% of the edges in a network while retaining all the information.

2.6.3 Visualisations of PSD Protein Interaction Networks

The NRC/MASC and PSD-95 associated proteins complex interaction networks can be seen in Figs. 2.10 and 2.11 respectively. Different illustrations of a network allow the researcher to focus on different features or properties of a network. Take for example the illustrations of NRC/MASC and PSD-95 associated proteins complex interaction networks in Figs. 2.12, where nodes involved in schizophrenia are highlighted. Just by examining them its easy to spot the correlations of the disease with specific clusters in the networks. These observations are also corroborated by the statistical analysis.

The concept of a collapsed visual representation using meta-networks is illustrated in Fig. 2.13. The latter figure also illustrates how while the NRC/MASC protein interaction network has similar representation of all proteins families except Enzymes, interconnected with a meta-edges of a similar weight, the PSD-95 associated proteins complex interaction network is slightly more dominated by Siganlling molecules and Enzymes, Cytoskeletal/Structural/Cell adhesion, Adaptor/Regulatory and Receptors/Channels/Transporters families, the self-interactions of the latter two families as well as interactions of the Kinases family with the Receptors/Channels/Transporters family.

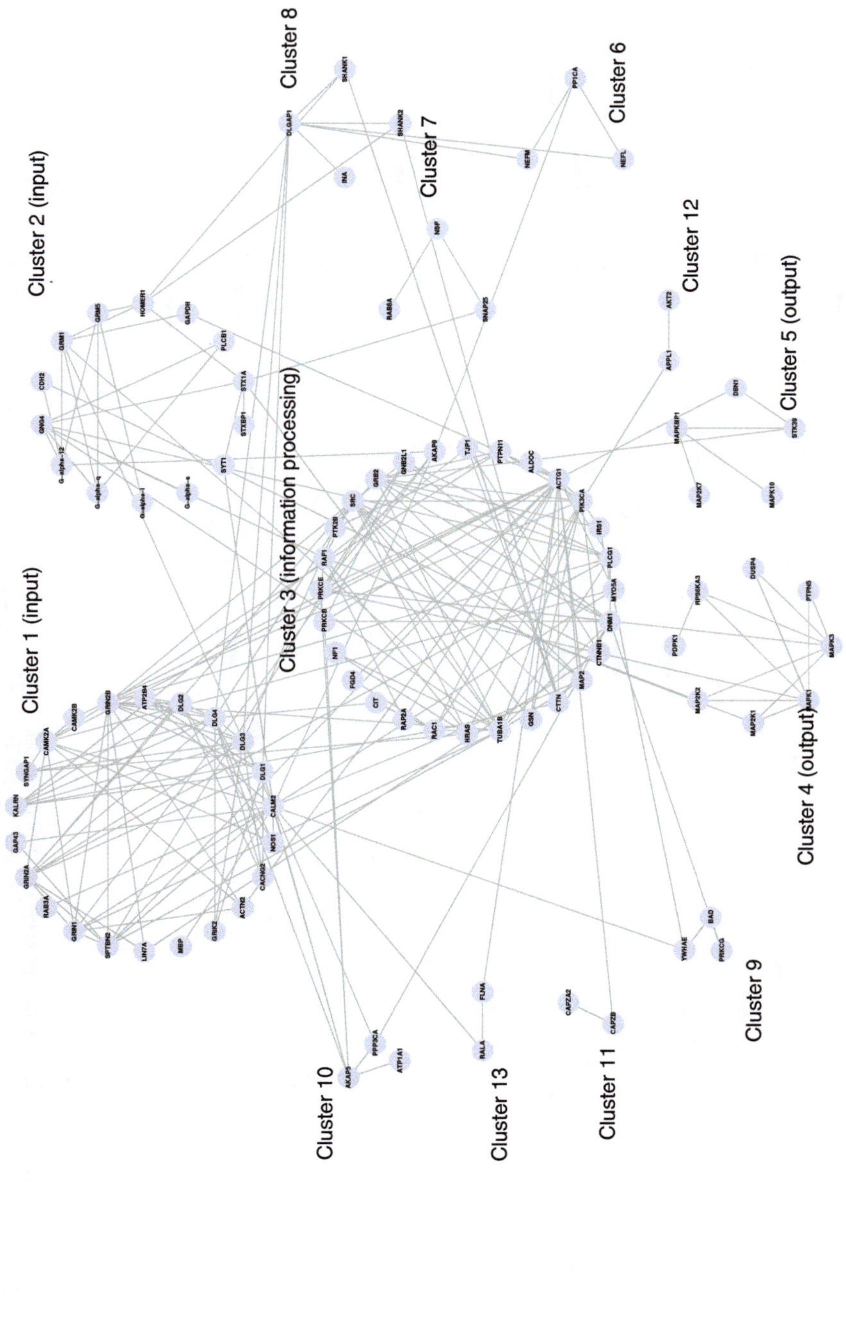

Fig. 2.10 The NRC/MASC interaction network

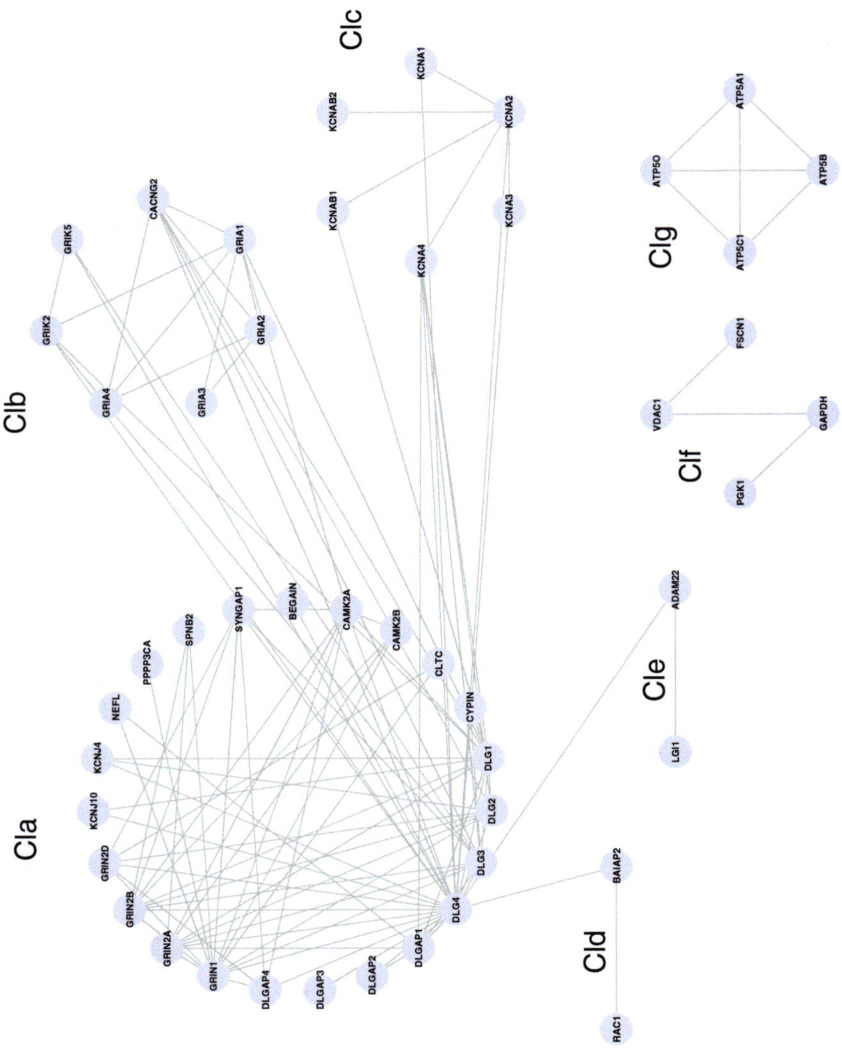

Fig. 2.11 The PSD-95 associated proteins interaction network

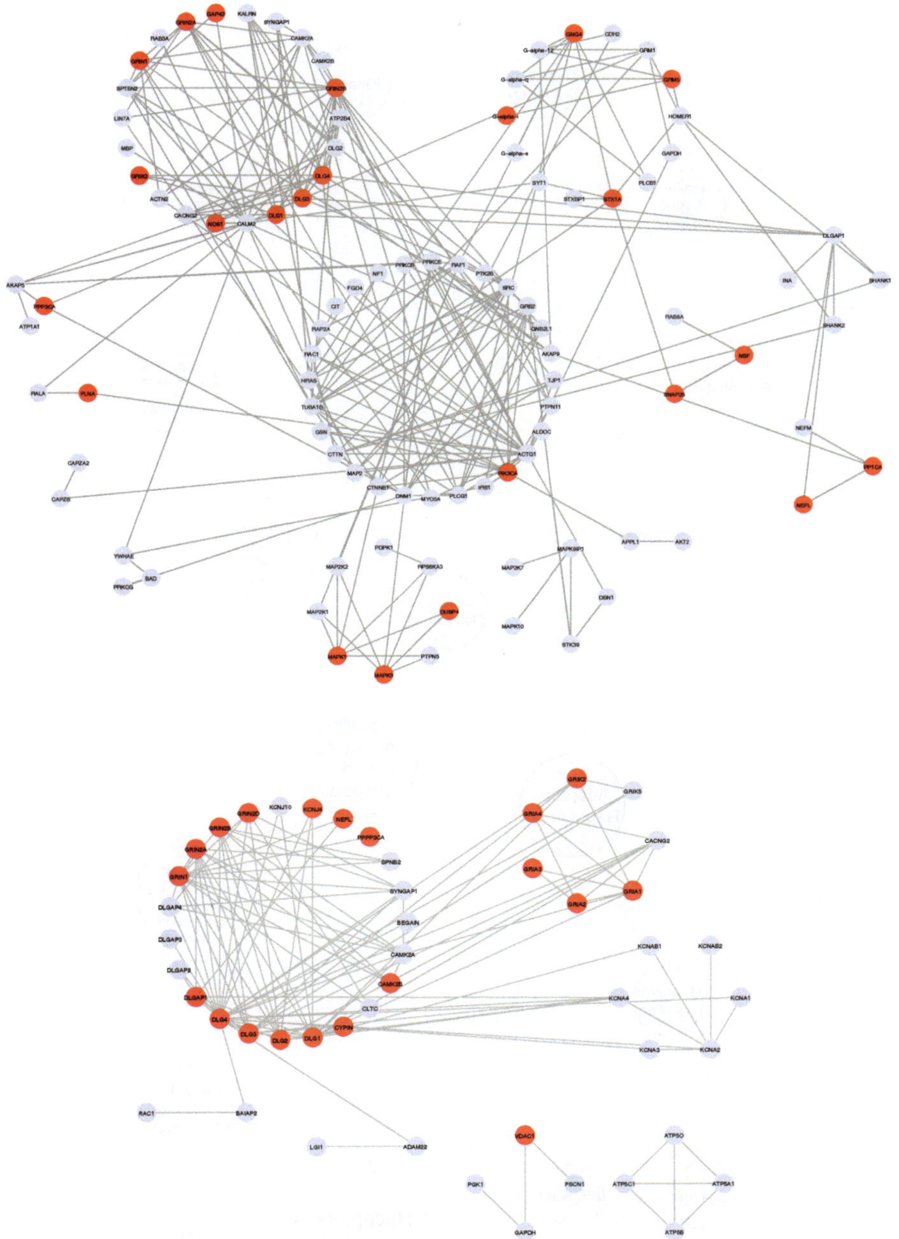

Fig. 2.12 Proteins implicated in schizophrenia (highlighted in *red*) in the NRC/MASC (*top*) and PSD-95 associated proteins complex (*bottom*) interaction networks

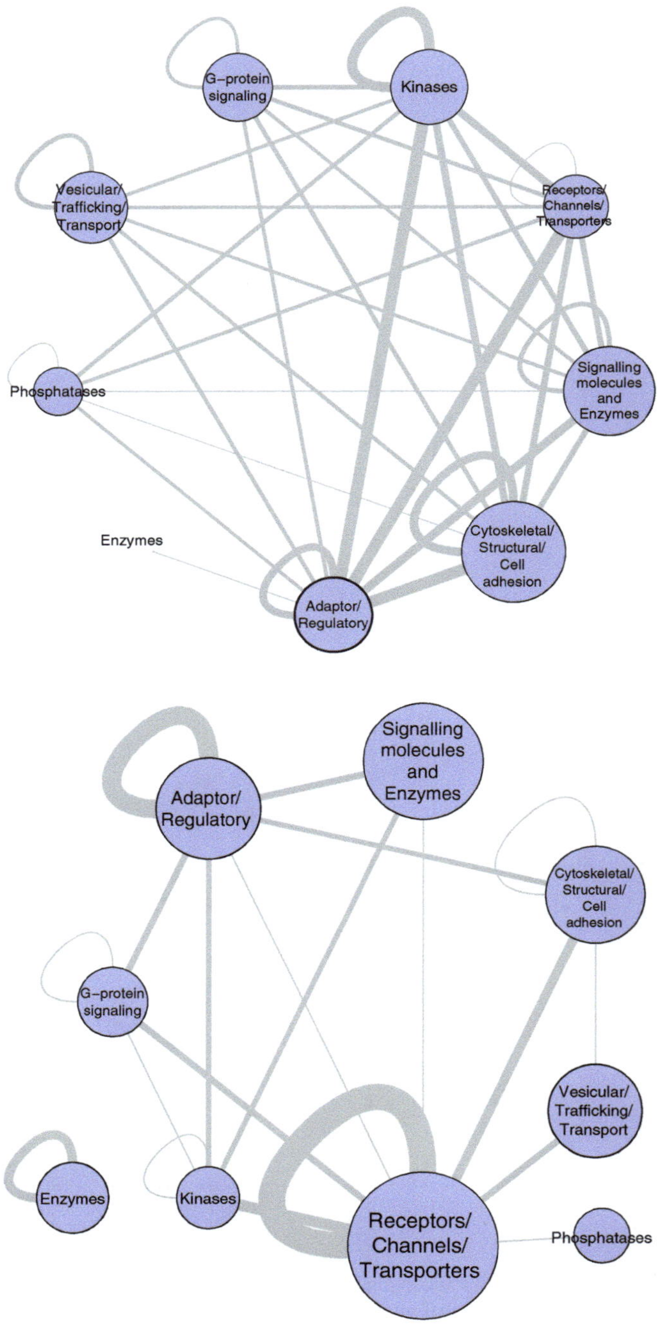

Fig. 2.13 NRC/MASC (*top*) and PSD-95 associated proteins complex (*bottom*) interaction networks collapsed by protein family. The node size and edge width are according to the percentage of nodes in the datasets and the number of interactions between nodes of two types respectively

2.7 Discussion

Systems Biology (and Systems Medicine) analysis of the brain aims to provide a framework upon which we can understand it at all levels of its complexity from bio-molecular events at synapses through complex networks of neurons, brain regions and systems and ultimately to individual and social behaviour. Here we have focussed on strategies and methods that help us capture, explore and understand molecular complexes identified from primary biochemical analysis of neural tissues. Protein interaction network models provide a powerful scaffold for knowledge integration and hypothesis generation. By combining data annotation, analysis and protein interaction network reconstruction researchers can model and investigate proteomics datasets. Utilising the resulting models as integrative descriptive tools can also offer an overview of the major constituent parts of molecular machines and also insight on how these parts get combined to give rise to the properties of a complex system such as the receptor signalling complexes embedded in synapse proteomes. Closer examination of these models can also be used to explore the validity of disease hypotheses. In the case of the PSD for example using models like the ones described here one can see that the primary interactors of genes associated with disease, e.g. schizophrenia, are spread throughout many modules within the network. This suggests that the overall network and its various clusters might play a role in schizophrenia, and while enriched, the glutamate receptors are not the entire story, as was might be considered in the "glutamate hypothesis" of schizophrenia (Greene 2001; Coyle 2006; Lisman et al. 2008). We can also start to query common mechanisms that might be shared across multiple diseases. For example, the reconstructed PSD models uncover 43 proteins linked by various lines of evidence (all individually weak) to schizophrenia of which 20 are also implicated in other diseases as well (bipolar disorder, depression mental retardation). As confidence in the data underpinning these models increases with time, these methods will start to deliver on their potential. However, as mentioned throughout, one must always bear in mind the significant limitations of these models when making any decisions based upon them. The list of components and their interactions are first and foremost limited by the sensitivity and the accuracy of the methods used to identify them and these are all developing technologies. Perhaps more importantly, we have described the use of a static modelling framework to analyse what is fundamentally an adaptive and highly dynamic structure. The spatio-temporal characteristics of the synapse are what gives it its function and while we can map some high level aspects of this onto static models as annotation, we are acutely aware that in the longer term we need to look to more dynamic modelling approaches. Dynamic models will require even more parameters (e.g. binding site affinities, quantitative proteomics etc.) and so extending from reduced pathway models of a few molecules to the scale of models described here (>100s of molecules) will take a significant investment in both modelling and biochemical characterisation.

References

Albert R, Jeong H, Barabási A (2000) Error and attack tolerance of complex networks. Nature 406(6794):378–382

Alex B, Haddow B, Grover C (2007) Recognising nested named entities in biomedical text. In: Proceedings of the workshop on BioNLP 2007: biological, translational, and clinical language processing, Prague, pp 65–72

Alex B, Grover C, Haddow B, Kabadjov M, Klein E, Matthews M, Tobin R, Wang X (2008) Automating curation using a natural language processing pipeline. Genome Biol 9(Suppl 2):S10

Allen NC, Bagade S, McQueen MB, Ioannidis JPA, Kavvoura FK, Khoury MJ, Tanzi RE, Bertram L (2008) Systematic meta-analyses and field synopsis of genetic association studies in schizophrenia: the szgene database. Nat Genet 40(7):827–834

Alon U (2007) Network motifs: theory and experimental approaches. Nat Rev Genet 8(6):450–461

Al-Shahrour F, Díaz-Uriarte R, Dopazo J (2004) Fatigo: a web tool for finding significant associations of gene ontology terms with groups of genes. Bioinformatics 20(4):578–580

Altaf-Ul-Amin M, Shinbo Y, Mihara K, Kurokawa K, Kanaya S (2006) Development and implementation of an algorithm for detection of protein complexes in large interaction networks. BMC bioinformatics 7(1):207

Antonov A, Schmidt T, Wang Y, Mewes H (2008) Profcom: a web tool for profiling the complex functionality of gene groups identified from high-throughput data. Nucleic Acids Res 36(suppl 2):W347

Arnau V, Mars S, Marín I (2005) Iterative cluster analysis of protein interaction data. Bioinformatics 21(3):364

Ashburner M, Ball C, Blake J, Botstein D, Butler H, Cherry J, Davis A, Dolinski K, Dwight S, Eppig J (2000) Gene ontology: tool for the unification of biology. The gene ontology consortium. Nat Genet 25(1):25

Asur S, Ucar D, Parthasarathy S (2007) An ensemble framework for clustering protein–protein interaction networks. Bioinformatics 23(13):i29

Bader G, Hogue C (2003) An automated method for finding molecular complexes in large protein interaction networks. BMC Bioinformatics 4(1):2

Bai F, Witzmann F (2007) Synaptosome proteomics. In: Bertrand E, Faupel MD (eds) Subcellular proteomics. Springer, Dordrecht, pp 77–98

Barabási A, Albert R (1999) Emergence of scaling in random networks. Science 286(5439):509

Barabási A, Dezso Z, Ravasz E, Yook S, Oltvai Z (2003) Scale-free and hierarchical structures in complex networks. AIP Conf Proc 661:1

Barnard G (1945) A new test for 2 × 2 tables. Nature 156(3954):177–177

Barrett T, Troup DB, Wilhite SE, Ledoux P, Rudnev D, Evangelista C, Kim IF, Soboleva A, Tomashevsky M, Marshall KA, Phillippy KH, Sherman PM, Muertter RN, Edgar R (2009) Ncbi geo: archive for high-throughput functional genomic data. Nucleic Acids Res 37(Database issue):D885–D890

Barrett T, Troup DB, Wilhite SE, Ledoux P, Evangelista C, Kim IF, Tomashevsky M, Marshall KA, Phillippy KH, Sherman PM, Muertter RN, Holko M, Ayanbule O, Yefanov A, Soboleva A (2011) Ncbi geo: archive for functional genomics data sets—10 years on. Nucleic Acids Res 39(Database issue):D1005–D1010

Barrios-Rodiles M, Brown KR, Ozdamar B, Bose R, Liu Z, Donovan RS, Shinjo F, Liu Y, Dembowy J, Taylor IW, Luga V, Przulj N, Robinson M, Suzuki H, Hayashizaki Y, Jurisica I, Wrana JL (2005) High-throughput mapping of a dynamic signaling network in mammalian cells. Science 307(5715):1621–1625

Bayés A, Grant S (2009) Neuroproteomics: understanding the molecular organization and complexity of the brain. Nat Rev Neurosci 10(9):635–646

Bayés A, van de Lagemaat L, Collins M, Croning M, Whittle I, Choudhary J, Grant S (2010) Characterization of the proteome, diseases and evolution of the human postsynaptic density. Nat Neurosci 14(1):19–21

Bécamel C, Alonso G, Galéotti N, Demey E, Jouin P, Ullmer C, Dumuis A, Bockaert J, Marin P (2002) Synaptic multiprotein complexes associated with 5-ht(2c) receptors: a proteomic approach. EMBO J 21(10):2332–2342

Bellman R (1958) On a routing problem. Q Appl Math 16(1):87–90

Benjamini Y, Hochberg Y (1995) Controlling the false discovery rate: a practical and powerful approach to multiple testing. J R Stat Soc Ser B (Methodological) 57(1):289–300

Benjamini Y, Yekutieli D (2001) The control of the false discovery rate in multiple testing under dependency. Ann Stat 29(4):1165–1188

Berglund AC, Sjölund E, Ostlund G, Sonnhammer ELL (2008) Inparanoid 6: eukaryotic ortholog clusters with inparalogs. Nucleic Acids Res 36(Database issue):D263–D266

Berkel S, Marshall CR, Weiss B, Howe J, Roeth R, Moog U, Endris V, Roberts W, Szatmari P, Pinto D, Bonin M, Riess A, Engels H, Sprengel R, Scherer SW, Rappold GA (2010) Mutations in the shank2 synaptic scaffolding gene in autism spectrum disorder and mental retardation. Nat Genet 42(6):489–491

Berriz GF, King OD, Bryant B, Sander C, Roth FP (2003) Characterizing gene sets with funcassociate. Bioinformatics 19(18):2502–2504

Bertram L, McQueen MB, Mullin K, Blacker D, Tanzi RE (2007) Systematic meta-analyses of alzheimer disease genetic association studies: the alzgene database. Nat Genet 39(1):17–23

Blake JA, Richardson JE, Bult CJ, Kadin JA, Eppig JT, Group MGD (2002) The mouse genome database (mgd): the model organism database for the laboratory mouse. Nucleic Acids Res 30(1):113–115

Bouwmeester T, Bauch A, Ruffner H, Angrand P, Bergamini G, Croughton K, Cruciat C, Eberhard D, Gagneur J, Ghidelli S (2004) A physical and functional map of the human tnf-a/nf-kb signal transduction pathway. Nat Cell Biol 6(2):97–105

Brandes U, Delling D, Gaertler M, Goerke R, Hoefer M, Nikoloski Z, Wagner D (2006) Maximizing modularity is hard. arXiv physics.data-an

Bu D, Zhao Y, Cai L, Xue H, Zhu X, Lu H, Zhang J, Sun S, Ling L, Zhang N (2003) Topological structure analysis of the protein–protein interaction network in budding yeast. Nucleic Acids Res 31(9):2443

Bult CJ, Eppig JT, Kadin JA, Richardson JE, Blake JA, Group MGD (2008) The mouse genome database (mgd): mouse biology and model systems. Nucleic Acids Res 36(Database issue):D724–D728

Burré J, Beckhaus T, Schägger H, Corvey C, Hofmann S, Karas M, Zimmermann H, Volknandt W (2006) Analysis of the synaptic vesicle proteome using three gel-based protein separation techniques. Proteomics 6(23):6250–6262

Carlin RK, Grab DJ, Cohen RS, Siekevitz P (1980) Isolation and characterization of postsynaptic densities from various brain regions: enrichment of different types of postsynaptic densities. J Cell Biol 86(3):831–845

Ceol A, Aryamontri AC, Licata L, Peluso D, Briganti L, Perfetto L, Castagnoli L, Cesareni G (2010) Mint, the molecular interaction database: 2009 update. Nucleic Acids Res 38(Database issue):D532–D539

Cesareni G, Chatr-aryamontri A, Licata L, Ceol A (2008) Searching the mint database for protein interaction information. Curr Protoc Bioinform Chapter 8:Unit 8.5

Chatr-aryamontri A, Ceol A, Palazzi LM, Nardelli G, Schneider MV, Castagnoli L, Cesareni G (2007) Mint: the molecular interaction database. Nucleic Acids Res 35(Database issue):D572–D574

Chatr-aryamontri A, Zanzoni A, Ceol A, Cesareni G (2008) Searching the protein interaction space through the mint database. Methods Mol Biol 484:305–317

Cho YR, Hwang W, Zhang A (2006) Efficient modularization of weighted protein interaction networks using k-hop graph reduction. BioInformatics and bioEngineering, 2006 BIBE 2006 sixth IEEE symposium on, Arlington, Virginia, USA, pp 289–298

Choudhary J, Grant SGN (2004) Proteomics in postgenomic neuroscience: the end of the beginning. Nat Neurosci 7(5):440–445

Clauset A, Newman M, Moore C (2004) Finding community structure in very large networks. Phys Rev E 70(6):066111
Clauset A, Shalizi C, Newman M (2007) Power-law distributions in empirical data. Arxiv preprint arXiv:07061062
Cline MS, Smoot M, Cerami E, Kuchinsky A, Landys N, Workman C, Christmas R, Avila-Campilo I, Creech M, Gross B, Hanspers K, Isserlin R, Kelley R, Killcoyne S, Lotia S, Maere S, Morris J, Ono K, Pavlovic V, Pico AR, Vailaya A, Wang PL, Adler A, Conklin BR, Hood L, Kuiper M, Sander C, Schmulevich I, Schwikowski B, Warner GJ, Ideker T, Bader GD (2007) Integration of biological networks and gene expression data using cytoscape. Nat protoc 2(10):2366–2382
Coba MP, Valor LM, Kopanitsa MV, Afinowi NO, Grant SGN (2008) Kinase networks integrate profiles of n-methyl-d-aspartate receptor-mediated gene expression in hippocampus. J Biol Chem 283(49):34101–34107
Coba MP, Pocklington AJ, Collins MO, Kopanitsa MV, Uren RT, Swamy S, Croning MDR, Choudhary JS, Grant SGN (2009) Neurotransmitters drive combinatorial multistate postsynaptic density networks. Sci Signal 2(68):ra19
Collins MO, Husi H, Yu L, Brandon JM, Anderson CNG, Blackstock WP, Choudhary JS, Grant SGN (2006) Molecular characterization and comparison of the components and multiprotein complexes in the postsynaptic proteome. J Neurochem 97(Suppl 1):16–23
Conant GC, Wagner A (2003) Convergent evolution of gene circuits. Nat Genet 34(3):264–266
Coyle JT (2006) Glutamate and schizophrenia: beyond the dopamine hypothesis. Cell Mol Neurobiol 26(4–6):365–384
Craig AM, Kang Y (2007) Neurexin-neuroligin signaling in synapse development. Curr Opin Neurobiol 17(1):43–52
Croning MDR, Marshall MC, McLaren P, Armstrong JD, Grant SGN (2009) G2cdb: the genes to cognition database. Nucleic Acids Res 37(Database issue):D846–D851
Dennis G, Sherman BT, Hosack DA, Yang J, Gao W, Lane HC, Lempicki RA (2003) David: database for annotation, visualization, and integrated discovery. Genome Biol 4(5):P3
Dijkstra E (1959) A note on two problems in connexion with graphs. Numer Math 1(1):269–271
Dittrich M, Klau G, Rosenwald A, Dandekar T, Muller T (2008) Identifying functional modules in protein–protein interaction networks: an integrated exact approach. Bioinformatics 24(13):i223–i231
Doniger SW, Salomonis N, Dahlquist KD, Vranizan K, Lawlor SC, Conklin BR (2003) Mappfinder: using gene ontology and genmapp to create a global gene-expression profile from microarray data. Genome Biol 4(1):R7
Doyle JP, Dougherty JD, Heiman M, Schmidt EF, Stevens TR, Ma G, Bupp S, Shrestha P, Shah RD, Doughty ML, Gong S, Greengard P, Heintz N (2008) Application of a translational profiling approach for the comparative analysis of cns cell types. Cell 135(4):749–762
Dunn R, Dudbridge F, Sanderson C (2005) The use of edge-betweenness clustering to investigate biological function in protein interaction networks. BMC Bioinformatics 6:39
Emes RD, Pocklington AJ, Anderson CNG, Bayes A, Collins MO, Vickers CA, Croning MDR, Malik BR, Choudhary JS, Armstrong JD, Grant SGN (2008) Evolutionary expansion and anatomical specialization of synapse proteome complexity. Nat Neurosci 11(7):799–806
Enright A, Ouzounis C (2001) Biolayout—an automatic graph layout algorithm for similarity visualization. Bioinformatics 17(9):853
Enright A, Dongen SV, Ouzounis C (2002) An efficient algorithm for large-scale detection of protein families. Nucleic Acids Res 30(7):1575
Eppig JT, Bult CJ, Kadin JA, Richardson JE, Blake JA, Anagnostopoulos A, Baldarelli RM, Baya M, Beal JS, Bello SM, Boddy WJ, Bradt DW, Burkart DL, Butler NE, Campbell J, Cassell MA, Corbani LE, Cousins SL, Dahmen DJ, Dene H, Diehl AD, Drabkin HJ, Frazer KS, Frost P, Glass LH, Goldsmith CW, Grant PL, Lennon-Pierce M, Lewis J, Lu I, Maltais LJ, McAndrews-Hill M, McClellan L, Miers DB, Miller LA, Ni L, Ormsby JE, Qi D, Reddy TBK, Reed DJ, Richards-Smith B, Shaw DR, Sinclair R, Smith CL, Szauter P, Walker MB, Walton

DO, Washburn LL, Witham IT, Zhu Y, Group MGD (2005) The mouse genome database (mgd): from genes to mice – a community resource for mouse biology. Nucleic Acids Res 33(Database issue):D471–D475

Eppig JT, Blake JA, Bult CJ, Kadin JA, Richardson JE, Group MGD (2007) The mouse genome database (mgd): new features facilitating a model system. Nucleic Acids Res 35(Database issue):D630–D637

Erdös P, Szckeres G (1987) A combinatorial problem in geometry. In: Gessel I, Rota G.-C (eds) Classic papers in combinatorics. Birkhaüser, Boston, pp 49–56

Ewing RM, Chu P, Elisma F, Li H, Taylor P, Climie S, McBroom-Cerajewski L, Robinson MD, O'Connor L, Li M, Taylor R, Dharsee M, Ho Y, Heilbut A, Moore L, Zhang S, Ornatsky O, Bukhman YV, Ethier M, Sheng Y, Vasilescu J, Abu-Farha M, Lambert JP, Duewel HS, Stewart II, Kuehl B, Hogue K, Colwill K, Gladwish K, Muskat B, Kinach R, Adams SL, Moran MF, Morin GB, Topaloglou T, Figeys D (2007) Large-scale mapping of human protein–protein interactions by mass spectrometry. Mol Syst Biol 3:89

Farr CD, Gafken PR, Norbeck AD, Doneanu CE, Stapels MD, Barofsky DF, Minami M, Saugstad JA (2004) Proteomic analysis of native metabotropic glutamate receptor 5 protein complexes reveals novel molecular constituents. J Neurochem 91(2):438–450

Fernández E, Collins M, Uren R, Kopanitsa M, Komiyama N, Croning M, Zografos L, Armstrong J, Choudhary J, Grant S (2009) Targeted tandem affinity purification of psd-95 recovers core postsynaptic complexes and schizophrenia susceptibility proteins. Mol Syst Biol 5:269

Finger JH, Smith CM, Hayamizu TF, McCright IJ, Eppig JT, Kadin JA, Richardson JE, Ringwald M (2011) The mouse gene expression database (gxd): 2011 update. Nucleic Acids Res 39(Database issue):D835–D841

Finn RD, Mistry J, Tate J, Coggill P, Heger A, Pollington JE, Gavin OL, Gunasekaran P, Ceric G, Forslund K, Holm L, Sonnhammer ELL, Eddy SR, Bateman A (2010) The pfam protein families database. Nucleic Acids Res 38(Database issue):D211–D222

Fisher R (1922) On the interpretation of chi squared from contingency tables, and the calculation of p. J R Stat Soc 85(1):87–94

Frankland PW, O'Brien C, Ohno M, Kirkwood A, Silva AJ (2001) Alpha-camkii-dependent plasticity in the cortex is required for permanent memory. Nature 411(6835):309–313

Franks KM, Sejnowski TJ (2002) Complexity of calcium signaling in synaptic spines. Bioessays 24(12):1130–1144

Gavin A, Bösche M, Krause R, Grandi P, Marzioch M, Bauer A, Schultz J, Rick J, Michon A, Cruciat C (2002) Functional organization of the yeast proteome by systematic analysis of protein complexes. Nature 415(6868):141–147

Gavin A, Aloy P, Grandi P, Krause R, Boesche M, Marzioch M, Rau C, Jensen L, Bastuck S, Dümpelfeld B (2006) Proteome survey reveals modularity of the yeast cell machinery. Nature 440(7084):631–636

Gelbart W, Crosby M, Matthews B, Rindone W, Chillemi J, Twombly S, Emmert D, Ashburner M, Drysdale R, Whitfield E (1997) Flybase: a drosophila database. The flybase consortium. Nucleic Acids Res 25(1):63

Gould CM, Diella F, Via A, Puntervoll P, Gemünd C, Chabanis-Davidson S, Michael S, Sayadi A, Bryne JC, Chica C, Seiler M, Davey NE, Haslam N, Weatheritt RJ, Budd A, Hughes T, Pas J, Rychlewski L, Travé G, Aasland R, Helmer-Citterich M, Linding R, Gibson TJ (2010) Elm: the status of the 2010 eukaryotic linear motif resource. Nucleic Acids Res 38(Database issue):D167–D180

Grant S (2003) Synapse signalling complexes and networks: machines underlying cognition. Bioessays 25(12):1229–1235

Greene R (2001) Circuit analysis of nmdar hypofunction in the hippocampus, in vitro, and psychosis of schizophrenia. Hippocampus 11(5):569–577

Greene D, Cagney G, Krogan N, Cunningham P (2008) Ensemble non-negative matrix factorization methods for clustering protein–protein interactions. Bioinformatics 24(15):1722

Grønborg M, Pavlos NJ, Brunk I, Chua JJE, Münster-Wandowski A, Riedel D, Ahnert-Hilger G, Urlaub H, Jahn R (2010) Quantitative comparison of glutamatergic and gabaergic synaptic vesicles unveils selectivity for few proteins including mal2, a novel synaptic vesicle protein. J Neurosci 30(1):2–12

Groth P, Pavlova N, Kalev I, Tonov S, Georgiev G, Pohlenz HD, Weiss B (2007) Phenomicdb: a new cross-species genotype/phenotype resource. Nucleic Acids Res 35(Database issue): D696–D699

Guimerà R, Amaral LAN (2005) Functional cartography of complex metabolic networks. Nature 433(7028):895–900

Hahn CG, Banerjee A, Macdonald ML, Cho DS, Kamins J, Nie Z, Borgmann-Winter KE, Grosser T, Pizarro A, Ciccimaro E, Arnold SE, Wang HY, Blair IA, Fox D (2009) The post-synaptic density of human postmortem brain tissues: an experimental study paradigm for neuropsychiatric illnesses. PLoS ONE 4(4):e5251

Hall M, Frank E, Holmes G, Pfahringer B (2009) The weka data mining software: an update. ACM SIGKDD Explor 11:10–18

Hamdan FF, Daoud H, Piton A, Gauthier J, Dobrzeniecka S, Krebs MO, Joober R, Lacaille JC, Nadeau A, Milunsky JM, Wang Z, Carmant L, Mottron L, Beauchamp MH, Rouleau GA, Michaud JL (2011) De novo syngap1 mutations in nonsyndromic intellectual disability and autism. Biol Psychiatry 69(9):898–901

Han JDJ, Bertin N, Hao T, Goldberg DS, Berriz GF, Zhang LV, Dupuy D, Walhout AJM, Cusick ME, Roth FP, Vidal M (2004) Evidence for dynamically organized modularity in the yeast protein–protein interaction network. Nature 430(6995):88–93

Han JDJ, Dupuy D, Bertin N, Cusick ME, Vidal M (2005) Effect of sampling on topology predictions of protein–protein interaction networks. Nat Biotechnol 23(7):839–844

Hanisch D, Zien A, Zimmer R, Lengauer T (2002) Co-clustering of biological networks and gene expression data. Bioinformatics 18(Suppl 1):S145–S154

Harrison PJ, Weinberger DR (2005) Schizophrenia genes, gene expression, and neuropathology: on the matter of their convergence. Mol Psychiatry 10(1):40–68; image 5

Hart P, Nilsson N, Raphael B (1968) A formal basis for the heuristic determination of minimum cost paths. IEEE Trans Syst Sci Cybern SSC4 4(2):100–107

Hartuv E, Shamir R (2000) A clustering algorithm based on graph connectivity. Inf Process Lett 76:175

Heintz N (2004) Gene expression nervous system atlas (gensat). Nat Neurosci 7(5):483

Hermjakob H, Montecchi-Palazzi L, Lewington C, Mudali S, Kerrien S, Orchard S, Vingron M, Roechert B, Roepstorff P, Valencia A, Margalit H, Armstrong J, Bairoch A, Cesareni G, Sherman D, Apweiler R (2004) Intact: an open source molecular interaction database. Nucleic Acids Res 32(Database issue):D452–D455

Hinman VF, Davidson EH (2007) Evolutionary plasticity of developmental gene regulatory network architecture. Proc Natl Acad Sci USA 104(49):19404–19409

Hinman VF, Nguyen AT, Cameron RA, Davidson EH (2003) Developmental gene regulatory network architecture across 500 million years of echinoderm evolution. Proc Natl Acad Sci USA 100(23):13356–13361

Ho Y, Gruhler A, Heilbut A, Bader GD, Moore L, Adams SL, Millar A, Taylor P, Bennett K, Boutilier K, Yang L, Wolting C, Donaldson I, Schandorff S, Shewnarane J, Vo M, Taggart J, Goudreault M, Muskat B, Alfarano C, Dewar D, Lin Z, Michalickova K, Willems AR, Sassi H, Nielsen PA, Rasmussen KJ, Andersen JR, Johansen LE, Hansen LH, Jespersen H, Podtelejnikov A, Nielsen E, Crawford J, Poulsen V, Sørensen BD, Matthiesen J, Hendrickson RC, Gleeson F, Pawson T, Moran MF, Durocher D, Mann M, Hogue CWV, Figeys D, Tyers M (2002) Systematic identification of protein complexes in saccharomyces cerevisiae by mass spectrometry. Nature 415(6868):180–183

Hosack DA, Dennis G, Sherman BT, Lane HC, Lempicki RA (2003) Identifying biological themes within lists of genes with ease. Genome Biol 4(10):R70

Huang DW, Sherman BT, Lempicki RA (2009) Bioinformatics enrichment tools: paths toward the comprehensive functional analysis of large gene lists. Nucleic Acids Res 37(1):1–13

Hunter S, Apweiler R, Attwood TK, Bairoch A, Bateman A, Binns D, Bork P, Das U, Daugherty L, Duquenne L, Finn RD, Gough J, Haft D, Hulo N, Kahn D, Kelly E, Laugraud A, Letunic I, Lonsdale D, Lopez R, Madera M, Maslen J, McAnulla C, McDowall J, Mistry J, Mitchell A, Mulder N, Natale D, Orengo C, Quinn AF, Selengut JD, Sigrist CJA, Thimma M, Thomas PD, Valentin F, Wilson D, Wu CH, Yeats C (2009) Interpro: the integrative protein signature database. Nucleic Acids Res 37(Database issue):D211–D215

Husi H, Grant SG (2001) Isolation of 2000-kda complexes of n-methyl-d-aspartate receptor and postsynaptic density 95 from mouse brain. J Neurochem 77(1):281–291

Husi H, Ward MA, Choudhary JS, Blackstock WP, Grant SG (2000) Proteomic analysis of nmda receptor-adhesion protein signaling complexes. Nat Neurosci 3(7):661–669

Hwang W, Cho Y, Zhang A, Ramanathan M (2006) A novel functional module detection algorithm for protein–protein interaction networks. Algorithms Mol Biol 1(1):24

Hwang W, Cho YR, Zhang A, Ramanathan M (2008) Cascade: a novel quasi all paths-based network analysis algorithm for clustering biological interactions. BMC Bioinformatics 9:64

Ideker T, Ozier O, Schwikowski B, Siegel AF (2002) Discovering regulatory and signalling circuits in molecular interaction networks. Bioinformatics 18(Suppl 1):S233–S240

Inlow JK, Restifo LL (2004) Molecular and comparative genetics of mental retardation. Genetics 166(2):835–881

Jamain S, Quach H, Betancur C, Råstam M, Colineaux C, Gillberg IC, Soderstrom H, Giros B, Leboyer M, Gillberg C, Bourgeron T, Study PARIS (2003) Mutations of the x-linked genes encoding neuroligins nlgn3 and nlgn4 are associated with autism. Nat Genet 34(1):27–29

Jansen R, Greenbaum D, Gerstein M (2002) Relating whole-genome expression data with protein–protein interactions. Genome Res 12(1):37

Jeong H, Mason S, Barabasi A, Oltvai Z (2001) Lethality and centrality in protein networks. Nature 411(6833):41–42

Jiang T, Keating A (2005) Avid: an integrative framework for discovering functional relationships among proteins. BMC bioinformatics 6(1):136

Jones R (2011) Tuning into places. PLoS Biol 9:e1001042

Jones AR, Overly CC, Sunkin SM (2009) The allen brain atlas: 5 years and beyond. Nat Rev Neurosci 10(11):821–828

Jordan BA, Fernholz BD, Boussac M, Xu C, Grigorean G, Ziff EB, Neubert TA (2004) Identification and verification of novel rodent postsynaptic density proteins. Mol Cell Proteomics 3(9):857–871

Jung S, Jang W, Hur H, Hyun B, Han D (2008) Protein complex prediction based on mutually exclusive interactions in protein interaction network. Genome Inform 21:77–88

Kahng G, Oh E, Kahng B, Kim D (2003) Betweenness centrality correlation in social networks. Phys Rev E 67:01710–01711

Kahraman A, Avramov A, Nashev LG, Popov D, Ternes R, Pohlenz HD, Weiss B (2005) Phenomicdb: a multi-species genotype/phenotype database for comparative phenomics. Bioinformatics 21(3):418–420

Kanehisa M, Goto S, Furumichi M, Tanabe M, Hirakawa M (2010) Kegg for representation and analysis of molecular networks involving diseases and drugs. Nucleic Acids Res 38(Database issue):D355–D360

Karayiorgou M, Simon TJ, Gogos JA (2010) 22q11.2 microdeletions: linking dna structural variation to brain dysfunction and schizophrenia. Nat Rev Neurosci 11(6):402–416

Kashtan N, Itzkovitz S, Milo R, Alon U (2004) Efficient sampling algorithm for estimating subgraph concentrations and detecting network motifs. Bioinformatics (Oxford, England) 20(11):1746–1758

Kim E, Sheng M (2004) Pdz domain proteins of synapses. Nat Rev Neurosci 5(10):771–781

Kim J, Jung SC, Clemens AM, Petralia RS, Hoffman DA (2007) Regulation of dendritic excitability by activity-dependent trafficking of the a-type k+ channel subunit kv4.2 in hippocampal neurons. Neuron 54(6):933–947

Kim H, Kishikawa S, Higgins A, Seong I, Donovan D, Shen Y, Lally E, Weiss L, Najm J, Kutsche K (2008) Disruption of neurexin 1 associated with autism spectrum disorder. Am J Hum Genet 82(1):199–207

Kirov G, Rujescu D, Ingason A, Collier DA, O'Donovan MC, Owen MJ (2009a) Neurexin 1 (nrxn1) deletions in schizophrenia. Schizophr Bull 35(5):851–854

Kirov G, Zaharieva I, Georgieva L, Moskvina V, Nikolov I, Cichon S, Hillmer A, Toncheva D, Owen MJ, O'Donovan MC (2009b) A genome-wide association study in 574 schizophrenia trios using dna pooling. Mol Psychiatry 14(8):796–803

Köhler J, Baumbach J, Taubert J, Specht M, Skusa A, Rüegg A, Rawlings C, Verrier P, Philippi S (2006) Graph-based analysis and visualization of experimental results with ondex. Bioinformatics 22(11):1383–1390

Komiyama NH, Watabe AM, Carlisle HJ, Porter K, Charlesworth P, Monti J, Strathdee DJC, O'Carroll CM, Martin SJ, Morris RGM, O'Dell TJ, Grant SGN (2002) Syngap regulates erk/mapk signaling, synaptic plasticity, and learning in the complex with postsynaptic density 95 and nmda receptor. J Neurosci 22(22):9721–9732

Krogan NJ, Cagney G, Yu H, Zhong G, Guo X, Ignatchenko A, Li J, Pu S, Datta N, Tikuisis AP, Punna T, Peregrín-Alvarez JM, Shales M, Zhang X, Davey M, Robinson MD, Paccanaro A, Bray JE, Sheung A, Beattie B, Richards DP, Canadien V, Lalev A, Mena F, Wong P, Starostine A, Canete MM, Vlasblom J, Wu S, Orsi C, Collins SR, Chandran S, Haw R, Rilstone JJ, Gandi K, Thompson NJ, Musso G, Onge PS, Ghanny S, Lam MHY, Butland G, Altaf-Ul AM, Kanaya S, Shilatifard A, O'Shea E, Weissman JS, Ingles CJ, Hughes TR, Parkinson J, Gerstein M, Wodak SJ, Emili A, Greenblatt JF (2006) Global landscape of protein complexes in the yeast saccharomyces cerevisiae. Nature 440(7084):637–643

Lagomarsino MC, Jona P, Bassetti B, Isambert H (2007) Hierarchy and feedback in the evolution of the escherichia coli transcription network. Proc Natl Acad Sci USA 104(13):5516–5520

Li KW, Jimenez CR (2008) Synapse proteomics: current status and quantitative applications. Expert Rev proteomics 5(2):353–360

Li KW, Hornshaw MP, Schors RCVD, Watson R, Tate S, Casetta B, Jimenez CR, Gouwenberg Y, Gundelfinger ED, Smalla KH, Smit AB (2004) Proteomics analysis of rat brain postsynaptic density. Implications of the diverse protein functional groups for the integration of synaptic physiology. J Biol Chem 279(2):987–1002

Li M, Wang J, Chen J (2008) A fast agglomerate algorithm for mining functional modules in protein interaction networks. In BioMedical engineering and informatics, 2008 BMEI 2008 international conference on, Sanya, Hainan, China, vol 1, pp 3–7

Lim J, Hao T, Shaw C, Patel A, Szabó G, Rual J, Fisk C, Li N, Smolyar A, Hill D (2006) A protein–protein interaction network for human inherited ataxias and disorders of purkinje cell degeneration. Cell 125(4):801–814

Lisman JE, Coyle JT, Green RW, Javitt DC, Benes FM, Heckers S, Grace AA (2008) Circuit-based framework for understanding neurotransmitter and risk gene interactions in schizophrenia. Trends Neurosci 31(5):234–242

Liu H, Hu ZZ, Zhang J, Wu C (2006) Biothesaurus: a web-based thesaurus of protein and gene names. Bioinformatics (Oxford, England) 22(1):103–105

Lu H, Shi B, Wu G, Zhang Y, Zhu X, Zhang Z, Liu C, Zhao Y, Wu T, Wang J (2006) Integrated analysis of multiple data sources reveals modular structure of biological networks. Biochem Biophys Res Commun 345(1):302–309

Lubovac Z, Gamalielsson J, Olsson B (2006) Combining functional and topological properties to identify core modules in protein interaction networks. Proteins 64(4):948–959

Maere S, Heymans K, Kuiper M (2005) Bingo: a cytoscape plugin to assess overrepresentation of gene ontology categories in biological networks. Bioinformatics 21(16):3448–3449

Magdaleno S, Jensen P, Brumwell CL, Seal A, Lehman K, Asbury A, Cheung T, Cornelius T, Batten DM, Eden C, Norland SM, Rice DS, Dosooye N, Shakya S, Mehta P, Curran T (2006) Bgem: an in situ hybridization database of gene expression in the embryonic and adult mouse nervous system. PLoS Biol 4(4):e86

Magrane M, Consortium U (2011) Uniprot knowledgebase: a hub of integrated protein data. Database (Oxford) 2011:bar009

Maraziotis IA, Dimitrakopoulou K, Bezerianos A (2007) Growing functional modules from a seed protein via integration of protein interaction and gene expression data. BMC Bioinformatics 8:408

Matos S, Arrais JP, Maia-Rodrigues J, Oliveira JL (2010) Concept-based query expansion for retrieving gene related publications from medline. BMC Bioinformatics 11:212

Matthews L, Gopinath G, Gillespie M, Caudy M, Croft D, de Bono B, Garapati P, Hemish J, Hermjakob H, Jassal B, Kanapin A, Lewis S, Mahajan S, May B, Schmidt E, Vastrik I, Wu G, Birney E, Stein L, D'Eustachio P (2009) Reactome knowledgebase of human biological pathways and processes. Nucleic Acids Res 37(Database issue):D619–D622

Matys V, Fricke E, Geffers R, Gössling E, Haubrock M, Hehl R, Hornischer K, Karas D, Kel AE, Kel-Margoulis OV, Kloos DU, Land S, Lewicki-Potapov B, Michael H, Münch R, Reuter I, Rotert S, Saxel H, Scheer M, Thiele S, Wingender E (2003) Transfac: transcriptional regulation, from patterns to profiles. Nucleic Acids Res 31(1):374–378

McKusick V (2007) Mendelian inheritance in man and its online version, omim. Am J Hum Genet 80(4):588

McKusick V, Amberger J (1994) The morbid anatomy of the human genome: chromosomal location of mutations causing disease. J Med Genet 31(4):265–279

Mering CV, Krause R, Snel B, Cornell M, Oliver S, Fields S, Bork P (2002) Comparative assessment of large-scale data sets of protein–protein interactions. Nature 417(6887):399–403

Meshi O, Shlomi T, Ruppin E (2007) Evolutionary conservation and over-representation of functionally enriched network patterns in the yeast regulatory network. BMC Syst Biol 1:1

Mi H, Guo N, Kejariwal A, Thomas P (2006) Panther version 6: protein sequence and function evolution data with expanded representation of biological pathways. Nucleic Acids Res 35(suppl 1):D247

Migaud M, Charlesworth P, Dempster M, Webster LC, Watabe AM, Makhinson M, He Y, Ramsay MF, Morris RG, Morrison JH, O'Dell TJ, Grant SG (1998) Enhanced long-term potentiation and impaired learning in mice with mutant postsynaptic density-95 protein. Nature 396(6710):433–439

Milenković T, Lai J, Pržulj N (2008) Graphcrunch: a tool for large network analyses. BMC Bioinformatics 9(1):70

Millar JK, Wilson-Annan JC, Anderson S, Christie S, Taylor MS, Semple CA, Devon RS, Clair DMS, Muir WJ, Blackwood DH, Porteous DJ (2000) Disruption of two novel genes by a translocation co-segregating with schizophrenia. Hum Mol Genet 9(9):1415–1423

Milo R, Shen-Orr S, Itzkovitz S, Kashtan N, Chklovskii D, Alon U (2002) Network motifs: simple building blocks of complex networks. Science 298(5594):824–827

Moessner R, Marshall CR, Sutcliffe JS, Skaug J, Pinto D, Vincent J, Zwaigenbaum L, Fernandez B, Roberts W, Szatmari P, Scherer SW (2007) Contribution of shank3 mutations to autism spectrum disorder. Am J Hum Genet 81(6):1289–1297

Morciano M, Burré J, Corvey C, Karas M, Zimmermann H, Volknandt W (2005) Immunoisolation of two synaptic vesicle pools from synaptosomes: a proteomics analysis. J Neurochem 95(6):1732–1745

Morciano M, Beckhaus T, Karas M, Zimmermann H, Volknandt W (2009) The proteome of the presynaptic active zone: from docked synaptic vesicles to adhesion molecules and maxi-channels. J Neurochem 108(3):662–675

Newman MEJ (2004) Analysis of weighted networks. arXiv cond-mat.stat-mech

Newman M, Girvan M (2004) Finding and evaluating community structure in networks. Phys Rev E 69(2):1–15

Noble WS (2009) How does multiple testing correction work? Nat Biotechnol 27(12):1135–1137

Nourry C, Grant SGN, Borg JP (2003) Pdz domain proteins: plug and play! Sci Signal 2003(179):RE7

Ostlund G, Schmitt T, Forslund K, Köstler T, Messina DN, Roopra S, Frings O, Sonnhammer ELL (2010) Inparanoid 7: new algorithms and tools for eukaryotic orthology analysis. Nucleic Acids Res 38(Database issue):D196–D203

Park D, Lee S, Bolser D, Schroeder M, Lappe M, Oh D, Bhak J (2005) Comparative interactomics analysis of protein family interaction networks using psimap (protein structural interactome map). Bioinformatics 21(15):3234

Parkinson H, Kapushesky M, Kolesnikov N, Rustici G, Shojatalab M, Abeygunawardena N, Berube H, Dylag M, Emam I, Farne A, Holloway E, Lukk M, Malone J, Mani R, Pilicheva E, Rayner TF, Rezwan F, Sharma A, Williams E, Bradley XZ, Adamusiak T, Brandizi M, Burdett T, Coulson R, Krestyaninova M, Kurnosov P, Maguire E, Neogi SG, Rocca-Serra P, Sansone SA, Sklyar N, Zhao M, Sarkans U, Brazma A (2009) Arrayexpress update—from an archive of functional genomics experiments to the atlas of gene expression. Nucleic Acids Res 37(Database issue):D868–D872

Paulo JA, Brucker WJ, Hawrot E (2009) Proteomic analysis of an alpha7 nicotinic acetylcholine receptor interactome. J Proteome Res 8(4):1849–1858

Peng J, Kim MJ, Cheng D, Duong DM, Gygi SP, Sheng M (2004) Semiquantitative proteomic analysis of rat forebrain postsynaptic density fractions by mass spectrometry. J Biol Chem 279(20):21003–21011

Peri S, Navarro JD, Amanchy R, Kristiansen TZ, Jonnalagadda CK, Surendranath V, Niranjan V, Muthusamy B, Gandhi TKB, Gronborg M, Ibarrola N, Deshpande N, Shanker K, Shivashankar HN, Rashmi BP, Ramya MA, Zhao Z, Chandrika KN, Padma N, Harsha HC, Yatish AJ, Kavitha MP, Menezes M, Choudhury DR, Suresh S, Ghosh N, Saravana R, Chandran S, Krishna S, Joy M, Anand SK, Madavan V, Joseph A, Wong GW, Schiemann WP, Constantinescu SN, Huang L, Khosravi-Far R, Steen H, Tewari M, Ghaffari S, Blobe GC, Dang CV, Garcia JGN, Pevsner J, Jensen ON, Roepstorff P, Deshpande KS, Chinnaiyan AM, Hamosh A, Chakravarti A, Pandey A (2003) Development of human protein reference database as an initial platform for approaching systems biology in humans. Genome Res 13(10):2363–2371

Persico M, Ceol A, Gavrila C, Hoffmann R, Florio A, Cesareni G (2005) Homomint: an inferred human network based on orthology mapping of protein interactions discovered in model organisms. BMC Bioinformatics 6(Suppl 4):S21

Pillet V, Zehnder M, Seewald AK, Veuthey AL, Petrak J (2005) Gpsdb: a new database for synonyms expansion of gene and protein names. Bioinformatics 21(8):1743–1744

Pinto D, Pagnamenta AT, Klei L, Anney R, Merico D, Regan R, Conroy J, Magalhaes TR, Correia C, Abrahams BS, Almeida J, Bacchelli E, Bader GD, Bailey AJ, Baird G, Battaglia A, Berney T, Bolshakova N, Bölte S, Bolton PF, Bourgeron T, Brennan S, Brian J, Bryson SE, Carson AR, Casallo G, Casey J, Chung BHY, Cochrane L, Corsello C, Crawford EL, Crossett A, Cytrynbaum C, Dawson G, de Jonge M, Delorme R, Drmic I, Duketis E, Duque F, Estes A, Farrar P, Fernandez BA, Folstein SE, Fombonne E, Freitag CM, Gilbert J, Gillberg C, Glessner JT, Goldberg J, Green A, Green J, Guter SJ, Hakonarson H, Heron EA, Hill M, Holt R, Howe JL, Hughes G, Hus V, Igliozzi R, Kim C, Klauck SM, Kolevzon A, Korvatska O, Kustanovich V, Lajonchere CM, Lamb JA, Laskawiec M, Leboyer M, Couteur AL, Leventhal BL, Lionel AC, Liu XQ, Lord C, Lotspeich L, Lund SC, Maestrini E, Mahoney W, Mantoulan C, Marshall CR, McConachie H, McDougle CJ, McGrath J, McMahon WM, Merikangas A, Migita O, Minshew NJ, Mirza GK, Munson J, Nelson SF, Noakes C, Noor A, Nygren G, Oliveira G, Papanikolaou K, Parr JR, Parrini B, Paton T, Pickles A, Pilorge M, Piven J, Ponting CP, Posey DJ, Poustka A, Poustka F, Prasad A, Ragoussis J, Renshaw K, Rickaby J, Roberts W, Roeder K, Roge B, Rutter ML, Bierut LJ, Rice JP, Salt J, Sansom K, Sato D, Segurado R, Sequeira AF, Senman L, Shah N, Sheffield VC, Soorya L, Sousa I, Stein O, Sykes N, Stoppioni V, Strawbridge C, Tancredi R, Tansey K, Thiruvahindrapduram B, Thompson AP, Thomson S, Tryfon A, Tsiantis J, Engeland HV, Vincent JB, Volkmar F, Wallace S, Wang K, Wang Z, Wassink TH, Webber C, Weksberg R, Wing K, Wittemeyer K, Wood S, Wu J, Yaspan BL, Zurawiecki D, Zwaigenbaum L, Buxbaum JD, Cantor RM, Cook EH, Coon H, Cuccaro ML, Devlin B, Ennis S, Gallagher L, Geschwind DH, Gill M, Haines JL, Hallmayer J, Miller J, Monaco AP, Nurnberger JI Jr, Paterson AD, Pericak-Vance MA, Schellenberg GD, Szatmari P, Vicente AM, Vieland VJ, Wijsman EM,

Scherer SW, Sutcliffe JS, Betancur C (2010) Functional impact of global rare copy number variation in autism spectrum disorders. Nature 466:368–372

Pocklington AJ, Cumiskey M, Armstrong JD, Grant SGN (2006) The proteomes of neurotransmitter receptor complexes form modular networks with distributed functionality underlying plasticity and behaviour. Mol Syst Biol 2:1–14

Prifti E, Zucker J, Clement K, Henegar C (2008) Funnet: an integrative tool for exploring transcriptional interactions. Bioinformatics 24(22):2636

Przulj N (2007) Biological network comparison using graphlet degree distribution. Bioinformatics 23(2):e177–e183

Przulj N, Wigle DA, Jurisica I (2004) Functional topology in a network of protein interactions. Bioinformatics 20(3):340–348

Przulj N, Corneil DG, Jurisica I (2006) Efficient estimation of graphlet frequency distributions in protein–protein interaction networks. Bioinformatics 22(8):974–980

Puig O (2001) The tandem affinity purification (tap) method: a general procedure of protein complex purification. Methods 24(3):218–229

Purcell S, Wray N, Stone J, Visscher P, O'Donovan M, Sullivan P, Sklar P, Ruderfer D, Mcquillin A, Morris D (2009) Common polygenic variation contributes to risk of schizophrenia and bipolar disorder. Nature 460(7256):748–752

Radicchi F, Castellano C, Cecconi F, Loreto V, Parisi D (2004) Defining and identifying communities in networks. Proc Natl Acad Sci USA 101(9):2658–2663

Rain JC, Selig L, Reuse HD, Battaglia V, Reverdy C, Simon S, Lenzen G, Petel F, Wojcik J, Schächter V, Chemama Y, Labigne A, Legrain P (2001) The protein–protein interaction map of helicobacter pylori. Nature 409(6817):211–215

Ravasz E, Somera AL, Mongru DA, Oltvai ZN, Barabási AL (2002) Hierarchical organization of modularity in metabolic networks. Science 297(5586):1551–1555

Redon R, Ishikawa S, Fitch KR, Feuk L, Perry GH, Andrews TD, Fiegler H, Shapero MH, Carson AR, Chen W, Cho EK, Dallaire S, Freeman JL, González JR, Gratacòs M, Huang J, Kalaitzopoulos D, Komura D, MacDonald JR, Marshall CR, Mei R, Montgomery L, Nishimura K, Okamura K, Shen F, Somerville MJ, Tchinda J, Valsesia A, Woodwark C, Yang F, Zhang J, Zerjal T, Zhang J, Armengol L, Conrad DF, Estivill X, Tyler-Smith C, Carter NP, Aburatani H, Lee C, Jones KW, Scherer SW, Hurles ME (2006) Global variation in copy number in the human genome. Nature 444(7118):444–454

Ringwald M, Davis G, Smith A, Trepanier L, Begley D, Richardson J, Eppig J (1997) The mouse gene expression database gxd. Semin Cell Dev Biol 8(5):489–497

Ringwald M, Eppig JT, Begley DA, Corradi JP, McCright IJ, Hayamizu TF, Hill DP, Kadin JA, Richardson JE (2001) The mouse gene expression database (gxd). Nucleic Acids Res 29(1): 98–101

Royer L, Reimann M, Andreopoulos B, Schroeder M (2008) Unraveling protein networks with power graph analysis. PLoS Comput Biol 4(7):e1000108

Rual J, Venkatesan K, Hao T, Hirozane-Kishikawa T, Dricot A, Li N, Berriz G, Gibbons F, Dreze M, Ayivi-Guedehoussou N (2005) Towards a proteome-scale map of the human protein–protein interaction network. Nature 437(7062):1173–1178

Ruan J, Zhang W (2007) Identifying network communities with a high resolution. arXiv physics.soc-ph

Ruiz-Cañada C, Koh YH, Budnik V, Tejedor FJ (2002) Dlg differentially localizes shaker k+-channels in the central nervous system and retina of drosophila. J Neurochem 82(6):1490–1501

Ryan TJ, Grant SGN (2009) The origin and evolution of synapses. Nat Rev Neurosci 10(10): 701–712

Rzhetsky A, Seringhaus M, Gerstein, M (2008) Seeking a new biology through text mining. Cell 134(1):9–13. doi:10.1016/j.cell.2008.06.029

Salter MW, Kalia LV (2004) Src kinases: a hub for nmda receptor regulation. Nat Rev Neurosci 5(4):317–328

Salwinski L, Miller CS, Smith AJ, Pettit FK, Bowie JU, Eisenberg D (2004) The database of interacting proteins: 2004 update. Nucleic Acids Res 32(Database issue):D449–D451

Sandelin A, Alkema W, Engström P, Wasserman WW, Lenhard B (2004) Jaspar: an open-access database for eukaryotic transcription factor binding profiles. Nucleic Acids Res 32(Database issue):D91–D94

Sasaki Y, McNaught J, Ananiadou S (2010) The value of an in-domain lexicon in genomics qa. J Bioinform Comput Biol 8(1):147–161

Schuemie MJ, Kang N, Hekkelman ML, Kors JA (2010) Genee: gene and protein query expansion with disambiguation. Bioinformatics 26(1):147–148

Schwikowski B, Uetz P, Fields S (2000) A network of protein–protein interactions in yeast. Nat Biotechnol 18(12):1257–1261

Segal E, Wang H, Koller D (2003) Discovering molecular pathways from protein interaction and gene expression data. Bioinformatics 19(Suppl 1):i264–i271

Selimi F, Cristea IM, Heller E, Chait BT, Heintz N (2009) Proteomic studies of a single cns synapse type: the parallel fiber/purkinje cell synapse. PLoS Biol 7(4):e83

Shannon P, Markiel A, Ozier O, Baliga NS, Wang JT, Ramage D, Amin N, Schwikowski B, Ideker T (2003) Cytoscape: a software environment for integrated models of biomolecular interaction networks. Genome Res 13(11):2498–2504

Sheng M, Hoogenraad CC (2007) The postsynaptic architecture of excitatory synapses: a more quantitative view. Annu Rev Biochem 76:823–847

Shen-Orr SS, Milo R, Mangan S, Alon U (2002) Network motifs in the transcriptional regulation network of escherichia coli. Nat Genet 31(1):64–68

Shimoyama M, Smith JR, Hayman T, Laulederkind S, Lowry T, Nigam R, Petri V, Wang SJ, Dwinell M, Jacob H, Team R (2011) Rgd: a comparative genomics platform. Hum Genomics 5(2):124–129

Simpson TI, Armstrong JD, Jarman AP (2010) Merged consensus clustering to assess and improve class discovery with microarray data. BMC Bioinformatics 11(1):590

Smith CL, Eppig JT (2009) The mammalian phenotype ontology: enabling robust annotation and comparative analysis. Wiley Interdiscip Rev Syst Biol Med 1(3):390–399

Smith CM, Finger JH, Hayamizu TF, McCright IJ, Eppig JT, Kadin JA, Richardson JE, Ringwald M (2007) The mouse gene expression database (gxd): 2007 update. Nucleic Acids Res 35(Database issue):D618–D623

Spirin V, Mirny LA (2003) Protein complexes and functional modules in molecular networks. Proc Nal Acad Sci USA 100(21):12123

Stark C, Breitkreutz BJ, Reguly T, Boucher L, Breitkreutz A, Tyers M (2006) Biogrid: a general repository for interaction datasets. Nucleic Acids Res 34(Database issue):D535–D539

Stein L, Sternberg P, Durbin R, Thierry-Mieg J, Spieth J (2001) Wormbase: network access to the genome and biology of caenorhabditis elegans. Nucleic Acids Res 29(1):82–86

Stone J, O'Donovan M, Gurling H, Kirov G, Blackwood D, Corvin A, Craddock N, Gill M, Hultman C, Lichtenstein P (2008) Rare chromosomal deletions and duplications increase risk of schizophrenia. Nature 455(7210):237–241

Stuart LM, Boulais J, Charriere GM, Hennessy EJ, Brunet S, Jutras I, Goyette G, Rondeau C, Letarte S, Huang H, Ye P, Morales F, Kocks C, Bader JS, Desjardins M, Ezekowitz RAB (2007) A systems biology analysis of the drosophila phagosome. Nature 445(7123):95–101

Stumpf MPH, Wiuf C, May RM (2005) Subnets of scale-free networks are not scale-free: sampling properties of networks. Proc Natl Acad Sci USA 102(12):4221–4224

Subramanian A, Tamayo P, Mootha VK, Mukherjee S, Ebert BL, Gillette MA, Paulovich A, Pomeroy SL, Golub TR, Lander ES, Mesirov JP (2005) Gene set enrichment analysis: a knowledge-based approach for interpreting genome-wide expression profiles. Proc Natl Acad Sci USA 102(43):15545–15550

Takamori S, Holt M, Stenius K, Lemke EA, Grønborg M, Riedel D, Urlaub H, Schenck S, Brügger B, Ringler P, Müller SA, Rammner B, Gräter F, Hub JS, Groot BLD, Mieskes G, Moriyama Y, Klingauf J, Grubmüller H, Heuser J, Wieland F, Jahn R (2006) Molecular anatomy of a trafficking organelle. Cell 127(4):831–846

te Velthuis AJW, Admiraal JF, Bagowski CP (2007) Molecular evolution of the maguk family in metazoan genomes. BMC Evol Biol 7:129

Theocharidis A, van Dongen S, Enright AJ, Freeman TC (2009) Network visualization and analysis of gene expression data using biolayout express(3d). Nat Protoc 4(10):1535–1550

Thomas GM, Huganir RL (2004) Mapk cascade signalling and synaptic plasticity. Nat Rev Neurosci 5(3):173–183

Thomas P, Campbell M, Kejariwal A, Mi H, Karlak B, Daverman R, Diemer K, Muruganujan A, Narechania A (2003) Panther: a library of protein families and subfamilies indexed by function. Genome Res 13(9):2129

Torii M, Hu Z, Wu CH, Liu H (2009) Biotagger-gm: a gene/protein name recognition system. J Am Med Inform Assoc 16(2):247–255

Trinidad J, Thalhammer A, Specht C, Lynn A, Baker P, Schoepfer R, Burlingame A (2008) Quantitative analysis of synaptic phosphorylation and protein expression. Mol Cell Proteomics 7(4):684

Uetz P, Giot L, Cagney G, Mansfield T, Judson R, Knight J, Lockshon D, Narayan V, Srinivasan M, Pochart P (2000) A comprehensive analysis of protein–protein interactions in saccharomyces cerevisiae. Nature 403(6770):623–627

Ulitsky I, Shamir R (2009) Identifying functional modules using expression profiles and confidence-scored protein interactions. Bioinformatics 25(9):1158–1164

Vêncio R, Shmulevich I (2007) Probcd: enrichment analysis accounting for categorization uncertainty. BMC bioinformatics 8(1):383

Venkatesan K, Rual JF, Vazquez A, Stelzl U, Lemmens I, Hirozane-Kishikawa T, Hao T, Zenkner M, Xin X, Goh KI, Yildirim MA, Simonis N, Heinzmann K, Gebreab F, Sahalie JM, Cevik S, Simon C, de Smet AS, Dann E, Smolyar A, Vinayagam A, Yu H, Szeto D, Borick H, Dricot A, Klitgord N, Murray RR, Lin C, Lalowski M, Timm J, Rau K, Boone C, Braun P, Cusick ME, Roth FP, Hill DE, Tavernier J, Wanker EE, Barabási AL, Vidal M (2009) An empirical framework for binary interactome mapping. Nat Method 6(1):83–90

Vidalain PO, Boxem M, Ge H, Li S, Vidal M (2004) Increasing specificity in high-throughput yeast two-hybrid experiments. Methods 32(4):363–370

Vilella AJ, Severin J, Ureta-Vidal A, Heng L, Durbin R, Birney E (2009) Ensemblcompara genetrees: complete, duplication-aware phylogenetic trees in vertebrates. Genome Res 19(2): 327–335

Walhout AJ, Sordella R, Lu X, Hartley JL, Temple GF, Brasch MA, Thierry-Mieg N, Vidal M (2000) Protein interaction mapping in c. Elegans using proteins involved in vulval development. Science 287(5450):116–122

Walhout AJM, Reboul J, Shtanko O, Bertin N, Vaglio P, Ge H, Lee H, Doucette-Stamm L, Gunsalus KC, Schetter AJ, Morton DG, Kemphues KJ, Reinke V, Kim SK, Piano F, Vidal M (2002) Integrating interactome, phenome, and transcriptome mapping data for the c. Elegans germline. Curr Biol 12(22):1952–1958

Walikonis RS, Jensen ON, Mann M, Provance DW, Mercer JA, Kennedy MB (2000) Identification of proteins in the postsynaptic density fraction by mass spectrometry. J Neurosci 20(11): 4069–4080

Walsh T, McClellan JM, McCarthy SE, Addington AM, Pierce SB, Cooper GM, Nord AS, Kusenda M, Malhotra D, Bhandari A, Stray SM, Rippey CF, Roccanova P, Makarov V, Lakshmi B, Findling RL, Sikich L, Stromberg T, Merriman B, Gogtay N, Butler P, Eckstrand K, Noory L, Gochman P, Long R, Chen Z, Davis S, Baker C, Eichler EE, Meltzer PS, Nelson SF, Singleton AB, Lee MK, Rapoport JL, King MC, Sebat J (2008) Rare structural variants disrupt multiple genes in neurodevelopmental pathways in schizophrenia. Science 320(5875):539–543

Wang J, Li M, Deng Y, Pan Y (2010) Recent advances in clustering methods for protein interaction networks. BMC Genomics 11(Suppl 3):S10

Wernicke S, Rasche F (2006) Fanmod: a tool for fast network motif detection. Bioinformatics (Oxford, England) 22(9):1152–1153

Wooten EC, Huggins GS (2011) Mind the dbgap: The application of data mining to identify biological mechanisms. Mol Interv 11(2):95–102

Wu K, Carlin R, Siekevitz P (1986) Binding of l-[3h]glutamate to fresh or frozen synaptic membrane and postsynaptic density fractions isolated from cerebral cortex and cerebellum of fresh or frozen canine brain. J Neurochem 46(3):831–841

Wuchty S, Oltvai ZN, Barabási AL (2003) Evolutionary conservation of motif constituents in the yeast protein interaction network. Nat Genet 35(2):176–179

Yasuyama K, Meinertzhagen IA, Schürmann FW (2002) Synaptic organization of the mushroom body calyx in drosophila melanogaster. J Comp Neurol 445(3):211–226

Yoshimura Y, Yamauchi Y, Shinkawa T, Taoka M, Donai H, Takahashi N, Isobe T, Yamauchi T (2004) Molecular constituents of the postsynaptic density fraction revealed by proteomic analysis using multidimensional liquid chromatography-tandem mass spectrometry. J Neurochem 88(3):759–768

Young KH (1998) Yeast two-hybrid: so many interactions, (in) so little time. Biol Reprod 58(2):302–311

Yu H, Kim PM, Sprecher E, Trifonov V, Gerstein M (2007) The importance of bottlenecks in protein networks: correlation with gene essentiality and expression dynamics. PLoS Comput Biol 3(4):e59

Yu H, Braun P, Yildirim MA, Lemmens I, Venkatesan K, Sahalie J, Hirozane-Kishikawa T, Gebreab F, Li N, Simonis N, Hao T, Rual JF, Dricot A, Vazquez A, Murray RR, Simon C, Tardivo L, Tam S, Svrzikapa N, Fan C, Smet ASD, Motyl A, Hudson ME, Park J, Xin X, Cusick ME, Moore T, Boone C, Snyder M, Roth FP, Barabasi AL, Tavernier J, Hill DE, Vidal M (2008a) High-quality binary protein interaction map of the yeast interactome network. Science 322(5898):104–110

Yu J, Pacifico S, Liu G, Finley RL (2008b) Droid: the drosophila interactions database, a comprehensive resource for annotated gene and protein interactions. BMC Genomics 9:461

Yu W, Wulf A, Liu T, Khoury MJ, Gwinn M (2008c) Gene prospector: an evidence gateway for evaluating potential susceptibility genes and interacting risk factors for human diseases. BMC bioinformatics 9:528

Zapala MA, Hovatta I, Ellison JA, Wodicka L, Rio JAD, Tennant R, Tynan W, Broide RS, Helton R, Stoveken BS, Winrow C, Lockhart DJ, Reilly JF, Young WG, Bloom FE, Lockhart DJ, Barlow C (2005) Adult mouse brain gene expression patterns bear an embryologic imprint. Proc Natl Acad Sci USA 102(29):10357–10362

Zeeberg BR, Feng W, Wang G, Wang MD, Fojo AT, Sunshine M, Narasimhan S, Kane DW, Reinhold WC, Lababidi S, Bussey KJ, Riss J, Barrett JC, Weinstein JN (2003) Gominer: a resource for biological interpretation of genomic and proteomic data. Genome Biol 4(4):R28

Zhang W, Zhang Y, Zheng H, Zhang C, Xiong W, Olyarchuk JG, Walker M, Xu W, Zhao M, Zhao S, Zhou Z, Wei L (2007) Syndb: a synapse protein database based on synapse ontology. Nucleic Acids Res 35(Database issue):D737–D741

Zhang B, Park BH, Karpinets T, Samatova NF (2008) From pull-down data to protein interaction networks and complexes with biological relevance. Bioinformatics (Oxford, England) 24(7):979–986

Zhong S, Storch KF, Lipan O, Kao MCJ, Weitz CJ, Wong WH (2004) Gosurfer: a graphical interactive tool for comparative analysis of large gene sets in gene ontology space. Appl Bioinformatics 3(4):261–264

Chapter 3
Using Chemical Kinetics to Model Neuronal Signalling Pathways

Lukas Endler, Melanie I. Stefan, Stuart J. Edelstein, and Nicolas Le Novère

Abstract Understanding the physical principles and mechanisms underlying biochemical reactions allows us to create mechanistic mathematical models of complex biological processes, such as those occurring during neuronal signal transduction. In this chapter we introduce basic concepts of chemical and enzyme kinetics, and reaction thermodynamics. Furthermore, we show how the temporal evolution of a reaction system can be described by ordinary differential equations that can be numerically solved on a computer. Finally we give a short overview of different approaches to modelling cooperative binding to, and allosteric control of, receptors and ion channels.

The transduction of neuronal signals and their effects on the behaviours and phenotypes of neurons involve many biochemical entities that interact, diffuse and transform, with different intensities and on different timescales. To understand those biological processes, dynamical and quantitative descriptions are necessary. However, a mere reproduction of experimental observations by non-mechanistic models often is not sufficient, as many experimental results represent averages over time. Furthermore, as the observables are emerging from complex biological systems, in general their behaviour can only be predicted and fully understood by considering the underlying reactions and biophysical processes.

Numerical simulations of models founded in chemical kinetics have been used successfully to describe neuronal signalling for a few decades. Early models concentrated on a single given entity, such as the models of acetylcholine receptors at the neuromuscular junction by Land et al. (1981). Later, models of complex pathways were designed that made full use of the wealth of data accumulated in the field of molecular neurobiology (Bhalla and Iyengar 1999).

L. Endler • M.I. Stefan • S.J. Edelstein • N. Le Novère (✉)
EMBL-EBI, Hinxton, UK
e-mail: lukas.endler@gmx.at; melanie.stefan@cantab.net; stuart.edelstein@unige.ch; lenov@ebi.ac.uk

N. Le Novère (ed.), *Computational Systems Neurobiology*,
DOI 10.1007/978-94-007-3858-4_3,

In this chapter we introduce basic concepts of chemical and enzyme kinetics, and show how the temporal evolution of a reaction system can be described by ordinary differential equations. Finally we give an overview of different approaches to model cooperative binding to, and allosteric control of, receptors and ion channels.

3.1 Introduction to Chemical Kinetics

A neuron, as any living cell, is built up as a series of compartments of various dimensions. The post-synaptic membrane is an example of a bi-dimensional compartment surrounding the cytosol of the dendritic spine, which is itself a tri-dimensional compartment. Microtubules are examples of uni-dimensional compartments. These compartments can be considered both as containers—we can count the number of instances of a certain type of entity present in, or attached to, a compartment—and as diffusional landscapes—the movements of the entities within the compartment depend on its properties. Within the compartments, the entities can move and react with each other. The object of chemical kinetics is to study the temporal evolution of the positions and quantities of the entities contained in a compartment, sometimes called a reactor.

In this chapter, we will not deal with the displacement of the chemical entities within a compartment. This question will be treated in Chap. 5. We will assume that an entity-pool, that is a set of entities that are indistinguishable as far as the model is concerned, is distributed homogeneously within the compartment. This hypothesis is known as the *well-stirred approximation* (Fig. 3.1). This approximation is based on the assumption that there is no diffusional anisotropy in the compartment, i.e. the molecules move randomly in any dimension.

3.1.1 Chemical Reactions

A chemical reaction is the transformation of one set of substances called reactants into another set called products. At a microscopic scale, such a transformation is in general reversible, although there are many cases in which the reverse reaction is of negligible importance compared to the forward one. In all cases, a reversible reaction can be split into forward and reverse reactions. For a given reaction, reactants generally combine in discrete and fixed ratios to form products. These ratios indicate the amount of each substance involved in the reaction. The amounts consumed or produced in one reaction event are called the stoichiometric coefficients or numbers, ν_X, and are positive for products, and negative for reactants. If a substance is neither consumed or produced by a reaction, its stoichiometric coefficient is 0. Equation 3.1 depicts a general reaction, in which A and B are reactants combining to form the product P. ν_A would be $-a$, $\nu_B = -b$ and $\nu_P = p$. The list $\{-a, -b, p\}$ is also called the stoichiometry of the reaction.

$$\mathrm{aA} + \mathrm{bB} \longrightarrow \mathrm{pP} \tag{3.1}$$

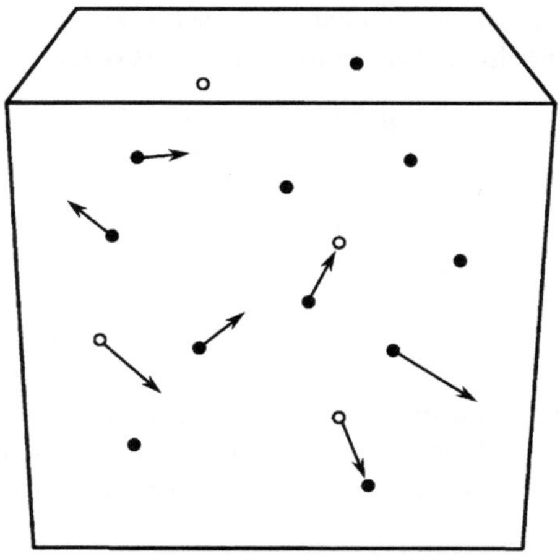

Fig. 3.1 Representation of a well-stirred container with two types of entities, represented by empty and filled *circles*. The *arrows* represent the direction and speed of their movements

$$P(\text{reaction } \bullet) = P(\text{find } \bullet) \times P(\bullet \text{ reacts})$$

$$P(\text{reaction } \bullet + \bullet) = P(\text{find } \bullet) \times P(\text{find another } \bullet) \times P(\bullet \text{ reacts with } \bullet)$$

$$P(\text{reaction } \bullet + \circ) = P(\text{find } \bullet) \times P(\text{find } \circ) \times P(\bullet \text{ reacts with } \circ)$$

$$\text{and } P(\text{find } \bullet) \propto \frac{n(\bullet)}{V} = [\bullet]$$

where V is the volume of the container

In many cases in biology only an overall transformation consisting of many sequential reactions is experimentally observable. In the finest grained form these reactions are also known as elementary reactions. An elementary reaction is defined as a minimal irreversible reaction with no stable intermediary products. The lumped stoichiometric coefficients of the overall reaction consist of the sums of the stoichiometric coefficients for each reactant over all elementary reactions.

Chemical kinetics is concerned with the velocity of such transformations, the rates with which substances are consumed and produced. As the rate of change for a reagent depends on its stoichiometric coefficients, it can be different for individual substances. Therefore it is convenient to define the reaction rate, v, as the rate of change of a substance divided by its stoichiometric coefficient. This effectively represents the number of reaction events taking place per unit of time and unit of compartment size.

$$v = \frac{1}{-a}\frac{d[\text{A}]}{dt} = \frac{1}{-b}\frac{d[\text{B}]}{dt} = \frac{1}{p}\frac{d[\text{P}]}{dt}$$

Therefore, we can compute the change of each substance as the product of the reaction rate and its stoichiometric coefficient for this reaction.

$$\frac{d[\mathrm{A}]}{dt} = -a \times v,$$
$$\frac{d[\mathrm{B}]}{dt} = -b \times v,$$
$$\frac{d[\mathrm{P}]}{dt} = p \times v$$

Reaction rates depend on many factors, and can effectively take any form for the purpose of modelling. In the following subsections, we will describe the simple cases where the reaction rates depend solely on the concentrations of the reacting substances.

3.1.2 Mass-Action Kinetics

For a chemical reaction to take place, the participants have to collide, or come into close vicinity of each other. The probability of such collisions depends, among other parameters, on the local density of the reactants, and hence, in well stirred environments, on their concentrations.[1] This relationship was first described by Guldberg and Waage in the second half of the nineteenth century in a series of articles on the dynamical nature of the chemical equilibrium (Waage and Guldberg 1864). They assumed that at equilibrium both the forward and backward reaction forces or velocities were equal, and that these velocities where proportional to the concentrations of the reactants to the power of their stoichiometric coefficients. The relationship of reaction velocities and concentrations is called the "Law of Mass-Action", and rate expressions equivalent to the ones employed in their articles are sometimes referred to as "Mass-Action Kinetics".[2]

The rates of simple unidirectional chemical reactions are usually proportional to the product of the concentrations of the reactants to the power of constant exponents, called *partial reaction orders* or n_X. The sum of all partial orders is called the *order*

---

[1]Under non-ideal conditions, as found in biology, activities instead of concentrations should actually be used both for describing rate equations and equilibria. As this is not common practise in biological modelling, we do not distinguish between activities and concentrations in the following. It should be noted, though, that activities can differ significantly from concentrations in cellular environments.

[2]The term *mass-action* stems from the proportionality of the so called reaction "force" to the mass of a substance in a fixed volume, which is proportional to the molar concentration of a substance.

n of a reaction, and the proportionality factor is called the *rate constant* k. For example, for the reaction described in Eq. 3.1 assuming mass action kinetics the reaction rate appears as follows:

$$v = k \times [\mathrm{A}]^{n_A} \times [\mathrm{B}]^{n_B}$$

The reaction has an order of $n = n_A + n_B$. In general, the order of elementary reactions is equal to the number of molecules interacting, also known as the *molecularity*. A *unimolecular* reaction $A \rightarrow P$ for example would have an order of one, a *bimolecular* reaction, such as $2A \rightarrow P$ or $A + B \rightarrow P$ would be a second order reaction etc. However, this equivalence is not always true, and anisotropy or crowding of the reaction environments may affect the motion of molecules, resulting in different, and sometimes non-integral, reaction orders.

While mass-action kinetics are only strictly valid for elementary reactions, they are widely and successfully applied in various fields of mathematical modelling in biology. Especially for large and vaguely defined reaction networks, as found in signal transduction, mass action kinetics are commonly employed as a very general initial approach. Most often, the partial orders are taken to be identical to the stoichiometric coefficients. The rate constants can either be calculated from separately measured equilibrium constants and characteristic times, or computationally fitted to reproduce experimental results.

3.1.2.1 Zeroth Order Reactions

Reactions of order zero have a reaction rate that does not depend on any reactant. Zeroth order reactions can be used for instance to represent constant creations from boundary condition reactants, such as:

$$\mathrm{X} \xrightarrow{k} \mathrm{P}$$

where X represent a set of source reactants that are not depleted by the reaction. The reaction rate is then equal to:

$$v = k \times [\mathrm{X}]^0 = k$$

in which k is the rate constant, and has the units of a concentration per time. The solution describing the evolution of P is of course a monotonic increase:

$$[\mathrm{P}](t) = [\mathrm{P}_0] + kt$$

3.1.2.2 First Order Reactions

In general unimolecular reactions are modelled using first order mechanisms. In irreversible first order reactions, the reaction rate linearly depends on the concentration of the reactant. Many decay processes show such kinetics, for example, radioactive decay, dissociation of complexes, or denaturation of proteins. For a simple reaction:

$$A \xrightarrow{k} P$$

the following rate law applies:

$$v = k \times [A]$$

in which k is the first order rate constant, and has the units of a reciprocal time, $[1/time]$. If this is the only reaction affecting the concentration of A in a system, the change of [A] equals the negative reaction rate. Similarly, the change of [P] equals the reaction rate.

$$\frac{d[A]}{dt} = -v = -k[A]$$

$$\frac{d[P]}{dt} = v = +k[A]$$

The first equation above can be easily rearranged and analytically solved, assuming an initial concentration $[A_0]$ at time $t = 0$. Furthermore, since $[P]_t + [A]_t = [P_0] + [A_0]$:

$$[A]_t = [A_0] \times e^{-kt}$$

$$[P]_t = [P_0] + [A_0] \times (1 - e^{-kt})$$

The rate constant in first order kinetics is directly related to some characteristic times of substances, which are often readily available. For example the average life time of the reactant, τ, and the time it takes for its concentration to halve, the half-life $t_{\frac{1}{2}}$, can be derived as (see Fig. 3.2):

$$\tau = \frac{1}{k}$$

$$t_{\frac{1}{2}} = \frac{ln2}{k}$$

3.1.2.3 Second Order Reactions

Second order reactions are often used to model bimolecular reactions, either between different types of molecules or between two instances of the same molecule. Examples are complex formation and dimerisation reactions. For a simple reaction:

$$A + B \xrightarrow{k} P$$

the following rate law applies:

$$v = k \times [A] \times [B]$$

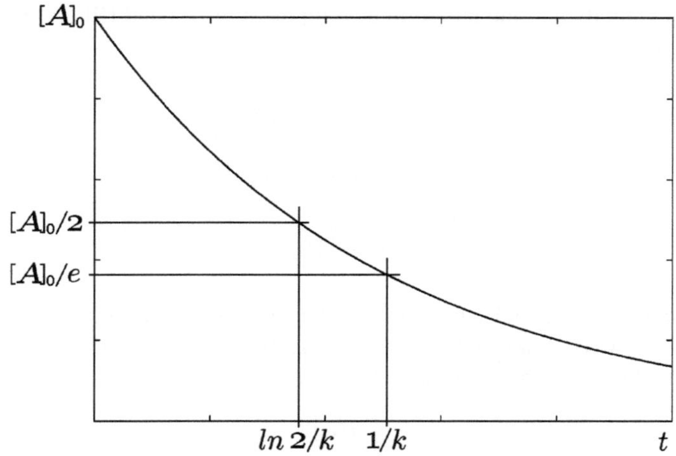

Fig. 3.2 Decay of a reactant A, that is consumed by a first order reaction with a constant k from an initial concentration of $[A_0]$. The average lifetime of a given molecule of A is given by $1/k$. [A] tends toward 0 while [P] tends towards $[A_0] + [P_0]$

in which k is the second order rate constant, and has the unit of $[1/(time \times concentration)]$. The change of [P] with time is described by the following differential equation:

$$\frac{d[P]}{dt} = v = k \times [A] \times [B]$$

Integration of the above expression using the initial concentrations $[A_0]$, $[B_0]$ and $[P_0]$ leads to a hyperbolic time dependency:

$$[P](t) = [P_0] + [A_0][B_0]\frac{e^{-kt[B_0]} - e^{-kt[A_0]}}{[A_0]e^{-kt[B_0]} - [B_0]e^{-kt[A_0]}}$$

Different from first order reactions, the characteristic times in second order reactions are not independent of the initial conditions, but depend on both the rate constant and the initial concentrations of the reactants. The half life of the limiting reactant, that is B in the case that $[A_0] \geq [B_0]$, is given by the following expression:

$$t_{\frac{1}{2}} = \frac{\ln\left(1 + \frac{[A_0]-[B_0]}{[A_0]}\right)}{([A_0] - [B_0])k}$$

A special case of bimolecular reaction, in which two reactant molecules of the same type react to form the product, occurs quite commonly in biology, for example in protein dimerisation reactions. For the general reaction:

$$2A \xrightarrow{k} P$$

the reaction velocity and the temporal development of [A] and [P] are given by the following equations:

$$v = k \times [\mathrm{A}] \times [\mathrm{A}]$$

$$\frac{d[\mathrm{A}]}{dt} = -2v = -2k[\mathrm{A}]^2$$

$$\frac{d[\mathrm{P}]}{dt} = v = k[\mathrm{A}]^2$$

Again, these differential equations can be integrated and, assuming the initial concentrations to be $[\mathrm{A}_0]$ and $[\mathrm{P}_0]$, resulting time courses for [A] and [P] are described by the following hyperbolic functions:

$$[\mathrm{A}](t) = \frac{[\mathrm{A}_0]}{2k[\mathrm{A}_0]t + 1}$$

$$[\mathrm{P}](t) = [\mathrm{P}_0] + \frac{2[\mathrm{A}_0]kt}{2k[\mathrm{A}_0]t + 1}$$

The half life, $t_{\frac{1}{2}}$ of [A] again depends on the initial concentration:

$$t_{\frac{1}{2}} = \frac{1}{2k[\mathrm{A}_0]}$$

3.1.3 The Thermodynamics of Reactions

The field of thermodynamics is concerned with the interconversion of different forms of energy, subsumed mainly under the notions of heat and work, and relates them to changes in observable properties of a system, such as temperature, electrochemical potentials, osmotic pressure, and concentrations of substances. The tools provided by chemical thermodynamics allow us to explore the energetics of a biochemical system and to determine the direction of coupled reactions and processes, such as the transport of ions across a membrane coupled to an electrochemical potential.

3.1.3.1 Energetics and Equilibrium

Central in chemical thermodynamics is the notion of the chemical equilibrium, a state in which all concentrations stay constant over time. While this means that all net reaction fluxes are zero, forward and reverse reactions can still occur, but simply cancel out. Solutions of reacting compounds in a closed environment tend towards a state of equilibrium at which the time evolution of their concentrations stops. In their work on the *dynamical* equilibrium Guldberg and Waage found that at equilibrium a certain ratio of the products and reactants, the so called mass action ratio, Γ, was constant for specific conditions. This value is called the equilibrium constant, K_{eq}. For a reversible reaction, Γ is defined as the product of all product concentrations, divided by the product of the reactants, with each concentration taken to the power of their stoichiometric coefficients. For the general reaction described in Eq. 3.1 Γ appears as follows:

$$\Gamma = \frac{[\mathrm{P}]^p}{[\mathrm{A}]^a \times [\mathrm{B}]^b}$$

and at equilibrium:

$$\Gamma_{eq} = K_{eq} = \frac{[\mathrm{P_{eq}}]^p}{[\mathrm{A_{eq}}]^a \times [\mathrm{B_{eq}}]^b} \tag{3.2}$$

The *disequilibrium ratio*, $\rho = \Gamma/K_{eq}$, gives the direction of a reaction. For $\rho < 1$, a reaction tends towards the products, while for $\rho > 1$ the reverse reaction rate is greater than that of the forward reaction.

The original derivation of K_{eq} by Guldberg and Waage was based on setting the forward rate of a reaction equal to the backward rate under the assumption of mass action kinetics. While this approach strictly speaking is only valid for simple elementary reactions, the derived expression for the equilibrium constant, which today is also called the *Law of Mass action* (3.2), is still valid under the caveat that under non–ideal conditions activities, rather than concentrations, have to be used. For the following reaction with mass action kinetics

$$\begin{aligned} &\mathrm{A} + \mathrm{B} \rightleftharpoons \mathrm{C} \\ \text{with :}\quad &\nu_f = k_f \times [\mathrm{A}] \times [\mathrm{B}] \\ &\nu_r = k_r \times [\mathrm{C}] \end{aligned}$$

the following relationship between the rate constants can be derived, by setting the forward and reverse rate equal:

$$\begin{aligned} \nu_f =& k_f \times [\mathrm{A_{eq}}] \times [\mathrm{B_{eq}}] = \nu_r = k_r \times [\mathrm{C_{eq}}] \\ K_{eq} =& \frac{[\mathrm{C_{eq}}]}{[\mathrm{A_{eq}}][\mathrm{B_{eq}}]} = k_f/k_r \end{aligned}$$

K_{eq} is related to the Gibb's free energy G, which describes the potential of a system to perform usable work, or equivalently, to undergo spontaneous change. The change of G, ΔG, accompanying a process, indicates whether this process is spontaneous and how much non-expansive work can be obtained during this change. Non-expansive work can be, for example, the movement of ions to create an electrochemical potential, fuelling of other non-spontaneous processes, such as synthesis of ATP, or mechanical work, such as muscle contraction. At equilibrium, the Gibb's free energy of a system is minimal and $\Delta G = 0$ for all processes.

The change of G for a reaction can be defined independent of the reactants' stoichiometries, so that $\Delta_r G$ is the change of G per defined amount of reaction turnover, e.g. 1 mol. These so called reaction Gibb's free energies can be calculated by subtracting the sum of the reactants' free energies times their stoichiometric coefficients from the products' free energies. For the general reaction described above (3.1) this would mean:

$$\Delta_r G = pG_P - (aG_A + bG_B)$$

In the literature, in general, free energies are given for standard conditions, such as room temperature and substance concentrations of 1 M. These standard reaction free energies, $\Delta_r G^0$, can be used to calculate $\Delta_r G$ for other concentrations of substances, simply by using their mass action ratios Γ. In general, if a reaction has a $\Delta_r G_X$ at a state X with a mass action ratio of Γ_X, then $\Delta_r G_Y$ at state Y with Γ_Y can be written as

$$\Delta_r G_Y = \Delta_r G_X + RT\ ln(\Gamma_Y / \Gamma_X)$$

In this R is the universal gas constant, and T stands for the absolute temperature. As at equilibrium the reaction free energy, $\Delta_r G_{eq}$, equals 0 and for standard reaction free energies, $\Delta_r G^0$, the mass action ratio, Γ^0, is in general equal to 1, the following relation between K_{eq} and $\Delta_r G^0$ can be derived:

$$\Delta_r G_{eq} = 0 = \Delta_r G^0 + RT\ ln(K_{eq}/\Gamma^0)$$
$$\Delta_r G^0 = -RTlnK_{eq}$$

It is also possible to calculate a reaction Gibb's free energy of a solution from the reaction's mass action ratio Γ and the equilibrium constant:

$$\Delta_r G = RT\ ln\frac{\Gamma}{K_{eq}} = RT\ ln\rho$$

For coupled reactions, the free energy changes, ΔG, of the individual reactions add up to give the overall change. As reaction free energies are proportional to the logarithm of the equilibrium constants, the overall equilibrium constant can be obtained as the product of the different individual reaction K_{eq}s. In the case of cyclic reaction systems that are not driven by an external energy source, the overall reaction free energy, $\Delta_r G$, is zero, and therefore the product of all equilibrium constants has to equal unity. As the product of all equilibrium constants K_i in a reaction cycle has to equal unity, the same holds true for the product of the ratio of the reverse and the forward rate constants k_{-i} and k_i, respectively. For a cycle of n reactions this leads to the following relationship, also known as detailed balance relation or Wegscheider's cyclicity condition (Heinrich and Schuster 1996):

$$\prod_{i=1}^{n} K_i = 1 = \prod_{i=1}^{n} \frac{k_{-i}}{k_i}$$

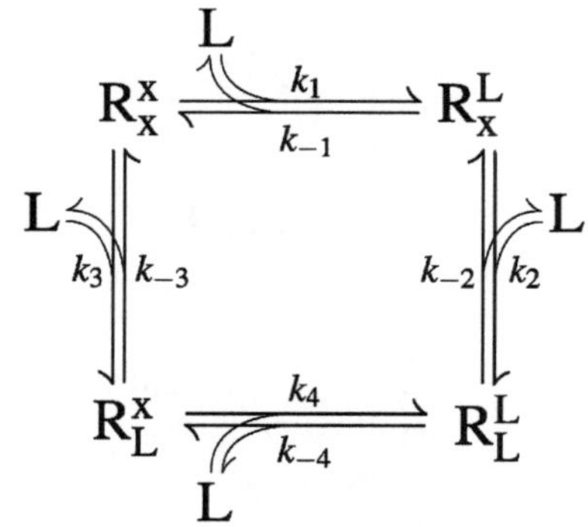

$$K_1 = \frac{k_1}{k_{-1}}; K_2 = \frac{k_2}{k_{-2}}; K_3 = \frac{k_3}{k_{-3}}; K_4 = \frac{k_4}{k_{-4}};$$
$$K_1 \cdot K_2 = K_3 \cdot K_4 \implies K_1 = \frac{K_3 \cdot K_4}{K_2}$$

Fig. 3.3 Reaction diagram for binding of a ligand L to two distinct sites on a receptor R. Assuming detailed balance allows us to express one equilibrium constant as a function of the others (After Colquhoun et al. 2004)

Applications of this principle to ions binding to a receptor are shown in Colquhoun et al. (2004). Figure 3.3 illustrates the method for binding of a ligand L to two, distinct binding sites on a receptor R. The detailed balance relation allows us to express one of the equilibria, or one of the rate constants through the other ones.

For binding reactions the inverse of the equilibrium constant, K_{eq}, also known as dissociation constant, K_{diss} is commonly used. In case of a simple complex formation reaction of a receptor R and a ligand L the dissociation constant would be defined as follows:

$$\mathrm{R} + \mathrm{L} \underset{k_{off}}{\overset{k_{on}}{\rightleftharpoons}} \mathrm{RL}$$

$$K_{diss} = \frac{1}{K_{eq}} = \frac{[\mathrm{R_{eq}}][\mathrm{L_{eq}}]}{[\mathrm{RL_{eq}}]} = \frac{k_{off}}{k_{on}}$$

with k_{on} and k_{off} being the complex association and dissociation rate constants.

3.1.3.2 Transition State and Temperature Dependence of Reaction Rates

Rate constants, in general, show a strong positive temperature dependence, that is, they increase with rising temperatures. This relation was studied in detail in the latter part of the nineteenth century first by Jacobus van't Hoff and Svante Arrhenius. Arrhenius derived an empirical expression for the temperature dependence of the rate constant, k, and postulated a general mechanism underlying this relationship. He assumed that a reaction could only occur if the reacting molecules possessed enough internal energy to overcome a threshold termed activation energy, E_a, and that the proportion of such molecules was given by a Boltzmann distribution.

$$k = Ae^{-E_a/RT}$$

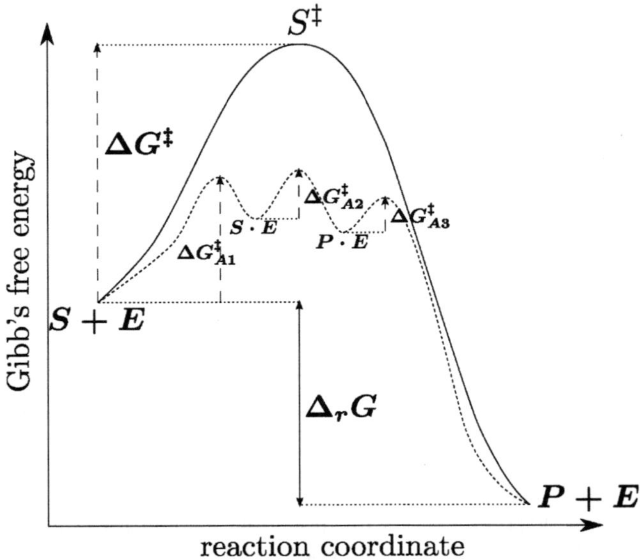

Fig. 3.4 Schematic free energy diagram for the reaction $S \rightarrow P$ without (*solid*) and with a catalyst (*dashed*), E. The reaction coordinate is a one-dimensional abstraction of the progress of the reaction. S spontaneously reacts to P via a transition state $S^\ddagger$. The catalyst E binds S and stabilises the transition state, leading to intermediate steps with a much smaller activation free energy $\Delta G^\ddagger$, thereby accelerating the reaction. The reaction energy $\Delta_r G$ is the same in both cases

In this relation, called Arrhenius equation, A is simply called the *pre-exponential* or *frequency factor*. It can be interpreted as the total frequency of reactant collisions in the correct constellation to react, but not necessarily possessing sufficient energy. A later and more detailed theory is the Transition State Theory, TST, initially pioneered by Henry Eyring and Michael Polanyi in the 1930s. Its basic assumption is that an elementary reaction runs over an unstable activated or transition state (see Fig. 3.4), with a free energy of $G^\ddagger$. Therefore, to reach the transition state, the participating reactants need at least an activation free energy of $\Delta G\ddagger = G^\ddagger - G_{gs}$, with G_{gs} being the free energy of the ground state of the reactants. The Eyring equation derived using TST relates k to the free activation energy $\Delta G\ddagger$ and the temperature similar to the Arrhenius equation:

$$k = \kappa \frac{k_B T}{h} e^{-\Delta G\ddagger / RT}$$

Here κ is called the transmission coefficient and indicates the proportion of transition states reacting to give products, k_B is the Boltzmann and h the Planck constant.

Catalysts, such as enzymes, work by reducing the free activation energy $\Delta G\ddagger$. One common possibility for this is to stabilise the transition state. It should always be kept in mind, though, that the reaction free energy, $\Delta_r G$, and with it the equilibrium constant, K_{eq}, is not affected by enzymes.

3.1.4 Representing the Evolution of Multi-reaction Systems

In the sections above, we only derived expressions describing the temporal evolution of species altered by single reactions. In biological systems, substances are involved in many different processes, leading to complex ordinary differential equation systems, that normally can only be solved numerically and with the help of computers.

3.1.4.1 Reconstruction of a System of Ordinary Differential Equations

Having carefully designed the elementary processes composing the system, reconstructing the differential equations representing the evolution of the different substances is a systematic and easy procedure. We already saw in Sect. 3.1.2.2 that the reaction:

$$\mathrm{A} \xrightarrow{k} \mathrm{P}$$

could be modelled by the system:

$$\frac{d[\mathrm{A}]}{dt} = -1v = -1k[\mathrm{A}]$$
$$\frac{d[\mathrm{P}]}{dt} = 1v = +1k[\mathrm{A}]$$

If the reaction is reversible, such as:

$$\mathrm{A} \underset{k_r}{\overset{k_f}{\rightleftharpoons}} \mathrm{P}$$

then we can consider it as a combination of two irreversible reactions, the rates of which depend on $[A]$ and $[P]$:

$$v_f = k_f \times [\mathrm{A}]$$
$$v_r = k_r \times [\mathrm{P}]$$

The evolution of both substances therefore depends on the forward and reverse reaction rates. A is consumed by the forward reaction and produced by the reverse reaction. It is the other way around for P.

$$\frac{d[\mathrm{A}]}{dt} = -1v_f + 1v_r = -1k_f[\mathrm{A}] + 1k_r[\mathrm{P}]$$
$$\frac{d[\mathrm{P}]}{dt} = +1v_f - 1v_r = +1k_f[\mathrm{A}] - 1k_r[\mathrm{P}]$$

To understand how to handle non-unity stoichiometric numbers, consider the following dimerisation:

$$2\mathrm{A} \underset{k_r}{\overset{k_f}{\rightleftharpoons}} \mathrm{P}$$

The forward reaction will be modelled using second-order kinetics, and the rates will therefore be:

$$v_f = k_f \times [\mathrm{A}]^2$$
$$v_r = k_r \times [\mathrm{P}]$$

As above the evolution of both substances therefore depends on the forward and reverse reaction rates. But this time two molecules of A are consumed by each forward reaction and produced by each reverse reaction. Therefore:

$$\frac{d[\mathrm{A}]}{dt} = -2v_f + 2v_r = -2k_f[\mathrm{A}]^2 + 2k_r[\mathrm{P}]$$
$$\frac{d[\mathrm{P}]}{dt} = +1v_f - 1v_r = +1k_f[\mathrm{A}]^2 - 1k_r[\mathrm{P}]$$

This approach can then be extended, independently of the size of the system considered. An ODE system will contain (at most) one differential equation for each substance. This equation will contain components representing the involvement of the substance in the different reactions of the system. For the substance S_n, involved in a system containing r reactions, the differential equation takes the following form:

$$\frac{d[\mathrm{S_n}]}{dt} = \sum_{i=1}^{r} \nu_{ni} v_i$$

ν_{ni} denotes the stoichiometric coefficient of S_n in reaction i, v_i the rate of this reaction. The resulting ODE system can also be represented in matrix notation, by introducing the stoichiometric matrix, **N**, and the reaction rate vector, **v**. The stoichiometric matrix, **N**, contains a row for each of the n species in the system, and a column for each of the r reactions. Its entries, N_{ij}, are the stoichiometric coefficients, ν_{ni}, of substance i in reaction j. **v** is a column vector with each element v_i indicating the rate of the ith reaction. Using the above, the change of the concentration vector **S** over time is described by:

$$\frac{d[\mathbf{S}]}{dt} = \mathbf{N} \cdot \mathbf{v}$$

3.1.4.2 Numerical Integration of ODE Models

Besides the most elementary systems containing only few well-behaved reactions, we cannot generally solve a system of ordinary differential equations analytically.

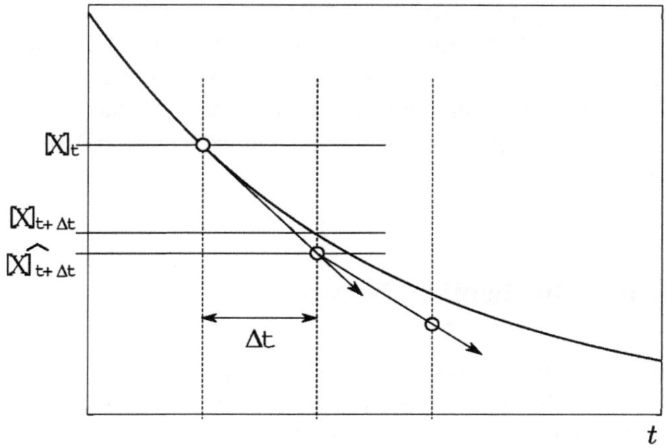

Fig. 3.5 Graphical representation of the forward Euler method to integrate ordinary differential equations. The *thick curve* represents $[X] = f(t)$, and the vectors its derivative. Note the progressive error introduced by the coarse time discretization

We have to resort to numerical integration, a method that goes back to the origin of differential calculus, where we approximate the current values of the variables based on the knowledge we have of their values in the (close) past. Many approximations have been developed. The simplest and easiest to grasp is the forward Euler rule. If we discretize the time, one can make the following approximation:

$$\frac{d[\mathrm{X}]}{dt} \approx \frac{\Delta[\mathrm{X}]}{\Delta t} = \frac{([\mathrm{X}]_{t+\Delta t} - [\mathrm{X}]_t)}{\Delta t}$$

We can rearrange the equation above and extract the concentration as follows:

$$[\mathrm{X}]_{t+\Delta t} \approx [\mathrm{X}]_t + \frac{d[\mathrm{X}]}{dt}(t) \times \Delta t$$

We know $d[X]/dt$ as a function of the vector of concentrations, obtained with the method described in Sect. 3.1.4.1, and can therefore compute the difference introduced during one Δt. This procedure is represented in Fig. 3.5. We can see on the figure that a systematic error is introduced by the time discretization. Such an error becomes larger for more complex dynamics, such as non-monotonic behaviours, or systems with fast and slow components. One can address the error by using tiny time steps but at the expense of computational efficacy. Many methods have been developed over the years to address this problem. A good introduction is given in LeMasson and Maex (2001) and a more comprehensive survey of the field by Hairer et al. (1993) and Hairer and Wanner (1996).

Biological modelling tools such as COPASI (Hoops et al. 2006), JDesigner/Jarnac (Sauro 2000), E-Cell (Takahashi et al. 2003) or CellDesigner (Funahashi et al. 2008) have their own in-built numerical ODE solver. They also generate the system of ODE to be solved automatically, so the user input that is required usually consists of a list of chemical reactions in some defined format and of the parameters governing those reactions.

3.2 Modelling Biochemical Networks

Modelling the biochemical pathways involved in neuronal function does not require much more than what has been presented in Sect. 3.1. The only complexity we will introduce in the following sections are slightly more complex expressions for the reaction rates.

3.2.1 Basal Level and Homoeostasis

Before modelling the effect of perturbations, such as extracellular signals, it is important to set up the right basal level for the substances that we will consider in the model. This basal level is obtained when the processes producing the substance and the processes consuming it are compensating each other. We then reach a steady-state, where input and output are equal. To illustrate this, we will build the simplest system possible that permits to have a steady basal concentration of calcium. The system is made up of a continuous creation of calcium, for instance due to leaky channels in the plasma membrane or in the internal stores, modelled as a zeroth order reaction (see Sect. 3.1.2.1). The calcium is then removed from the system for instance by pumps or buffers in excess, modelled as a first order reaction (see Sect. 3.1.2.2).

$$\emptyset \xrightarrow{k_{in}} \mathrm{Ca}^{2+} \xrightarrow{k_{out}} \emptyset$$

The instantaneous changes of calcium concentration then result from the combination of the two reaction rates (Fig. 3.6).

$$\frac{d[Ca^{2+}]}{dt} = k_{in} - k_{out}[Ca^{2+}]$$

The steady-state level is reached when the changes are null, that is $[Ca^{2+}] = k_{in}/k_{out}$. If the concentration of calcium is higher than this ratio, the second term wins and the concentration decreases. In contrast, if the concentration of calcium is lower than this ratio, the first term wins and the concentration increases. k_{out}

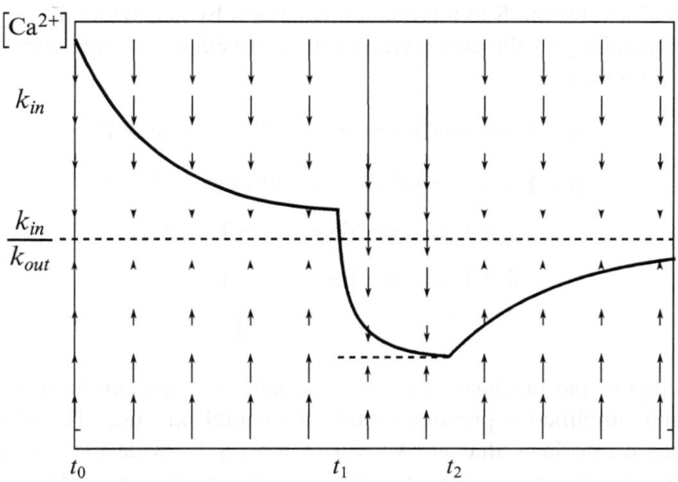

Fig. 3.6 Evolution of calcium concentration over time. Between t_0 and t_1, the extrusion is stronger than the creation. At t_1, k_{in} strongly decreases, for instance by a block of leak channels, and the concentration is brought to a lower steady-state value. At t_2 the block is removed. The creation becomes stronger than extrusion, and brings back the concentration to the initial steady-state. *Vertical arrows* represent the intensity and direction of the reaction's flux for a given concentration of calcium

can be estimated from the decay observed after stimulation. k_{in} can therefore be computed from the steady-state. Changing k_{in} in a discrete manner is a simple way of modelling the opening or closing of calcium channels.

Such a homoeostatic control is extremely simple. More complex schemes can be designed, with control loops such as negative feed-backs on the creation steps and positive feedforwards on the extrusion steps.

3.2.2 *Representing Enzymatic Reactions*

In order to accelerate chemical reactions and select among different isomers, cells use enzymes, which are protein-based catalysts. They can increase reaction rates to a tremendous degree and often are essential to making reactions occur at a measurable rate. Enzyme catalysed reactions tend to follow complex sequences of reaction steps, and the exact reaction mechanisms are generally unknown. The single reaction steps can be contracted into an overall description with lumped stoichiometries. However, since the detailed reaction mechanisms are most often unknown, and also parameters for each of these steps are hard to come by, such reactions can rarely be modelled considering each single step and using mass action kinetics. Depending on how much detail is known, an enzyme catalysed reaction can be described on different levels. The reaction equations for a simple conversion of a substrate S to a product P catalysed by an enzyme E, for example, can vary depending on the consideration of intermediate enzyme complexes and reaction reversibility:

$$
\begin{gathered}
S + E \rightleftharpoons ES \rightleftharpoons EP \rightleftharpoons E + P \\
S + E \rightleftharpoons ES \longrightarrow EP \rightleftharpoons E + P \\
S + E \rightleftharpoons ES \rightleftharpoons E + P \\
S + E \rightleftharpoons ES \longrightarrow E + P \\
S + E \longrightarrow E + P
\end{gathered} \tag{3.3}
$$

Knowledge of the mechanism of an enzymatic reaction can be used to derive compact and simplified expressions fitting the overall kinetics. The alternative is to use generic rate laws that are known to loosely fit wide classes of reaction mechanisms, and to choose the ones that seem most appropriate for the reaction in question. The kinetics of the overall reaction are determined by the reaction mechanisms of the elementary steps, but exact derivations can become quite complex and cumbersome to handle. In general it is safer and more convenient to use approximate expressions in biological modelling, even more so as exact mechanisms are rarely known.

Two assumptions are available to simplify complex enzymatic reaction descriptions. The more general one is the quasi steady state approximation, QSSA. The QSSA considers that some, or all, of the intermediary enzyme substrate complexes tend to a near constant concentration shortly after the reaction starts. The other widely used assumption, called the rapid equilibrium assumption, is that some steps are much faster than the overall reaction, meaning that the participating enzyme forms are virtually at equilibrium and that their concentrations can be expressed using equilibrium constants. This approach is often used to model fast reactant or modulator binding to the enzyme. The application of these techniques depends very much on how much of the reaction mechanism is known. An excellent introduction into enzyme kinetics is given by Cornish-Bowden (2004). For a more exhaustive treatment with detailed derivations of rate laws for a multitude of mechanisms please refer to the standard work by Segel (1993).

3.2.2.1 Henri-Michaelis-Menten Kinetics

At the beginning of the twentieth century, Henri (1902) proposed a reaction scheme and an accompanying expression for describing the rate of sucrose hydrolysis catalysed by invertase. This reaction showed a deviation from normal second-order kinetics and tended to a maximal velocity directly proportional to the enzyme concentration. Making use of the existence of an intermediary substrate-enzyme complex, ES, and assuming that the substrate S and the enzyme E were in a rapid binding equilibrium with the complex, he could derive an expression fitting the experimental observations. A similar approach was taken and expanded in 1913 by Michaelis and Menten (1913), who proposed the current form of the reaction rate based on a rapid equilibrium between enzyme and substrate.

$$\mathrm{E} + \mathrm{S} \underset{k_{-1}}{\overset{k_1}{\rightleftharpoons}} \mathrm{ES} \xrightarrow{k_2} \mathrm{E} + \mathrm{P}$$

(k_2 is the catalytic constant, or turnover number, and often called k_{cat}.)

A more general derivation, using the QSSA, was proposed by Briggs and Haldane (1925). The substrate binding and dissociation, as well as the product formation step, lead to the following expression for the time dependence of [ES]:

$$\frac{d[\mathrm{ES}]}{dt} = k_1[E][S] - k_{-1}[ES] - k_2[ES]$$

At steady state, the concentration of the intermediate complex,[ES], is constant hence $d[\mathrm{ES}]/dt = 0$. Rearranging this equation and setting $K_M = \frac{k_{-1}+k_2}{k_1}$, we obtain $[E] = [ES] \times K_M/[S]$. Furthermore, because the concentration of enzyme is constant, we have $[E] = [E_t] - [ES]$. Equating both, we obtain:

$$v = \frac{d[\mathrm{P}]}{dt} = k_2[\mathrm{ES}] = k_2[\mathrm{E_t}]\frac{[\mathrm{S}]}{K_M + [\mathrm{S}]} \tag{3.4}$$

$k_2 \times [E_t]$ is sometimes called the maximal velocity v_{max}.

This rate expression is often used—and abused—when modelling biochemical processes for which the exact mechanisms are unknown. However, one has to realise that it only holds true if the concentration of the enzyme-substrate complex stays constant, which in turns implies that the concentration of substrate is in large excess. Those conditions are very rarely met in signal transduction systems, resulting in many artifacts.

Plotting the reaction velocity, v, against the substrate concentration, [S], gives a rectangular hyperbolic curve (see Fig. 3.7). The parameter K_M has the unit of a concentration and is of central importance in describing the form of the substrate dependence of the reaction velocity. As can be seen by inserting K_M for [S] in Eq. 3.4, it denotes the substrate concentration at which the reaction speed is half of the limiting velocity. If $[\mathrm{S}] \ll K_M$, then [S] in the denominator can be disregarded and the reaction becomes linear with regard to S, showing first order characteristics:

$$[\mathrm{S}] \ll K_M \Rightarrow v \approx \frac{v_{max}}{K_M} \times [\mathrm{S}]$$

On the other extreme, for high substrate concentrations, $[\mathrm{S}] \gg K_M$, the reaction speed becomes virtually independent of [S] and tends toward v_{max}.

$$[\mathrm{S}] \gg K_M \Rightarrow v \approx v_{max} = k_{cat} \times [\mathrm{E_t}]$$

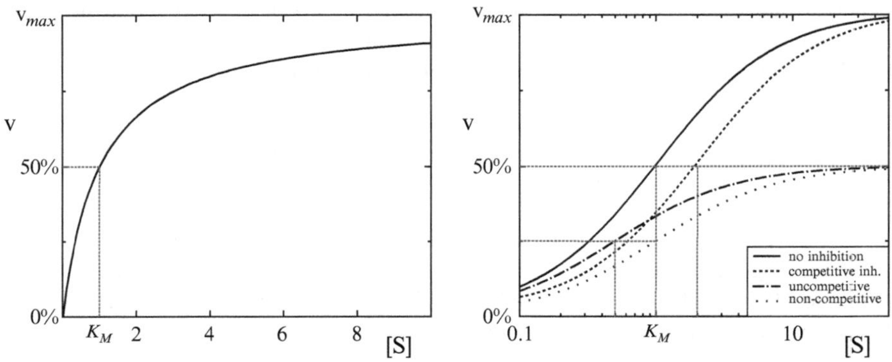

Fig. 3.7 Dependence of the reaction velocity, v, of the irreversible Michaelis Menten equation on the concentration of the substrate, S. The *left graph* shows the uninhibited case. On the *right* various forms of inhibition are shown in a semi-logarithmic plot. The *horizontal dotted lines* indicate the apparent half maximal velocities, the *vertical* ones the apparent K_Ms. Competitive inhibition does not alter the maximal velocity, but shifts the K_M to higher values, while non-competitive inhibition simply decreases the apparent v_{max}. The special case of uncompetitive inhibition leads to an apparent increase of substrate affinity of the enzyme, that is a lower K_M, but a reduction of the apparent v_{max}. Mechanistically this is due to the unproductive enzyme-substrate-inhibitor complex ($K_M = 1$; [I]= 1; comp., uncomp. and non-comp. inhib.: $K_I = 1$)

Most enzyme catalysed reactions show a similar rate behaviour inasmuch as they exhibit first or higher order dependencies on the substrate at lower substrate concentrations and tend to a limiting rate depending only on the enzyme concentration when the reactant concentrations are high.

While the original Michaelis–Menten equation was derived to describe the initial velocity of the enzymatic reaction in absence of product, allowing the reverse reaction to be neglected, the QSSA can also be used to derive a reversible Michaelis–Menten equation describing the most extensive reaction scheme in Eq. 3.3.

Using the same procedure as above, the following expression for the reaction velocity in dependence of E_T, S and P can be derived:

$$v = \frac{v_{fwd}\dfrac{[S]}{K_{MS}} - v_{rev}\dfrac{[P]}{K_{MP}}}{1 + \dfrac{[S]}{K_{MS}} + \dfrac{[P]}{K_{MP}}} \qquad (3.5)$$

As the net rate of a reversible reaction has to vanish at equilibrium, one of the parameters of Eq. 3.5 can be expressed using the equilibrium constant by setting the numerator of the expression to zero. The so called Haldane relationship connects kinetic and thermodynamic parameters of an enzymatic reaction. While some mechanisms lead to much more complicated expressions, at least one Haldane relationship exists for every reversible reaction.

$$K_{eq} = \frac{v_{fwd} K_{MP}}{v_{rev} K_{MS}} = \frac{k_2 K_{MP}}{k_{-1} K_{MS}}$$

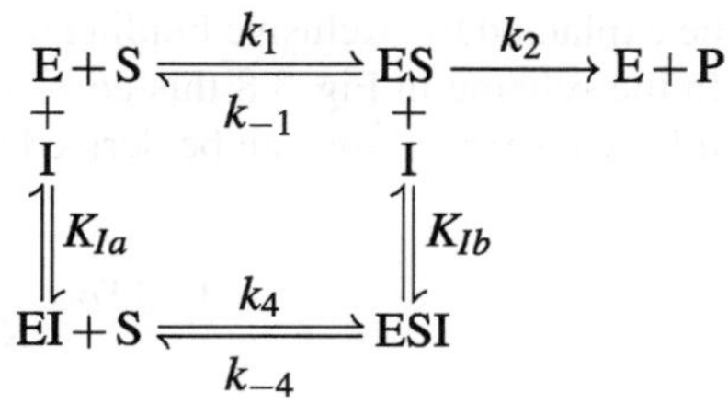

Fig. 3.8 Reversible inhibition of an enzyme E by binding of an inhibitor I. Depending on the values of the dissociation constants K_{Ia} and K_{Ib}, the inhibition can be of competitive, uncompetitive or mixed type

3.2.2.2 Enzyme Regulation

An important aspect of enzyme catalysed reactions is the regulation of enzyme activity by effectors or modifications. There are many possible mechanisms for both activation and inhibition of enzymes, often leading to complicated and unwieldy mathematical expressions. Luckily, for modelling purposes crude approximations can be sufficient in many cases.

Alteration of an enzyme's activity by covalent modifications, such as phosphorylation of cyclin dependent kinases or cleavage in the case of caspases often have to be modelled directly using explicit differential equations for each state. As binding processes are normally much faster, regulation by reversible binding of effectors is more amenable to using rapid equilibrium and steady state assumptions and deriving compact mathematical expressions.

Inhibition can be either *irreversible* or *reversible*, depending on whether the inhibitor disrupts enzyme activity permanently or not. For reversible inhibition, three basic cases can be distinguished, *competitive*, *uncompetitive* and *noncompetitive* inhibition. For these cases, minimal mechanisms can be assumed in combination with the Michaelis Menten scheme as depicted in Fig. 3.8. In this scheme, inhibitor binding is characterised by using the dissociation constants, K_{Ia} and K_{Ib} of the enzyme inhibitor complexes.

In *competitive* inhibition, the inhibitor does not alter the limiting rate, but increases the effective Michaelis constant, K_M, the concentration of substrate needed to reach half maximal activity. One possible explanation for this behaviour is that the inhibitor competes with the substrate for the enzyme by binding the same site and blocking it. In the scheme in Fig. 3.8 this corresponds to the inhibitor exclusively binding the free enzyme, $\frac{1}{K_{Ib}} = 0$ and $k_4 = k_{-4} = 0$. Under a quasi steady-state assumption the following dependence of the velocity on the substrate, S, and inhibitor, I, concentrations can be derived:

$$v = v_{max} \frac{[\mathrm{S}]}{K_M \left(1 + \frac{[\mathrm{I}]}{K_{Ia}}\right) + [\mathrm{S}]}$$

The effect of *uncompetitive* inhibitors on the other hand cannot be counteracted by higher substrate concentrations. They alter both the apparent limiting rate as well as the effective Michaelis constant. In the case of the simple Michaelis Menten mechanism, both v_{max} and K_m are altered by the same factor. This behaviour can be explained by exclusive binding of the inhibitor to the enzyme substrate complex. In the scheme in Fig. 3.8 this corresponds to $\frac{1}{K_{Ia}} = 0$ and $k_4 = k_{-4} = 0$ and the following expression can be derived for the velocity:

$$v = v_{max} \frac{[S]}{K_M + [S]\left(1 + \frac{[I]}{K_{Ib}}\right)}$$

Noncompetitive inhibition is a rarely occurring case, in which the inhibitor only alters the apparent limiting velocity. A possible mechanism would be that inhibitor binds the enzyme independent of the substrate, totally abolishing the enzyme's activity. In the scheme in Fig. 3.8 this would correspond to $K_I = K_{Ia} = K_{Ib}$ and $k_4 = k_1, k_{-4} = k_{-1}$ and an expression for the velocity of the form:

$$v = v_{max} \frac{[S]}{K_M + [S]} \frac{1}{\left(1 + \frac{[I]}{K_{Ib}}\right)}$$

The more realistic scenario, in which inhibitor binding depends on substrate binding, is called *mixed* inhibition. With this form of inhibition both the apparent limiting rate and K_M are altered by the inhibitor. This scenario subsumes all three other forms of inhibition as special cases with the proper K_{Ia} and K_{Ib}. An expression for the mixed type mechanism with the scheme in Fig. 3.8 can be derived by using the steady state assumptions equivalent to those used for derivation of the irreversible Michaelis–Menten equation (3.4) and considering all possible enzyme states. This gives an expanded conservation relation for the total enzyme concentration, $[E_T] = [E] + [ES] + [EI] + [EIS]$. Taking a rapid equilibrium approach for inhibitor binding, [EI] and [EIS] can be expressed using their dissociation constants, K_{Ia} and K_{Ib} respectively, and the concentrations of free enzyme and inhibitor:

$$[EI] = \frac{[E] \times [I]}{K_{Ia}} \quad \text{and} \quad [EIS] = \frac{[ES] \times [I]}{K_{Ib}}$$

Proceeding as for Eq. 3.4 the rate law for the simple mixed type inhibition mechanism in Fig. 3.8 results as:

$$v = k_{cat}[E_T] \frac{[S]}{K_M\left(1 + \frac{[I]}{K_{Ia}}\right) + [S]\left(1 + \frac{[I]}{K_{Ib}}\right)}$$

3.2.3 *Modelling Simple Transport Processes*

Compartmentalisation of molecular species and transport across membranes is of great importance in biological systems, and often needs to be implicitly accounted for or explicitly included into models.

Transport across membranes can either occur passively by simple diffusion, or be coupled to another reaction to actively move molecules against a chemical potential gradient. In the simplest form of passive diffusion, molecules just directly pass through a membrane or an open channel or pore. As the connected compartments in general have differing volumes, the change of concentration of a substance flowing from one compartment to another is not equal in both compartments. Therefore the rate of translocation is commonly described by the flux, j, of a substance, that is the amount of a substance crossing a unit area per time unit. In case of no other influences on the translocation, but simple diffusion, the flux of a substance S into a cell through a membrane follows a variant of Fick's first law:

$$[\mathrm{S_{out}}] \rightleftharpoons [\mathrm{S_{in}}]$$

$$j_S = p_S([\mathrm{S_{out}}] - [\mathrm{S_{in}}])$$

in which $[\mathrm{S_{out}}]$ and $[\mathrm{S_{in}}]$ are the concentrations of S on the exterior and inside the cell, respectively. p_S denotes the permeability of the membrane for S. The permeability for direct diffusion is proportional to the diffusion coefficient of S and, for pores or channels, to the number of open channels per area.

To derive an expression of the change of concentration of S, it is important to consider that the flux is given as amount per area and time and not as concentration per time. Therefore the volumes of the exterior and the cell have to be included in the differential expressions of concentrations. The overall rate of translocation, v_t, depends on the surface area, A, of the membrane, and the permeability and area can be contracted to a transport rate constant, $k_S = p_S \times A$. For the change of $[\mathrm{S_{out}}]$ and $[\mathrm{S_{in}}]$, respectively, the following expressions can be derived:

$$\frac{d[\mathrm{S_{out}}]}{dt} = -\frac{v_T}{V_{out}} = -\frac{k_S}{V_{out}}([\mathrm{S_{out}}] - [\mathrm{S_{in}}])$$

$$\frac{d[\mathrm{S_{in}}]}{dt} = \frac{v_T}{V_{in}}$$

with V_{out} and V_{in} being the volumes of the exterior and the cell.

In the case of a molecule that does not simply diffuse through a membrane or pore, but needs to bind a carrier to be translocated from one compartment to the other, the kinetic expressions depend on the exact mechanism of translocation. The simplest case of facilitated, or carrier-mediated, diffusion consists of a carrier with a single binding site, C, which can bind a substance A with equal affinity on each side of the membrane, and flips from one side of the membrane to the other. Using the steady state approach the following expression can be derived for the translocation rate:

$$v_t = \frac{v_{max}([A_{out}] - [A_{in}])}{K_M + [A_{out}] + [A_{in}] + \frac{K_i[A_{out}][A_{in}]}{K_M}}$$

In this equation v_{max} is the limiting rate of translocation and depends mostly on the amount of carrier. K_M is the concentration of A on one side at half maximal translocation in case of zero concentration on the other side of the membrane, and K_i, called the interactive constant, depends on the relative mobility of the free and loaded carrier (for details see Kotyk 1967).

3.3 Modelling Cooperative Modulation of Dynamical Processes

Reactions in biological systems are not only regulated by the availability of reactants and catalysts, but also by compounds modulating the activity of channels and enzymes, often without any direct involvement in the specific reactions. Examples are neurotransmitters, such as acetylcholine and gamma-aminobutyric acid, that alter the flow of ions through channels, without direct involvement in the translocation process.

Often, these processes display cooperativity. Intuitively, one can imagine a cooperative scenario as one where the modulating effect of a compound depends on its concentration in a non-linear manner, where *the whole is more (or less) than the sum of its parts*. In this section, we will first introduce useful measures of ligand binding and conformational state, and then examine how cooperativity can be modelled using different frameworks.

3.3.1 Binding of Modulators and Conformational State

The activities of receptors, channels, and enzymes are often regulated by ligands binding to them. One important characteristic of such binding processes is the *fractional occupancy*, $\bar{Y}$, of the bound compound. It is defined as the number of binding sites occupied by a ligand, divided by the total number of binding sites. For a ligand X binding to a single binding site of a protein P, we can express [PX] and $\bar{Y}$ as follows, using the dissociation constant $K_{diss} = \frac{k_{off}}{k_{on}}$ and the total protein concentration $[\mathrm{P_t}] = [\mathrm{P}] + [\mathrm{PX}]$:

$$\mathrm{P} + \mathrm{X} \underset{k_{off}}{\overset{k_{on}}{\rightleftharpoons}} \mathrm{PX}$$

$$[\mathrm{PX}] = \frac{[\mathrm{P_t}][\mathrm{X}]}{K_{diss} + [\mathrm{X}]}$$

$$\bar{Y} = \frac{[\mathrm{PX}]}{[\mathrm{P_t}]} = \frac{[\mathrm{X}]}{K_{diss} + [\mathrm{X}]}\,. \tag{3.6}$$

Equation 3.6, also known as the Hill-Langmuir equation, is very similar to the Michaelis–Menten equation. Like [S] in Eq. 3.4, [X] stands for the concentration of *free* ligand, but can be substituted with the total ligand concentration $[X_t] = [X] + [PX]$ in case that $[X_t] \gg [P_T]$.

Often, a protein can exist in various distinct states, only one of which can perform a specific function. Many enzymes, for example, have an inactive state, in which their active site is blocked and an active state, in which this block is relieved. Ion channels can be open and closed. Some proteins exist in two (or more) distinct structural conformations that favour distinct binding partners. We call such a conformation of interest the R state for reasons that will become apparent later in this chapter. Since not all proteins in a population of protein P are necessarily in the same conformation, it is useful to define a *fractional conformational state*, analogous to the previous definition of the *fractional occupancy*. We denote fractional conformational state by $\bar{R}$ and define it as follows:

$$\bar{R} = \frac{[R]}{[P_t]}$$

It is important to note that fractional occupancy and fractional conformational state do not necessarily coincide. Occupancy is usually easier to measure, but the conformational state might be more relevant (and, indeed, sufficient) in some modelling scenarios. Both ligand binding and conformational change can display cooperative behaviour.

In the case of multiple ligand binding sites on a protein, *cooperativity* can arise if the binding of a ligand to one site influences binding to the others. If the binding of a ligand increases the affinity to other ligands, the binding is said to exhibit *positive*, if it decreases the affinity, *negative* cooperativity. Effects of ligands binding to a protein on an activity physically separated from their binding sites are called *allosteric*. They often occur in the regulation of channels by ligands that are structurally unrelated to the transported compounds. Depending on the kind of ligands that influence each others' binding, allosteric and cooperative effects are called *homotropic*, if a ligand influences the binding of ligands of the same kind, or *heterotropic*, if it influences the affinity to ligands of a different kind.

3.3.2 *The Hill Equation*

The first description of cooperative binding to a multi-site protein was developed by Hill (1910). Drawing on observations of oxygen binding to hæmoglobin, Hill suggested the following formula for the fractional occupancy $\bar{Y}$ of a protein with several ligand binding sites:

$$\bar{Y} = \frac{\frac{[X]^h}{K_H}}{1 + \frac{[X]^h}{K_H}} = \frac{[X]^h}{K_H + [X]^h}$$

where [X] denotes ligand concentration, K_H is an apparent dissociation constant (with the unit of a concentration to the power of h) and h is the *Hill coefficient*, which need not be an integer. The *Hill coefficient h* indicates the degree of cooperativity, and in general is different from the number of ligand binding sites, n. While n is an upper bound for h, it is not possible to estimate the binding sites from measurements of the Hill coefficient alone. This is exemplified in Hill's original analysis, in which he found exponents ranging from $h = 1.6$ to 3.2 for the binding of oxygen to hæmoglobin, while the heterotetrameric protein possesses four binding sites for O_2. The Hill equation can show positive and negative cooperativity, for exponent values of $h > 1$ and $0 < h < 1$, respectively. In case of $h = 1$ it shows hyperbolic binding behaviour. With increasing exponents, the ligand binding curve becomes more and more sigmoid, with the limit of a step function with a threshold value of $\sqrt[h]{K_H}$. The number $K_h = \sqrt[h]{K_H}$ gives the ligand concentration at which half the binding sites are occupied, or, in purely phenomenological uses, activation or inhibition by the effector is half maximal.

It is important to note that the above formula, known as the *Hill* equation is a purely phenomenological description of Hills observations of oxygen binding to hæmoglobin. It does not offer a mechanistic description of the underlying processes. Because it is a purely phenomenological description, however, it can be used to describe the cooperativity of conformational change as a function of ligand concentration just as well as it can be used to describe the cooperativity of ligand binding. It is enough to replace K_H by a phenomenological constant pertaining to conformational change (the physical equivalent of which we need not worry about) and h by an appropriate Hill coefficient that fits the data for conformational change. Note, however, that the Hill framework does not offer a way of relating ligand saturation and conformational change.

Because the Hill framework is not concerned with a mechanistic explanation of cooperative ligand binding, all binding sites are treated as equal and cooperativity itself does not change with ligand saturation. In other words: Cooperativity in the Hill model is solely a property of ligand molecules, rather than a property of binding sites.

3.3.2.1 Using Hill Functions to Model the Regulation of Biochemical Processes

Hill functions can easily be adapted for use in modelling and to describe interactions with little prior knowledge. Let us assume a channel C transporting a substance S, that is regulated in a nonlinear fashion by a ligand A, for example by direct binding. If the channel is activated with increasing concentrations of A, sometimes this behaviour can be approximated using a Hill function:

$$v_T([\mathrm{C}], [\mathrm{A}], [\mathrm{S}]) = v_{Tmax}([\mathrm{C}], [\mathrm{S}]) \frac{[\mathrm{A}]^h}{K_h^h + [\mathrm{A}]^h}$$

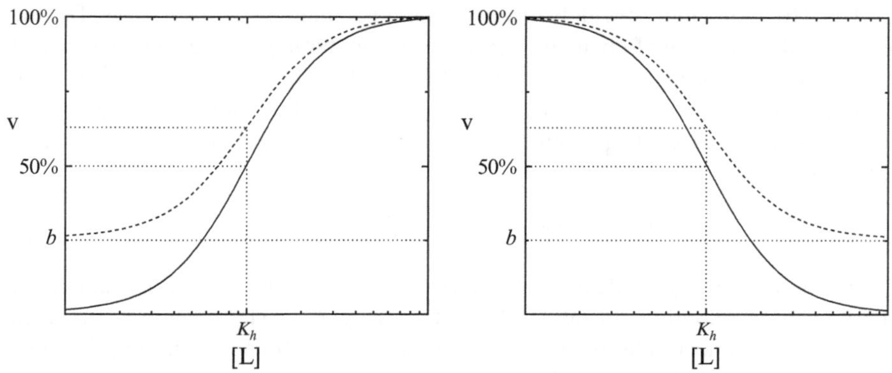

Fig. 3.9 Activation (*left*) and inhibition (*right*) modelled using Hill functions with a Hill exponent, n_h of 2. The concentration of the ligand is shown in units of the concentration of half maximal activation or inhibition, respectively, K_h on a logarithmic scale, the velocity v in percent of the fully activated or uninhibited velocity, v_{max}. The *dashed line* shows cases with a basal rate, v_{bas}, of 25% of v_{max} ($b = \frac{v_{bas}}{v_{max}} = 0.25$)

Here v_T is the actual flux rate of S through the channel C. v_{Tmax} indicates the maximal flux rate at high concentrations of A. K_h and h indicate the ligand concentration of half maximal activation and the Hill coefficient.

An inhibitory effect of a ligand I on the flux through a channel C can often be described using a similar expression:

$$v_T([\mathrm{C}], [\mathrm{I}], [\mathrm{S}]) = v_{Tmax}([\mathrm{C}], [\mathrm{S}]) \frac{1}{K_h^h + [\mathrm{I}]^h}$$

In this equation K_h stands for the concentration at which the ligand I shows half maximal inhibition.

In case of non-essential activation or leaky inhibition, a process still proceeds at a basal rate v_{bas} in absence of the activator or at high concentrations of the inhibitor. This can be accounted for by using the relative basal rate, $b = \frac{v_{bas}}{v_{max}}$:

$$v = v_{max}(b + (1 - b)\gamma([\mathrm{X}]))$$

in which $\gamma([\mathrm{X}])$ is a function describing the relative activity in dependence of the concentration of the regulating ligand X (Fig. 3.9). In the simplest case for an activating ligand A or an inhibitory ligand I, γ takes the following form:

$$\gamma([\mathrm{A}]) = \frac{[\mathrm{A}]^h}{K_h^h + [\mathrm{A}]^h}$$

$$\gamma([\mathrm{I}]) = \frac{1}{K_h^h + [\mathrm{I}]^h}$$

The Hill equation is widely used in neuronal modelling, especially for the kinetics of ligand-gated channels. An example containing two different types of activation is given in Borghans et al. (1997) for the Ca^{2+} induced Ca^{2+} release (CICR) via the inositol triphosphate ($InsP_3$) receptor. Equation 19 in that article describes the release of calcium from a calcium sensitive pool with a flux rate given by:

$$v_{InsP3R} = v_{max} \frac{[\mathrm{Ca_p}]^2}{K_1^2 + [\mathrm{Ca_p}]^2} \frac{[\mathrm{Ca_c}]^2}{K_2^2 + [\mathrm{Ca_c}]^2}$$

In this equation v_{max} denotes the maximal release rate, and $[\mathrm{Ca_p}]$ and $[\mathrm{Ca_c}]$ the Ca^{2+} concentrations in the pool and the cytoplasm. The release is regulated by the Ca^{2+} concentrations on both sides of the membrane separating the pool and the cytosol, and K_1 and K_2 stand for the threshold concentrations for these activations.

Parthimos et al. (2007) used an even more complex expression for the CICR from the sarcoplasmic reticulum via the $InsP_3$ receptor. The receptor was modelled to be both activated and inactivated by cytosolic Ca^{2+}, Ca_c, using two Hill functions involving Ca_c. A possible mechanistic explanation for this form would be the existence of independent activation and inhibition sites, with different affinities and degrees of cooperativity for Ca^{2+}. In the flux rate through the $InsP_3$ receptor

$$v_{InsP3R} = v_{max} \frac{[\mathrm{Ca_s}]^2}{K_1^2 + [\mathrm{Ca_s}]^2} \frac{[\mathrm{Ca_c}]^4}{K_2^4 + [\mathrm{Ca_c}]^4} \frac{K_3^4}{K_3^4 + [\mathrm{Ca_c}]^4} \tag{3.7}$$

K_2 and K_3 indicate the cytosolic Ca^{2+} concentrations at which activation and inhibition of CICR, respectively, are half maximal. If they are chosen in such a way that $K_2 < K_3$, the flux rate through the receptor reaches a maximum for concentration values between the values of the two constants and vanishes for higher cytosolic Ca^{2+} concentrations (see Fig. 3.10), creating a complex on-off behaviour of the $InsP_3$ receptor in dependence of the Ca^{2+} concentration.

3.3.3 The Adair-Klotz Framework

Adair (1925) and Klotz (1946) (reviewed in Klotz 2004) further explored the notion of cooperative binding. According to their framework, cooperativity was no longer fixed, but dependent on saturation: Binding of the first ligand molecule would alter the affinity of the protein for the following ligand molecules.

This type of cooperative binding can be elucidated in the simplest case of a protein possessing two identical ligand binding sites. Assuming that the first ligand molecule, X can bind either site of P with a dissociation constant K_1 to give the complex PX and the second molecule with a dissociation constant K_2 to give the complex PX_2:

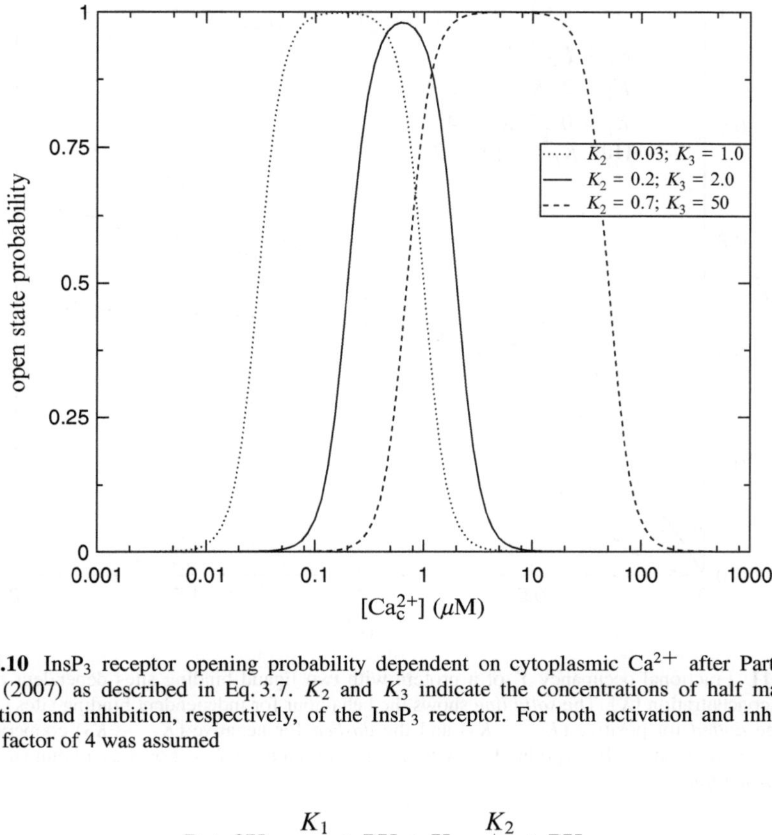

Fig. 3.10 $InsP_3$ receptor opening probability dependent on cytoplasmic Ca^{2+} after Parthimos et al. (2007) as described in Eq. 3.7. K_2 and K_3 indicate the concentrations of half maximal activation and inhibition, respectively, of the $InsP_3$ receptor. For both activation and inhibition a Hill factor of 4 was assumed

$$P + 2X \underset{}{\overset{K_1}{\rightleftharpoons}} PX + X \underset{}{\overset{K_2}{\rightleftharpoons}} PX_2$$

As the first ligand can choose from two binding sites, a factor 2 has to be included in the expression for [PX]. For the concentrations of the complexes the following relations follow:

$$[PX] = 2\frac{[P][X]}{K_1} \quad \text{and} \quad [PX_2] = \frac{[P][X]^2}{K_1 K_2}$$

or for the fractional saturation $\bar{Y}$:

$$\bar{Y} = \frac{[PX] + 2[PX_2]}{2\left([P] + [PX] + [PX_2]\right)} = \frac{\frac{[X]}{K_1} + \frac{[X]^2}{K_1 K_2}}{1 + 2\frac{[X]}{K_1} + \frac{[X]^2}{K_1 K_2}}$$

The two binding affinities, K_1 and K_2 determine the form of cooperativity exhibited by the binding process. If the binding of the ligand to both sites is completely independent, that is $K_1 = K_2$, the protein exhibits hyperbolic binding. On the other hand, if binding of the first ligand leads to an increased affinity, ie.

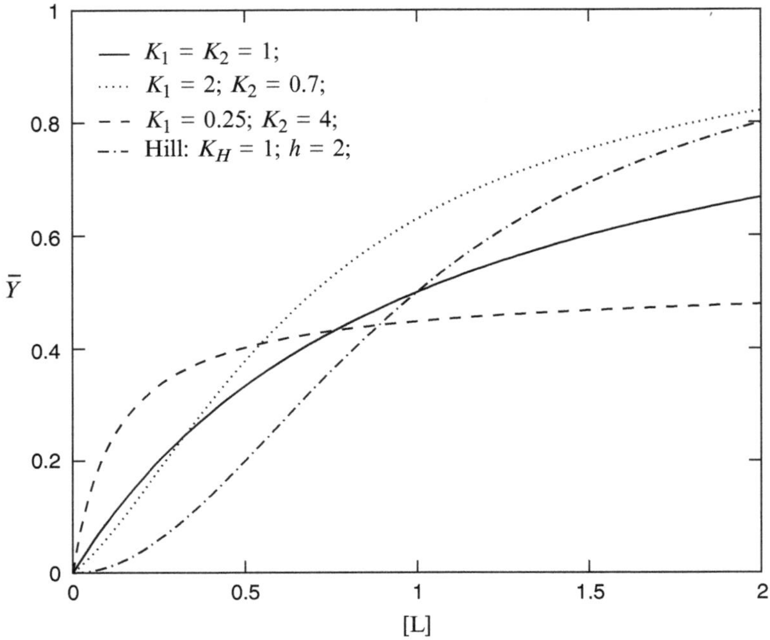

Fig. 3.11 Fractional occupancy $\bar{Y}$ of a protein with two ligand binding sites dependent on the ligand concentration $[X]$. The *solid line* shows the behaviour for independent binding sites ($K_1 = K_2$), the *dotted* for positive ($K_1 < K_2$) and the *dashed* for negative ($K_1 > K_2$) cooperativity between binding sites. Hill type binding with a Hill coefficient of 2 is shown as a boundary case (*dot-dashed line*)

decreased dissociation constant for the second site, $K_1 > K_2$, the protein exhibits positive cooperativity. In case of negative cooperativity, the binding of the first ligand decreases the affinity of the second site, $K_1 < K_2$, and the sensitivity of the protein to the ligand concentration decreases faster than with hyperbolic binding. Figure 3.11 shows different forms of cooperativity for this binding process.

In the case of $K_1 \gg K_2$, the concentration of the intermediary can be neglected and it can be assumed that the binding occurs in a single step, with both ligands binding at the same time. In this case, the above equation reduces to a Hill equation with an appropriate phenomenological dissociation constant K_H.

The Adair-Klotz framework gives a sequence of binding constants, exactly as many as there are binding sites on protein P for ligand X. It is worth noting that these constants do not relate to individual binding sites. They describe *how many* binding sites are occupied, rather than *which ones*. In that sense, the reported dissociation constants are phenomenological. At the same time, they are easily observable by fitting an Adair-Klotz equation to data on protein saturation as a function of ligand concentration. They are therefore widely used by experimentalists to describe measurements of ligand binding in terms of sequential apparent binding constants.

Note that the Adair-Klotz equation cannot be used to describe conformational change, nor is there an easy way to deduce a conformational state from a fractional occupancy. For some applications, this might not be a problem, because conformational information might not be needed, or it might be a good enough approximation to add a simple assignment such as, for instance, equating full ligand saturation with the active state. It needs to be borne in mind, though, that this approximation does not always hold and that some signalling proteins with subtle regulation patterns need to be described using models that account for their conformational state as well as their ligand saturation.

3.3.4 Allosteric Models

By the middle of the twentieth century, there was an increased interest in models that would not only describe binding curves phenomenologically, but offer an underlying biochemical mechanism. Koshland (1958) and Koshland et al. (1966) offered a tentative biochemical explanation of the mechanism described by Adair (1925) and Klotz (1946) for proteins made of identical subunits with one ligand binding site per subunit. The Koshland, Nemethy and Filmer (KNF) model assumes that each subunit can exist in one of two conformations: active or inactive. Ligand binding to one subunit would induce an immediate conformational change of that subunit from the inactive to the active conformation, a mechanism described as *induced fit*. Cooperativity, according to the KNF model, would arise from interactions between the subunits, the strength of which varies depending on the relative conformations of the subunits involved. This *sequential* model directly links saturation to the conformational state of a subunit. Importantly, it posits that not all subunits of a protein need to be in the same conformational state at the same time.

3.3.4.1 The MWC Model

In contrast, The Monod-Wyman-Changeux (MWC) model of concerted allosteric transitions (Monod et al. 1965) assumes that all subunits in the enzyme undergo conformational change together, a concept known as *concerted* transition. The probability of transition between two conformational states of the proteins, termed the *tense* (T) and the *relaxed* (R) state, depends on the binding of ligands that have different affinities for each of these two states. A schema of an MWC-type protein is shown in Fig. 3.12.

In the absence of a ligand, for instance, the T state prevails, but as more ligand molecules bind, the R state (which has higher affinity for the ligand) becomes more and more populated. Remembering the discussion in Sect. 3.1.3.2, we can describe the situation in terms of free energy: In the absence of ligand, the T state has a lower free energy than the R state and is therefore the preferred state. As more and more ligand binds, however, the R state becomes the energetically favoured conformation (see Fig. 3.13).

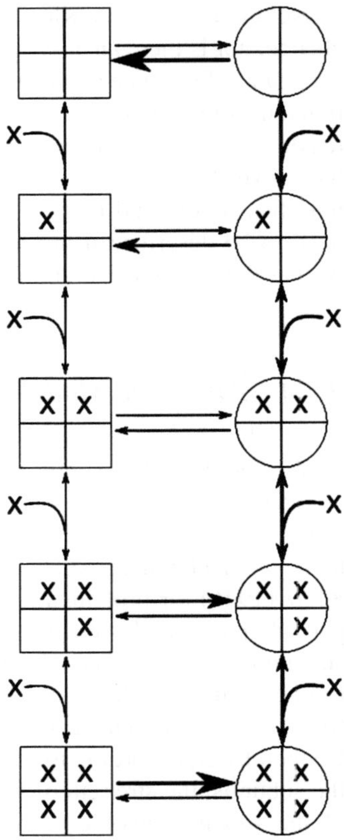

Fig. 3.12 Schematic view of an MWC protein with four subunits. The T state is shown as a *square*, the R state as a *circle*. In this case, the R state has a higher affinity for the ligand L; ligand binding thus stabilises the R state over the T state

A few new parameters have to be introduced in order to conveniently describe an MWC protein. The allosteric isomerisation constant L describes the equilibrium between both states when no ligand molecule is bound: $L = \frac{[T_0]}{[R_0]}$. If L is very large, most of the protein exists in the tense state in the absence of ligand. If L is small (close to one), the R state is nearly as populated as the T state. While the Adair-Klotz framework traditionally operates with association constants, the MWC framework has traditionally been described using dissociation constants. The ratio of dissociation constants for the R and T states is described by the constant c: $c = \frac{K_d^R}{K_d^T}$. If c = 1, both R and T states have the same ligand affinity and the ligand does not affect isomerisation. The value of c also indicates how much the equilibrium between T and R states changes upon ligand binding: the smaller c, the more the equilibrium shifts towards the R state. According to the MWC model (Monod et al. 1965), fractional occupancy is described as follows:

$$\bar{Y} = \frac{\frac{[X]}{K_d^R}\left(1+\frac{[X]}{K_d^R}\right)^{n-1} + Lc\frac{[X]}{K_d^R}\left(1+c\frac{[X]}{K_d^R}\right)^{n-1}}{\left(1+\frac{[X]}{K_d^R}\right)^{n} + L\left(1+c\frac{[X]}{K_d^R}\right)^{n}} \quad (3.8)$$

with K_d^R, L and c as described in the paragraph above.

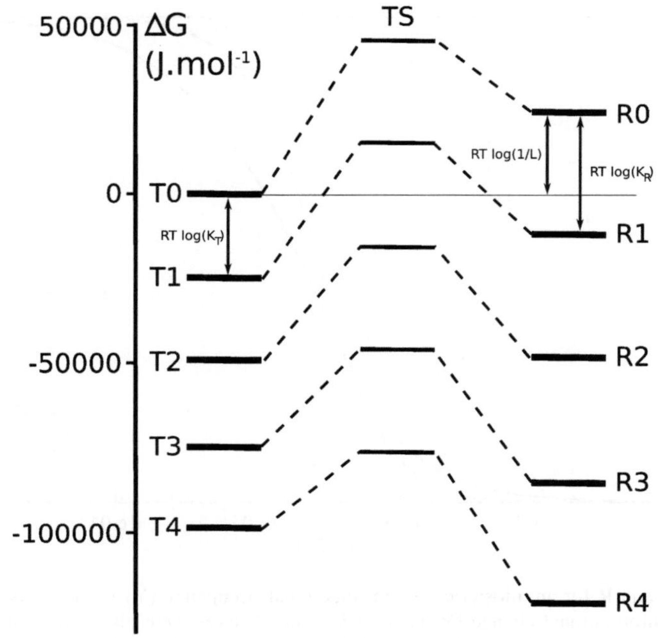

Fig. 3.13 Free energy diagram for an allosteric protein with four binding sites. Energy levels (in J/mol) were computed as in (Edelstein et al. 1996) using an allosteric model of calmodulin (Stefan et al. 2008). Each level of energy represents all the forms carrying the same number of ligand ions. Free energy differences between the T state and the corresponding R state relate to the allosteric isomerisation constant. Between corresponding T and R states, a hypothetical transition state is depicted based on estimates of rate constants. T state is shown on the *left*, R state on the *right* and the transition state in the *middle*

The degree of conformational change is described by the state function $\bar{R}$, which denotes the fraction of protein present in the R state. As the energy diagram illustrates, $\bar{R}$ increases as more ligand molecules bind. The expression for $\bar{R}$ according to the MWC model (Monod et al. 1965) is:

$$\bar{R} = \frac{\left(1 + \frac{[X]}{K_d^R}\right)^n}{\left(1 + \frac{[X]}{K_d^R}\right)^n + L\left(1 + c\frac{[X]}{K_d^R}\right)^n} \quad (3.9)$$

Thus, the MWC model can express both ligand binding and conformational change as a function of ligand concentration, and the relationship between the two is well defined because both expressions rely on the same set of microscopic parameters. It is important to note that the curves for $\bar{Y}$ and $\bar{R}$ do not overlap (Rubin and Changeux 1966), i. e. fractional saturation is not a direct indicator of conformational state (and hence, of activity). This is illustrated in Fig. 3.14.

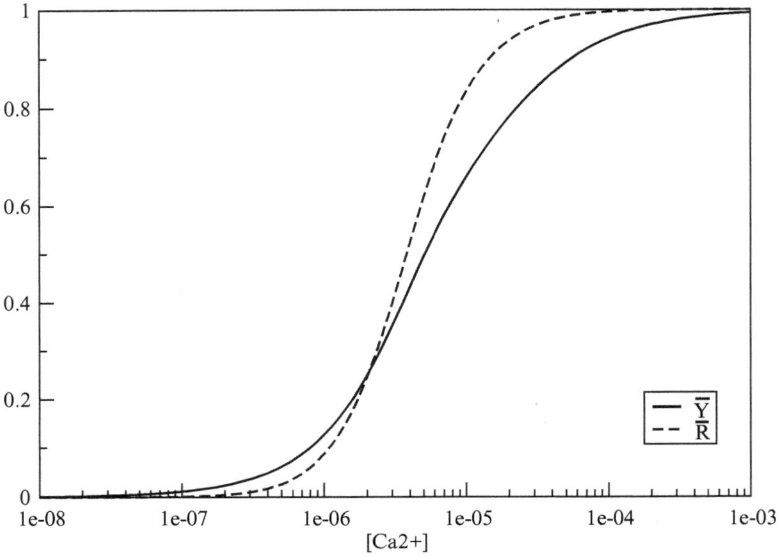

Fig. 3.14 $\bar{Y}$ and $\bar{R}$ for an allosteric protein. Fractional occupancy ($\bar{Y}$) is shown as a *solid line*; fractional conformational change ($\bar{R}$) as a *dashed line*. Curves were obtained using a model of calmodulin (Stefan et al. 2008) with a calmodulin concentration of 2×10^{-7} m

Within the MWC model, the function of an allosteric protein can be modulated by an allosteric effector: An effector that binds preferentially to the R state and hence stabilises it is called an allosteric activator, while an effector that prefers the T state is called an allosteric inhibitor (Monod et al. 1965).

Generalisation and extensions of the MWC framework have been presented to account for various additional scenarios. Other generalisations of the MWC framework have been presented to account for allosteric proteins with multiple states (Edelstein et al. 1996), for proteins that bind to different types of ligand (Mello and Tu 2005), proteins that bind to several ligands and multiple allosteric activators or inhibitors (Najdi et al. 2006) and proteins with non-identical binding sites for the same ligand (Stefan et al. 2009).

The conformational spread model by Duke et al. (2001) is a general allosteric model that encompasses both the KNF model and the MWC model as special cases.

3.3.5 Which Framework to Use

In theory, the same system can be described using either of the frameworks presented above (and a nice example for data interpreted both using the Adair-Klotz framework and the MWC framework is given in Yonetani et al. 2002), and simpler models arise as special cases from the more complicated ones. However, for the purposes of computational modelling, it is important to bear in mind that the different frameworks describing cooperativity have different scopes, drawbacks and advantages.

The Hill function is quick and easy to implement, features few unknown parameters that can readily be derived by fitting to experimental data, and can be used to describe either ligand binding or activation. On the other hand, it is a purely phenomenological description that will not offer a better mechanistic understanding of the protein-ligand system in question and does not allow for subtle effects such as a change of cooperativity as a function of saturation.

The Adair-Klotz framework is used widely in experimental work on ligand binding to protein. Therefore, dissociation constants found in the literature can often be plugged directly into an Adair-Klotz equation without the need for parameter conversion or estimation. The Adair-Klotz framework is wider in scope than the Hill equation and has more mechanistic relevance in that the association constants are related to real binding events. However, the Adair-Klotz framework itself is strictly limited to ligand binding and disregards conformational change.

The MWC framework accounts for both ligand binding and conformational change and therefore allows for the modelling of rather subtle effects, especially when the two do not coincide. It offers the greatest level of mechanistic detail and is therefore very powerful. However, allosteric parameters such as L and c are rarely found in the experimental literature and are harder to measure than the apparent Adair constants, so the demands on data analysis and parameter estimation are higher.

Note that while all frameworks provide an assigment rule that allows for ligand saturation at equilibrium to be computed from concentration of free ligand, only the Adair-Klotz framework and the MWC framework allow for a separate formulation of forward and reverse reactions, and hence for a representation of kinetic effects.

Also note that the expressions for $\bar{Y}$ in all three frameworks and for $\bar{R}$ in the MWC model only hold if the concentration of free ligand equals that of total ligand, i. e. if ligand supply is unlimited. In biological systems, this is not always the case, which means that the real dose-response curve can differ from the theoretical one. A discussion of this phenomenon, called ligand depletion, is given in Edelstein et al. (2010). An explicit simulation, in which ligand supply is not unlimited and ligand is consumed as the reactions proceed, offers a more realistic approach, although it might be more tedious to implement.

Whatever model is used will depend on a number of factors, including data availability, computational cost, scale of the model, and the biological question under investigation.

3.4 Further Reading

Biophysical chemistry, James P. Allen. This is a complete and concise presentation of the physical and chemical basis of life (Allen 2008).

Computational Cell Biology, Christopher P. Fall, Eric S. Marland, John M. Wagner, John J. Tyson. Also known as "the yellow book", this is an excellent introduction to modelling cellular processes. It contains chapters dedicated to ion channels, transporters, biochemical oscillations, molecular motors and more (Fall et al. 2002).

Enzyme kinetics, Irwin H. Segel and *Fundamentals of Enzyme Kinetics*, Athel Cornish-Bowden. Also known as "the black book" and the "the red book", these are the two reference books if one wants to know how to model an enzymatic reaction, regardless of its complexity.

Solving Ordinary Differential Equations I and II, Ernst Hairer, Syvert P. Norsett, Gerhard Wanner. Extensive coverage of the domain of ordinary differential equations, from Newton and Leibniz to the most advanced techniques using implicit solvers.

References

Adair GS (1925) The hemoglobin system. IV. The oxygen dissociation curve of hemoglobin. J Biol Chem 63:529–545

Allen JP (ed) (2008) Biophysical chemistry. Wiley-Blackwell, Oxford

Bhalla U, Iyengar R (1999) Emergent properties of networks of biological signaling pathways. Science 283(5400):381–387

Borghans JM, Dupont G, Goldbeter A (1997) Complex intracellular calcium oscillations. A theoretical exploration of possible mechanisms. Biophys Chem 66(1):25–41

Briggs GE, Haldane JB (1925) A note on the kinetics of enzyme action. Biochem J 19(2):338–339

Colquhoun D, Dowsland KA, Beato M, Plested AJR (2004) How to impose microscopic reversibility in complex reaction mechanisms. Biophys J 86(6):3510–3518

Cornish-Bowden A (2004) Fundamentals of enzyme kinetics. Portland Press, London

Duke TA, Le Novère N, Bray D (2001) Conformational spread in a ring of proteins: a stochastic approach to allostery. J Mol Biol 308(3):541–553

Edelstein SJ, Schaad O, Henry E, Bertrand D, Changeux JP (1996) A kinetic mechanism for nicotinic acetylcholine receptors based on multiple allosteric transitions. Biol Cybern 75(5):361–379

Edelstein SJ, Stefan MI, Le Novère N (2010) Ligand depletion in vivo modulates the dynamic range and cooperativity of signal transduction. PLoS One 5(1):e8449

Fall C, Marland E, Wagner J, Tyson J (eds) (2002) Computational cell biology. Springer, New York

Funahashi A, Matsuoka Y, Jouraku A, Morohashi M, Kikuchi N, Kitano H (2008) CellDesigner 3.5: a versatile modeling tool for biochemical networks. Proc IEEE 96:1254–1265

Hairer E, Wanner G (1996) Solving ordinary differential equation II: stiff and differential-algebraic problems. Springer, Berlin

Hairer E, Nørsett SP, Wanner G (1993) Solving ordinary differential equations: nonstiff problems. Springer, Berlin

Heinrich R, Schuster S (1996) The regulation of cellular systems. Springer, New York

Henri V (1902) Theorie generale de l'action de quelques diastases. C R Acad Sci. Paris 135:916–919

Hill AV (1910) The possible effects of the aggregation of the molecules of hæmoglobin on its dissociation curves. J Physiol 40:i–vii

Hoops S, Sahle S, Gauges R, Lee C, Pahle J, Simus N, Singhal M, Xu L, Mendes P, Kummer U (2006) COPASI–a complex pathway simulator. Bioinformatics 22(24):3067–3074

Klotz IM (1946) The application of the law of mass action to binding by proteins. Interactions with calcium. Arch Biochem 9:109–117

Klotz IM (2004) Ligand-receptor complexes: origin and development of the concept. J Biol Chem 279(1):1–12

Koshland DE (1958) Application of a theory of enzyme specificity to protein synthesis. Proc Natl Acad Sci USA 44(2):98–104

Koshland DJ, Nmethy G, Filmer D (1966) Comparison of experimental binding data and theoretical models in proteins containing subunits. Biochemistry 5(1):365–385

Kotyk A (1967) Mobility of the free and of the loaded monosaccharide carrier in saccharomyces cerevisiae. Biochim Biophys Acta 135(1):112–119

Land BR, Salpeter EE, Salpeter MM (1981) Kinetic parameters for acetylcholine interaction in intact neuromuscular junction. Proc Natl Acad Sci USA 78(11):7200–7204

LeMasson G, Maex R (2001) Computational neuroscience. In: De Schutter E (ed) Chapter Introduction to equation solving and parameter fitting, CRC Press, London, pp 1–23

Mello BA, Tu Y (2005) An allosteric model for heterogeneous receptor complexes: understanding bacterial chemotaxis responses to multiple stimuli. Proc Natl Acad Sci USA 102(48):17354–17359

Michaelis L, Menten M (1913) Die Kinetik der Invertinwirkung. Biochem Z 49:333–369

Monod J, Wyman J, Changeux JP (1965) On the nature of allosteric transitions: a plausible model. J Mol Biol 12:88–118

Najdi TS, Yang CR, Shapiro BE, Hatfield GW, Mjolsness ED (2006) Application of a generalized MWC model for the mathematical simulation of metabolic pathways regulated by allosteric enzymes. J Bioinform Comput Biol 4(2):335–355

Parthimos D, Haddock RE, Hill CE, Griffith TM (2007) Dynamics of a three-variable nonlinear model of vasomotion: comparison of theory and experiment. Biophys J 93(5):1534–1556

Rubin MM, Changeux JP (1966) On the nature of allosteric transitions: implications of non-exclusive ligand binding. J Mol Biol 21(2):265–274

Sauro H (2000) Jarnac: a system for interactive metabolic analysis. In: Hofmeyr JHS, Rohwer JM, Snoep JL (eds) Animating the cellular map 9th international BioThermoKinetics meeting, chap. 33. Stellenbosch University Press, Stellenbosch, pp 221–228

Segel IH (ed) (1993) Enzyme kinetics. Wiley, New York

Stefan MI, Edelstein SJ, Le Novère N (2008) An allosteric model of calmodulin explains differential activation of PP2B and CaMKII. Proc Natl Acad Sci USA 105(31):10768–10773

Stefan MI, Edelstein SJ, Le Novère N (2009) Computing phenomenologic Adair-Klotz constants from microscopic MWC parameters. BMC Syst Biol 3:68

Takahashi K, Ishikawa N, Sadamoto Y, Sasamoto H, Ohta S, Shiozawa A, Miyoshi F, Naito Y, Nakayama Y, Tomita M (2003) E-cell 2: multi-platform E-cell simulation system. Bioinformatics 19(13):1727–1729

Waage P, Guldberg C (1864) Studies concerning affinity. Forhandlinger: Videnskabs-Selskabet i Christiana 35

Yonetani T, Park SI, Tsuneshige A, Imai K, Kanaori K (2002) Global allostery model of hemoglobin. Modulation of O2 affinity, cooperativity, and Bohr effect by heterotropic allosteric effectors. J Biol Chem 277(37):34508–34520

Chapter 4
Breakdown of Mass-Action Laws in Biochemical Computation

Fidel Santamaria, Gabriela Antunes, and Erik De Schutter

Abstract The objective of this chapter is to describe conditions where the classical laws of mass action and diffusion no longer apply to biological systems, particularly neurons and other types of cells. This type of phenomena typically takes place at the nano- to micro-scale levels. An increasing number of studies show that the classical diffusion process dominated by Brownian motion cannot be directly applied to cells. Instead, a process called anomalous diffusion seems to be fundamental to the propagation of biochemical signals. Anomalous diffusion implies an increase in the correlation of movement among the diffusing molecules, which is the basis of the deviation from classical diffusion phenomena. Such a process has important consequences not only on the diffusion of molecules inside cells but also on their reaction rates. We first describe structural causes of anomalous diffusion and stochastic simulation algorithms that can be used to computationally simulate its effects. We end the chapter by describing another cause of anomalous diffusion, molecular crowding, and speculations on the significance of these phenomena for neural function.

4.1 Neuronal Structure Causes a Breakdown of the Classical Diffusion Process

Most intracellular signaling pathways require the diffusion of molecules. This is particularly important in neurons, the most morphologically polarized cells in Nature, where proteins are mainly produced in the cell body and have to find

F. Santamaria (✉)
Department of Biology, University of Texas at San Antonio, San Antonio 78249, TX, USA

G. Antunes • E. De Schutter
Computational Neuroscience Unit, Okinawa Institute of Science and Technology, Okinawa 904-0411, Japan

E. De Schutter
Laboratory of Theoretical Neurobiology, University of Antwerp, Antwerp 2610, Belgium

N. Le Novère (ed.), *Computational Systems Neurobiology*,
DOI 10.1007/978-94-007-3858-4_4,

their way to specific sites of large dendritic and axonal arborizations. The dendritic trees of neurons are particularly important because they contain most of the area and volume of the cell. Large dendritic trees contain structures spanning multiple spatial scales. The principal components of dendrites are branches, which vary in length from 10 to 1,000 μm. Dendritic branches can have different lengths, changes in diameter and bifurcations. Another important morphological component of dendrites are dendritic spines. Dendritic spines are small protrusions of about 1 fL where most excitatory synaptic contacts take place (Fiala et al. 2008). Spines are composed of a head and a neck, attached to the surface of dendrites. Dendrites can be covered – spiny – or not – smooth – with varying densities of dendritic spines.

Diffusion over branched dendritic trees that bifurcate and taper is basically determined by the classical laws of diffusion (Santamaria et al. 2006). The length of dendritic segments over which molecules diffuse (tens of micrometers) is at least an order of magnitude larger than their diameter (~1 μm) (Fiala et al. 2008). Therefore, differences in concentration in the radial direction of the dendrite are rapidly equilibrated compared to diffusion along the dendritic axis. Figure 4.1a shows a three-dimensional (3D) model of diffusion in a smooth dendrite. Clearly, a diffusing molecule in such a structure follows a random walk along the dendrite. The mean square displacement (MSD) along the length of the dendrite of all the simulated molecules is given by

$$MSD(t) = \frac{1}{N}\sum_{i=1}^{N} r_i^2 \tag{4.1}$$

where r_i is the relative position to their origin at $t = 0$ and N the number of molecules diffusing. Plotting the MSD vs t results in the well known linear dependence of normal diffusion

$$MSD \propto Dt \tag{4.2}$$

where D is the diffusion coefficient.

The structural properties of dendritic spines are well known and rather variable. Molecules diffusing along dendrites can be expected to enter spines, remain trapped inside the spine head and then escape to continue their axial diffusion (Fig. 4.1b). The natural distribution of spine shapes might give rise to a distribution of waiting times inside spines that, in turn, could slow down axial diffusion along the dendrite. It might be expected that the slowing down of the diffusing particle together with the high viscosity of the cytosol would reduce the diffusion rate of a particle but remain linear

$$MSD \propto D_{app}t \tag{4.3}$$

where $D > D_{app} = constant$, with D_{app} called the apparent diffusion coefficient. Such a reduced normal diffusion process has been extensively studied in neurons in the

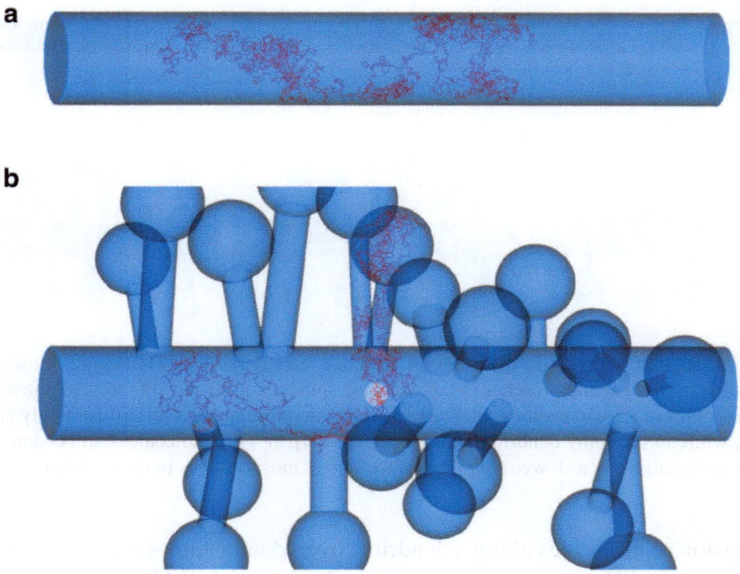

Fig. 4.1 Simulated 3D random walks in a smooth dendrite (**a**) and a spiny dendrite (**b**). Spines are modeled by a neck (a thin tube) connecting the head (a large sphere) to the dendritic shaft. Once a molecule enters a spine head it becomes trapped inside for variable amounts of time, resulting in a decreased lateral movement of molecules inside a spiny dendrite

context of calcium buffering (Wagner and Keizer 1994; Zador and Koch 1994; Naraghi and Neher 1997). However, the trapping of molecules by dendritic spines results in a quite different diffusion process.

Figure 4.1b shows a 3D model of a dendrite covered with multiple spines. The sizes of the spines are varied to mimic the heterogeneous distribution of morphological properties of spines. Diffusing molecules can become trapped in the spine head: once they enter this relatively large volume they can leave only when they hit the small diameter neck entry. We have previously shown (Santamaria et al. 2006) that the trapping time is related to the ratio of the spine head to spine neck diameters, which can be quite variable (Fiala et al. 2008). Diffusion along a spiny dendrite is therefore better described by a power law:

$$MSD \propto D_{app} t^{\alpha} \tag{4.4}$$

This type of diffusion process is termed anomalous diffusion (ben-Avraham and Havlin 2000) and α is the *anomalous exponent*. It is easy to see that while under normal diffusion D_{app} is constant ($D_{app} = MSD/t$) during anomalous diffusion its value depends on time ($D_{app} = D\, t^{\alpha-1}$). This anomalous diffusion is fundamentally different than other reduced structural forms of slowed diffusion such as tortuosity in which the D_{app} remains constant (Valiullin and Skirda 2001; Lacks 2008). Thus,

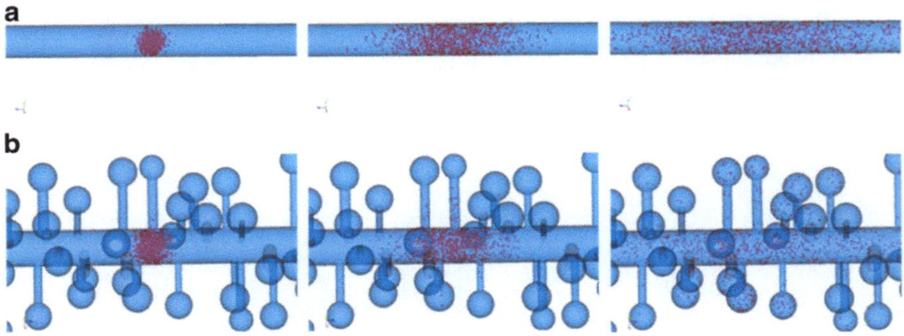

Fig. 4.2 Simulated anomalous diffusion in Purkinje cell dendrites. The figure illustrates the diffusion in absence (**a**) and presence (**b**) of dendritic spines at 1, 20 and 200 ms after the release of 1,000 molecules in a 3 μm center spot. For the smooth dendrite, molecules diffuse freely along the dendrite, while in the spiny dendrite (density = 6 spines/μm^2) the molecules can become trapped in the spines resulting in a slower lateral spread which is most obvious in the mid panels

the diffusion of molecules along a dendrite covered with spines cannot be described with a constant diffusion coefficient. The lack of a single diffusion coefficient that characterizes the diffusional process suggests that some molecules can undergo free diffusion while others experience different degrees of obstructed diffusion due to trapping by spines. If there is not a single diffusion coefficient at any given spatial compartment then the classical assumptions of homogeneous distribution of mass and a single diffusion coefficient for a given material required in Fick's law cannot be applied, resulting in a breakdown of the classical laws of diffusion. The trapping of molecules in dendritic spines also causes an increase in the correlation of particle displacement (Dagdug et al. 2007; Fedotov and Mendez 2008).

Anomalous diffusion processes can be described phenomenologically by multiple waiting times models (Hrabetova et al. 2003; Dagdug et al. 2007; Campos et al. 2008; Fedotov and Mendez 2008) as long as the characteristic length of diffusion is larger than the size of the traps. The emergence of anomalous diffusion in spiny dendrites depends on the relative scales imposed by the shape of spines and length of the dendrite. In the neuromuscular junction, at short observation times diffusion is anomalous, while at long times diffusion can return to normal (Lacks 2008). In our case, if the time spent in the spine is negligible, or the length of the dendritic segment is short, then no anomalous diffusion is observed (Sen et al. 1994).

We have shown that Purkinje cells (Fig. 4.2) (Santamaria et al. 2006), and more recently, pyramidal cells (Santamaria et al. 2011), have anomalous diffusion of intracellular signals. We tested this idea by photolyzing a caged inert compound in different parts of the dendrite. We found that the diffusion of molecules in smooth dendrite is almost normal ($\alpha = 1$), while in spiny dendrites it is anomalous. Detailed 3D simulations (Fig. 4.3) predicted that the value of the anomalous exponent will decrease linearly with the spine density, which was confirmed for the pyramidal neurons.

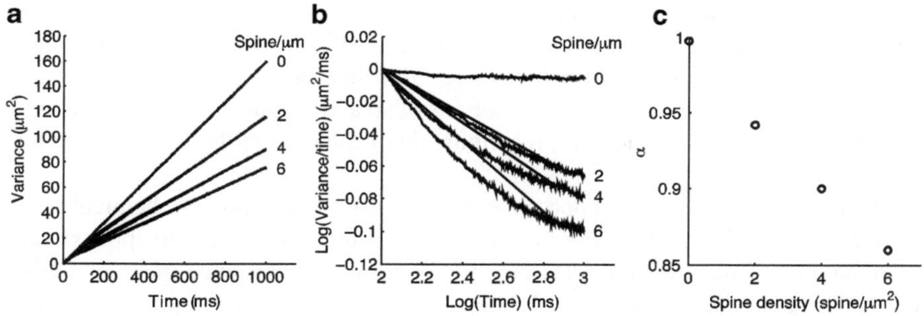

Fig. 4.3 Simulated anomalous diffusion in Purkinje cell dendrites with different spine densities. (**a**) The spatial variance (equivalent to MSD, Eq. 4.1) decreased non-linearly with increasing spine density. (**b**) Logarithmic transform of spatial variance values calculated in Purkinje cell dendrites at the indicated spine densities. (**c**) Relationship between α and spine density

We also tested second messengers, in particular Ca^{2+} and IP_3. These two second messengers are very important in the initiation of long term depression, a type of synaptic plasticity in Purkinje cells believed to underlie associative learning (Ito 2001). We found that Ca^{2+} did not show anomalous diffusion in spiny dendrites. This lack of structural influence on diffusion was due to the fast buffering and pumping of Ca^{2+} ions after they were photolyzed. In fact, the fast time constant of Ca^{2+} decay only allowed Ca^{2+} ions to move about 5 μm away from the site of release before being excluded from the cytosol; thus the characteristic length of diffusion was shorter than the size of the traps. The other molecule we tested was IP_3. This molecule showed anomalous diffusion in spineless dendrites also. This was probably caused by the biochemical reaction of IP_3 binding to receptors. In this case, the delay between IP_3 binding and unbinding from the IP_3 receptor causes the temporal trapping that reduces the spread of the signal along the dendrite. However, IP_3 diffusion in spiny dendrites showed much stronger levels of anomalous diffusion due to the presence of spines.

4.2 Methods: The Gillespie Stochastic Simulation Algorithm

Simulations presented in Figs. 4.1 and 4.2 were made with the STEPS software (Wils and De Schutter 2009) which implements the Gillespie Stochastic Simulation Algorithm (SSA) in tetrahedral meshes. We first describe the SSA before giving an overview of the STEPS software functionality.

The basis of chemical reactions is a random process in which the reactants must collide with the appropriate direction and energy (Connors 1990; Alexander and Zare 1998). If the system contains a small molecular population of some of its reactants the discreteness of the considered system generates dynamics different from the predictions of the corresponding deterministic model (Bhalla 2004). Thus, the mass action law breaks down. Such cases are frequent in cellular environments, leading to fluctuation in the biochemical reactions with considerable effects (Kuthan 2001; Elowitz et al. 2002). In order to accurately predict the evolution of the system it then becomes necessary to track each reaction instead of computing the mean behavior using ordinary differential equations (Cao and Samuels 2009). With a high number of molecular species and reactions, this problem can be computationally challenging.

There are several different approaches to stochastic simulation of chemical reactions (Bhalla and Wils 2010). For mesoscopic systems, it is possible to approximate the dynamics of biochemical reactions by assuming the existence of a volume in which the chemical population is homogeneously distributed in space (Gillespie 1976, 1977), also known as a well-mixed system. Thus, the position and velocity of the individual chemical entities can be ignored (Gillespie 2007). Under these conditions it becomes possible to calculate the probabilistic behavior of chemical systems over the discrete state space of entities (Gillespie 1992). This is the central principle enclosed in the chemical master equation (CME) (McQuarrie 1967).

We consider a system with N species S_i and a set of reaction rules R_j. Each R_j describes an irreversible reaction. Reversible reactions can always be described by two irreversible reactions. The state of this system is described by a state vector $\boldsymbol{x}$ of size N, with each integer component X_i corresponding to the number of molecules of species S_i present. The number of molecules X_i can be converted to concentrations if the volume of the system is known. The system is then characterized by an initial state $\boldsymbol{x}_0$ and its evolution over time is described by which reactions R_j take place and how these reactions change $\boldsymbol{x}$. The state change vector $\boldsymbol{v}_j$ (size N) represents the change induced by one reaction R_j. For large systems most integer values of $\boldsymbol{v}_j$ will be zero as only a limited number of molecules participate in any reaction. If the system is in the state $\mathbf{x}$ and one R_j occurs, the system changes to state $\boldsymbol{x} + \boldsymbol{v}_j$.

To simulate the behavior of this system one needs to know which reactions R_j occur at what time. This is determined by an important property of reactions, called the propensity function a_j, defined so that $a_j(x)dt$ is the probability that one R_j will occur in the next infinitesimal time interval $[t;\ t + dt)$, for a specific $\boldsymbol{x}_t$. The propensity function reflects the fundamental characteristics that stochastic chemical kinetics are dependent on the number of reactants available and on reaction rate constants. So, for a given R_j, this implies the existence of a reaction rate constant c_j, defined such that $c_j dt$ is the probability that an arbitrarily chosen combination of R_j reactant molecules will react in the next infinitesimal time dt. Therefore, the propensity function can be defined as $a_j(x) = c_j x_1$, for the unimolecular reaction $S_1 \rightarrow$ products(t), and as $a_j(x) = c_j x_1 x_2$ for second order reactions of the form $S_1 + S_2 \rightarrow$ products(t) (Gillespie 2007). The propensity functions are the basis for the CME for $X(t)$ by using the laws of probability:

$$\frac{\partial P\left(x, t \mid x_0, t_0\right)}{\partial t} = \sum_{j=1}^{M} \left[a_j\left(x - v_j\right) P\left(x - v_j, t \mid x_0, t_0\right) - a_j(x) P\left(x, t \mid x_0, t_0\right)\right] \tag{4.5}$$

The probability that the system is in a state x at time t, knowing that it was in x_0 at the initial time t_0, depends on the difference between the probability that it enters x from a closely related state (first term of the summation in Eq. 4.5) and the probability that it was already in state x but has changed (second term), with the summation taken over all possible reactions.

Compared to deterministic kinetics approaches, the CME is more physically rigorous and based on thermodynamics concepts (Gillespie 1992). However, Eq. 4.5 corresponds to a large system of coupled ODEs that is analytically intractable for most biologically relevant models. Therefore, to simulate cellular processes stochastically, numerical approaches should be used.

During the 1970s, Daniel Gillespie (1976, 1977) published an exact stochastic simulation algorithm for chemical kinetics, which has become the most common algorithm used for stochastic simulation of biochemical systems. The SSA is a procedure for generating statistically exact time-evolution trajectories of a finite well-stirred population of individuals distributed over a finite set of states. It is exact in the sense that the generated sample paths are precisely distributed according to the solution of the CME, but because it scales with the number of reactions in the system it can be quite slow for large simulations.

In the SSA, as occurs in the CME, the reactions are defined as events that change the number of individuals in each state following chemical rules, altering the molecular populations by integer numbers. The SSA is based on computing iteratively the time when the next chemical reaction of a given system will occur and then determining which reaction it will be (Gillespie 1976, 1977). This is based on a new function that calculates the probability, given $X(t) = \boldsymbol{x}$, that the next reaction of the system will occur in the infinitesimal time interval $[t + \tau; t + \tau + d\tau)$ and will be of type R_j:

$$p(\tau, j|x, t) = a_j(x)e^{-a_0(x)\tau} \tag{4.6}$$

with $a_0(x) = \sum_{j=1}^{M} a_j(x)$ called the zero propensity, the probability that any reaction will occur.

Gillespie suggested several equivalent procedures for constructing exact numerical realizations of the SSA. STEPS utilizes the so-called Direct Method. In order to advance the system from state $\mathbf{x}$ at time t, the Direct Method generates two random numbers r_1 and r_2 from the uniform distribution in the unit interval. The time until the next reaction τ is then determined by:

$$\tau = \frac{1}{a_0(x)} \ln\left(\frac{1}{r_1}\right) \tag{4.7}$$

The reaction j that occurs is selected with the second random number so that j is the smallest integer satisfying $\sum_{i=1}^{j} a_i(x) > r_2 a_0(x)$. The state $\mathbf{x}$ is then updated to $\boldsymbol{x} + \boldsymbol{v}_j$ and t to $t + \tau$ and the procedure is repeated until some predefined time or state is reached.

4.3 The STEPS Simulator

In this section, we briefly describe the software STEPS (STochastic Engine for Pathway Simulation), available at http://steps.sourceforge.net/ (Wils and De Schutter 2009). STEPS is a platform for stochastic simulation of coupled reaction–diffusion systems in complex 3D boundary conditions. The Python-based user interface allows for a flexible description of the model and its initial conditions (Wils and De Schutter 2009). By invoking a different solver the same model can be simulated either deterministically, using a fixed time step Runge–Kutta method or stochastically using SSA (Fig. 4.4). The accuracy of STEPS has been verified extensively. An online user manual is available to get the novice user started.

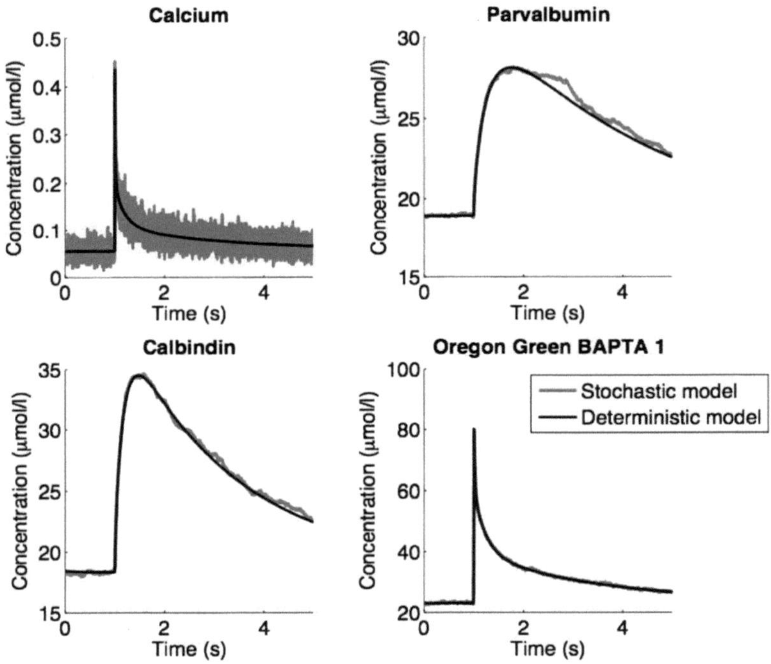

Fig. 4.4 An example of a stochastic and a deterministic model of Ca^{2+} dynamics performed with STEPS. In spines of Purkinje cells, transient changes in Ca^{2+} concentration are involved in numerous processes, including the induction of synaptic plasticity. The kinetics of Ca^{2+} transients is fundamental in determining its action, and is regulated by several factors. In this example (Adapted from Schmidt et al. 2003) mechanisms of Ca^{2+} extrusion, exogenous (Oregon green BAPTA 1), and endogenous buffers (parvalbumin and calbindin) were responsible for shaping the transients within the spines. Due to the small volume of the spine, only the stochastic simulation (*grey* line) is able to represent the discrete fluctuations in the Ca^{2+} concentration. However, the high concentration buffers exhibit a similar behavior in both stochastic and deterministic simulations

An important limitation of the Gillespie algorithm is that it can be only be applied to chemical homogeneous environments, which is not an appropriated assumption when considering the cellular milieu. Cells are characteristically chemical inhomogeneous systems, presenting several mechanisms of source and sink of some key entities taking place at specific positions (Mazel et al. 2009). Furthermore, most of the cellular biochemical reactions happens in aqueous environment or in fluid lipid membranes, where the chemical entities are allowed to move (Singer and Nicolson 1972; Saffman and Delbruck 1975), making their diffusion an important event regulating biochemical process (Kholodenko 2003; Bhalla 2004).

In order to simulate stochastic diffusion, STEPS uses a modification of the Gillespie algorithm. Space inhomogeneity is incorporated in the SSA by discretizing a model into tetrahedral sub-volumes, also called voxels, considered to be separate reaction volumes. STEPS supports importing tetrahedral meshes from several popular mesh generation packages, which is a very important advantage in comparison to other biochemical simulators (Ander et al. 2004; Hattne et al. 2005). To satisfy the requirements of the SSA, each tetrahedron must have a sufficiently small volume allowing the chemical population inside of it to be considered spatially homogeneous distributed. Diffusion is simulated as the transition of molecules between neighboring elements. These transitions are expressed in the form of a set of reactions, which are added to the chemical system. The rate constant (k) for the diffusion between two voxels (i, j) for a given entity is calculated through the expression:

$$k_{i \to j} = \frac{D_z A_j}{dx_i V_i} \tag{4.8}$$

where $k_{i,j}$ is the transition from the voxel i to j, D_z is the diffusion coefficient for the entity z, A_j is the cross-sectional area of the triangle separating the tetrahedrons i and j, V_i is the volume of the voxel i, and dx_i is defined as the barycenter-to-barycenter distance between tetrahedron i and its neighbor j.

4.4 Anomalous Diffusion Caused by Molecular Crowding

There is increasing evidence that the membranes and cytosolic content of cells cannot be treated as well-mixed compartments at the temporal and spatial scales in which important biological function occurs (Winckler et al. 1999; Fukano et al. 2004; Santamaria et al. 2006; Ehlers et al. 2007). The source of this deviation could be due to the structure of the cell (Santamaria et al. 2006) as explained in the first section, or from the high concentration of macromolecules found in the cytosol or anchored in the membrane (Saxton 1994; Ritchie et al. 2005; Guigas et al. 2007a, b; Guigas and Weiss 2008), a condition referred to as **molecular crowding** (Schnell and Turner 2004).

Without any obstacles to diffusion the displacement of a particle in 2D or 3D is described with the classical law of diffusion. However, macromolecules, molecules that weigh several kilo Daltons, can occupy up to 40% of the intracellular volume of a cell (Record et al. 1998). At such concentrations, the influence of steric interactions among diffusing molecules cannot be assumed to be negligible (Minton 2006). Theoretical, modeling and experimental results show that molecular crowding caused by macromolecules significantly reduces the spread of biochemical signals in the cytosol (Weiss et al. 2004; Banks and Fradin 2005), transforming the classical diffusion equation again to a power law (Saxton 1994; Deverall et al. 2005). There is an increasing amount of experimental evidence suggesting that many cytosolic and membrane molecules show anomalous diffusion (Tang and Edidin 2003; Ritchie et al. 2005; Wawrezinieck et al. 2005), although this has been recently challenged (Dix and Verkman 2008). It is important to note that molecular crowding specially arises when the distribution of molecules is random (Minton 2006). Furthermore, molecular crowding seems to be a property that cells tend to maintain independent of volume (Guigas et al. 2007a).

Molecular crowding may also play an important role in neurotransmitter receptor trafficking in the postsynaptic density (PSD) (Santamaria et al. 2010). Modeling suggests that the high concentration and excluded volume caused by anchored PSD molecules result in molecular crowding and makes diffusion of glutamate receptors (AMPAR) in the PSD anomalous. Anomalous diffusion of AMPAR results in retention of these receptors inside the PSD for periods ranging from minutes to several hours in the absence of strong binding of receptors to PSD molecules. The distribution of time AMPARs spend in the PSD reproduces multiple experimental results (Ehlers et al. 2007; Frischknecht et al. 2009; Petrini et al. 2009; Renner et al. 2009). Trapping of receptors in the PSD by crowding effects was very sensitive to the concentration of PSD molecules, showing a switch-like behavior for retention of receptors. Interestingly, non-covalent binding of AMPAR to anchored PSD molecules allowed the synapse to become well-mixed, resulting in normal diffusion of AMPAR. Binding also allowed the exchange of receptors in and out of the PSD. This modeling work suggests that molecular crowding could be an important biophysical mechanism to maintain homeostatic synaptic concentrations of AMPARs in the PSD without the need of energetically expensive biochemical reactions. In this context, binding of AMPAR with PSD molecules could collaborate with crowding to maintain synaptic homeostasis but could also allow synaptic plasticity by increasing the exchange of these receptors with the surrounding extra-synaptic membrane.

In general, because under anomalous diffusion there is no constant diffusion coefficient, the space in which the molecule is diffusing cannot be considered a well-mixed environment (Saxton 2007). The break-down of such a fundamental mass transport law has not only implications on how far and fast a biochemical signal propagates, but also changes the probability of activating downstream biochemical reactions (Chebotareva et al. 2004; Guigas et al. 2007a; Lizana et al. 2008; Nicolau and Burrage 2008). In fact, reaction rates, which under normal diffusion are constant, depend on the amount of molecular crowding (ben-Avraham and

Havlin 2000; Chebotareva et al. 2004). If diffusion is anomalous then reaction rates also follow a power law (ben-Avraham and Havlin 2000) and vary as a function of molecular crowding (Chebotareva et al. 2004). Most theories of biochemical signaling in cells assume that the dynamics of the system is reaction-limited; however, if anomalous diffusion is present then the system is diffusion-limited.

4.5 Significance

Anomalous diffusion due to dendritic spines or by molecular crowding could be a basic regulatory mechanism used by neurons to control the propagation and reaction of any biochemical signal in the cytosol or membrane. The shape and content of dendrites changes with multiple biological processes such as development, learning, aging, and illnesses (Harris et al. 1992; Maletic-Savatic et al. 1999; Lendvai et al. 2000; Rocher et al. 2008). As a neuron modifies its shape, normal biochemical signaling might be affected. This may require a re-assessment of the biophysical mechanisms used by neurons to process biochemical information (Holcman et al. 2004; Cornelisse et al. 2007; Schmidt et al. 2007a, b). Furthermore, the fundamental theories used to interpret experimental results have to be reformulated to include such effects. In that context, new theoretical, computational, and experimental work is needed to understand how synaptic plasticity takes place in the complex environment of cells and not in a test-tube.

Acknowledgments FS was supported by NSF-HDR 0923339 and EF-1137897. GA and ADS were supported by OISTPC.

References

Alexander AJ, Zare RN (1998) Anatomy of elementary chemical reactions. J Chem Educ 75:1105–1118

Ander M, Beltrao P, Di Ventura B, Ferkinghoff-Borg J, Foglierini M, Kaplan A, Lemerle C, Tomas-Oliveira I, Serrano L (2004) SmartCell, a framework to simulate cellular processes that combines stochastic approximation with diffusion and localisation: analysis of simple networks. Syst Biol 1:129–138

Banks DS, Fradin C (2005) Anomalous diffusion of proteins due to molecular crowding. Biophys J 89:2960–2971

ben-Avraham D, Havlin S (2000) Diffusion and reactions in fractals and disordered systems. Cambridge University Press, Cambridge

Bhalla US (2004) Signaling in small subcellular volumes. I. Stochastic and diffusion effects on individual pathways. Biophys J 87:733–744

Bhalla US, Wils S (2010) Reaction–diffusion modeling. In: De Schutter E (ed) Computational modeling methods for neuroscientists. The MIT Press, Cambridge

Campos D, Fedotov S, Mendez V (2008) Anomalous reaction-transport processes: the dynamics beyond the law of mass action. Phys Rev E Stat Nonlin Soft Matter Phys 77(6):061130

Cao Y, Samuels DC (2009) Discrete stochastic simulation methods for chemically reacting systems. Methods Enzymol 454:115–140

Chebotareva NA, Kurganov BI, Livanova NB (2004) Biochemical effects of molecular crowding. Biochemistry (Mosc) 69:1239–1251

Connors KA (1990) Chemical kinetics: the study of reaction rates in solution. VCH Publishers, New York

Cornelisse LN, van Elburg RA, Meredith RM, Yuste R, Mansvelder HD (2007) High speed two-photon imaging of calcium dynamics in dendritic spines: consequences for spine calcium kinetics and buffer capacity. PLoS One 2:e1073

Dagdug L, Berezhkovskii AM, Makhnovskii YA, Zitserman VY (2007) Transient diffusion in a tube with dead ends. J Chem Phys 127(22):224712-1–224712-9

Deverall MA, Gindl E, Sinner EK, Besir H, Ruehe J, Saxton MJ, Naumann CA (2005) Membrane lateral mobility obstructed by polymer-tethered lipids studied at the single molecule level. Biophys J 88:1875–1886

Dix JA, Verkman AS (2008) Crowding effects on diffusion in solutions and cells. Annu Rev Biophys 37:247–263

Ehlers MD, Heine M, Groc L, Lee MC, Choquet D (2007) Diffusional trapping of GluR1 AMPA receptors by input-specific synaptic activity. Neuron 54:447–460

Elowitz MB, Levine AJ, Siggia ED, Swain PS (2002) Stochastic gene expression in a single cell. Science 297:1183–1186

Fedotov S, Mendez V (2008) Non-Markovian model for transport and reactions of particles in spiny dendrites. Phys Rev Lett 101:218102

Fiala JC, Spacek J, Harris KM et al (2008) Dendrite structure. In: Stuart G (ed) Dendrites, 2nd edn. Oxford University Press, Oxford

Frischknecht R, Heine M, Perrais D, Seidenbecher CI, Choquet D, Gundelfinger ED (2009) Brain extracellular matrix affects AMPA receptor lateral mobility and short-term synaptic plasticity. Nat Neurosci 12:897–904

Fukano T, Hama H, Miyawaki A (2004) Similar diffusibility of membrane proteins across the axon-soma and dendrite-soma boundaries revealed by a novel FRAP technique. J Struct Biol 147:12–18

Gillespie DT (1976) General method for numerically simulating stochastic time evolution of coupled chemical-reactions. J Comput Phys 22:403–434

Gillespie DT (1977) Exact stochastic simulation of coupled chemical-reactions. Abstr Pap Am Chem Soc 173:128–128

Gillespie DT (1992) A rigorous derivation of the chemical master equation. Phys A 188: 404–425

Gillespie DT (2007) Stochastic simulation of chemical kinetics. Annu Rev Phys Chem 58: 35–55

Guigas G, Weiss M (2008) Sampling the cell with anomalous diffusion – the discovery of slowness. Biophys J 94:90–94

Guigas G, Kalla C, Weiss M (2007a) The degree of macromolecular crowding in the cytoplasm and nucleoplasm of mammalian cells is conserved. FEBS Lett 581:5094–5098

Guigas G, Kalla C, Weiss M (2007b) Probing the nanoscale viscoelasticity of intracellular fluids in living cells. Biophys J 93:316–323

Harris KM, Jensen FE, Tsao B (1992) Three-dimensional structure of dendritic spines and synapses in rat hippocampus (CA1) at postnatal day 15 and adult ages: implications for the maturation of synaptic physiology and long-term potentiation. J Neurosci 12:2685–2705

Hattne J, Fange D, Elf J (2005) Stochastic reaction–diffusion simulation with MesoRD. Bioinformatics 21:2923–2924

Holcman D, Schuss Z, Korkotian E (2004) Calcium dynamics in dendritic spines and spine motility. Biophys J 87:81–91

Hrabetova S, Hrabe J, Nicholson C (2003) Dead-space microdomains hinder extracellular diffusion in rat neocortex during ischemia. J Neurosci 23:8351–8359

Ito M (2001) Cerebellar long-term depression: characterization, signal transduction, and functional roles. Physiol Rev 81:1143–1195

Kholodenko BN (2003) Four-dimensional organization of protein kinase signaling cascades: the roles of diffusion, endocytosis and molecular motors. J Exp Biol 206:2073–2082

Kuthan H (2001) Self-organisation and orderly processes by individual protein complexes in the bacterial cell. Prog Biophys Mol Biol 75:1–17

Lacks DJ (2008) Tortuosity and anomalous diffusion in the neuromuscular junction. Phys Rev E Stat Nonlin Soft Matter Phys 77:041912

Lendvai B, Stern EA, Chen B, Svoboda K (2000) Experience-dependent plasticity of dendritic spines in the developing rat barrel cortex in vivo. Nature 404:876–881

Lizana L, Bauer B, Orwar O (2008) Controlling the rates of biochemical reactions and signaling networks by shape and volume changes. Proc Natl Acad Sci U S A 105:4099–4104

Maletic-Savatic M, Malinow R, Svoboda K (1999) Rapid dendritic morphogenesis in CA1 hippocampal dendrites induced by synaptic activity. Science 283:1923–1927

Mazel T, Raymond R, Raymond-Stintz M, Jett S, Wilson BS (2009) Stochastic modeling of calcium in 3D geometry. Biophys J 96:1691–1706

McQuarrie DA (1967) Stochastic approach to chemical kinetics. J Appl Probab 4:413–478

Minton AP (2006) How can biochemical reactions within cells differ from those in test tubes? J Cell Sci 119:2863–2869

Naraghi M, Neher E (1997) Linearized buffered Ca2+ diffusion in microdomains and its implications for calculation of [Ca2+] at the mouth of a calcium channel. J neurosci 17: 6961–6973

Nicolau DV, Burrage K (2008) Stochastic simulation of chemical reactions in spatially complex media. Comput Math Appl 55:1007–1018

Petrini EM, Lu J, Cognet L, Lounis B, Ehlers MD, Choquet D (2009) Endocytic trafficking and recycling maintain a pool of mobile surface AMPA receptors required for synaptic potentiation. Neuron 63:92–105

Record MT Jr, Courtenay ES, Cayley S, Guttman HJ (1998) Biophysical compensation mechanisms buffering *E. coli* protein-nucleic acid interactions against changing environments. Trends Biochem Sci 23:190–194

Renner ML, Cognet L, Lounis B, Triller A, Choquet D (2009) The excitatory postsynaptic density is a size exclusion diffusion environment. Neuropharmacology 56:30–36

Ritchie K, Shan XY, Kondo J, Iwasawa K, Fujiwara T, Kusumi A (2005) Detection of non-Brownian diffusion in the cell membrane in single molecule tracking. Biophys J 88:2266–2277

Rocher AB, Kinson MS, Luebke JI (2008) Significant structural but not physiological changes in cortical neurons of 12-month-old Tg2576 mice. Neurobiol Dis 32:309–318

Saffman PG, Delbruck M (1975) Brownian motion in biological membranes. Proc Natl Acad Sci U S A 72:3111–3113

Santamaria F, Wils S, De Schutter E, Augustine GJ (2006) Anomalous diffusion in Purkinje cell dendrites caused by spines. Neuron 52:635–648

Santamaria F, Gonzalez J, Augustine GJ, Raghavachari S (2010) Quantifying the effects of elastic collisions and non-covalent binding on glutamate receptor trafficking in the post-synaptic density. PLoS Comput Biol 6:e1000780

Santamaria F, Wils S, De Schutter E, Augustine GJ (2011) The diffusional properties of dendrites depend on the density of dendritic spines. Eur J Neurosci 34:561–568.

Saxton MJ (1994) Anomalous diffusion due to obstacles: a Monte Carlo study. Biophys J 66: 394–401

Saxton MJ (2007) A biological interpretation of transient anomalous subdiffusion. I. Qualitative model. Biophys J 92:1178–1191

Schmidt H, Stiefel KM, Racay P, Schwaller B, Eilers J (2003) Mutational analysis of dendritic Ca2+ kinetics in rodent Purkinje cells: role of parvalbumin and calbindin D28k. J Physiol 551(1):13–32

Schmidt H, Arendt O, Brown EB, Schwaller B, Eilers J (2007a) Parvalbumin is freely mobile in axons, somata and nuclei of cerebellar Purkinje neurones. J Neurochem 100:727–735

Schmidt H, Kunerth S, Wilms C, Strotmann R, Eilers J (2007b) Spino-dendritic cross-talk in rodent Purkinje neurons mediated by endogenous Ca2+ −binding proteins. J Physiol 581:619–629
Schnell S, Turner TE (2004) Reaction kinetics in intracellular environments with macromolecular crowding: simulations and rate laws. Prog Biophys Mol Biol 85:235–260
Sen PN, Schwartz LM, Mitra PP, Halperin BI (1994) Surface relaxation and the long-time diffusion coefficient in porous media: periodic geometries. Phys Rev B: Condens Matter 49:215–225
Singer SJ, Nicolson GL (1972) The fluid mosaic model of the structure of cell membranes. Science 175:720–731
Tang Q, Edidin M (2003) Lowering the barriers to random walks on the cell surface. Biophys J 84:400–407
Valiullin R, Skirda V (2001) Time dependent self-diffusion coefficient of molecules in porous media. J Chem Phys 114:452–458
Wagner J, Keizer J (1994) Effects of rapid buffers on Ca2+ diffusion and Ca2+ oscillations. Biophys J 67:447–456
Wawrezinieck L, Rigneault H, Marguet D, Lenne PF (2005) Fluorescence correlation spectroscopy diffusion laws to probe the submicron cell membrane organization. Biophys J 89:4029–4042
Weiss M, Elsner M, Kartberg F, Nilsson T (2004) Anomalous subdiffusion is a measure for cytoplasmic crowding in living cells. Biophys J 87:3518–3524
Wils S, De Schutter E (2009) STEPS: modeling and simulating complex reaction–diffusion systems with Python. Front Neuroinformatics 3:15
Winckler B, Forscher P, Mellman I (1999) A diffusion barrier maintains distribution of membrane proteins in polarized neurons. Nature 397:698–701
Zador A, Koch C (1994) Linearized models of calcium dynamics: formal equivalence to the cable equation. J Neurosci 14:4705–4715

Chapter 5
Spatial Organization and Diffusion in Neuronal Signaling

Sherry-Ann Brown, Raquell M. Holmes, and Leslie M. Loew

Abstract The intricate architecture of neuronal cells suggests that morphology plays a key role in cellular function. Yet descriptions and analyses of neuronal signaling systems often focus solely on biochemical reaction pathways. Models developed from these data implicitly assume that reactions occur in well mixed homogenous environments with instantaneous diffusion. However, if we have any hope of building truly predictive quantitative models, the intricate geometries and large length scales of neurons compel us to explicitly account for molecular diffusion and spatial organization.

The workdescribed in this chapter makes use of the Virtual Cell (VCell) modeling and simulation software (accessible at http://www.vcell.org) to examine the significance of spatial parameters on signaling events in neuronal systems. As a specific example to focus these concepts, we use the PIP2-IP3-Ca release cascade initiated by binding of glutamate to the metabotropic glutamate receptor in the Purkinje neuron spine. The data for these simulations were obtained with live cell imaging and biochemical techniques (Xu et al. 2003a; Wang et al. 2000). By combining imaging and biochemical data with computational frameworks for spatial modeling, we are able to develop quantitative understandings of the role of cellular geometries on reaction–diffusion systems. In our approach, we have employed a variety of models using geometrical descriptions of differing dimensions to elucidate the relationship of cellular geometry to calcium signaling in the dendritic spine.

S.-A. Brown • R.M. Holmes • L.M. Loew (✉)
University of Connecticut Health Center, Center for Cell Analysis & Modeling, 263 Farmington Avenue, CT 06030-6406, Farmington
e-mail: SABrown@student.uchc.edu; rholmes@uchc.edu; les@volt.uchc.edu

N. Le Novère (ed.), *Computational Systems Neurobiology*,
DOI 10.1007/978-94-007-3858-4_5,

5.1 Virtual Cell

The model building process in VCell makes it quite straightforward to explore such geometrical features, approximations and assumptions (Slepchenko and Loew 2010; Moraru et al. 2008; Schaff et al. 1997). Figure 5.1a is a screenshot of the VCell user interface showing panes that provide (clockwise from upper left):

the overall model tree (labeled Navigation),
a full view of the selected model component (in this case the reaction diagram),
a detailed specification of properties for the selected model element, in this case a reaction rate expression and its parameters, and
available VCell and external database resources for finding models or model elements.

The first step is to define the system "Physiology", which includes all the molecular species, their locations in cellular compartments (e.g. cytosol, nucleus, endoplasmic reticulum . . .) or membranes, their reactions (including stoichiometries and kinetic rate expressions) and any of their membrane transport mechanisms. Once the Physiology is fully specified, it can be used in multiple "Applications", which specify initial conditions and geometrical features of the system. Applications also can be used to clamp certain species, turn reactions off or impose protocols or events on the evolving system; in other words, the Applications can be considered virtual experiments that probe the behavior of the Physiology.

The completed Application contains sufficient information to automatically generate a mathematical description of a model which can then be simulated in Virtual Cell using a number of methods and solvers. Once the surface area and volume of each compartment are specified, the software can automatically generate either ordinary differential equation (ODE) or stochastic simulations. The "Physiology" can also be associated with explicit geometries in an Application, either analytical or derived from experimental images. Simulations then require solution of partial differential equations (PDE) with all mobile species assigned diffusion coefficients. In this way, as we move from compartmental models to spatial models, we are able to evaluate and examine the relationship of geometry to biochemical behaviors in implicit and explicit ways. In the works for the next version of VCell is a new capability to do spatial stochastic modeling via the Smoldyn algorithm (Andrews et al. 2010); this will allow the Physiology to be simulated as molecules that are represented as discrete particles both moving and reacting stochastically within the geometry.

Figure 5.1b is a screenshot showing the selection of the simulation component within one Application entitled "3D PIP2_experimentally derived geometry". Shown over the main window is the simulation results window for one of four simulations that were run in this Application. This illustrates several additional Features of Virtual Cell, such as the ability to perform parameter scans at the simulation level and the ability to visualize complex 3D simulation results. This

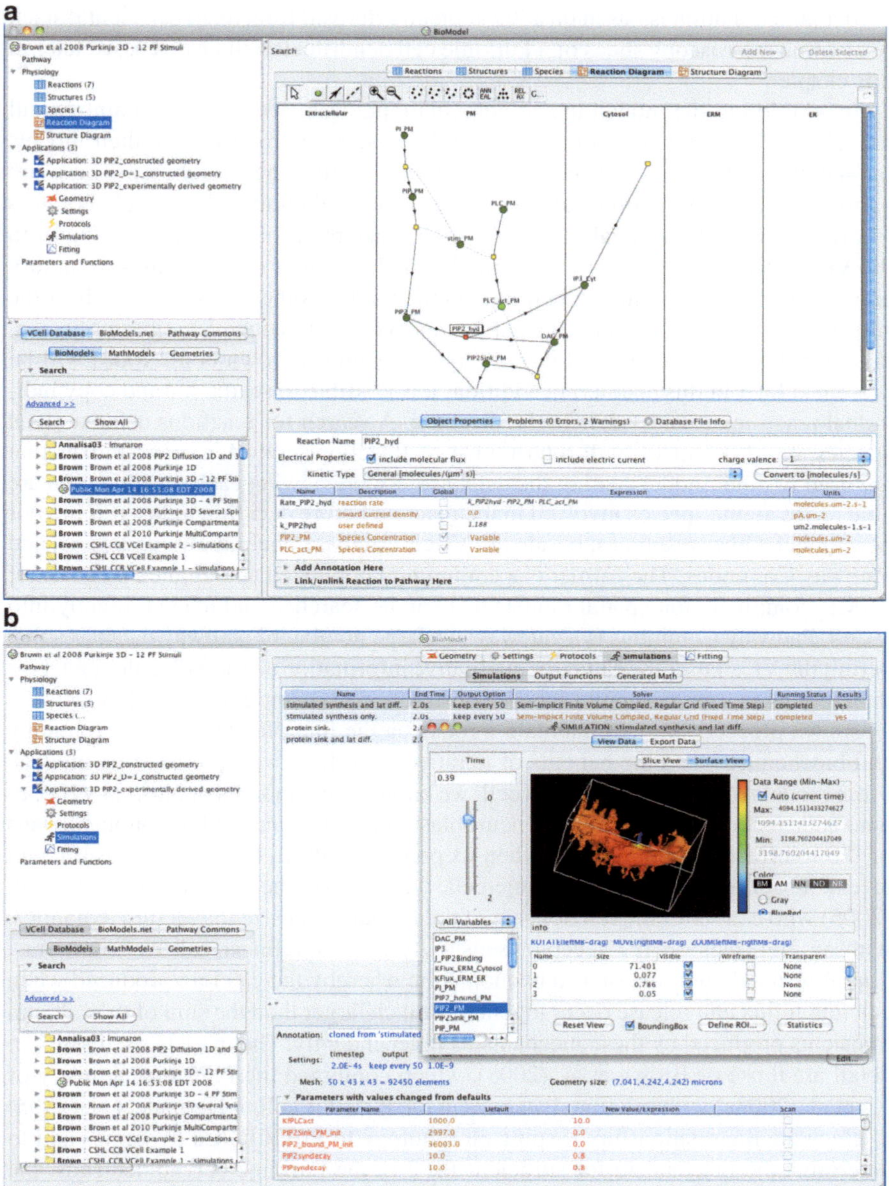

Fig. 5.1 VCell Screenshots. (**a**) The main VCell user interface window with four panes labeled according to their general functionality. (**b**) VCell user interface with a simulation component selected. A simulation results window is show on top of the main VCell window in the surface rendering view. The first of four simulations has been selected and the Properties pane indicates that several default parameters have been altered for this simulation

particular simulation is based on a 3D geometry derived from an experimental stack of electron microscopy slices of a Purkinje cell dendrite. It will be discussed later in this chapter.

In addition to the modeling and computational framework, VCell maintains all models in an internet-accessible database. Users can control access to their models by specifying them as private, shared, public or published (archive). These are various levels of permission that allow a user to share their model with collaborators (shared) or with the general public. As such, readers of this chapter may log in to the Virtual Cell, view, run or copy a public model to their own accounts to modify and test their own hypothesis, building on the work of others. The lower left of the screenshot in Fig. 5.1 shows that a public model by user "Brown" has been selected; this model for PIP2 dynamics in Purkinje cell spines (Brown et al. 2008) will be presented later in this chapter. In addition, it is possible to search for and copy individual reactions from models in the database. A search for reactions using a given species, e.g. PIP2 or PI, will return a list of Reactions and Membrane Fluxes that are used in other models within the database. By previewing each reaction, you are able to see all the species involved in the reaction before deciding to insert it in your new model. Other resources, beside the VCell Database, include Biomodels.net and Pathway Commons. The former is a community populated and curated database of SBML-compliant non-spatial models that can be searched and loaded directly into VCell. Pathway Commons is a gateway to a large number of pathway databases that permit the VCell user to search for qualitative information about molecules and their interactions as a starting point for the construction of quantitative VCell models.

As we proceed, we will maintain a focus on the interplay between morphology and biochemistry and the benefits of iteratively working with pre-existing models. From models of the neuroblastoma cell, we build on the findings of Xu et al. (2003a) who address the question: how do neuroblastoma cells generate high enough levels of IP3 to effect calcium release, when its precursor, PIP2, is present at insufficient levels? From models of the Purkinje cell, we draw on the work of Hernjak et al. (2005) who examines the relationship of spine shape and size to calcium dynamics, with an emphasis on IP3 receptor sensitivity and supralinear calcium transients. Supralinear calcium release is a phenomenon whereby the calcium produced when calcium influx and release occur together is much larger than the sum of the calcium transients produced by these mechanisms independently. The models we discuss in detail are those of Brown et al. (2008), which combined the PIP2 biochemistry of Xu et al. (2003a) with calcium dynamics in the Purkinje cell of Hernjak et al. (2005) to elucidate the relationship of spine morphology to PIP2/IP3/Ca dynamics and to determine the mechanism for sufficient PIP2 in the Purkinje cell spine. Thus, the primary focus is the impact of spine morphology on the dynamics of IP3 and PIP2 involved in mGluR-mediated calcium release, with an emphasis on the diffusive barrier created by the spine neck.

We attempt to clarify the process of integrating model components (topology, kinetics and output) from one set of computational experiments to another in hopes that the reader will be able to make use of this approach to enhance their own capacity to explore neuronal systems through compartmental as well as spatial simulation of reaction diffusion systems.

5.2 Making Use of Pre-Existing Models, Linking Across Dimensions

Each Purkinje neuron receives input on its distal dendrites from more than 150,000 granule cell axons, called *parallel fibers (PFs)*, each of which activates an individual synapse on dendritic spines. When a PF stimulates a spine at a distal dendrite of the Purkinje neuron, it releases a neurotransmitter, glutamate, which binds to the metabotropic glutamate receptor (mGluR), on the plasma membrane of the Purkinje spine (Takechi et al. 1998; Finch and Augustine 1998). This leads to the activation of PLC, an enzyme that hydrolyzes PIP2 to produce IP3, which, in turn, mediates calcium release from the endoplasmic reticulum. A more global activation is produce by climbing fiber (CF) inputs which depolarize the dendrite to open voltage-sensitive calcium channels; this results in calcium influx. These components of the spine signaling pathway are summarized in Fig. 5.2. Experiments in Purkinje cells in which caged IP3 is released via photoactivation have consistently revealed that $20 - 70\ \mu M$ IP3 is required for a calcium response that can ultimately lead to long term depression (LTD) of the synapse (Khodakhah and Ogden 1993; Finch and Augustine 1998). In non-neuronal cells, IP3-meditated calcium release requires 2 orders of magnitude less IP3. Yet IP3-mediated calcium release is critical for coincidence detection between parallel and climbing fiber synaptic inputs and LTD of synaptic inputs. Our previous work calculated a density of PIP2 in the plasma membrane of 4,000 molecules/μm^2 (Xu et al. 2003a). Simulations that use this value as an initial condition of PIP2 were unable to achieve sufficient IP3 generation. Given that the Purkinje cell physiology requires that high transient concentrations of IP3 be generated in the spine, how can the system supply sufficient substrate PIP2 during synaptic PF activity? We have examined three potential mechanisms (diffusion, synthesis and sequestration) for supplying sufficient PIP2 in the spine head for IP3 signal.

Stimulated synthesis hypothesizes that the stimulus that activates PIP2 hydrolysis also effects a burst of PIP2 production that will allow it to keep up with the demand for IP3. Local sequestration hypothesizes that there are local binding sites, perhaps as part of the post synaptic density in the spine, which can maintain a higher concentration of PIP2 in the spine membrane than in the rest of the dendrite. Lateral diffusion posits that as PIP2 is depleted in the spine, diffusion from neighboring regions of the dendrite is sufficiently rapid to resupply it. A primary focus of the study by Brown et al. (2008) was to explore these three mechanisms and the likelihood that they could produce sufficient IP3 for the requisite calcium release. To do this, a 3D model incorporating PIP2 lateral membrane diffusion was required. The reaction mechanism and kinetics for stimulated synthesis were derived from Xu et al. (2003a), and parameters modified to be appropriate for Purkinje cells rather than neuroblastoma cells. The Hernjak et al. (2005) physiology was modified to include the stimulated synthesis from Xu et al. (2003a) and reactions newly added by Brown et al. (2008) for PIP2 sequestration. Lateral diffusion of PIP2 is, of course, a constitutive mechanism that is treated directly in any spatial model (but

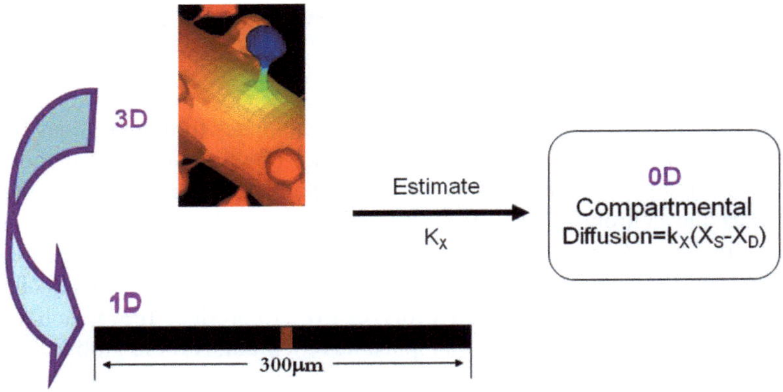

Fig. 5.2 Relationships between 3D, 1D and 0D geometrical models. 3D models consider diffusion in all regions, including on the membrane surface, explicitly. In a 1D geometry (*straight line*), diffusion down the dendrite can be modeled explicitly within a long linear geometry and diffusion between spine and dendrite is modeled implicitly according to Eq. 5.4. Various segments of the straight line in 1D can be designated by x coordinates to span a single activated spine unit, highlighted in gray in the center of the straight line. The 0D (compartmental) model uses ordinary differential equations to mathematically describe a single spine where the geometries are taken into account by considering the relative volumes and surface to volume ratios of the spine and dendrite compartments. Again, the diffusion of species between spine and dendrite must be modeled by Eq. 5.4 and an additional term is also included for the decay of the dendritic species due to the equilibration between the region just under the activated spine and the steady state concentration further way in the dendritic shaft (see Hernjak et al. (Slepchenko and Loew 2010) for full details of the structure of the compartmental model) (Reprinted from Brown et al. (2008). Copyright 2008, with permission from Elsevier)

can be turned off, in a virtual experiment, to assess its importance). In this series of models, the common physiological linkage is the IP3 signal. Xu et al. (2003a) looked at the ability of PIP2 to generate IP3. Hernjak et al. (2005) examined the relationship of spine shape and size on IP3 stimulated calcium dynamics using parameters that matched experimental conditions of parallel fiber or climbing fiber stimulation (Wang et al. 2000). Then, the results of the 3D models in Brown et al. (2008) on PIP2 and IP3 dynamics were used as inputs to the Hernjak et al. (2005) compartmental model to determine whether they were sufficient to produce similar calcium transients. The public models associated with these studies can be found and accessed by their names provided in Table 5.1.

As part of this flow of inputs and outputs, we had to consider how to link models from one biological system to another across spatial scales and dimensionality. These ideas are illustrated in Fig. 5.3 and Table 5.2. Geometries with two or three dimensions can represent the intricacies of neuronal geometry accurately, but reaction–diffusion simulations in any region larger than a small segment of dendrite would be computationally too intense to be practical. The challenge for a small segment of dendrite is to properly account for boundary conditions; for example, reactions that produce or destroy molecules in a localized region such as a spine

Table 5.1 The papers and models that are used to elucidate the relationship between spine morphology and neuronal signaling via IP3 are listed with the cell type that was modeled, the dimensionality of models and what species were measured or tracked in the models

Papers	Cell type	VCell models[a]	0D	1D	2D	3D
(1)	Neuroblastoma	BioModel: les: PIP2 Hydrolysis JCB Submission-update	PIP2, IP3		PIP2, IP3	
(8)	Purkinje spines	BioModel: hernjak: purkinje_2D BioModel: hernjak: purkinje_compartmental MathModel:hernjak:purkinje_1D	IP3, Ca (implicit diffusion)	IP3,Ca	IP3, Ca	
(7)	Purkinje spines	BioModel: Brown: Brown et al. 2008	IP3, Ca (implicit diffusion)	PIP2, IP3		PIP2, IP3

[a]The Virtual Cell Models are listed with Virtual Cell model type (BioModel), the user name (les, hernjak, Brown) and model name

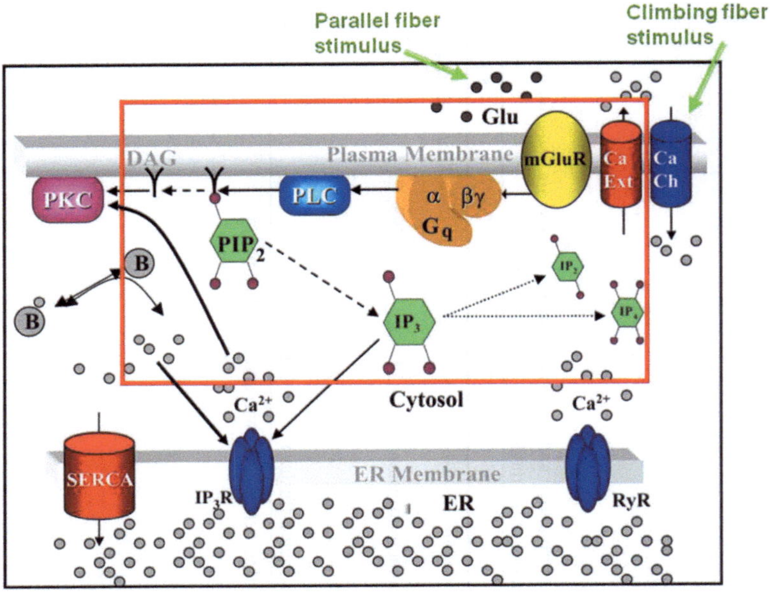

Fig. 5.3 Cartoon representation of PIP2 mediated signaling in Purkinje cell spines (Reprinted from Brown et al. (2008). Copyright 2008, with permission from Elsevier)

could produce time varying species concentrations at the open edges of the dendrite segment. In the 1D model, activation confined to a spine at the center of a long enough dendrite could be efficiently simulated so that boundary conditions could be modeled as fixed initial concentrations. However, diffusion between the spine and the adjacent dendrite in the ID model could not be explicitly simulated, but had to approximated as a flux between compartments using the expression shown in Table 5.2. These 1D results could then be used to fit an empirical time-dependent expression as an approximation for the change in the concentration of species X at the boundary of a 2D or 3D geometry, permitting a fully explicit model of dynamics in the spine.

Brown et al. (2008) simulated the biochemical reactions of the Purkinje cell physiology in both experimentally derived and analytically constructed 3D geometries. The simulations results were sufficiently similar to allow for further simulations to be carried out in the analytic geometries with confidence. The analytic 3D geometry was constructed based on the average values for spine morphological parameters determined by Harris and Stevens (1988) and also used by Hernjak et al. (2005). The same average values for the spines were used to construct a 1D geometry and included as terms for length, area and volume in the compartmental models. Thus, although the mathematical representations of geometry differ, the values for the spatial characteristics did not.

Table 5.2 Linkage points between simulations in models of different dimensions

Dimension	Parallel fiber stim	Climbing fiber stim	Diffusion	Spatial parameters
3D	Parallel Fibers (PLC activation)	NA	Explicit	Within average values (constructed geometry); explicit image-based geometry
1D	IP3 signal	Calcium influx	Explicit along dendrite; between spine and adjacent dendrite: $D_X(A_{neck}/l_{neck} V_s)$	Average values
0D	IP3 signal	Calcium influx	Between spine and adjacent dendrite: $D_X(A_{neck}/l_{neck} V_s)$; *Adjacent to distal dendrite:* k_x	Average values

Parallel fiber stimulation was represented in all models, yet that representation differed structurally in each. Despite explicit inclusion of PLC activation in 3D models, the IP3 signal in all three types of models was significantly the same (see Fig. 5.5)

5.3 Diffusive Barriers and Spine Geometry

The Purkinje neuron is often admired for its dendritic arbor. A good majority of its hundreds of dendrites are decorated with spines. On average, the Purkinje neuron has approximately 14 spines per micron of dendrite (Harris and Stevens 1988). The spine is attached to the dendrite branchlet by its *neck*, which varies in its diameter and length. It has long been known that the spines have a diffusive barrier that limits the ability of calcium to diffuse out of the spine head (Holmes and Rall 1995). The particular geometry of the spine neck shapes the dynamics of diffusion, and thus *reaction–diffusion*, in the spiny dendrite (Brown et al. 2008; Santamaria et al. 2006). As identified by Santamaria, Antunes, and Schutter in chapter 5, spines have an affect on the diffusion of cytosolic molecules within the dendrite creating anomalous diffusion which can affect biochemical reactions and neuronal signaling. They use stochastic simulations to examine cytoplasmic diffusion within dendrites with spines at various densities, while we examine the role of dendritic spine shape on reaction–diffusion of PIP2-IP3-Calcium dynamics.

When the spine is activated, the diffusive barrier of the spine neck leads to the localized accumulation of the cytoplasmic phosphoinositol IP3, and cytosolic calcium. Though lipids and receptors in the plasma membrane typically diffuse more slowly than molecules in the cytosol, we have shown that both *lateral membrane* and volumetric cytoplasmic diffusion are influenced by spine neck geometry (Brown et al. 2008). Biochemical signaling can be limited by substrate availability, which, in turn, is controlled by reactions (synthesis, degradation) and diffusion. For example, local PIP2 availability for cleavage into IP3 is determined, in part, by its diffusion in the membrane to the region that has been stimulated. Thus, examining the influence of geometry on lateral PIP2 diffusion and the effect of lateral diffusion on signaling is critical to understanding underlying biochemical processes of Purkinje signaling.

The experimentally derived geometry used by Brown et al. (2008) is a reconstruction of an actual Purkinje neuron dendrite branchlet, based on a stack of 3D electron microscope images.[1] The use of the analytic geometric shapes allows for more systematic investigation of the impact of spatial properties on signaling events. In the Virtual Cell, a regular grid (mesh) of points is created that define the geometric shape and within which the differential equations are solved. Within each mesh element, molecules are assumed to be uniformly distributed. Virtual Cell has a choice of semi-implicit or fully- implicit finite volume solvers to numerically solve the set of differential equations created to describe the reaction–diffusion system.

---

[1]Courtesy of Maryanne Martone, Mark Ellisman and Masako Terada of the National Center for Microscopy and Imaging Research in San Diego, CA.

5.3.1 PIP2 Lateral Diffusion in Distinct Spine Neck Geometries

In this section, we will analyze lateral diffusion of PIP2 in more detail in order to determine if lateral diffusion of PIP2 displays anomalous diffusion characteristics. In our model simulations, we initially depleted PIP2 from three spine heads (Fig. 5.4a, spines 1–3) with differing morphologies. We also examined the lateral diffusion of PIP2 into a selected patch of dendrite, indicated with an asterisk in Fig. 5.4a.

To initialize the concentration of PIP2 in each of the four selected regions in the Fig. 5.3, we set it to 0 molecules/μm^2 compared to a basal surface density of 4,000 molecules/μm^2 in the rest of the dendrite. We do this in the Virtual Cell by writing an expression that sets x, y, z coordinates for a 3D region (e.g., shaded 1 in Fig. 5.4) outside of which the basal concentration of PIP2 is 4,000 molecules/μm^2 and inside of which the density of PIP2 is 0 molecules/μm^2. The coordinates are set in relation to the entire domain of the geometry. Thus the Boolean expression below says that the value of PIP2 is 4,000 when x is less than 13.33 and greater than 14.75 and, y is less than 2.22 and greater than and z is less than 2.5 and greater than 3.2:

$$(4000.0^*((x < 13.3)||(x > 14.75)||(y < 2.22)||(y > 2.4)||(z < 2.5)||(z > 3.2))).$$

Virtual Cell then uses the assigned diffusion rate to generate a differential equation of the form:

$$\frac{d[PIP2\_PM]}{dt} = PIP2\,Diffusion\,rate \times \nabla^2[PIP2\_PM] \tag{5.1}$$

where [PIP2_PM] represents the concentration, i.e. surface density, of PIP2, and is the Laplacian operator.

The simulation begins with the PIP2 distribution defined above, and we track the restoration of the basal concentration of PIP2 in the center of each initially depleted region. This allows us to investigate the dynamics of PIP2 lateral diffusion in the spine membrane in comparison to the dendritic shaft membrane. Figure 5.3b shows the change in PIP2 surface density with time. In the patch of dendrite (*), PIP2 recovers rapidly to 4,000 molecules/μm^2. In spine 2 which has a very short and broad neck, PIP2 recovers quickly, but not as rapidly as in the patch of dendrite. This suggests that the presence of the spine neck alters the PIP2 diffusion dynamics. Spines 1 and 3, whose neck geometries are narrower, show even slower recoveries. This demonstrates that the overall geometry of the spine neck dynamically restricts lateral diffusion, such that the rates of recovery are slower in the thinner longer necks than shorter wider necks.

The inset in Fig. 5.4b further analyzes the PIP2 diffusion pattern. For conventional, simple diffusion along a surface, time is proportional to the inverse of the molecular concentration, such that in our case:

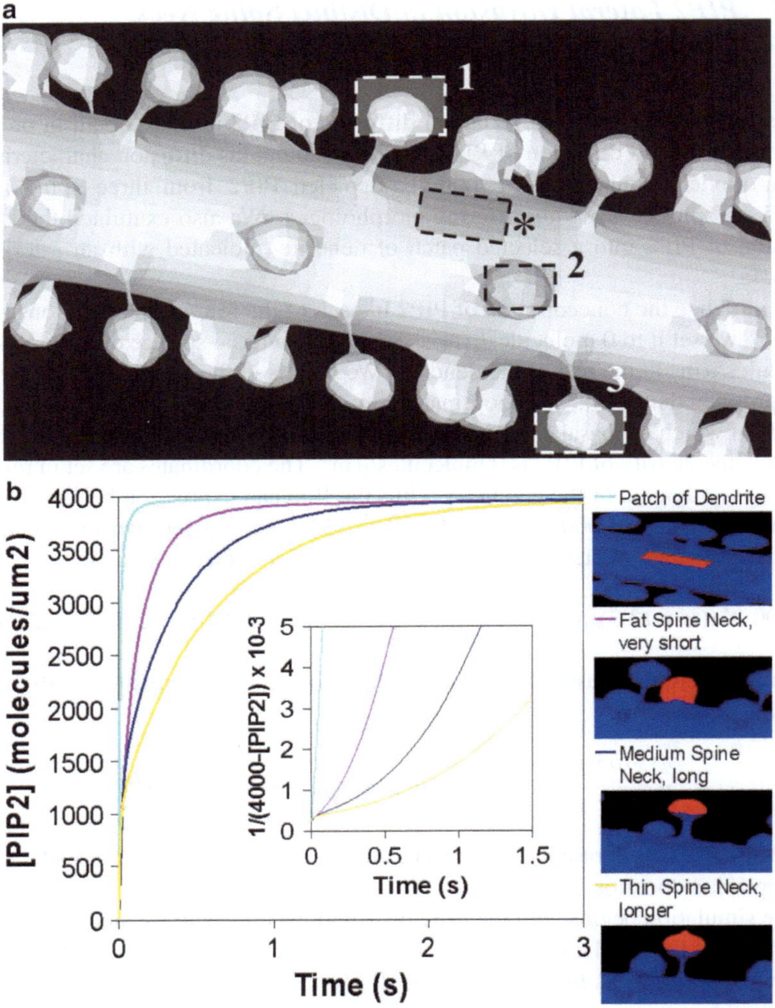

Fig. 5.4 PIP2 lateral diffusion in selected spines and a patch of dendrite from our constructed 3D geometry. (**a**) Constructed 3D geometry with three types of spines, determined by three different, specific spine geometries – a spine with a neck of intermediate proportions (Xu et al. 2003a), a spine with a very short, broad neck (Wang et al. 2000), a spine with a long, narrow neck (Slepchenko and Loew 2010); an aspiny patch of dendrite (*). (**b**) PIP2 lateral diffusion in the four selected regions from a. The chart shows that PIP2 concentration recovers much more rapidly in the aspiny patch of dendrite (*) than in the spines (Xu et al. 2003a; Wang et al. 2000; Slepchenko and Loew 2010). *Inset*. Analysis of PIP2 lateral diffusion. For conventional surface diffusion, time α 1/[4000 – PIP2]. The graph shows that PIP2 rapidly diffuses into the patch of dendrite (*). PIP2 diffuses into the spines more slowly. The arrows indicate lateral PIP2 diffusion into spine 1, through the spine neck. The white portion of each spine indicates the area where PIP2 is instantaneously depleted (Reprinted from Brown et al. (2008). Copyright 2008, with permission from Elsevier)

$$time\,\alpha\,\frac{1}{(4000 - [PIP2\_PM])} \tag{5.2}$$

where 4,000 molecules/μm^2 is the basal concentration of PIP2 in the model. For simple diffusion, plotting 1/(4000-[PIP2_PM]) against time should give us a straight line. For the patch of dendrite (*) in the insert in Fig. 5.4, the slope is indeed a straight line, and is very steep. This suggests that within the first few seconds after depletion, PIP2 diffuses normally and rapidly into the patch of dendrite. However, non-linear curves are obtained for the lateral diffusion of PIP2 into the spine heads to replenish the initially depleted PIP2. The shortest and broadest of the three spines (spine 2 in Fig. 5.4) gives a curve that is closest to linearity. This suggests that the short and broad spine neck provides the least restriction to lateral diffusion. The spine with a long neck of intermediate width (spine 1 in Fig. 5.4) provides more restriction, as evidenced by slower recovery and further departure from linearity. The spine with the longest, narrowest neck (spine 3 in Fig. 5.4) has the most non-linear curve and is therefore most restrictive to diffusion. The departure from linearity exhibited by all three spines is suggestive of 'anomalous' diffusion due to the influence of the spine neck geometry (Brown et al. 2008).

The time courses for replenishment of PIP2 by lateral diffusion for the three cases shown in Fig. 5.4 are too slow to replace the rapidly hydrolyzed PIP2 in the spine at a rate that would be required for the requisite IP3 signal in any of the spines. This is because the same geometric feature, the constricted spine neck, which impedes the lateral diffusion of PIP2, is also required to assure retention of the IP3. Thus even though the stubby spine in Fig. 5.4 may allow for rapid diffusion of PIP2, that spine loses its IP3 too quickly to the dendrite for there to be any significant accumulation. On the other hand, the spines labeled 2 and 3 in the Figure are able to retain IP3 for a longer period but the slowed lateral diffusion of PIP2 effectively isolates the small pool of PIP2 initially available in the spine head; this pool is insufficient for the requisite levels of IP3 (Brown et al. 2008).

Clearly geometry can be used by the cell to regulate the rate of diffusion and thereby compartmentalize molecules within different regions. To quantify further the relationship between shape/size and diffusion, we turn later in the chapter to 1D and compartmental models in which we can rapidly perturb the spatial parameters and measure their effect on the dynamics of PIP2 biochemistry, IP3 generation and calcium transients.

5.3.2 PIP2 Supplied by Stimulated Synthesis or Sequestration

In the previous section, we showed how a diffusive barrier prevents lateral diffusion of PIP2 from supplying sufficient PIP2 to generate the high concentration of IP3 required to activate the calcium transient. The other two hypotheses are (1) PIP2 could be rapidly synthesized concurrently with hydrolysis upon stimulation, as

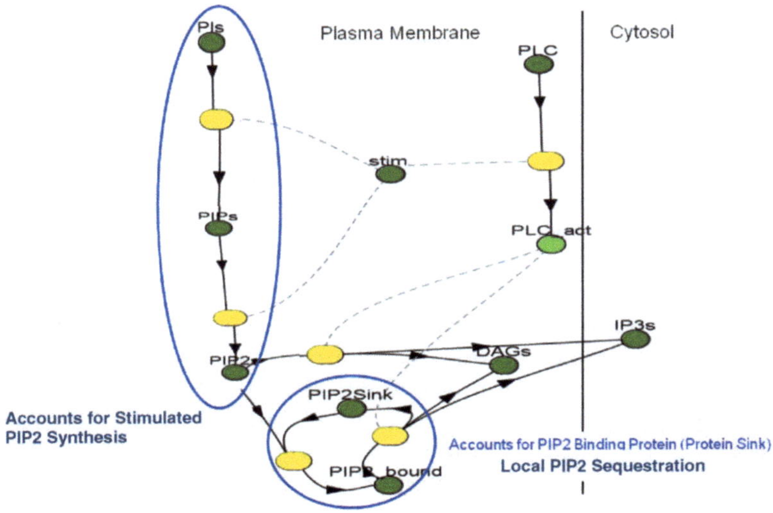

Fig. 5.5 Reaction diagrams for local stimulated synthesis and sequestration (Reprinted from Brown et al. (2008). Copyright 2008, with permission from Elsevier)

established for the N1E-115 mouse neuroblastoma cell line (Xu et al. 2003a) or (2) PIP2 could be bound to some sequestering protein at the post-synaptic membrane, providing a locally high concentration of bound PIP2. The 3D models we used to distinguish between the two possible sources of sufficient PIP2 differ only in the reactions (synthesis or unbinding) that lead to the production of PIP2 (Fig. 5.5). The remaining topology of the model is kept the same, yet the production of IP3 from hydrolysis of PIP2 is fit to achieve approximately the same amplitude as that required for peak calcium release (~25 μM). Individual equations associated with stimulated synthesis and local sequestration can be viewed in the public VCell models (Brown et al. 2008).

We enter a localized stimulus in our 3D model in the same way that we established regions of depleted PIP2 in the previous Section: by defining a spatial region in xyz that evaluates to 1 for a specific spine head. This is done by including a spatially localized stimulus (*stim* in Fig. 5.5).

The variable *stim_PM* is defined in initial conditions to encompass about half of a spine head using a Boolean expression of the Cartesian coordinates; for example the expression for stimulation of spine 1 in the 3D constructed geometry (Fig. 5.4) is:

$$Stim\_PM = ((x > 13.25)\ \&\&\ (x < 14.6)\ \&\&\ (y < 1.18)\ \&\&\ (z > 2.6)\ \&\&\ (z < 3.9)) \quad (5.3)$$

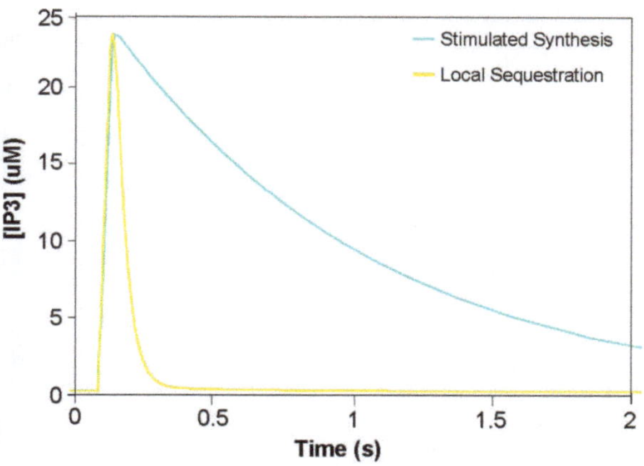

Fig. 5.6 IP3 signal with stimulated PIP2 synthesis or local PIP2 sequestration in 3D model. The equivalent of a train of four parallel fiber stimuli was applied to the Purkinje neuron spine. The IP3 obtained with stimulated PIP2 synthesis lasted for much longer than the signal obtained with local PIP2 sequestration. The sources of PIP2 gave different signal durations although they had the same amplitudes (Reprinted from Brown et al. (2008). Copyright 2008, with permission from Elsevier)

Details of the rate expressions for the stimulated synthesis and local sequestration pathways (Fig. 5.5) can be found in the original paper (Brown et al. 2008) and the public VCell model. The resulting IP3 signals for these mechanisms are shown in Fig. 5.6. The IP3 signal from the hydrolysis of PIP2 concomitant with stimulated PIP2 synthesis lasts for much longer than the signal generated with local PIP2 sequestration. Although the amplitudes are the same between the models, the decay time for stimulated synthesis was similar to that used by Hernjak et al. (2005) in their studies of spine shape regulation of calcium transients. This provides an obvious biochemical linkage between the hypothesized synthesis mechanism in the 3D models and the required IP3 signal in the compartmental model. Given this, we will use the same IP3 signal obtained from stimulated synthesis in the 3D model to further explore the affects of spine shape on IP3 diffusion and achieving supralinear calcium transients.

5.3.3 IP3 Dynamics with Stimulated PIP2 Synthesis

In this section, we will discuss the influence of the spine neck geometry on the cytosolic diffusion of IP3 (produced from PIP2 hydrolysis) out of the spine head and into the dendritic reservoir. Recall, that the geometry of the spine neck influences the lateral diffusion of PIP2. 3D geometries that support fast diffusion of PIP2 correlate with loss of IP3 in vivo and in models that do not account for increased supplies of PIP2. The timecourse of IP3 generated by stimulated synthesis closely resembles

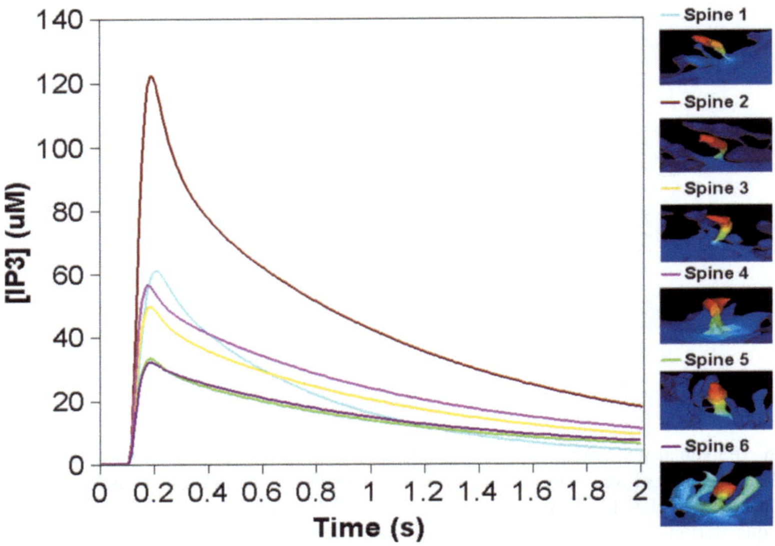

Fig. 5.7 Spine compartmentalization of IP3, due to restricted diffusion through the spine neck. a. IP3 in six spines of our experimentally derived 3D geometry. A train of four parallel fiber stimulus pulses were applied to the Purkinje spine. The graph shows that the IP3 signal obtained with stimulated synthesis depends on the unique geometry of each spine neck

the IP3 profile used by Hernjak et al. (2005), thus we will focus on results that use stimulated PIP2 synthesis as a source of sufficient PIP2 in our 3D models (Nishizuka 1988). Diffusion of the IP3 from the spine into the dendrite is retarded due to the restricted diffusion through the spine neck, as suggested by Santamaria et al. (2006), Brown et al. (2008). This retardation allows IP3 to build up in the spine and is critical to neuronal signaling, as large amounts of IP3 are needed for significant calcium release in the Purkinje neuron (Khodakhah and Ogden 1993).

Figure 5.7 shows the IP3 signal obtained with stimulated PIP2 synthesis in the experimental 3D Purkinje dendrite geometry (see the full geometry in the screenshot from the VCell interface of Fig. 5.1b) in six different spines with geometries that vary in length, width and curvature. The IP3 signal varies greatly from spine to spine. For example, note that the amplitude of the IP3 curve in spines 5 and 6 is almost half the amplitude of the IP3 produced in spine 4. This is remarkable given that the IP3 reaction kinetics are identical in each spine.

In order to rapidly examine the relationship between spine size parameters (spine head area and spine neck length) and IP3 accumulation, we turned to a compartmental model of the same biology (Brown and Loew 2012). In the compartmental model adapted from Hernjak et al. (2005), diffusion and geometry are included parametrically (Hernjak et al. 2005). To model diffusion and spatial features in compartmental models, the geometrical features are reduced to measurements of length, area and volume; each species is taken to be uniformly distributed within each compartment (i.e., rapid diffusion) and geometry-based expression is used to model the diffusion rate between compartments. Thus, in compartmental models, diffusion of molecules or species from one cellular region to another can be modeled implicitly via a rate equation such as

$$dXS/dt = -kX(|XS| - |XD|) \tag{5.4}$$

Where kx is given by the geometry and diffusion coefficient in cytosol as discussed in connection with Table 5.2.

Average values for the radius of the spine head (0.29 μm) and the radius and length of the spine neck (r = 0.1; l = 0.66) are based on an experimental study by Harris and Stevens (1988). Harris et al. noted that in examined Purkinje neuron dendrites, shorter spine necks associated with larger spine heads, and inversely, longer spine necks associated with smaller spine heads. Here, we explore the effect of this association by examining the influence of spine head radius on compartmentalization and comparing this directly with the influence of spine neck length. Average values from Harris and Stevens (1988) were used for one simulation (average values) and subsequent simulations either halved (neck length halved, spine head radius halved) or doubled (neck length doubled, spine head radius doubled) to investigate the influence of these changes on IP3 accumulation and diffusion from the spine. (See Table 5.2 for diffusion expression including spine head volume and spine neck length, recall the relationship between spine head volume and spine head radius; Brown and Loew 2012).

Figure 5.8a shows that changes in IP3 concentration were greater in response to changes in spine head radius than spine neck length. This suggests that the compartmentalization in the spine head may be more sensitive to changes in spine head radius than to changes in the length of the spine neck. Interestingly, the range of diameters of the spine head (0.43 – 0.68 μm) determined experimentally by Harris and Stevens (1988) is much smaller than the range of lengths of the spine neck (0.12 – 2.18 μm). Of note, the sensitivity of IP3 concentration to the size of the spine could have contributions from both the volume of the spine to be occupied and the area of membrane contributing the PIP2 precursor. Further, the rate of IP3 decay in the spine depends on its diffusion through the spine neck, which in turn depends on the size of the spine head. Therefore, we performed simulations to understand the relative contribution of the volume of the spine to the amplitude and duration of the IP3 signal. The results are shown in Fig. 5.8b. The figure shows that halving or doubling the spine neck length has similar effects on IP3 concentration as halving or doubling the spine head volume, respectively. Examining the expression for IP3 diffusion through the spine neck (see Table 5.2) shows that an equal manipulation of either the spine neck length or the spine head volume would lead to the same effect on this expression. Figure 5.8b suggests that IP3 concentration, and compartmentalization, is exquisitely sensitive to IP3 diffusion though the spine neck, apparently more so than to total spine PIP2.

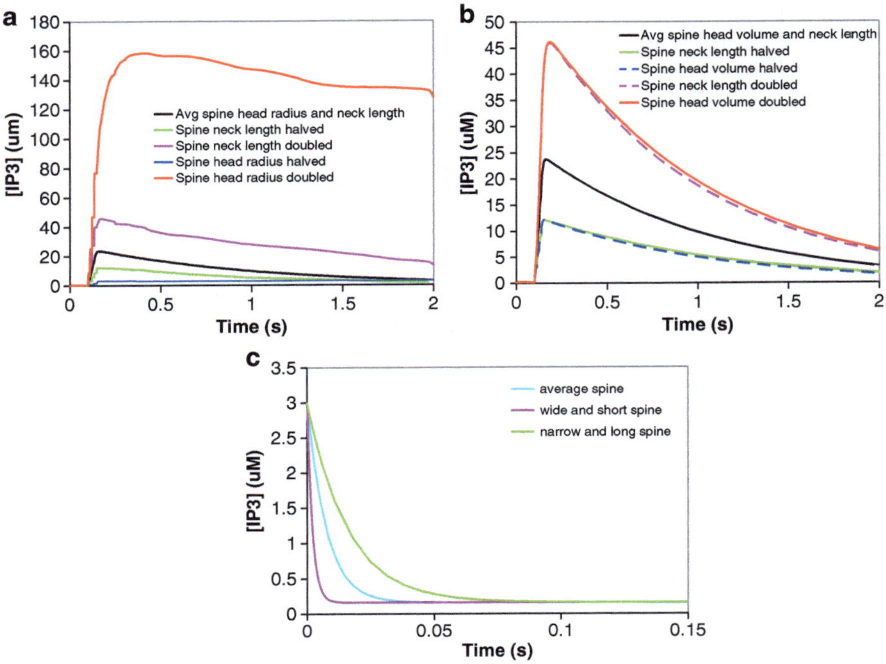

Fig. 5.8 Compartmental model. Spine compartmentalization of IP3, due to restricted diffusion through the spine neck and to spine head geometry. (**a**) IP3 concentrations are plotted from a single stimulated spine in which average values of spine head radius and spine neck lengths are compared by halved and doubled values. (**b**) IP3 concentration changes similarly when spine neck length and spine head volume are halved or doubled. (**c**) Uncaged IP3 concentration decreases most slowly in a long and narrow spine, and most quickly in a wide and short spine

Taken together, these findings suggest that increasing spine size increases compartmentalization of IP3, while decreasing spine neck length has an opposite effect. Thus, one could speculate that the inverse association between spine neck length and spine head size, as discovered by Harris and Stevens (1988), could function to maintain a normalized level of compartmentalization across spines of various geometries. However, Fig. 5.7 shows that compartmentalization is not maintained at a constant level across spines in the experimentally derived 3D geometry. It may therefore be that although the inverse association serves to ensure compartmentalization and appropriate IP3 production, based on the particular synaptic activity and biochemical needs for each individual spine, it does not suggest constant compartmentalization across spines.

This same compartmental model was used in new simulations for this chapter to examine the behavior of photoactivated caged-IP3. In each case we instantaneously uncaged 3 μM IP3 within the full spine volume. These simulations, therefore, allow us to eliminate the effect of PIP2 availability on the spine membrane and focus only on the kinetics of IP3 decay as a function of spine geometry. We demonstrated that IP3 diffusion out of the spine head occurs more slowly in a spine with a long and narrow spine neck (Fig. 5.8c) and more quickly out of a spine with a wide and short spine neck than a spine with a neck of intermediate size. This is consistent with and expands on the spine compartmentalization studies reported by Hernjak et al. (2005) for calcium, and Santamaria et al. (2006) and Brown et al. (2008) for calcium and IP3. It also parallels PIP2 lateral diffusion results from Brown et al. (2008).

IP3 is also removed from the spine by another method: degradation, which is modeled by an expression such as:

$$\left(kIP3deg^{*}\left([IP3\_Cyt] - [IP3\_basal]\right)\right) \tag{5.5}$$

where *kIP3deg* is the rate constant for degradation of IP3, *[IP3_Cyt]* is the cytosolic concentration of IP3, and *[IP3_basal]* is the basal concentration of IP3. This rate expression was used in all simulations and is based on the measured degradation rate in neuroblastoma cells. That rate has a half time of ca. 8 s. Thus, changes in degradation are not expected to explain differences in accumulation from one geometry to another.

To understand the need for such large amounts of IP3 (Khodakhah and Ogden 1993), we need to consider unique properties of the Purkinje IP3R1. The cerebellar Purkinje neuron possesses significantly more IP3 receptor Type 1 than any other cell type (De Smedt et al. 1997). However, the IP3R1 in the Purkinje neuron is less sensitive to IP3 than in any other cell type (Khodakhah and Ogden 1993; Fujiwara et al. 2001). This interesting mix of high abundance and low sensitivity is important for proper neuronal signaling in the Purkinje cell (Hernjak et al. 2005), the unitary output cell for the cerebellum. Reduced sensitivity of IP3R1 to IP3 is one reason that the Purkinje neuron needs a large amount of IP3 for significant calcium release from internal stores. This large, local, and high concentration of IP3 is facilitated by spine neck-dependent compartmentalization, as previously discussed. Without compartmentalization by the spine neck, IP3 would quickly diffuse out of the spine head as seen in Fig. 5.8b and spine 5 of Fig. 5.7.

5.3.4 Fine Tuning Supralinear Calcium Transients

In this section, we describe briefly the dynamics of the calcium transient resulting from IP3 binding its receptor on the surface of the ER, as well as from the opening of voltage-gated calcium channels in the Purkinje neuron plasma membrane (Ito et al. 1982). Just as with IP3, the geometry of the spine neck creates a diffusive barrier that allows for compartmentalization of the Ca signal which enters the cytosol as a calcium flux from the ER and Ca influx from extracellular space. The experimental system of Wang et al. (2000) used four PF pulses followed by a CF pulse 50 ms later to stimulate a spine and subsequently measure the change in calcium using

Magnesium Green. Hernjak et al. (2005) was able to match the calcium transients by modeling PF stimuli as increases in IP3 at a frequency matching the Wang protocol (Wang et al. 2000) for PF frequency and followed by a CF pulse, modeled as a time constrained flux of external calcium into the cytosol.

In cerebellar Purkinje neurons, intracellular calcium transients can also be produced by influx across the plasma membrane, through voltage-gated calcium channels. This occurs when the Purkinje neuron receives afferent stimulation from the single *climbing fiber (CF)* that innervates the cell body and proximal dendrites (Ito et al. 1982). The CF stimulus depolarizes the entire neuron membrane, thus opening the voltage-gated calcium channels. The rapid and transient flow of calcium into the cell due to CFs can be represented by (Hernjak et al. 2005):

$$\left(Jch^{*}\ (t > t1)^{*}\ (t < t2)^{*}\ (Ca\_Extracellular - Ca\_Cytosol)\right) \tag{5.6}$$

where *Jch* adjusts the amplitude of the rate of calcium influx, *Ca_Extracellular* is the calcium concentration outside the cell, and *Ca_Cytosol* is the calcium concentration inside the cell; the influx begins at time = t1, and ends when time = t2 (Hernjak et al. 2005). The combined calcium response due to release from internal stores and to influx from outside the cell can produce supralinear spine calcium transients (Wang et al. 2000), i.e. a calcium signal that is greater than the sum of the calcium responses produced by each stimulus independently.

It has been suggested that the vast amount of calcium coming into the cytosol during *coincident activation* may overwhelm the cytosolic buffers resulting in supralinear calcium (Hernjak et al. 2005). Calcium is quickly taken up by various buffers in solution, including calbindin, parvalbumin, and any fluorescent indicators (Hernjak et al. 2005). Calcium buffers modify the spatiotemporal patterns of calcium transients (Schwaller et al. 2002). This means that even while total calcium is rising, whether due to calcium influx from outside the cell or calcium release from the ER, calcium-binding proteins act to buffer and thereby limit the rise in free calcium concentration. The computational model includes a species as a calcium indicator with the low affinity of Magnesium Green (kd = 19 uM) (Brown et al. 2008). We have found that binding reactions for indicators must be included in order to accurately predict the behavior of molecules that are buffered by the same molecules used to measure their concentrations, such as calcium dyes. Figure 5.9 shows the resulting change in fluorescence as represented by the concentration of calcium bound to the Mg Green.

The resulting spike in fluorescence has the same amplitude for both sources of PIP2: stimulated synthesis and local sequestration. However, stimulated PIP2 synthesis provides a longer calcium signal; the calcium transient from local PIP2 sequestration decays more quickly. This is consistent with Fig. 5.6, in which the IP3 signal from stimulated PIP2 synthesis is of lasting duration, whereas the IP3 signal with local PIP2 sequestration is of short duration. The effect of IP3 signal duration on the dynamics of the calcium transient is explored further below, after we consider calcium influx.

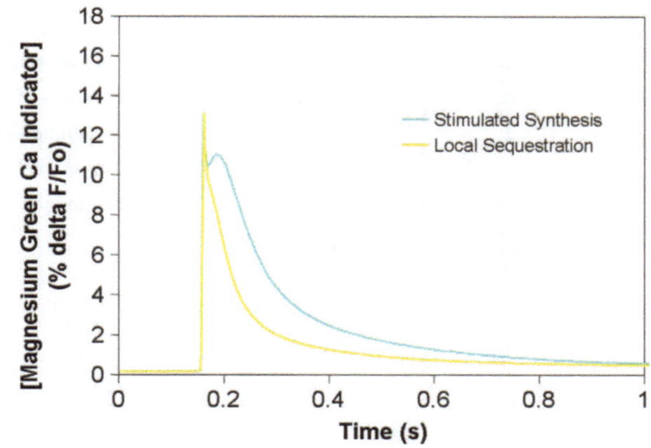

Fig. 5.9 Calcium transients with stimulated PIP2 synthesis and local PIP2 sequestration. The calcium release obtained from a signal of IP3 equivalent to a train of four parallel fiber stimuli (based on an experimental protocol by Wang et al.) is more robust with stimulated synthesis than with local sequestration

One potential mechanism for the supralinear calcium (Wang et al. 2000) is the IP3 receptor sensitivity to the concentration of cytosolic calcium (Bezprozvanny et al. 1991). Calcium entering the cell can bind to the IP3 receptor, thus potentiating IP3-dependent calcium release from the ER (Fiala et al. 1996). Consequently, the greatest calcium response occurs when the initiation of the PF stimuli immediately precedes the CF stimulus (Wang et al. 2000; Hernjak et al. 2005; Sarkisov and Wang 2008). This allows IP3 concentration to begin to rise, and allows for a significant amount of IP3 to bind its receptor before the voltage-gated calcium channels open in the plasma membrane, inducing calcium influx from outside the cell. The resulting calcium response from this *coincident detection* of the PF and CF stimuli is greater than the sum of the individual calcium transients obtained from isolated PF and CF stimuli.

Depending on the source of sufficient PIP2 for hydrolysis, the supralinear calcium response in our simulations can be sensitive to the timing between PF and CF stimuli (Brown et al. 2008). Figure 5.10 displays simulation results indicating that the supralinear calcium response from stimulated PIP2 synthesis is not very sensitive to timing, whereas the supralinear calcium response from local PIP2 sequestration is. Both conditions show supralinear calcium when the PF stimulus occurs 50 ms before the CF stimulus. Figure 5.10a shows that the supralinear calcium response obtained when the CF stimulus occurs 150 ms after the initial PF stimulus is almost identical to the 50 ms pairing. However, when the timing between the CF and PF stimuli in our sequestration model is increased from 50 to 150 ms (Fig. 5.10b), the supralinear calcium response disappears! This suggests that local PIP2 sequestration would provide the cerebellar Purkinje neuron with a means of fine-tuning *coincidence detection* (Brown et al. 2008). It is reasonable to expect that this fine-tuning would be crippled without the spine neck geometry-dependent compartmentalization of cytosolic molecules (IP3 and calcium) and restriction of diffusion of the surface molecules involved (i.e., PIP2). This effect on timing is similar to experimental results from Wang et al. (2000) and Sarkisov and Wang

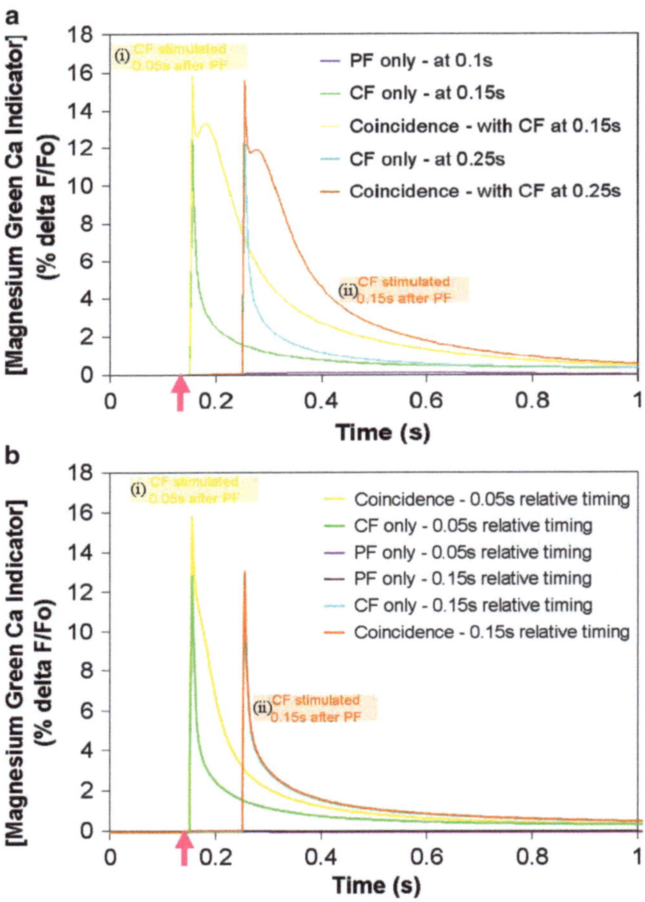

Fig. 5.10 Effect of timing on supralinear calcium release. (**a**) The supralinear calcium release obtained with stimulated PIP2 synthesis is almost identical when the initiation of the four parallel fiber stimuli (*arrow*, 100 ms) precedes the single climbing fiber stimulus at (i) 150 ms and (ii) 250 ms. The supralinear calcium release. (**b**) obtained with local PIP2 sequestration is sensitive to timing. Whereas the supralinear calcium release is obtained when the climbing fiber stimulus follows the initial PF stimulus (*arrow*, 100 ms) by 150 ms, it is no longer obtained when the timing is increased to 250 ms. The modeling is based on experimental protocols by Wang et al. (2000)

(2008). Both of these experimental studies suggest that there is an optimal *timing window*, during which closely timed pairing of parallel fiber and climbing fiber stimuli will yield a supralinear calcium response. Outside of the timing window, the calcium response is reduced, as recapitulated by the modeling results in Fig. 5.10.

5.3.5 Physiological Relevance

The role of dendritic spines in Purkinje neurons is critical for the central task of the cerebellum in coordination of movement, as well as motor learning and conditioned reflexes. *Ataxia* is a medical condition referring to lack of motor coordination (Goetz 2003), and long-term depression is a biological process underlying motor learning (Ito 2001). Mouse models and humans with mutations in various molecules involved in calcium homeostasis exhibit ataxia (Alonso et al. 2005; Kim et al. 1997; Zhuchenko et al. 1997; Guida et al. 2001; Iwaki et al. 2008; Yue et al. 1997), and some of the mouse models also exhibit reduced long-term depression (Aiba et al. 1994; Inoue et al. 1998).

LTD is a form of synaptic plasticity: it involves the weakening of synapses between the Purkinje neuron dendrite and the presynaptic parallel fiber bouton (Ito et al. 1982; Ito 2001; Ito and Kano 1982). The structural connection largely remains unchanged, but the response of the Purkinje neuron to stimulation by the parallel fibers decreases. This is because other glutamate receptors, namely the α-amino-3-hydroxy-5-methyl-4-isoxazolepropionic acid receptors (AMPARs), are downregulated in the Purkinje dendrite membrane (Neuroscience. Sinauer Associates 2999; Chung et al. 2003; Chung et al. 2000). One way to produce LTD in the Purkinje neuron is by coincident activation by parallel fibers and a climbing fiber (Ito et al. 1982). Briefly, transiently high calcium in the cytosol binds IP3R1, increasing the probability of the receptor existing in an open state (Fiala et al. 1996). IP3 is also bound to its receptor, and this coincident binding leads to much more calcium release than with activation of IP3R1 by IP3 or calcium alone (Wang et al. 2000; Hernjak et al. 2005; Sarkisov and Wang 2008; Brown et al. 2008). Cytosolic calcium, along with DAG (produced with IP3 from PIP3 hydrolysis), activates protein kinase C (PKC) (Nishizuka 1984). PKC phosphorylates various molecules that are recruited to the plasma membrane and remove AMPA receptors from the membrane (Neuroscience. Sinauer Associates 2999; Chung et al. 2000, 2003). AMPAR internalization depresses the response of the Purkinje dendrite to stimulus by parallel fibers.

From the discussions in this chapter, we see that the geometrical features of the Purkinje cell spine may modulate the calcium signaling that leads to LTD. In addition, the IP3 receptors in the Purkinje cell have a higher density and a lower affinity for IP3 than other cell types, leading to the ability to exquisitely localize spine calcium signals (Hernjak et al. 2005). Further, the unique features of the calcium handling machinery in the spine allows for coincidence detection between the localized inputs of parallel fibers and the more global depolarization produced by climbing fibers. Mice completely lacking the IP3 receptor Type 1 exhibit motor discoordination (ataxia) (Matsumoto et al. 1996) and reduced long-term depression (LTD) (Inoue et al. 1998). Integration of the biochemical, electrophysiological and morphological data into computational models can produce simulations that explore the relationships between LTD, Ca++, IP3, IP3R1 and cerebellar synapses, as well as the influence of geometry on the signaling processes (Xu et al. 2003a; Brown et al.

2008; Hernjak et al. 2005; Zhou et al. 2008; Fink et al. 2000). Such models and simulations can now be directly applied to test "virtual therapies" for human patients suffering from IP3R1-associated ataxias (Brown and Loew 2009, 2012).

5.3.6 Next Steps

Now that we have discussed in detail the influence of spine geometry on biochemistry and related physiological processes, an open question that remains is the influence of the spine neck on electrophysiological properties, such as membrane potential and current propagation. Literature suggests that on one hand, the resistance of the spine neck may alter propagation of electrical membrane properties (Araya et al. 2006a, b; 2007; Segev et al. 1995); other studies indicate that the resistance of the spine neck is insufficient to modulate neuronal electrophysiology (Harris and Stevens 1988; Turner and Schwartzkroin 1983; Palmer and Stuart 2009). In addition to better experimental methods, solving this dichotomy will require a true integration of biochemical and electrophysiological modeling algorithms and software. It will be useful for us to create models that combine biochemistry with electrophysiology to more thoroughly study disorders such as IP3R1-associated ataxia (Brown et al. 2008; Brown and Loew 2009; Schorge et al. 2010), in which both intracellular calcium release and membrane electrophysiology are affected (Matsumoto et al. 1996). We have developed a merged biochemical-electrophysiological model in Virtual Cell in order to take advantage of its current reaction–diffusion capabilities and to drive further enhancements to its abilities for modeling electrophysiology ((Brown et al. 2011). Such a combined biochemical-electrophysiological model could allow for detailed modeling of the physiology and electrical responses of the Purkinje neuron. For example, large conductance calcium-activated voltage-gated potassium channels are implicated in various forms of ataxia in mice (Sausbier et al. 2004; Walter et al. 2006; Cheron et al. 2009), and are regulated by some of the same molecules (protein kinase A, PKA; protein kinase C, PKC) (Hall and Armstrong 2000; Widmer et al. 2003) that regulate IP3R1 (DeSouza et al. 2002; Ferris et al. 1991). Merging neuronal electrophysiology with neurobiology in this way provides an additional layer of interaction between computational neuroscience, systems biology and spatial simulations inherent in reaction–diffusion models of cellular systems.

Using mathematical models and computer simulations to explore details of the interplay between spine geometry, phosphoinositol and calcium signaling gives us insight into how these cellular processes work together as an efficient system. The Virtual Cell provides a robust computational framework for prototyping, revising, sharing and re-utilizing models and model components. Our work exemplifies and is a practical application of Computational Systems Neurobiology. Expanding such interaction between the computational neuroscience and systems biology research communities (De Schutter 2008; Le Novère 2007) is clearly an exciting opportunity to advance research in both fields.

5.3.7 Further Reading

1. Brown et al. (2008) Spatiotemporal analysis of PIP2 and IP3 signaling in the Purkinje neuron.
2. Hernjak et al. (2005) Analysis of calcium and IP3 signaling in the Purkinje neuron.
3. Santamaria et al. (2006) Experimental study on Anomalous IP3 and calcium diffusion.
4. Harris and Stevens (1988) Experimental study of the spine geometry in cerebellar Purkinje neurons.
5. Xu et al. (2003) Kinetic analysis of receptor-activated phosphoinositide turnover.

5.3.8 VCell Exercises

These exercises are designed to guide the reader through the exploration of diffusion restriction and compartmentalization afforded by three different spines with unique morphologies in our constructed and experimentally derived 3D geometries. Instructions on how to view, run and save models can be found in the Virtual Cell User Guides at http://www.vcell.org/.

1. *Reproducing PIP2 lateral diffusion in the experimentally derived 3D geometry.* In Virtual Cell, open the public BioModel under username Brown entitled 'Spatial Organization Models'.

   (a) *Create spatially defined PIP2 depletion.* Double click to open the application '3D PIP2 diffusion'. In the initial conditions view, click on PIP2, then enter the following initial concentration:

   $$\begin{aligned}&(4000.0*(((x < 1.75) \,||\, (x > 2.25) \,||\, (y < 2.2) \,||\, (y > 2.95) \,||\, (z > 0.65))\\&\quad \&\&((x < 2.8) \,||\, (x > 3.4) \,||\, (y < 3.13) \,||\, (y > 3.5) \,||\, (z > 0.86))\\&\quad \&\& ((x < 1.6) \,||\, (x > 2.36) \,||\, (y < 1.6) \,||\, (y > 2.25) \,||\, (z > 0.65))\\&\quad \&\& ((x > 5.35) \,||\, (x < 4.5) \,||\, (y > 1.25) \,||\, (y < 0.7) \,||\, (z > 0.45) \,||\, (z < 0.1))))\end{aligned}$$

   This expression creates a cube around three individual spine heads of the 3D geometry, as well as an aspiny patch of dendrite. Each cube defines the initial concentration of PIP2 at 4,000 molecules/um_2. Outside of each cube, the concentration of PIP2 is 0. In the simulation view, run "Simulation". At time 0, the simulation will instantaneously deplete PIP2 in spines 2, 3, and 5 of the geometry as well as in the patch of dendrite. This is consistent with Fig. 5.3 of the chapter.

   (b) *View time course of diffusion or recovery in the depleted spines.* View the simulation results. Select PIP2 in the variable list on the left of the window, then click the '+1' positive zoom button until the depleted spines and patch of dendrite are visible. In this view, basal levels of PIP2 are in red and depleted areas are in blue (in the 'Auto' values range view). Click the 'point icon' located below the arrow button on the right of the simulation window, then click on each blue spine (the depleted spines) as well as the blue area of dendrite. Click on 'Show Time Plot' at the bottom of the screen. A sub-window will appear displaying time plots of PIP2 concentration in the selected spines and patch of dendrite.

(c) *Export simulation results to view diffusion of PIP2 over time*. In the Results view, uncheck 'Auto' under 'Data Values Range'. Click 'Show Membrane Surfaces'. A sub-window will appear displaying a 3D rendering of the Purkinje neuron dendrite with basal levels of PIP2 in red and depleted areas in blue. Click on 'Make Movie', then 'OK'. Select a destination for your movie, save it with the extension '.mov', then open it using Quicktime. Play the movie to view PIP2 lateral diffusion in real time.

2. *IP3 production in the constructed 3D geometry*. In Virtual Cell, open the public BioModel under username Brown entitled 'Spatial Organization Models'.

(a) Double click to open the application 'IP3 production in 3D' and select the Initial Conditions view. Click on the variable 'stim', then enter the following initial condition:

$$((x > 1.75)\ \&\&\ (x < 2.25)\ \&\&\ (y < 2.2)\ \&\&\ (y < 2.95)\ \&\&\ (z < 0.65))$$
$$\&\&\ ((x > 2.8)\ \&\&\ (x < 3.4)\ \&\&\ (y > 3.13)\ \&\&\ (y < 3.5)\ \&\&\ (z < 0.86))$$
$$\&\&\ ((x > 1.6)\ \&\&\ (x < 2.36)\ \&\&\ (y > 1.6)\ \&\&\ (y < 2.25)\ \&\&\ (z < 0.65))$$

This expression localizes the production of IP3 only to spines 2, 3, and 5 as seen in Fig. 5.3 of the chapter.

(b) *View time course of IP3 production in the three spines*. In the Application, select the Simulation tab and view the simulation results. Select IP3 in the variable list on the left of the window. Type 0.2 (s) in the box for 'Time'. This will allow us to view IP3 concentrations once stimulated production has already begun. Click the '+1' positive zoom button until the three active spines are visible, with varying shades of red based on IP3 concentration. Click the 'point icon' located below the arrow button on the right of the simulation window, then click in the middle each of the three active spines. Click on 'Show Time Plot' at the bottom of the screen. A sub-window will appear displaying time plots of IP3 concentration in the three spines.

(c) *View time course of PIP2 concentration changes during IP3 production in the three spines*. Viewthe simulation results. Select PIP2 in the variable list on the left of the window. Change the Max value (box in upper right of screen) to 20,000 (molecules/μm^2). This will facilitate rises in PIP2 concentration due to stimulated synthesis concomitant with PIP2 hydrolysis, as hypothesized in the chapter. Before viewing membrane surfaces, type 0.2 (s) in the box for

'Time'. This will allow us to view PIP2 concentrations once IP3 production has already begun. Click 'Show Membrane Surfaces'. A sub-window will appear displaying a 3D rendering of the Purkinje neuron dendrite with various levels of PIP2, in each of the three spines, represented by the displayed color scale.

References

Aiba A, Kano M, Chen C, Stanton M, Fox G, Herrup K, Zwingman T, Tonegawa S (1994) Deficient cerebellar long-term depression and impaired motor learning in mGluR1 mutant mice. Cell 79:377–388

Alonso I, Costa C, Gomes A, Ferro A, Seixas A, Silva S, Cruz V, Coutinho P, Sequeiros J, Silveira I (2005) A novel H101Q mutation causes PKCgamma loss in spinocerebellar ataxia type 14. J Hum Genet 50:523–529

Andrews SS, Addy NJ, Brent R, Arkin AP (2010) Detailed simulations of cell biology with Smoldyn 2.1. PLoS Comput Biol 6:e1000705

Araya R, Jiang J, Eisenthal K, Yuste R (2006a) The spine neck filters membrane potentials. Proc Natl Acad Sci U S A 103:17961–17966

Araya R, Eisenthal K, Yuste R (2006b) Dendritic spines linearize the summation of excitatory potentials. Proc Natl Acad Sci U S A 103:18799–18804

Araya R, Nikolenko V, Eisenthal K, Yuste R (2007) Sodium channels amplify spine potentials. Proc Natl Acad Sci U S A 104:12347–12352

Bezprozvanny I, Watras J, Ehrlich B (1991) Bell-shaped calcium-response curves of Ins(1,4,5)P3- and calcium-gated channels from endoplasmic reticulum of cerebellum. Nature 351:751–754

Brown S-A, Loew LM (2009) Toward a computational model of IP3R1-associated ataxia. Biophys J 96((3:S1)):96a

Brown S-A, Loew LM (2012) Computational analysis of calcium signaling and membrane electrophysiology in cerebellar Purkinje neurons associated with ataxia. BMC Systems Biology (In Press)

Brown S, Morgan F, Watras J, Loew L (2008) Analysis of phosphatidylinositol-4,5-bisphosphate signaling in cerebellar Purkinje spines. Biophys J 95:1795–1812

Brown SA, Moraru II, Schaff J, Loew LM (2011) Virtual NEURON: a strategy for merged biochemical and electrophysiological modeling. J Comput Neurosci 31(2):385–400

Cheron G, Sausbier M, Sausbier U, Neuhuber W, Ruth P, Dan B, Servais L (2009) BK channels control cerebellar Purkinje and Golgi cell rhythmicity in vivo. PLoS One 4:e7991

Chung H, Xia J, Scannevin R, Zhang X, Huganir R (2000) Phosphorylation of the AMPA receptor subunit GluR2 differentially regulates its interaction with PDZ domain-containing proteins. J Neurosci 20:7258–7267

Chung H, Steinberg J, Huganir R, Linden D (2003) Requirement of AMPA receptor GluR2 phosphorylation for cerebellar long-term depression. Science 300:1751–1755

De Schutter E (2008) Why are computational neuroscience and systems biology so separate? PLoS Comput Biol 4:e1000078

De Smedt H, Missiaen L, Parys J, Henning R, Sienaert I, Vanlingen S, Gijsens A, Himpens B, Casteels R (1997) Isoform diversity of the inositol trisphosphate receptor in cell types of mouse origin. Biochem J 322(Pt 2):575–583

DeSouza N, Reiken S, Ondrias K, Yang Y, Matkovich S, Marks A (2002) Protein kinase a and two phosphatases are components of the inositol 1,4,5-trisphosphate receptor macromolecular signaling complex. J Biol Chem 277:39397–39400

Ferris C, Huganir R, Bredt D, Cameron A, Snyder S (1991) Inositol trisphosphate receptor: phosphorylation by protein kinase C and calcium calmodulin-dependent protein kinases in reconstituted lipid vesicles. Proc Natl Acad Sci U S A 88:2232–2235

Fiala J, Grossberg S, Bullock D (1996) Metabotropic glutamate receptor activation in cerebellar Purkinje cells as substrate for adaptive timing of the classically conditioned eye-blink response. J Neurosci 16:3760–3774

Finch E, Augustine G (1998) Local calcium signalling by inositol-1,4,5-trisphosphate in Purkinje cell dendrites. Nature 396:753–756

Fink C, Slepchenko B, Moraru I, Watras J, Schaff J, Loew L (2000) An image-based model of calcium waves in differentiated neuroblastoma cells. Biophys J 79:163–183

Fujiwara A, Hirose K, Yamazawa T, Iino M (2001) Reduced IP3 sensitivity of IP3 receptor in Purkinje neurons. Neuroreport 12:2647–2651

Goetz CG (2003) Textbook of clinical neurology. WB Saunders, St. Louis

Guida S, Trettel F, Pagnutti S, Mantuano E, Tottene A, Veneziano L, Fellin T, Spadaro M, Stauderman K, Williams M, Volsen S, Ophoff R, Frants R, Jodice C, Frontali M, Pietrobon D (2001) Complete loss of P/Q calcium channel activity caused by a CACNA1A missense mutation carried by patients with episodic ataxia type 2. Am J Hum Genet 68:759–764

Hall S, Armstrong D (2000) Conditional and unconditional inhibition of calcium-activated potassium channels by reversible protein phosphorylation. J Biol Chem 275:3749–3754

Harris K, Stevens J (1988) Dendritic spines of rat cerebellar Purkinje cells: serial electron microscopy with reference to their biophysical characteristics. J Neurosci 8:4455–4469

Harris K, Stevens J (2006) Anomalous diffusion in Purkinje cell dendrites caused by spines. Neuron 8:635–648

Hernjak N, Slepchenko B, Fernald K, Fink C, Fortin D, Moraru I, Watras J, Loew L (2005) Modeling and analysis of calcium signaling events leading to long-term depression in cerebellar Purkinje cells. Biophys J 89:3790–3806

Inoue T, Kato K, Kohda K, Mikoshiba K (1998) Type 1 inositol 1,4,5-trisphosphate receptor is required for induction of long-term depression in cerebellar Purkinje neurons. J Neurosci 18:5366–5373

Ito M (2001) Cerebellar long-term depression: characterization, signal transduction, and functional roles. Physiol Rev 81:1143–1195

Ito M, Kano M (1982) Long-lasting depression of parallel fiber-Purkinje cell transmission induced by conjunctive stimulation of parallel fibers and climbing fibers in the cerebellar cortex. Neurosci Lett 33:253–258

Ito M, Sakurai M, Tongroach P (1982) Climbing fibre induced depression of both mossy fibre responsiveness and glutamate sensitivity of cerebellar Purkinje cells. J Physiol 324:113–134

Iwaki A, Kawano Y, Miura S, Shibata H, Matsuse D, Li W, Furuya H, Ohyagi Y, Taniwaki T, Kira J, Fukumaki Y (2008) Heterozygous deletion of ITPR1, but not SUMF1, in spinocerebellar ataxia type 16. J Med Genet 45:32–35

Khodakhah K, Ogden D (1993) Functional heterogeneity of calcium release by inositol trisphosphate in single Purkinje neurones, cultured cerebellar astrocytes, and peripheral tissues. Proc Natl Acad Sci U S A 90:4976–4980

Kim D, Jun K, Lee S, Kang N, Min D, Kim Y, Ryu S, Suh P, Shin H (1997) Phospholipase C isozymes selectively couple to specific neurotransmitter receptors. Nature 389:290–293

Le Novère N (2007) The long journey to a Systems Biology of neuronal function. BMC Syst Biol 1:28

Matsumoto M, Nakagawa T, Inoue T, Nagata E, Tanaka K, Takano H, Minowa O, Kuno J, Sakakibara S, Yamada M, Yoneshima H, Miyawaki A, Fukuuchi Y, Furuichi T, Okano H, Mikoshiba K, Noda T (1996) Ataxia and epileptic seizures in mice lacking type 1 inositol 1,4,5-trisphosphate receptor. Nature 379:168–171

Moraru II, Schaff JC, Slepchenko BM, Blinov ML, Morgan F, Lakshminarayana A, Gao F, Li Y, Loew LM (2008) Virtual cell modelling and simulation software environment. IET Syst Biol 2:352–362

Neuroscience. Sinauer Associates, Inc

Nishizuka Y (1984) The role of protein kinase C in cell surface signal transduction and tumour promotion. Nature 308:693–698

Nishizuka Y (1988) The role of protein kinase C in cell surface signal transduction and tumour promotion. Nature 8:693–698

Palmer L, Stuart G (2009) Membrane potential changes in dendritic spines during action potentials and synaptic input. J Neurosci 29:6897–6903

Santamaria F, Wils S, De Schutter E, Augustine G (2006) Anomalous diffusion in Purkinje cell dendrites caused by spines. Neuron 52:635–648

Santamaria F, Watras J, Loew L (2008) Anomalous diffusion in Purkinje spines. Neuron 95: 635–648

Sarkisov D, Wang S (2008) Order-dependent coincidence detection in cerebellar Purkinje neurons at the inositol trisphosphate receptor. J Neurosci 28:133–142

Sausbier M, Hu H, Arntz C, Feil S, Kamm S, Adelsberger H, Sausbier U, Sailer C, Feil R, Hofmann F, Korth M, Shipston M, Knaus H, Wolfer D, Pedroarena C, Storm J, Ruth P (2004) Cerebellar ataxia and Purkinje cell dysfunction caused by Ca2+ −activated K + channel deficiency. Proc Natl Acad Sci U S A 101:9474–9478

Schaff J, Fink CC, Slepchenko B, Carson JH, Loew LM (1997) A general computational framework for modeling cellular structure and function. Biophys J 73:1135–1146

Schorge S, van de Leemput J, Singleton A, Houlden H, Hardy J (2010) Human ataxias: a genetic dissection of inositol triphosphate receptor (ITPR1)-dependent signaling. Trends Neurosci 33(5):211–219

Schwaller B, Meyer M, Schiffmann S (2002) 'New' functions for ''old' proteins: the role of the calcium-binding proteins calbindin D-28 k, calretinin and parvalbumin, in cerebellar physiology. Studies with knockout mice. Cerebellum 1:241–258

Segev I, Friedman A, White E, Gutnick M (1995) Electrical consequences of spine dimensions in a model of a cortical spiny stellate cell completely reconstructed from serial thin sections. J Comput Neurosci 2:117–130

Slepchenko BM, Loew LM (2010) Use of virtual cell in studies of cellular dynamics. Int Rev Cell Mol Biol 283:1–56

Takechi H, Eilers J, Konnerth A (1998) A new class of synaptic response involving calcium release in dendritic spines. Nature 396:757–760

Turner D, Schwartzkroin P (1983) Electrical characteristics of dendrites and dendritic spines in intracellularly stained CA3 and dentate hippocampal neurons. J Neurosci 3:2381–2394

Walter J, Alviña K, Womack M, Chevez C, Khodakhah K (2006) Decreases in the precision of Purkinje cell pacemaking cause cerebellar dysfunction and ataxia. Nat Neurosci 9:389–397

Wang S, Denk W, Häusser M (2000) Coincidence detection in single dendritic spines mediated by calcium release. Nat Neurosci 3:1266–1273

Widmer H, Rowe I, Shipston M (2003) Conditional protein phosphorylation regulates BK channel activity in rat cerebellar Purkinje neurons. J Physiol 552:379–391

Xu C, Watras J, Loew L (2003) Kinetic analysis of receptor-activated phosphoinositide turnover. J Cell Biol 161:779–791

Yue Q, Jen J, Nelson S, Baloh R (1997) Progressive ataxia due to a missense mutation in a calcium-channel gene. Am J Hum Genet 61:1078–1087

Zhou W, Yan P, Wuskell J, Loew L, Antic S (2008) Dynamics of action potential backpropagation in basal dendrites of prefrontal cortical pyramidal neurons. Eur J Neurosci 27:923–936

Zhuchenko O, Bailey J, Bonnen P, Ashizawa T, Stockton D, Amos C, Dobyns W, Subramony S, Zoghbi H, Lee C (1997) Autosomal dominant cerebellar ataxia (SCA6) associated with small polyglutamine expansions in the alpha 1A-voltage-dependent calcium channel. Nat Genet 15:62–69

Chapter 6
The Performance (and Limits) of Simple Neuron Models: Generalizations of the Leaky Integrate-and-Fire Model

Richard Naud and Wulfram Gerstner

Abstract The study of neuronal populations with regards to coding, computation and learning relies on its primary building bloc: the single neuron. Describing the activity of single neurons can be done by mathematical models of various complexity. In this chapter we start with the integrate-and-fire model, and then consider a set of enhancements so as to approach the behaviour of multiple types of real neurons.

The study of neuronal populations with regards to coding, computation and learning relies on its primary building bloc: the single neuron. Describing the activity of single neurons can be done by mathematical models of various complexity. On one hand there are complex biophysical models and on the other hand there are the simpler integrate-and-fire models. In order to relate with higher functionalities such as computation or coding, it is not necessary to model all the spatio-temporal details of ionic flow and protein interactions. There is a level of description that is simple, that bridges the gap to higher functionalities, and that is sufficiently complete to match real neurons. In this chapter we start with the integrate-and-fire model and then consider a set of enhancements so as to model the behaviour of multiple types of neurons.

The formalism considered here takes a stimulating current $I(t)$ as an input, to which the neuron responds with a voltage trace $V(t)$, which contain multiple spikes. The input current can be injected experimentally in vitro. In a living brain, the input comes from synapses, gap junctions or voltage-dependent ion channels of the neuron's membrane.

R. Naud (✉) • W. Gerstner
Ecole Polytechnique Fédérale de Lausanne, Lausanne, 1015-EPFL, Switzerland
e-mail: richard.naud@epfl.ch; wulfram.gerstner@epfl.ch

N. Le Novère (ed.), *Computational Systems Neurobiology*,
DOI 10.1007/978-94-007-3858-4_6,

6.1 Basic Threshold Models

Nerve cells communicate by action potentials – also called spikes. Each neuron gathers input from thousands of synapses to decide when to produce a spike. In the absence of input, the neurons would stay at a resting membrane potential around −70 mV. Most neurons will fire a spike when their membrane potential reaches a value around −55 to −50 mV. Action potential firing can be considered an all-or-nothing event. These action potentials are very stereotypical. This suggests that spikes can be replaced by unitary events that are generated by a threshold-crossing process.

6.1.1 The Perfect Integrate-and-Fire Model

Fundamentally, we can say that a single neuron accumulates electric charge supplied by an input current $I(t)$ on its membrane. This translates into an by increase of the membrane potential $V(t)$. When the voltage hits the threshold (V_T), a spike is said to be emitted and the voltage is reset to V_r. Mathematically we write:

$$\frac{dV}{dt} = \frac{1}{C} I(t) \tag{6.1}$$

$$\text{when } V(t) > V_T \text{ then } V(t) \to V_r. \tag{6.2}$$

Here, C is the total capacitance of the membrane. This equation is the first Kirchoff law for an impermeable membrane: the current injected can only load the capacitance. The greater the capacitance the greater the amount of current required to increase the potential by a given amount.

This system is called the (perfect or non-leaky) integrate-and-fire (IF) model. Solving the first-order differential equation shows the integration process explicitly; after a previous spike at t_0 the voltage at time t is given by:

$$V(t) = V_r + \frac{1}{C} \int_{t_0}^{t} I(s)ds. \tag{6.3}$$

Here a pulse of current will never be forgotten. In other words one could inject a small pulse of current every hour and their repercussions on the voltage will cumulate to eventually make the model neuron fire. This conflicts with the behaviour of real neurons, which are not perfect integrators but "leaky" ones. In fact, the IF model is used here for didactic purposes only because it summarizes the central idea: integrate and fire.

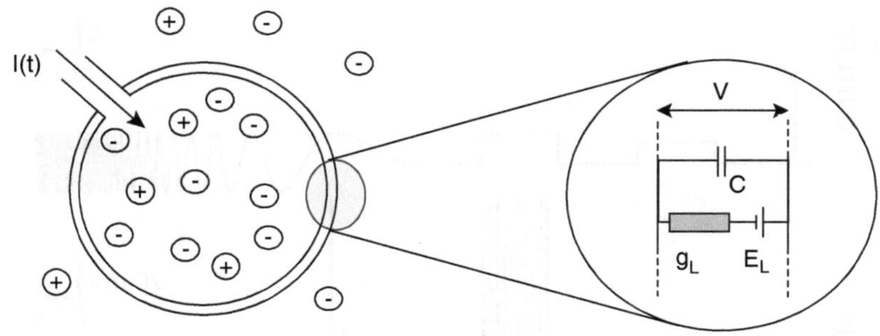

Fig. 6.1 Positively and negatively charged ions are distributed inside and outside the cell. Current ($I(t)$) entering the cell will modify the difference in electric potential between the exterior and the interior (V). The dynamics of the LIF correspond to a RC-circuit composed of a conductance (g_L) in parallel with a capacitance (C). The electric supply corresponds to the resting potential (E_L)

6.1.2 The Leaky Integrate-and-Fire Model

Real neuronal membranes are leaky. Ions can diffuse through the neuronal membrane. The membrane of neurons can be seen as providing a limited electrical conductance (g_L) for charges crossing the cellular membrane (Fig. 6.1). The difference in electric potential at equilibrium depends on the local concentration of ions and is often called the equilibrium (or resting) potential (E_0). This additional feature leads to the more realistic Leaky-Integrate-and-Fire (LIF) model (Lapicque 1907):

$$C\frac{dV}{dt} = -g_L(V - E_0) + I(t) \tag{6.4}$$

$$\text{when } V(t) > V_T \text{ then } V(t) \to V_r. \tag{6.5}$$

Again, this is the Kirchoff law for conservation of charge. The current injected can either leak out or accumulate on the membrane. The effect of a short current pulse will cause a transient increase in voltage. This can be seen by looking at the solution of the linear differential equation given a previous spike at time $\hat{t}_0$:

$$V(t) = E_0 + \eta_r(t - \hat{t}_0) + \int_0^{t-\hat{t}_0} \kappa(s) I(t-s) ds \tag{6.6}$$

$$\eta_r = (V_r - E_0)e^{-t/\tau}\Theta(t) \tag{6.7}$$

$$\kappa(t) = \frac{1}{C}e^{-t/\tau}\Theta(t) \tag{6.8}$$

where $\Theta(t)$ is the Heaviside function and $\tau = C/g_L$ is the membrane time constant. In Eq. 6.6, three terms the voltage. The first term is the equilibrium potential. The second term is the effect of voltage reset which acts as an initial condition

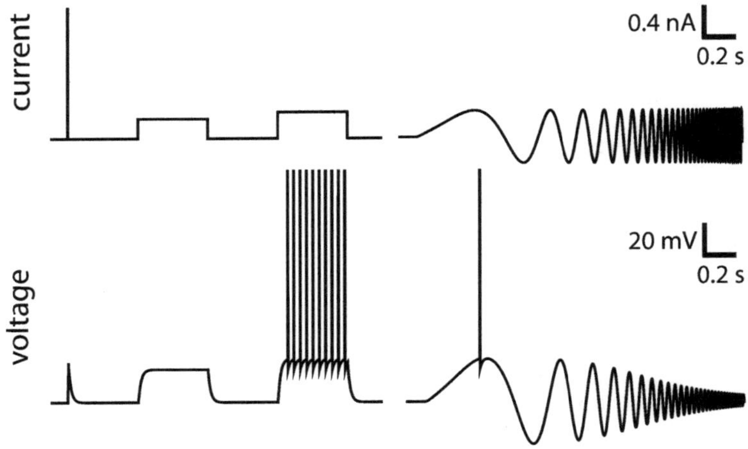

Fig. 6.2 The defining responses of the LIF model. A short but strong pulse will make a marked increased in potential which will then decay exponentially. A subthreshold step of current leads to exponential relaxation to the steady-state voltage, and to an exponential relaxation back to resting potential after the end of the step. A supra-threshold step of current leads to tonic firing. If a sinusoidal wave of increasing frequency is injected in the model, only the *lowest* frequencies will respond largely. The *higher* frequencies will be attenuated

for the integration of the differential equation and gives rise to the function η_r. Note that far away from the last spike ($\hat{t}_0$) this term vanishes. The last term – made of the convolution of the filter $\kappa(t)$ with the current – is the influence of input current on the voltage. We see that the voltage is integrating the current but the current at an earlier time has a smaller effect than current at later times. The membrane time constant of real neurons can vary between 10 and 50 ms. The theory of signal processing tells us that the membrane acts as a low-pass filter of the current (Fig. 6.2). In fact, input current fluctuating slowly is more efficient at driving the voltage than current fluctuating very rapidly.

There is another way to implement the reset. Instead resetting to fixed value, we assume that whenever the voltage equals V_T, there is a sudden decrease in voltage to V_r caused by a short negative pulse of current, which reflects the fact that the membrane loses charge when the neuron sends out a spike. We write the LIF equations differently, using the Dirac delta function ($\delta(t)$):

$$C\frac{dV}{dt} = -g_L(V - E_0) + I(t) - C(V_T - V_r)\sum_i \delta(t - \hat{t}_i) \tag{6.9}$$

where the sum runs on all spike times $\hat{t}_i \in \{\hat{t}\}$, defined as the times where $V(t) = V_T$. The integrated form is now:

$$V(t) = E_0 + \sum_i \eta_a(t - \hat{t}_i) + \int_0^\infty \kappa(s) I(t-s) ds \tag{6.10}$$

$$\eta_a(t) = -(V_T - V_r) e^{-t/\tau} \Theta(t). \tag{6.11}$$

The two different ways to implement the reset yield slightly different formalisms. Even though including the leak made the mathematical neuron model more realistic, it is not sufficient to describe experiments with real neurons. We also need a process to account for adaptation (Benda and Herz 2003; Rauch et al. 2003; Jolivet et al. 2006b), and this is the topic of the next section.

6.2 Refractoriness and Adaptation

Refractoriness prevents a second spike immediately after a first one was emitted. One can distinguish between an absolute and a relative refractory period. During the absolute refractory period, no spike can be emitted, no matter the strength of the stimulus. The duration of the spike is often taken as the absolute refractory period since it is impossible to emit a spike while one is being generated. During the relative refractory period it is possible to fire a spike, but a stronger stimulus is required. In this case the current required depends on the time since the last spike. Manifestly, the absolute refractory period always precedes the relative refractory period, and the absolute refractory period can be seen as a very strong relative refractory period.

Spike-frequency adaptation, on the other hand, is the phenomenon whereby a constant stimulus gives rise to a firing frequency that slows down in time (Fig. 6.3). Here it is the cumulative effect of previous spikes that prevents further spiking. In other words: the more a neuron has fired in the recent past, the less it is likely to spike again. How long can this history dependence be? Multiple studies have pointed out that spikes emitted one or even ten seconds earlier can still reduce the instantaneous firing rate (La Camera et al. 2004; Lundstrom et al. 2008).

Refractoriness and adaptation are two similar but distinct effects, and we need to define the difference in precise terms. Although refractoriness mostly affects the earliest times after a spike and adaptation the latest times, this distinction is not adequate: spike-triggered currents can cumulate even at small time scales. It is more convenient to distinguish the two processes based on the history-dependence whereby refractoriness prevents further spiking as a function of time since the last spike only, while adaptation implies a dependency on the time of all the previous spikes. In other words, adaptation is a refractoriness that cumulates over the spikes. Equivalently, refractoriness can be distinguished from adaptation by the type of reset: a fixed reset like in Eq. 6.5 leads to dependency on the previous spike only (see Eq. 6.7) and hence to no adaptation. A relative reset allows the effect of multiple spikes to cumulate and can lead to spike-frequency adaptation.

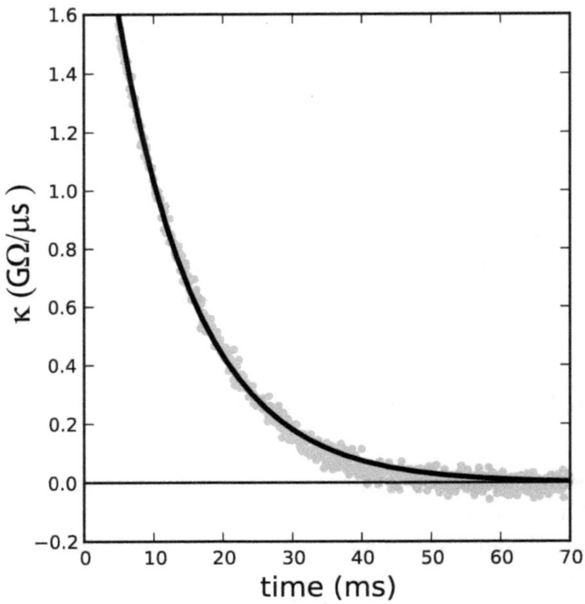

Fig. 6.3 Comparing two simple mechanism of spike-triggered adaptation. A step current (*top*) is injected in a model with spike-triggered hyperpolarizing current (*left*) or in a model with moving threshold (*right*). Both models can produce the same firing patterns, but the voltage trace differs qualitatively

In terms that are compatible with threshold models, both refractoriness and adaptation can be mediated either by a transient increase of the threshold after each spike or by the transient increase of a hyperpolarizing current. These will be discussed next.

6.2.1 Spike-Triggered Current

Some ion channels seem to be essential to the action potential but influence very weakly the subthreshold dynamics. Some other ion channels play no role in defining the shape of the spike, are partially activated during a spike, and their level of activation decays exponentially after the spike. These are ion channels that can mediate adaptation or refractoriness. Such ion channels have voltage-dependent sensitivity to the membrane potential: at high voltage they rapidly open, at low voltage they slowly close. Since a spike is a short period of high voltage this creates a short jump in the level of activation which will slowly decay after the spike. Such a situation can be implemented in the LIF by adding an hyperpolarizing current w which is incremented by b each time there is a spike and otherwise decays towards zero with a time constant τ_w (Baldissera et al. 1976; Izhikevich 2004; Benda and Herz 2003):

$$C\frac{dV}{dt} = -g_L(V - E_0) - w + I(t) - C(V_T - V_r)\sum_i \delta(t - \hat{t}_i) \quad (6.12)$$

$$\tau_w \frac{dw}{dt} = -w + b\tau_w \sum_i \delta(t - \hat{t}_i). \tag{6.13}$$

The above system of equations is a simple mathematical model for a neuron with spike-frequency adaptation. The current w is triggered by the spikes and will move the membrane potential away from the threshold when $b < 0$. This equation can be integrated to yield:

$$V(t) = E_0 + \sum_i \eta_a(t - \hat{t}_i) + \int_0^\infty \kappa(s) I(t - s) ds \tag{6.14}$$

$$\eta_a(t) = \frac{b\tau\tau_w}{C(\tau_w - \tau)} \left[e^{-t/\tau_w} - e^{-t/\tau}\right] \Theta(t) - (V_T - V_r) e^{-t/\tau} \Theta(t) \tag{6.15}$$

where $\kappa(t)$ is the same as in Eq. 6.9. The spike-triggered current that cumulates over the spikes is reflected in a stereotypical change in voltage η_a that can also cumulate over the spikes. Such a spike-triggered current can also make refractoriness if we replace its cumulative reset by a fixed reset :

$$\tau_w \frac{dw}{dt} = -w + (b - w)\tau_w \sum_i \delta(t - \hat{t}_i) \tag{6.16}$$

so that at each time instead of incrementing *by* b, we increment *to* b. In this case the amount w of refractory current depends only on the time since the last spike. Integration of this yields:

$$\eta_r(t) = \frac{b\tau\tau_w}{C(\tau_w - \tau)} \left[e^{-t/\tau_w} - e^{-t/\tau}\right] \Theta(t) \tag{6.17}$$

The shape of the spike after potential can be mediated by a handful of ion channels. Likely candidates for mediating a spike triggered current of the type described above must have a slow to medium activation at supra-threshold potentials and a very slow inactivation or de-activation at subthreshold potentials. An action potential will then induce a small increase in the number of open channels which could cumulate over the spikes. The time constant of the hyperpolarizing current τ_w relates to the time constant for the closure of the ion channels at subthreshold potentials. Typical examples are: slow potassium current I_K with de-activation time constant around 30–40 ms (Korngreen and Sakmann 2000), muscarinic potassium current, I_M, with de-activation time constant around 30–100 ms (Passmore et al. 2003) or the calcium-dependent potassium current $I_{K[Ca]}$ which can have a time constant in the order of multiple seconds (Schwindt et al. 1989). Finally, active dendritic processes can also induce current to flow back into the somatic compartment after each spike (Doiron et al. 2007). In this case the current is depolarizing rather than hyperpolarizing, leading to facilitation rather than adaptation.

6.2.2 Moving Threshold

Spike-triggered currents are not the only way to implement refractoriness and adaptation in neuron models. Multiple experiments have shown that the effective threshold of neurons is dynamic (Hill 1936; Fuortes and Mantegazzini 1962; Azouz and Gray 2000, 2003; Badel et al. 2008b). If instead of adding a spike-triggered current we let the threshold V_T be dynamic: the threshold can increase by δV_T each time there is a spike and decay exponentially with time constant τ_T to the minimum threshold $V_T^{(0)}$. This is summarized by the supplementary equations:

$$\tau_T \frac{dV_T}{dt} = -(V_T - V_T^{(0)}) \tag{6.18}$$

$$\text{when } V(t) \geq V_T(t) \text{ then } V(t) \rightarrow V_r \tag{6.19}$$

$$\text{and } V_T(t) \rightarrow V_T(t) + \delta V_T. \tag{6.20}$$

Again, the moving threshold can implement adaptation (as with Eqs. 6.18–6.20 above) or refractoriness if we replace the relative reset by a fixed reset: $V_T(t) \rightarrow V_T^{(0)} + \delta V_T$.

It is often possible to find parameters for which a model with a moving threshold will yield the same spikes times than a model with a spike-triggered current. Indeed moving the membrane potential away from the threshold with a spike-triggered current is equivalent to moving the threshold away from the membrane potential. In particular, when only the spike times are predicted, we can put Eqs. 6.18–6.20 in the form of Eq. 6.14 by keeping a fixed threshold and adding to $\eta_a(t)$:

$$\delta V_T e^{-t/\tau_T} \Theta(t). \tag{6.21}$$

It is not yet clear which biophysical mechanisms are responsible for moving thresholds. One likely candidate is the sodium channel inactivation (Azouz and Gray 2000). An increase in sodium channels inactivation can increase the voltage threshold for spike initiation. Inactivated sodium channels de-inactivate with a time constant of 2–6 ms (Huguenard et al. 1988). Furthermore, it is believed that sodium channels can de-inactivate on time scales as long as multiple seconds (Baker and Bostock 1998).

6.3 Linearized Subthreshold Currents

There are ion channels influencing principally the shape of the spike, some the refractoriness, and others the adaptation of neurons. However, there are also channels whose dynamics depends and influences only the subthreshold potentials. An example is the hyperpolarization activated cation current I_h which start to

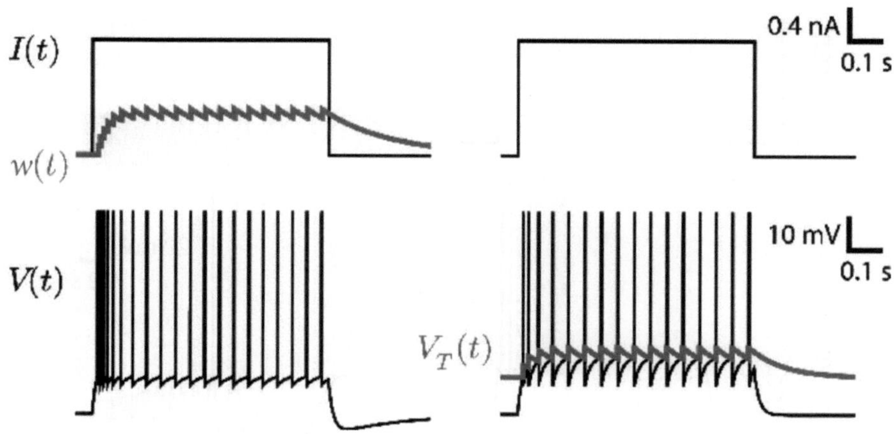

Fig. 6.4 The filter κ can be measured in real neurons (Jolivet et al. 2006a). Here the shape of the filter is shown as measured in the soma of a pyramidal neuron of the layer 5 of the cortex (*gray circles*). Data points before 5 ms are not shown because they bear a heavy artefact due to the electrode used for recording. In *black* is a fit of Eq. 6.8 with $C = 408$ pF and $\tau = 11.4$ ms

activate around −70 mV. The first-order effect of such currents can be added to the LIF equations (see Exercices 6.3 or Mauro et al. 1970 for details on the Taylor expansion):

$$C\frac{dV}{dt} = -g_L(V - E_0) - w + I(t) \tag{6.22}$$

$$\tau_w\frac{dw}{dt} = a(V - E_w) - w \tag{6.23}$$

$$\text{if } V(t) > V_T \text{ then } V(t) \to V_r. \tag{6.24}$$

Here a regulates the magnitude of the subthreshold current and τ_w rules the time constant of the coupling with the voltage. E_w should correspond to the average voltage, we will assume that $E_w = E_0$ for the following treatments. When a is negative, w is said to cause subthreshold facilitation. The response properties will resemble the LIF as (in Figs. 6.2 and 6.4) but with a longer impulse-response function. When a is positive, w is said to generate subthreshold adaptation. For a sufficiently strong positive a we see the emergence of resonance as shown in Fig. 6.5. This model can be called the resonate and fire (RF; Izhikevich 2001; Richardson et al. 2003).

This system of equations can be mapped to the equations of a damped oscillator with a driving force. It is a well-studied system that comes from the dynamics of a mass hanging on a spring in a viscous medium. We know that this system has three dynamical regimes:

- $4C\tau_w(g_L + a) < (g_L\tau_w + C)^2$ overdamped,

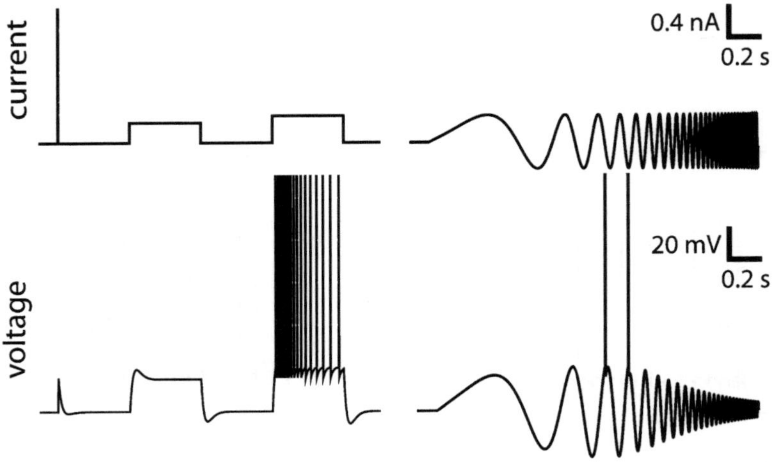

Fig. 6.5 The responses of the LIF with a single linearized current. A short but strong pulse will generate a jump in potential which will then relaxes with an undershoot. A subthreshold step of current leads to a characteristic overshoot after the onset and an undershoot after the offset of the current. A supra-threshold step of current leads to firing with spike-frequency adaptation. If a sinusoidal wave of increasing frequency is injected in the model, the lowest frequencies and the highest frequencies will be attenuated. In-between frequencies will yield the greatest response

- $4C\tau_w(g_L + a) = (g_L\tau_w + C)^2$ critically damped,
- $4C\tau_w(g_L + a) > (g_L\tau_w + C)^2$ underdamped.

Overdamped and critically damped systems have no resonating frequency. It is only when the system is underdamped that resonance appears. Such resonance is seen in multiple types of neurons, typically in some cortical interneurons (Markram et al. 2004), in mesencephalic V neurons (Wu et al. 2005) and in the apical dendrites of pyramidal neurons (Cook et al. 2007).

What are the main characteristics of a resonating membrane? Contrasting with the standard LIF, the response to a current pulse in Figs. 6.5 and 6.6 is not a single exponential. Instead, the voltage makes a short undershoot before relaxing to the resting potential. Similarly, when the input is a step current, there is a substantial overshoot at the onset and undershoot after the offset of the current. The neuron model resonates around a characteristic frequency for which it will respond with maximal amplitude:

$$\omega = \sqrt{\frac{g_L + a}{C\tau_w} - \frac{(g_L\tau_w + C)^2}{4C^2\tau_w^2}}. \tag{6.25}$$

Resonating membranes are bandpass filters as we can see from the response to a sinusoidal wave of increasing frequency (Fig. 6.5). The shape of the filter can be written as:

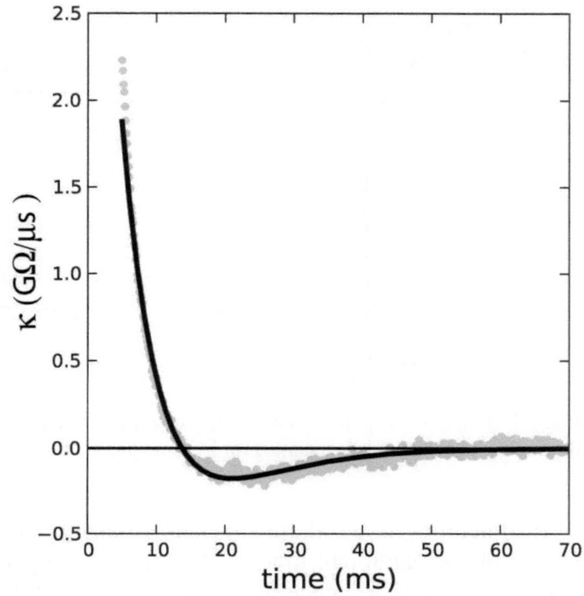

Fig. 6.6 The shape of the filter κ in the presence of resonance. Here the shape of the filter is shown as measured in the apical pyramidal neuron of the layer 5 of the cortex (*gray points*). Data points before 5 ms are not shown because they bear a heavy artefact from due to the measurement electrode. In *black* is shown a fit of Eqs. 6.22–6.23 ($a = 13.6$ nS, $g_L = 35.0$ nS , $C = 168$ pF, $\tau_w = 15.5$ ms) (Data a courtesy of Brice Bathellier)

$$\kappa(t) = \exp(-t/\tau_\omega)\left[\frac{1}{C}\cos\omega t + \frac{\tau_\omega \tau_w + 1}{g_L + a}\sin\omega t\right]\Theta(t) \qquad (6.26)$$

where the decay time constant is:

$$\tau_\omega = \frac{2\tau_w \tau}{\tau_w + \tau}. \qquad (6.27)$$

Figure 6.6 shows the shape of the filter, as measured in a neuron with resonance.

6.4 Nonlinear Integrate-and-Fire Models

Do neurons really have a voltage threshold? Imagine that a neuron was to receive a stimulus that brings its membrane potential to a value which triggers the spike but then the stimulus is stopped. The membrane potential would continue to increase even in the absence of stimulus, and produce the action potential. Can we say that a spike is produced whenever the membrane potential reaches this threshold voltage? No. At the earliest times of the action potential, a negative current can veto the spike even though the membrane potential was above the threshold for spike initiation. Another example that makes the conceptualization of a threshold dubious is shown in Fig. 6.7. Here the voltage threshold measured from a current pulse is significantly different from the voltage threshold measured with a current step.

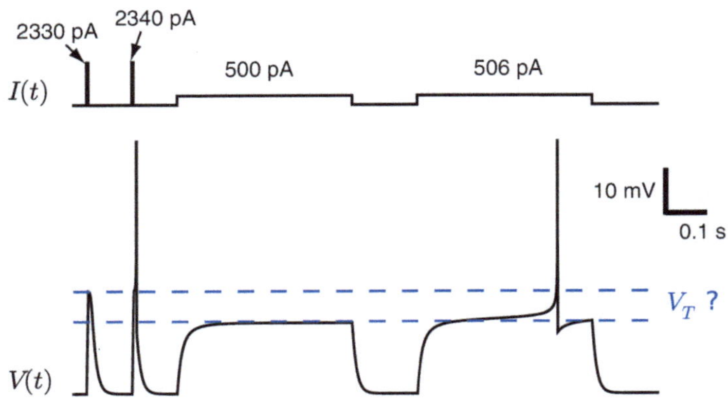

Fig. 6.7 Where is the voltage threshold? Current pulses and current steps show different voltage thresholds that can not be accounted for by a moving threshold

Real neurons have sodium ion channels that gradually open as a function of the membrane potential and time. If the channel is open, positive sodium ions flow into the cell, which increases the membrane potential even further. This positive feedback is responsible for the upswing of the action potential. Although this strong positive feedback is hard to stop, it can be stopped by a sufficiently strong hyperpolarizing current, thus allowing the membrane potential to increase above the activation threshold of the sodium current.

Sodium ion channels responsible for the upswing of the action potential react very fast. So fast that the time it takes to reach their voltage-dependent level of activation is negligible. Therefore these channels can be seen as currents with a magnitude depending nonlinearly on the membrane potential, which suggests that a simplified description of spike generation should be possible. This section explores the LIF augmented with a nonlinear term for smooth spike initiation.

6.4.1 The Exponential Integrate-and-Fire Model

Let us assume that the transmembrane current is some function of V, so that the membrane dynamics is an equation of the type:

$$C\frac{dV}{dt} = F(V) + I(t), \tag{6.28}$$

where $F(V)$ is the current flowing through the membrane. For the perfect IF it is zero ($F(V) = 0$), for the LIF it is linear with a negative slope ($F(V) = -g_L(V - E_0)$). We can speculate on the shape of the non-linearity. The simplest non-linearity would arguably be the quadratic : $F(V) = -g_L(V - E_0)(V - V_T)$

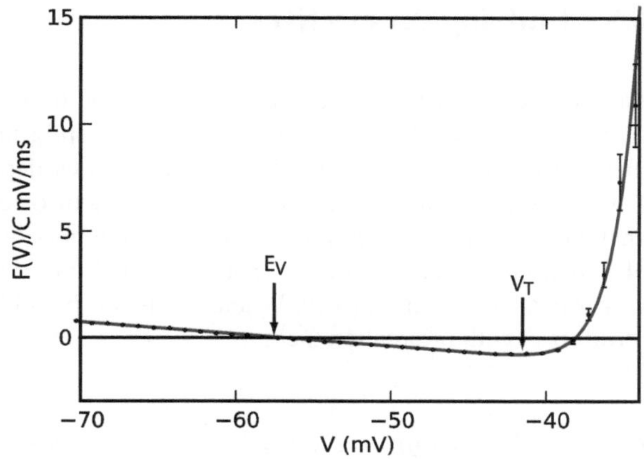

Fig. 6.8 Experimental measurement of the nonlinearity for spike initiation. Example of the measured $F(V)$ in *black circles* for a layer 5 pyramidal neuron of the barrel cortex. The errorbars indicate one standard error of the mean and the *blue line* is a fit of Eq. 6.29 (Data is a courtesy of Badel et al. (2008b))

(Quadratic Integrate and Fire, QIF; Latham et al. 2000). However this implies that the dynamics at hyperpolarized potentials is non-linear, which conflicts with experimental observations. Other possible models could be made with cubic or even quartic functions of V (Touboul 2008). An equally simple non-linearity is the exponential function:

$$F(V) = -g_L(V - E_0) + g_L \Delta_T \exp\left(\frac{V - V_T}{\Delta_T}\right) \qquad (6.29)$$

where Δ_T is called the slope factor that regulates the sharpness of the spike initiation. The first term on the right hand side of Eq. 6.29 is identical to that of the LIF. The second one accounts for nonlinearity in spike initiation.

The Exponential Integrate-and-Fire (EIF; Fourcaud-Trocme et al. 2003) model integrates the current according to Eqs. 6.28 and 6.29 and resets the dynamics to V_r (i.e. produces a spike) once the simulated potential reaches a value θ. The exact value of θ does not matter, as long as $\theta >> V_T + \Delta_T$. As in the LIF, we have to reset the dynamics once we have detected a spike. The value at which we stop the numerical integration should not be confused with the threshold for spike initiation. We reset the dynamics once we are sure the spike *has been* initiated. This can be at a membrane potential of 0, 10 mV or infinity. In fact, because of the exponential nonlinearity V goes from $V_T + \Delta_T$ to infinity in a negligible amount of time.

Which $F(V)$ is the best? This can be measured experimentally (provided that we can estimate membrane capacitance beforehand Badel et al. 2008a,b). Figure 6.8 shows the function $F(V)$ as measured in pyramidal neurons of the cortex. Choosing $F(V)$ as a linear plus exponential allow a good fit to the experimental data. Similar curves are observed in neocortex interneurons (Badel et al. 2008a).

6.5 Unifying Perspectives

We have seen that the LIF can be augmented with mechanisms for refractoriness, adaptation, subthreshold resonances and smooth spike initiation. Models combining these features can be classified in two categories depending on the presence of a non-linearity. This is because the linear system of equations can be integrated analytically, whereas integration is generally not possible for the non-linear systems of equations. When integration can be carried out, the dynamics can be studied with signal processing theory. When this is not possible, the dynamical system is scrutinized with bifurcation theory.

6.5.1 *The Adaptive Exponential Integrate-and-Fire Model*

One obvious way to combine all the features is to add to the EIF model several linearized currents with cumulative spike-triggered adaptation:

$$C\frac{dV}{dt} = -g_L(V - E_0) + g_L\Delta_T \exp\left(\frac{V - V_T}{\Delta_T}\right) + I(t) - \sum_{i=1}^{N} w_i \quad (6.30)$$

$$\tau_i \frac{dw_i}{dt} = a_i(V_i - E_0) - w_i \quad (6.31)$$

$$\text{if } V(t) > V_T \text{ then } V(t) \rightarrow V_r \quad (6.32)$$

$$\text{and } w_i(t) \rightarrow w_i(t) + b_i \quad (6.33)$$

where each additional current w_i can be tuned by adapting its subthreshold coupling constant a_i and its spike-triggered jump size b_i. The simplest version of this framework assumes $N = 1$ and it is known as the Adaptive Exponential Integrate-and-Fire (AdEx; Brette and Gerstner 2005; Gerstner and Brette 2009). This model compares very well with many types of real neurons, as we will see in Sect. 6.7.

6.5.2 *Integrated Models*

For some neurons the spike initiation is sharp enough and can be neglected. In fact, if the slope factor $\Delta \rightarrow 0$ in Eq. 6.30, then the AdEx turns into a linear model with a sharp threshold. As we have seen in Sect. 6.2, the solution to the linear dynamical system can be cast in the form:

$$V(t) = E_0 + \int_0^{\infty} \kappa(s)I(t-s)ds + \sum_i \eta_a(t - \hat{t}_i) \quad (6.34)$$

where $\kappa(t)$ is the input filter and $\eta_a(t)$ is the shape of the spike with its cumulative tail. The sum runs on all the spike times $\{\hat{t}\}$ defined as the times where the

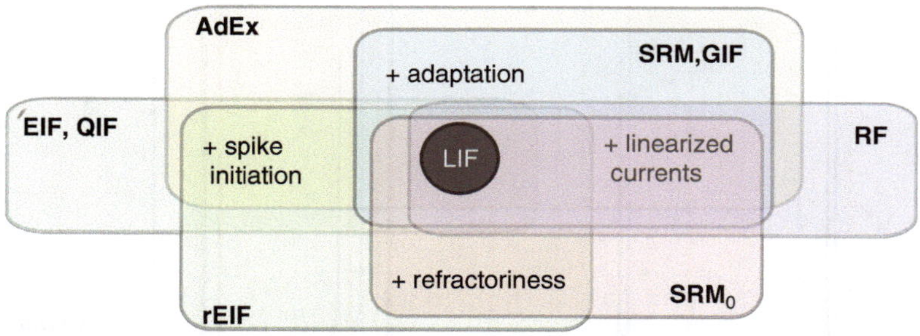

Fig. 6.9 Generalizations of the LIF include either refractoriness, adaptation, linearized currents, or smooth spike initiation. Various regroupments have various names. For instance, the refractory-Exponential-Integrate-and-Fire (rEIF) regroups refractoriness, smooth spike initiation and the features of a LIF (Badel et al. 2008b)

voltage crosses the threshold ($V > V_T$). To be consistent with the processes seen in the previous sections the functions $\kappa(t)$ and $\eta_a(t)$ must be sums of exponentials. However, for the sake of fitting a model to experimental data, any type of basis function can be used. For arbitrary shape of the kernels $\kappa(t)$ and $\eta_a(t)$, this model is known as the cumulative Spike Response Model (SRM, Gerstner et al. 1996; Gerstner 2008) or more recently as the Generalized Integrate-and-Fire (GIF, Paninski et al. 2004).

The simplified Spike Response Model (SRM_0; Gerstner 2008) is another related model worth pointing out. In the SRM_0 the sum in Eq. 6.34 extends to the last spike only. This makes a purely refractory model without spike-frequency adaptation.

Figure 6.9 summarizes the nomenclature for the combinations of generalizations. As we will see in the next section, the formalism of the SRM and SRM_0 models bridges the gap to a more general class of spiking models where the influence of noise is taken into account.

6.6 Noise

Variability is ubiquitous in the nervous system. In many ways this variability is seen as a feature rather than a defect (Faisal et al. 2008). It is important therefore not to ignore all the noise, but to listen carefully in order to understand what it is trying to communicate.

In view of the intrinsic probabilistic nature of the neurons, it is difficult to predict the exact spike times because a neuron will fire with some jitter around an average spike time (Fig. 6.10). Rather, the models of neuronal behaviour must predict the probability to emit a spike in a given time interval. The probability $p(t)$ of observing a spike in a given small interval of δt defines the instantaneous firing rate:

$$p(t) = r(t)\delta t. \tag{6.35}$$

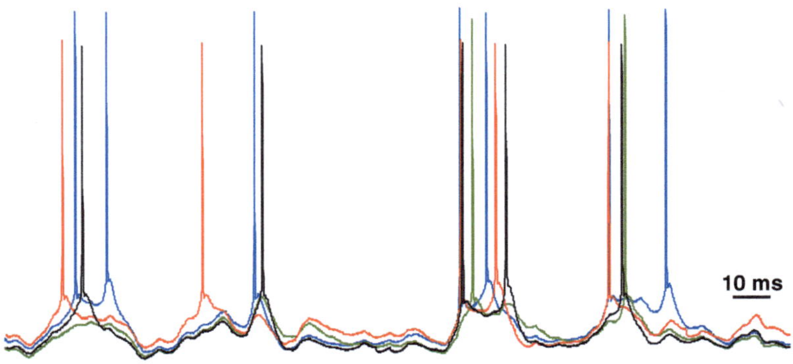

Fig. 6.10 Membrane potential recorded from four repetition of the same stimulus. Spikes are missed others are shifted and this variability is intrinsic to the neuron

Mathematically speaking, $r(t)$ is a stochastic intensity, a terminology borrowed from the theory of point-processes (Daley and Vere-Jones 1988). In the context of neuron models $r(t)$ is called the firing intensity or instantaneous firing rate. It is also related to the experimentalist concept of a Peri-Stimulus Time Histogram (PSTH).

Chapter 12 describes the implication of noise at the biophysical level. Here we describe how noise influences the level of description of the LIF model. Furthermore, we describe a simple framework through which models such as the SRM are related to models of the firing intensity $r(t)$.

6.6.1 Synaptic Noise

Synaptic noise deals with the input $I(t)$. In this chapter $I(t)$ refers to the current arriving in the region which is solely responsible for spike initiation. This signal can be seen as being noisy: because some synaptic events happened without a pre-synaptic spike, because the unreliability of axonal propagation prevented a pre-synaptic spike from getting to the synapse, or because the variability in vesicle release and receptor channel opening make the amplitude of the postsynaptic event variable. From yet another point of view, one may be interested in the current input coming from an identified subpopulation of pre-synaptic neurons. In this view the rest of the pre-synaptic neurons form a considerable background synaptic noise. In any case, the synaptic noise is added to a deterministic current:

$$I(t) \to I(t) + \xi(t, V) \tag{6.36}$$

where $\xi(t)$ is the synaptic noise. The dependance of the noise on V is added because synapses make changes in conductance which must be multiplied by V to yield the current. Despite that, the dependance on V can often be neglected and the synaptic noise is considered as a time-dependent current $\xi(t)$.

There are multiple stochastic models of synaptic noise. Maybe the simplest situation takes into account a random occurrence of synaptic events, $\hat{t}^{(pre)}$ each bringing in an exponentially decaying pulse of current:

$$\xi(t) = c \sum_{\hat{t}_i \in \{\hat{t}^{(pre)}\}} \exp\left(-\frac{t - \hat{t}_i}{\tau_s}\right) \theta(t - \hat{t}_i) \tag{6.37}$$

where τ_s is the time constant and c is the amplitude of the post-synaptic current decay. If the synaptic events occur randomly with frequency ν_e, the mean, the variance and the autocorrelation of the noise are (Stein 1965):

$$\mu \equiv \langle \xi \rangle = c\tau_s \nu_e \tag{6.38}$$

$$\sigma^2 \equiv \text{Var}[\xi] = \frac{c^2 \tau_s \nu_e}{2} \tag{6.39}$$

$$\langle [\xi(t) - \mu][\xi(t + s) - \mu] \rangle = \sigma^2 e^{-s/\tau_s}. \tag{6.40}$$

In the limit of small synaptic jumps ($c \to 0$) and large frequency of synaptic events ($\nu_e \to \infty$), the synaptic noise can be seen as a gaussian noise with exponential autocorrelation function. The dynamics of such a noise current is often written with the equation:

$$\xi(t + dt) = \xi(t) + \frac{(\mu - \xi(t))}{\tau_s} dt + \sigma G_t \sqrt{\frac{2dt}{\tau_s}} \tag{6.41}$$

where G_t is a random number taken from a standard normal distribution and dt is the step size. This is called an Ornstein-Uhlenbeck process, a similar equation rules the position of a particle undergoing Brownian motion in an attracting potential. Equation 6.41 is the diffusion approximation for synaptic inputs.

6.6.2 Electrical Noise

Electrical noise groups the thermal noise and the channel noise. The thermal noise (also called Nyquist or Johnson noise) adds fluctuation to the current passing through the membrane due to the thermal agitation of the ions. In this case the variance of the voltage fluctuations at rest is (Manwani and Koch 1999): $k_B T B / g_L$, where k_B is the Boltzmann constant, T is the temperature and B is the bandwidth of the system. Thermal noise is not the main source of electrical noise as it is three orders of magnitude smaller than the channel noise (Faisal et al. 2008).

Channel noise is due to the stochastic opening and closing of the ion channels (which itself is of thermal origin). This noise creates fluctuations in current that depend on the membrane potential following the activation profile of ion channels. Noise due to the Na-channels responsible for spike initiation can explain how the probability of firing depends on the amplitude of the stimulation when the stimulation consists of a short pulse of current (White et al. 2000). Noise due to Na ion channels is therefore seen as an important source of noise which adds variability to the threshold for spike initiation.

Next section will explore the idea of a stochastic threshold further. More details on stochastic models of ion channels can be found in Chap. 12.

6.6.3 Generalized Linear Models

Consider the noiseless dynamics for $V(t)$ (as given by the SRM_0 or SRM) and replace the fixed voltage threshold with a stochastic voltage threshold:

$$V_T \to \theta + \zeta(t) \tag{6.42}$$

where θ is the average – or deterministic – threshold, and $\zeta(t)$ is a zero-mean white noise. This type of noise is called '*escape noise*' (Gerstner and Kistler 2002) and relates to the escape rate in models of chemical reactions (van Kampen 1992). In this scenario, the probability of finding the membrane potential above the threshold depends solely on the instantaneous difference between the voltage and the average threshold. We can write in general terms that the firing intensity is a nonlinear function of the modelled voltage trace:

$$r(t) = f\,(V(t) - \theta). \tag{6.43}$$

The monotonically increasing nonlinear function f is the cumulative distribution function of $\zeta(t)$.

Models such as the SRM_0 or the SRM have an explicit formulation for $V(t)$ that we can substitute in Eq. 6.43. These formulations for $V(t)$ require the knowledge of the spiking history. In this case, the firing intensity is dependent on the knowledge of the previous spike times. We will label this intensity differently, $\lambda(t|\{\hat{t}\}_t)$, since it does not equal the PSTH observed experimentally anymore. Writing the convolution operation of Eq. 6.34 with an asterisk and substituting the voltage in Eq. 6.43 by an explicit formula, we have:

$$\lambda = f\left(\kappa * I + \sum_i \eta_a(t - \hat{t}_i) + E_0 - \theta\right). \tag{6.44}$$

When all kernel κ and η_a are expressed as a linear combination of basis functions, the firing intensity would be a linear model if f were linear. With the nonlinear link-function f, this is instead a Generalized Linear Models (GLM). GLMs have convenient properties in terms of finding the parameters of the model and estimating the validity of the estimates (McCullagh and Nelder 1989). One of these properties is that the likelihood of observing a given set of spike times is a convex function of the parameters (when the link function is strictly convex Paninski 2004). This means that it is always possible to find the best set of parameters to explain the data. It is a remarkable fact that only the knowledge of the spike times observed in response to a given stimulus is sufficient to estimate the filter κ and the shape of the adaptation function η.

The GLM model above depends on all the previous spikes, and therefore shows spike-frequency adaptation through the kernel $\eta_a(t)$. By excluding the possible influence of cumulative adaptation, it is possible to make a purely refractory stochastic model by dropping the dependence on all the previoius spikes but the last one. This framework allows the theorems of renewal theory (Cox 1962) to be applied, and to study the behaviour of networks of neurons analytically (Gerstner and Kistler 2002).

In the same vein, it is possible to ignore completely the refractoriness and consider only the filtering of the input:

$$\nu = f\left(\kappa * I\right). \tag{6.45}$$

Despite the fact that it appears as too crude an assumption, we can gain considerable knowledge on the functional relationship between external stimuli and neuronal response (Gazzaniga 2004). This model referred to as the Linear-Nonlinear Poisson model (LNP) was extensively used to describe the response of single neuron in the retina, thalamus or cortex as a function of the visual stimulus. When multiple neurons pave the way between the stimulus and the spikes generated by the LNP model, the filter function no longer represents the membrane filter of the cell but rather the linear filter corresponding to successive stages of processing before the signal reaches the neurons.

The mere fact that it is possible to make a decent firing-rate prediction with such a simple model makes a strong claim about the role of the neurons and the neural code. A claim that could be challenged by experiments pointing towards missing features. In the next section we further elaborate on the models described in the previous sections. We want to make sure we find the *simplest* description possible for a given set of experiments, but not *simpler*.

6.7 Advantages of Simple Models

How good are the simple models discussed in this chapter? Before addressing this question one needs to define what is meant by a good model. Suppose we focus on an experiment which injects time-varying current in a neuron that otherwise receives no input. What can be reproduced by the model? the average firing rate? the PSTH? the timing of each spike? The subthreshold potential? A model that is able to reproduce the averaged firing-rate may not be sufficient to underly the fast computational capabilities of neuron networks (Thorpe et al. 1996). On the other hand, modelling the exact time-course of the membrane potential may not be necessary given that a later neuron only receives the spikes, and no information about the membrane potential of the neuron is transmitted. Perhaps the most appropriate task of a model is to predict the occurrence of spikes of a neuron receiving in-vivo-like input. Before evaluating the performance at predicting the spike times, we assess the ability of simple models to reproduce qualitatively the firing patterns observed in various neuron types.

6.7.1 Variety of Firing Patterns

Neurons throughout the nervous system display various types of excitability. The diversity is best illustrated in experiments injecting a predefined stimulus in cells that otherwise receive no input. For example, if an identical step current is used to drive multiple cells any differences in the observed firing patterns between the cells must be attributed to intrinsic mechanisms rather than the stimulation pattern. It is common to classify neurons according to their initial response to the step current as well as according to different steady state responses (Markram et al. 2004). Consequently, the onset of firing is categorized as being either delayed, bursting or tonic. On the other hand the steady state can be tonic, adapting, bursting or irregular. Simple threshold models can reproduce all the firing patterns observed experimentally (Fig. 6.11). The study of excitability types in such simple models sheds light on the basic principles contributing to the neuronal diversity.

Delayed spiking. Delayed spiking can be due to smooth spike initiation as in the EIF or to linearized subthreshold currents. Indeed, the EIF can produce delayed spiking when the stimulating current is slightly greater than $g_L(V_T - E_0 - \Delta)$. Another possibility is that a subthreshold current slowly activates at higher voltage to depolarize the membrane further. For instance Eqs. 6.22–6.23 may lead to delayed spiking onset when $a < 0$ and $\tau_w > \tau$.

Bursting. Bursting can arise from many different causes and it is possible to define multiple types of bursting (Izhikevich 2007). Perhaps the simplest bursting model is the LIF with adaptation (Eqs. 6.12–6.13). The high firing-frequency during a burst increases the adaptation current to a point where the neuron can no longer spike until its level of adaptation has decreased substantially. In this case the inter-burst interval is related to the time constant of adaptation. A hallmark of this type of bursting is the slowing down of the firing during the burst.

Transient spiking. Upon the onset of the stimulus, the neuron fires one or multiple spikes and then remains quiescent, even if the stimulus is maintained for a very long time. Spike-triggered adaptation or a moving threshold cannot account for this pattern. Transient spiking is due to a resonance with a subthreshold current (Fig. 6.5).

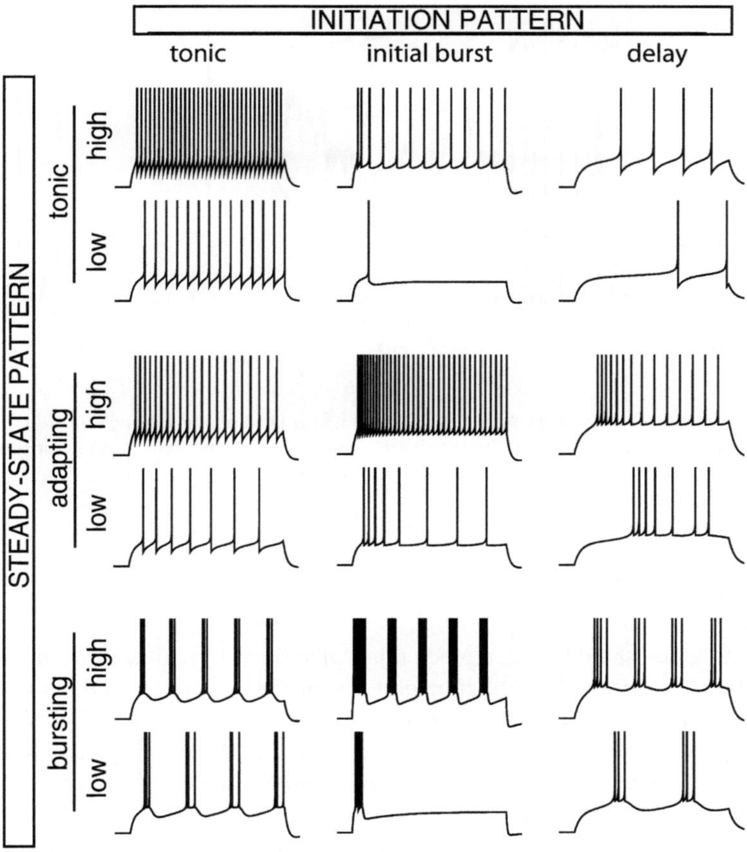

Fig. 6.11 Multiple firing patterns are reproduced by merely tuning the parameters of a simple threshold model. Here the AdEx fires with tonic bursts, initial burst, spike-frequency adaptation, or delay. For each set of parameters, the model is simulated on a step current with low (close to current threshold for firing) or high (well above the firing threshold)

Adaptation. We have seen that spike-frequency adaptation is brought by either hyperpolarizing currents or a moving threshold which cumulates over multiple spikes.

This brief description emphasizes only the most important firing patterns, we will discuss the analysis of the bursting and initial bursting firing patterns further in Sect. 6.7.3. The main observation is that a two-dimensional model such as the AdEx is capable of describing a large variety of different firing patterns. The question now is whether these models can describe not only qualitative features of firing but also quantitative ones. Concretely, can we predict the timing of spikes in experiments using simple threshold models?

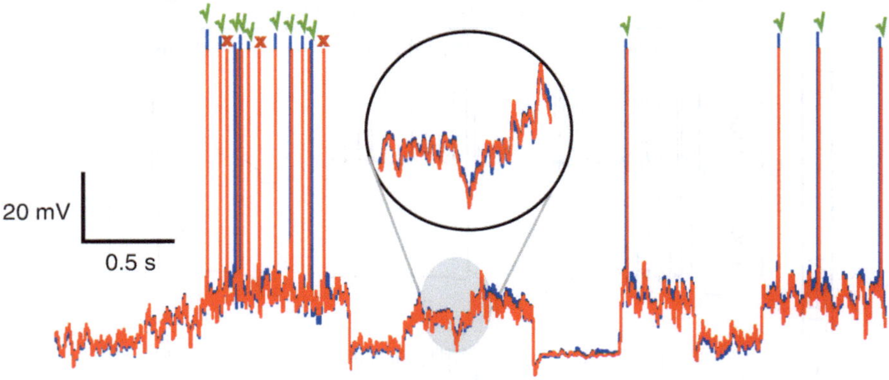

Fig. 6.12 Overlayed traces of an AdEx model (*red*) and a fast-spiking interneuron (*blue*). Most of the spikes are predicted correclty by the model (*green check marks*) but some extra spike are predicted (*red crosses*). The subthreshold voltage traces match very well (*inset*) (The data is a courtesy of Mensi et al. 2012)

6.7.2 Spike-Time Prediction

Models of the threshold type can predict the spike times of real neurons in vitro. It is important to focus on the prediction performance and not simply on reproducing those spike times that were used to calibrate the model parameters. Otherwise, a very complex model could reproduce perfectly the data used for fitting while it would fail completely to reproduce the response to novel stimulus (a phenomenon called overfitting). The correct procedure is therefore to separate the data into a training set and a test set. The first is used for finding the best model parameters and the second to test the performance.

In vitro it is possible to simulate realistic conditions by injecting a fluctuating current in the soma of the neuron. For instance the injected $I(t)$ can be taken to be as in Eq. 6.37 or 6.41 as it would be expected from a high number of synaptic events affecting the soma. This current when injected in the soma drives the neuron to fire spikes. 'Injecting' this current in the mathematical neuron models will give a similar voltage trace. After determining the model parameters that yield the best performance on the training set, the neuron model can be used to predict the voltage trace and the spike times on the test set. As we can see from Fig. 6.12, the AdEx is capable of predicting the subthreshold voltage and spike timings to an impressive degree of accuracy.

Deterministic models such as the SRM_0 with a dynamic threshold (Jolivet et al. 2006a), the AdEx (Jolivet et al. 2008), the rEIF (Badel et al. 2008b), the SRM (Mensi et al. 2012) or other similar models (Kobayashi et al. 2009) have been fitted to such in vitro experiments and their predictive performance evaluated. To evaluate the performance of deterministic models, the number of predicted spikes that fall within ± 4 ms of the observed spikes are counted. When discounting for the number of spikes that can be predicted by chance (Kistler et al. 1997), a coincidence factor is obtained ranging from zero (chance level) to one (perfect prediction). This coincidence factor ranged from 0.3 to 0.6 for pyramidal neurons of the layer 5, and from 0.5 to 0.8 for fast-spiking interneurons of the layer 5 of the cortex.

It turns out that these performances are very close to optimal, as we can see if we consider that a real neuron would not reproduce exactly the same spike times after multiple repetitions of the same stimulus (as shown in Fig. 6.10). The best coincidence factor achievable by the models is the intrinsic reliability R, which is the average of the coincidence factor across all pairs of experimental spike trains generated with the same stimulus. This value can be seen as an upper bound on the coincidence factor achievable by the mathematical models. Scaling the model-to-neuron coincidence factor by the intrinsic reliability and multiplying by 100 gives a measure of the percentage of the predictable spikes that were predicted. For models like the AdEx or the SRM, this number ranged from 60% to 82% for pyramidal neurons, and from 60% to 100% for fast-spiking interneurons. Simpler models do not share this predictive power: the LIF only accounts for 46–48% of the predictable portion of spikes.

Models from the GLM family have been used to predict the spike times of neurons in the retina, thalamus or cortex from the knowledge of the light stimulus. Almost perfect prediction can be obtained in the retina and in the thalamus (Pillow et al. 2005). Furthermore, it has been shown that refractoriness or adaptation is required for good prediction (Berry and Meister 1998; Pillow et al. 2005; Truccolo et al. 2010). Furthermore, the quality of the prediction can be improved by taking into account the coupling between adjacent cells (Pillow et al. 2008).

6.7.3 Ease of Mathematical Analysis

The greater simplicity of the neuron models discussed in this chapter compared to biophysical models of the Hodgkin–Huxley type have another advantage: the ease of mathematical analysis. This is particularly advantageous for studying neuron networks and for investigating synaptic plasticity. This is a vast field of study where many macroscopic properties of neuron networks can be scrutinized: learning, oscillations, synchrony, travelling waves, coding, and possibly others (see Hoppensteadt and Izhikevich 1997; Dayan and Abbott 2001; Gerstner and Kistler 2002 for introductions). Paving the way to these exciting fields, mathematical analysis has yielded important insights that relate the function of single neurons with that of networks.

The characteristic of the response to various types of stimulations can often be described in mathematical terms. For a constant current I, the firing frequency of the LIF is given by (see Exercices 6.5):

$$\nu = \left[\tau \ln\left(\frac{g_L(V_r - E_0) - I}{g_L(V_T - E_0) - I}\right)\right]^{-1}. \tag{6.46}$$

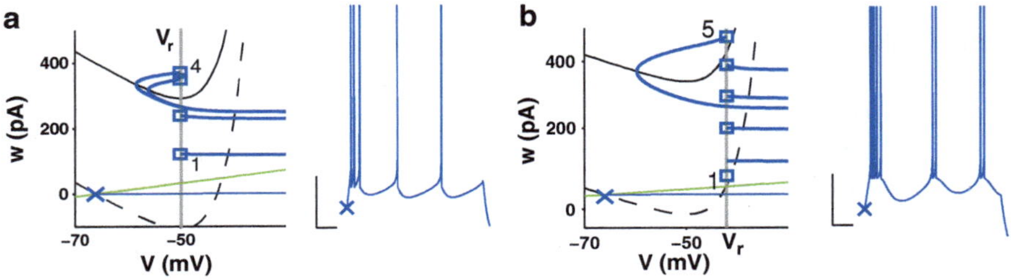

Fig. 6.13 Initially bursting (**a**) and regular bursting (**b**) trajectories in both phase plane and as a time-series. The initial state of the neuron model before the step current is marked with a cross and is associated with the V-nullcline shown as a *dash line*. After the step increase in current the V-nullcline shifts upwards (*full line*). The w-nullcline is shown in *green* and the trajectories are shown in *blue* with each subsequent reset marked with a *square*. The first and last resets are labeled with their *ordinal* number. Regular bursting is distinguished from initial bursting by the presence of two voltage resets between the trajectory of the fifth spike and the V-nullcline

The mean frequency when the current is an Ornstein-Uhlenbeck process (Eq. 6.41) can be written as an integral formula (Johannesma 1968):

$$\nu = \left[\tau \sqrt{\pi} \int_{\sqrt{\tau}(g_L(V_r - E_0) - \mu)/(\sigma\sqrt{2\tau_s})}^{\sqrt{\tau}(g_L(V_T - E_0) - \mu)/(\sigma\sqrt{2\tau_s})} e^{x^2} \left[1 + \mathrm{erf}(x)\right] dx \right]^{-1} \tag{6.47}$$

where erf(x) is the error function. When the non-linearity for spike initiation is taken into account, it is not possible to arrive at exact solutions anymore. Yet, for some parameter values there are approximations that can be worked in. This way, approximated solutions can be written for the EIF or the AdEx receiving gaussian white noise (Fourcaud-Trocme et al. 2003; Richardson 2009). When there is a strong effect of adaptation it is not possible to arrive at closed-form solutions for ν and one must rely on numerical integration of the appropriate Fokker-Planck equations can be used (Muller et al. 2007; Toyoizumi et al. 2009).

Mathematical analysis is equally successful to relate the observed types of firing patterns with the parameter values. We will illustrate the application of bifurcation theory with an example related to Fig. 6.11: distinguishing initial bursting from a regular bursting using phase plane analysis.

Repetitive bursting is created in the AdEx model by constantly alternating between slow and fast interspike intervals. When the injection is a step increase in current (as in Fig. 6.11) this corresponds to a sudden shift of the V-nullcline (i.e. the set of points where $\frac{dV}{dt} = 0$) in the phase plane. Before the step, the state of the neuron is at the stable fixed point which is sitting at the intersection between the V- and the w-nullcline. After the step, the stable fixed point has disappeared and this results in repetitive firing. The distinction between an initially bursting AdEx model and a regular bursting one is made by considering the location of the resets in the phase plane. Figure 6.13 shows that both types of bursting have spike resets above and below the V-nullcline. Resetting above the V-nullcline brings a much larger interspike interval than resetting below. To achieve this, a spike reset above the V-nullcline must be able to make at least one spike reset below the V-nullcline before being mapped above again. This is seen in Fig. 6.13b where there is sufficient space below the V-nullcline and above the trajectory of the fifth spike for at least one reset. This leads the repetitive bursting of two spikes. On the other hand, in Fig. 6.13a the fourth spike – which was reset above the V-nullcline – is followed by another reset above the V-nullcline. This leads to the end of the initial burst which is followed only by long interspike intervals. By observing at the phase planes in Fig. 6.13, one can conclude that a higher voltage reset helps to bring about repetitive bursting.

Similar conclusions can be drawn for all the firing patterns (Naud et al. 2008). Furthermore, it is sometimes possible to have analytical expressions relating the parameter values and the firing patterns (Touboul 2008). Such conclusions are possible mainly because the complexity of the model was kept low. Indeed, the structure of firing patterns have been studied in other simple models (Izhikevich 2007; Mihalaş and Niebur 2009).

6.8 Limits

The neuron models presented in this chapter present an idealized picture. The first idealization is that all these models consider the neuron as a point with no spatial structure. This point could represent the axon initial segment where the spike is initiated. However, real neurons receive input distributed not only in the soma but also in their dendritic arborizations. Dendrites bring multiple types of non-linear interactions between the inputs. Dendritic ion channels combine with basic properties of AMPA, GABA and NMDA synapses to make dendritic output to the soma dependent on the spatio-temporal pattern of excitation. Spatially extensive models (see Chap. 11) may be required to correctly translate input spikes into current arriving at the axon initial segment, but to what extent? This remains to be addressed experimentally.

Another central assumption prevalent throughout this chapter is that the spike does not change its length or shape with different stimuli. While this is seen as a good approximation for cortical neurons firing at low frequency (Bean 2007), some neurons have action potential shapes that vary strongly as a function of either stimulus, firing frequency or neuromodulation. For instance, the interneurons taking part in the motor pattern generation of flight in locusts display spikes that reduce by half their amplitude and width when part of a burst of activity (Wolf and Pearson 1989). Similarly, half-blown spikes are observed frequently in the lobster stomatogastric pattern generator (Clemens et al. 1998) or in complex spikes of cartwheel cells in the dorsal cochlear nucleus (Golding and Oertel 1997).

The simple models described in this chapter can be seen as a high-level description of the complex biophysical mechanisms mediated by ion channels, ion pumps, and various chemical reactions involving neurotransmitters. At this level of description the molecular cascades for the effect of – for example – acetylcholine are not modelled. Though the effect of neuromodulators can be calibrated and cast in stereotypical modification of the simple model (Slee et al. 2005), it is not an intrinsic feature and the calibration has to be performed for each scenario. Complex biochemical as well as biophysical models are the norm if pharmacological procedures are studied (see Chaps. 17 and 18).

To conclude, simple neuron models capture essential features that are required for transducing input into spikes. If we bear in mind the inherent limitations, the study of simple neuron models will continue bringing new insights about the role of single neurons. In particular, the roles of adaptation and variability are only starting to be considered.

6.9 Further Reading

Spiking neuron models. Uses single-neuron models as a building block for studying fundamental questions such as neuronal coding, signal transmission, dynamics of neuronal populations and synaptic plasticity (Gerstner and Kistler 2002).

Dynamical systems in neuroscience. Neuron models are made of systems of differential equations, this book provides a very didactic approach to the mathematical theory associated with such dynamical systems (Izhikevich 2007).

Exercices

1. With $E_w = E_0$, cast Eqs. 6.22–6.23 in the form of:

$$V(t) = E_0 + \int_0^\infty \kappa(s) I(t-s) ds + \sum_i \eta_a(t - \hat{t}_i). \tag{6.48}$$

Solution :

$$\eta_a = 0$$
$$\kappa(t) = k_+ e^{\lambda_+ t} + k_- e^{\lambda_- t}$$

where

$$\lambda_\pm = \frac{1}{2\tau\tau_w}\left(-(\tau+\tau_w) \pm \sqrt{(\tau+\tau_w) - 4\tau\tau_w(g_L + a)/g_L}\right)$$

and

$$k_\pm = \pm \frac{\lambda_\pm \tau_w + 1}{C\tau_w(\lambda_+ - \lambda_-)}.$$

2. If we connect a dendritic compartment to a LIF point neuron, the dynamics of the system obey:

$$C\frac{dV}{dt} = -G_L(V - V_L) + G_c(V_d - V) + I$$
$$C_d\frac{dV_d}{dt} = -G_d(V - E_d) + G_c(V - V_d)$$

where g_c is the conductance coupling the LIF compartment with membrane potential V to the dendritic compartment with potential V_d. The leak conductance, the capacitance and the equilibrium potential for the dendritic compartment are g_d, C_d and E_d, respectively. How should you define w, a, g_L, E_0, E_w and τ_w to show that this situation is equivalent to Eqs. 6.22–6.23?

Solution: $w = G_c V_d$, $g_L = G_L + G_c$, $E_0 = G_L V_L/(G_L + G_c)$, $a = G_d + G_c$, $E_w = E_d G_d/(G_c + G_d)$, $\tau_w = C_d/G_d$.

3. Reduce the Hodgkin–Huxley equations (see Chap. 11) to the AdEx model. Start with the single-compartment Hodgkin and Huxley equations:

$$C\frac{dV}{dt} = -g_m(V - E_m) - g_{Na}m^3h(V - E_{Na}) - g_K n^4(V - E_K), \quad (6.49)$$

$$\tau_x(V)\frac{dx}{dt} = -x + x_\infty(V), \quad (6.50)$$

with x being any of the gating variables m, h or n.

Solution :

(i) Assume that $\tau_m(V) << C/g_L$.
(ii) Assume that when the voltage reaches some relatively high value (e.g. $-10\,\text{mV}$) there is a spike being emitted and the voltage is reseted to V_r, while the potassium activation make a stereotypical change: $n \rightarrow n + \delta_n$.
(iii) Assume that the sodium inactivation variable plays no role sub-threshold and during the spike initiation: $h(t) = h_0$.
(iv) There remains only subthreshold dynamics and spike initiation, therefore the dynamics of h, n and I_K can be linearized around E_0 (multivariate taylor expansion). The resulting system is:

$$C\frac{dV}{dt} = I - g_L(V - E_0) - g_{Na}h_0 m_\infty^3(V)(V - E_{Na}) - w \quad (6.51)$$

$$\tau_w\frac{dw}{dt} = a_w(V - E_0) - w \quad (6.52)$$

$$\text{if } V > 0mV \text{ then } V = V_r,\ w = w + b. \quad (6.53)$$

with $w(t) = 4g_K(n(t) - n_\infty(E_0))n_\infty^3(E_0)(E_0 + E_K)$, $a = 4g_K n_\infty(E_0)^3 (E_0 + E_K) \left.\frac{\partial n_\infty}{\partial V}\right|_{E_0}$, $\tau_w = \tau_n(E_0)$, $b = 4g_K\delta_n n_\infty^3(E_0)(E_0 - E_K)$.

4. Computer aided exercice. Build a simple SRM model to reproduce all the firing patterns in Fig. 6.11. *Hint:* let kernels $\eta(t)$ and $\kappa(t)$ each be constituted of two exponentials.
5. Derive equation 6.46 starting from Eqs. 6.6 to 6.8 with a constant current. *Hint*: Call ν^{-1} the time it takes to go from the reset to the threshold. With I constant the integrals can be computed explicitly.

References

Azouz R, Gray C (2000) Dynamic spike threshold reveals a mechanism for coincidence detection in cortical neurons in vivo. Proc Natl Acad Sci USA 97:8110–8115

Azouz R, Gray CM (2003) Adaptive coincidence detection and dynamic gain control in visual cortical neurons in vivo. Neuron 37:513–523

Badel L, Lefort S, Berger T, Petersen C, Gerstner W, Richardson MJE (2008a) Extracting non-linear integrate-and-fire models from experimental data using dynamic i–v curves. Biol Cybern 99:361–370

Badel L, Lefort S, Brette R, Petersen C, Gerstner W, Richardson M (2008b) Dynamic i-v curves are reliable predictors of naturalistic pyramidal-neuron voltage traces. J Neurophysiol 99:656–666

Baker MD, Bostock H (1998) Inactivation of macroscopic late na+ current and characteristics of unitary late na+ currents in sensory neurons. J Neurophysiol 80(5):2538–2549

Baldissera F, Gustafsson B, Parmiggiani F (1976) A model for refractoriness accumulation and secondary range firing in spinal motoneurones. Biol Cybern 24(2):61–65

Bean BP (2007) The action potential in mammalian central neurons. Nat Rev Neurosci 8(6):451–65

Benda J, Herz A (2003) A universal model for spike-frequency adaptation. Neural Comput 15(11):2523–2564

Berry M, Meister M (1998) Refractoriness and neural precision. J Neurosci 18:2200–2211

Brette R, Gerstner W (2005) Adaptive exponential integrate-and-fire model as an effective description of neuronal activity. J Neurophysiol 94(5):3637–3642

Clemens S, Combes D, Meyrand P, Simmers J (1998) Long-term expression of two interacting motor pattern-generating networks in the stomatogastric system of freely behaving lobster. J Neurophysiol 79(3):1396–408

Cook EP, Guest JA, Liang Y, Masse NY, Colbert CM (2007) Dendrite-to-soma input/output function of continuous time-varying signals in hippocampal ca1 pyramidal neurons. J Neurophysiol 98(5):2943–2955

Cox DR (1962) Renewal theory. Methuen, London

Daley D, Vere-Jones D (1988) An introduction to the theory of point processes. Springer, New York

Dayan P, Abbott LF (2001) Theoretical neuroscience. MIT Press, Cambridge

Doiron B, Oswald A, Maler L (2007) Interval coding. II. Dendrite-dependent mechanisms. J Neurophysiol 97:2744–2757

Faisal A, Selen L, Wolpert D (2008) Noise in the nervous system. Nat Rev Neurosci 9(4):292

Fourcaud-Trocme N, Hansel D, Vreeswijk CV, Brunel N (2003) How spike generation mechanisms determine the neuronal response to fluctuating inputs. J Neurophysiol 23(37):11628–11640

Fuortes M, Mantegazzini F (1962) Interpretation of the repetitive firing of nerve cells. J Gen Physiol 45:1163–1179

Gazzaniga MS (2004) The cognitive neurosciences, 3rd edn. MIT Press, Cambridge

Gerstner W (2008) Spike-response model. Scholarpedia 3(12):1343

Gerstner W, Brette R (2009) Adaptive exponential integrate-and-fire model. Scholarpedia 4(6):8427
Gerstner W, Kistler W (2002) Spiking neuron models. Cambridge University Press, New York
Gerstner W, van Hemmen J, Cowan J (1996) What matters in neuronal locking? Neural Comput 8:1653–1676
Golding NL, Oertel D (1997) Physiological identification of the targets of cartwheel cells in the dorsal cochlear nucleus. J Neurophysiol 78(1):248–60
Hill A (1936) Excitation and accomodation in nerve. Proc R Soc B 119:305–355
Hoppensteadt FC, Izhikevich EM (1997) Weakly connected neural networks. Springer, New York
Huguenard JR, Hamill OP, Prince DA (1988) Developmental changes in na+ conductances in rat neocortical neurons: appearance of a slowly inactivating component. J Neurophysiol 59(3):778–95
Izhikevich E (2001) Resonate-and-fire neurons. Neural Netw 14:883–894
Izhikevich E (2004) Which model to use for cortical spiking neurons? IEEE Trans Neural Netw 15:1063–1070
Izhikevich EM (2007) Dynamical systems in neuroscience: the geometry of excitability and bursting. MIT Press, Cambridge
Johannesma P (1968) Diffusion models of the stochastic acticity of neurons. In: Caianiello ER (ed) Neural networks, Springer, Berlin, pp 116–144
Jolivet R, Rauch A, Lüscher HR, Gerstner W (2006a) Integrate-and-fire models with adaptation are good enough. In: Weiss Y, Schölkopf B, Platt J (eds) Advances in neural information processing systems 18, MIT Press, Cambridge, pp 595–602
Jolivet R, Rauch A, Luscher HR, Gerstner W (2006b) Predicting spike timing of neocortical pyramidal neurons by simple threshold models. J Comput Neurosci 21(1):35–49
Jolivet R, Kobayashi R, Rauch A, Naud R, Shinomoto S, Gerstner W (2008) A benchmark test for a quantitative assessment of simple neuron models. J Neuroscience Methods 169:417–424
Kistler WM, Gerstner W, van Hemmen JL (1997) Reduction of Hodgkin-Huxley equations to a single-variable threshold model. Neural Comput 9:1015–1045
Kobayashi R, Tsubo Y, Shinomoto S (2009) Made-to-order spiking neuron model equipped with a multi-timescale adaptive threshold. Front Comput Neurosci 3:9
Korngreen A, Sakmann B (2000) Voltage-gated k+ channels in layer 5 neocortical pyramidal neurones from young rats:subtypes and gradients. J Physiol 525(3):621–639
La Camera G, Rauch A, Lüscher HR, Senn W, Fusi S (2004) Minimal models of adpated neuronal responses to in-vivo like input currents. Neural Comput 16:2101–2104
Lapicque L (1907) Recherches quantitatives sur l'excitation electrique des nerfs traitée comme une polarization. J Physiol Pathol Gen 9:620–635. Cited in Tuckwell HC (1988) Introduction to theoretic neurobiology. Cambridge University Press, Cambridge
Latham PE, Richmond B, Nelson P, Nirenberg S (2000) Intrinsic dynamics in neuronal networks. I. Theory. J Neurophysiol 83:808–827
Lundstrom B, Higgs M, Spain W, Fairhall A (2008) Fractional differentiation by neocortical pyramidal neurons. Nat Neurosci 11(11):1335–1342
Manwani A, Koch C (1999) Detecting and estimating signals in noisy cable structures, I: Neuronal noise sources. Neural Comput 11:1797–1829
Markram H, Toledo-Rodriguez M, Wang Y, Gubta A, Silbrberg G, Wu C (2004) Interneurons of the neocortical inhibitory system. Nat Rev Neurosci 5:793–807
Mauro A, Conti F, Dodge F, Schor R (1970) Subthreshold behavior and phenomenological impedance of the squid giant axon. J Gen Physiol 55(4):497–523
McCullagh P, Nelder JA (1989) Generalized linear models, vol 37, 2nd edn. Chapman and Hall, London
Mensi S, Naud R, Pozzorini C, Averman M, Petersen CCH, Gerstner W (2012) Parameter extraction and classification of three cortical neuron types reveals two distinct adaptation mechanisms. J Neurophys 107(6):1756–1775
Mihalaş S, Niebur E (2009) A generalized linear integrate-and-fire neural model produces diverse spiking behaviors. Neural Comput 21(3):704–18

Muller E, Buesing L, Schemmel J, Meier K (2007) Spike-frequency adapting neural ensembles: beyond mean adaptation and renewal theories. Neural Comput 19(11):2958–3010

Naud R, Marcille N, Clopath C, Gerstner W (2008) Firing patterns in the adaptive exponential integrate-and-fire model. Biol Cybern 99:335–347

Paninski L (2004) Maximum likelihood estimation of cascade point-process neural encoding models. Netw Comput Neural Syst 15(4):243–262

Paninski L, Pillow J, Simoncelli E (2004) Maximum likelihood estimate of a stochastic integrate-and-fire neural encoding model. Neural Comput 16:2533–2561

Passmore G, Selyanko A, Mistry M, Al-Qatari M, Marsh SJ, Matthews EA, Dickenson AH, Brown TA, Burbidge SA, Main M, Brown DA (2003) Kcnq/m currents in sensory neurons: significance for pain therapy. J Neurosci 23(18):7227–7236

Pillow J, Paninski L, Uzzell V, Simoncelli E, EJChichilnisky (2005) Prediction and decoding of retinal ganglion cell responses with a probabilistic spiking model. J Neurosci 25:11003–11023

Pillow JW, Shlens J, Paninski L, Sher A, Litke AM, Chichilnisky EJ, Simoncelli EP (2008) Spatio-temporal correlations and visual signalling in a complete neuronal population. Nature 454(7207):995–1000

Rauch A, Camera GL, Luscher H, Senn W, Fusi S (2003) Neocortical pyramidal cells respond as integrate-and-fire neurons to in vivo-like currents. J Neurophysiol 90:1598–1612

Richardson MJE (2009) Dynamics of populations and networks of neurons with voltage-activated and calcium-activated currents. Phys Rev E 80(2 Pt.1):021928

Richardson M, Brunel N, Hakim V (2003) From subthreshold to firing-rate resonance. J Neurophysiol 89:2538-2554

Schwindt, Spain W, Crill W (1989) Long-lasting reduction of excitability by a sodium-dependent potassium current in cat cortical neurons. J Neurosci 61(2):233–244

Slee SJ, Higgs MH, Fairhall AL, Spain WJ (2005) Two-dimensional time coding in the auditory brainstem. J Neurosci 25(43):9978–9988

Stein RB (1965) A theoretical analysis of neuronal variability. Biophys J 5:173–194

Thorpe S, Fize D, Marlot C (1996) Speed of processing in the human visual system. Nature 381:520–522

Touboul J (2008) Bifurcation analysis of a general class of nonlinear integrate-and-fire neurons. SIAM J Appl Math 68(4):1045–1079

Toyoizumi T, Rad K, Paninski L (2009) Mean-field approximations for coupled populations of generalized linear model spiking neurons with markov refractoriness. Neural Comput 21(5):1203–1243

Truccolo W, Hochberg LR, Donoghue JP (2010) Collective dynamics in human and monkey sensorimotor cortex: predicting single neuron spikes. Nat Neurosci 13(1):105–111

van Kampen NG (1992) Stochastic processes in physics and chemistry, 2nd edn. North-Holland, Amsterdam

White JA, Rubinstein JT, Kay AR (2000) Channel noise in neurons. Trends Neurosci 23(3):1–7

Wolf H, Pearson K (1989) Comparison of motor patterns in the intact and deafferented flight system of the locust. III: patterns of interneuronal activity. J Comp Physiol A Sens Neural Behav Physiol 165(1):61–74

Wu N, Enomoto A, Tanaka S, Hsiao CF, Nykamp DQ, Izhikevich E, Chandler SH (2005) Persistent sodium currents in mesencephalic v neurons participate in burst generation and control of membrane excitability. J Neurophysiol 93(5):2710–22

Chapter 7
Multi-compartmental Models of Neurons

Upinder S. Bhalla

Abstract The electrical properties of single neurons can be acurately modeled using multicompartmental modeling. Such models are biologically motivated and have a close correspondence with the underlying biophysical properties of neurons and their ion channels. These multicompartment models are also important as building blocks for detailed network models. Finally, the compartmental modeling framework is also well suited for embedding molecular signaling pathway models which are important for studying synaptic plasticity. This chapter introduces the theory and practice of multicompartmental modeling.

Electrical models of neurons are one of the rather rare cases in Biology where a concise quantitative theory accounts for a huge range of observations and works well to predict and understand physiological properties. The mark of a successful theory is that people take it for granted and use it casually. Single neuronal models are no longer remarkable: with the theory well in hand, most interesting questions using models have moved to the networks of neurons in which they are embedded, and the networks of signalling pathways that are in turn embedded in neurons. Nevertheless, good single-neuron models are still rather rare and valuable entities, and it is an important goal in neuroinformatics (and this chapter) to make their generation a well-tuned process.

7.1 Frog Legs and Monks: A Perspective

Twitching frog legs and dancing monks were some of the earliest indications that behavior and motion were somehow coupled to electricity. In classical studies in the late 1700s, Galvani found that under certain circumstances, touching a metal object

U.S. Bhalla (✉)
National Centre for Biological Sciences, TIFR, Bangalore, India
e-mail: bhalla@ncbs.res.in

N. Le Novère (ed.), *Computational Systems Neurobiology*,
DOI 10.1007/978-94-007-3858-4_7,

to a stripped frog leg caused it to twitch. This was ascribed to 'animal electricity.' The role of electricity and its application to nerves was already inferred from these early experiments, though Galvani incorrectly believed it to have a specifically animal origin. The use of static electricity in causing humans to twitch was also well known; in one celebrated experiment Nollet used an early capacitor to cause a row of Carthusian monks to leap into the air. There this theory sat for a long time.

By the early 1900s, Bernstein and Cajal had pinned down the electrical and morphological significance of neurons in behavior. Bernstein recorded the first action potentials (Seyfarth 2006) and proposed his "membrane theory" that stated that ion fluxes across a semipermeable membrane accounted for nerve conduction. Cajal had begun to champion the role of neurons in behavior. He called them "the mysterious butterflies of the soul," and went so far as to suggest that their function held "... the secret of mental life." The 1940s saw a burst of activity in establishing how neurons worked, notably the experiments and theories of voltage-gated ion channels by Hodgkin and Huxley that still remains the benchmark for realistic physiological models (Hodgkin and Huxley 1952a, b). It is worth noting that these early theoretical excursions were met with significant hostility, even though they were superbly backed by experiment.

If the work of Hodgkin and Huxley defined the standards for studying the active properties of cells, Wilfrid Rall laid the theoretical foundations for studying their passive and computational roles from the 1960s. Till this time the dendrites had largely been disregarded: they were believed to be mere cellular extensions to plug synapses onto. Rall's work showed that dendrites were not 'small' and that their passive properties substantially defined the computational function of the cell (Rall 1959, 1967, reviewed in Segev 2006). In the conservative neuroscience tradition, this work too met with a great deal of resistance: first to the idea that dendrites did anything interesting, and then to his analysis of their functional and computational roles. Experiments eventually caught up with the theory, and today we take these concepts for granted.

It is now easy to grab a simulator such as NEURON or GENESIS, and run models ranging from the original Hodgkin-Huxley squid theory, to immensely detailed 3-D models of neurons. These models, and their rather obvious success in accounting for many neuronal properties, have substantially eased the acceptance of neuronal modeling in neuroscience. Detailed compartmental models are just one category of models in active use. More abstract versions range from integrate-and-fire models, which represent some of the passive and synaptic properties of neurons, but typically do away with the geometry and the active currents, to almost unrecognizable summation entities that form the basis of 'neural' networks. Each provides insights at different levels. However, to understand physiology, compartmental models are the way to go. They are based on physiology, they do a good job of replicating experimental findings, and they bridge the interesting range between molecular networks and networks of neurons which is the focus of this book.

7.2 Cable Theory

Cable theory describes passive propagation of charge through an insulated cable surrounded by a conductor. This turns out to be a pretty good description of a segment of a neuron: the cytoplasm surrounded by membrane is the cable, and it is surrounded by the extracellular medium which is a conductor. Electrical signal propagation through undersea telegraph cables is essentially the same problem, and was originally analyzed in the 1850s by Lord Kelvin. The equations are similar to those for heat conduction in a wire, which had been worked out still earlier by Fourier.

In this section we will see how small segments of neurons can be represented as equivalent electrical circuits, and use these to derive the cable equation. We will then follow Wilfrid Rall in exploring the implications of the cable equation for neuronal function and experiments.

7.2.1 *Compartments*

A common trick in analyzing a spatially extended object, such as a cable, is to subdivide it into compartments which are so small that the property of interest is uniform over the entire extent of the compartment. For neurons, this property of interest is the potential difference across the membrane (*Vm*). The compartment is a small cylinder with cytoplasm inside, membrane around it, and the extracellular medium around that. This forms a capacitor (*Cm*), as the membrane is a good insulator between two conductors. The large but finite resistance of the membrane (*Rm*) is in parallel with this capacitor. The ion gradients across the membrane add a battery (*Em*) to this little circuit (Fig. 7.1). This forms a compartmental circuit diagram and is the starting point for our analysis. The convention is to treat the extracellular medium as a common ground. As we shall see later, exactly this electrical equivalent is used to perform numerical simulations of neurons.

In order to complete the spatial picture, we need to connect compartments to each other. Here we consider the axial flow of current between compartments. As the cytoplasm is not a perfect conductor, we treat it as a resistance, *Ra*. The compartmental trick lets us collapse all the resistive effects into the junctions between cables. In this case we split the contribution into *Ra*/2, on either side (Fig. 7.1). Branches are readily represented as electrical branches.

It is certainly possible to use more complex compartmental forms, for example conical sections rather than cylinders. In the limit of very short compartments, they all give the same results (Jack et al. 1975). It is also possible to consider non-uniformities around the circumference of the cable: we discuss this later. For almost all detailed neuronal models, we are saved by the fact that the length-scale of variations along the axis of the cable is usually greater than the cable diameter. Thus we can get away with this very simple description of neurons as thin cylinders with branching.

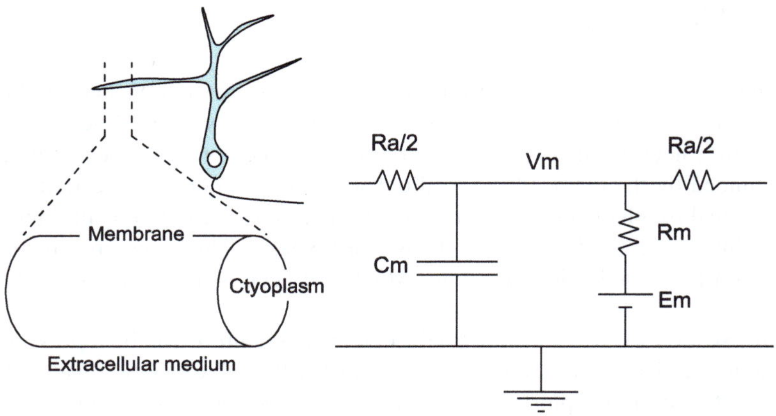

Fig. 7.1 Electrical equivalent circuit of a membrane compartment. Each segment of dendrite or axon is represented as a compartment, which is internally uniform. The electrical properties of this compartment can be mapped onto equivalent electrical circuit elements

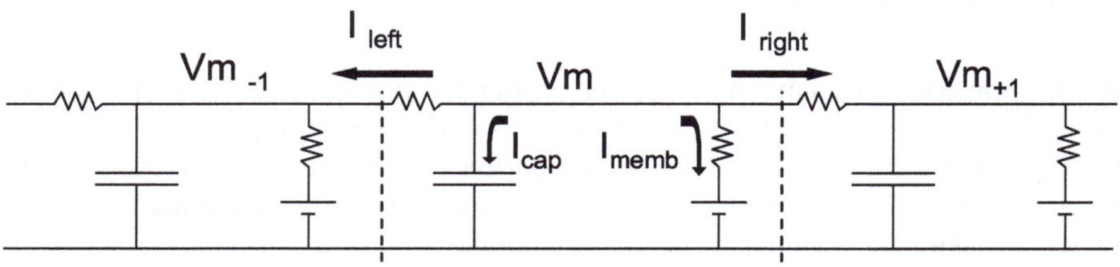

Fig. 7.2 Series of electrical compartments in a cable, used to derive the cable equation

7.2.2 Derivation and Implications

To derive the cable equation, we consider three adjacent compartments on a uniform cable (Fig. 7.2). Instead of absolute electrical values, we will use r_a, r_m and c_m, which are per unit length. We set Em to zero for this derivation. The middle compartment is our reference, and the one to the left and right are numbered -1 and $+1$.

Assume that they are separated by a distance *Dx*. We set the total current leaving the compartment to zero. This includes terms for current leaving from the left, the right, through the membrane resistance, and through charging the capacitor:

$$\frac{Vm - Vm_{-1}}{r_a.Dx} + \frac{Vm - Vm_{+1}}{r_a.Dx} + \frac{Vm.Dx}{r_m} + c_m.Dx.\frac{dVm}{dt} = 0 \tag{7.1}$$

The first two terms refer to the voltage gradient on either side of the compartment:

$$\frac{dVm/dx\,|_{-1} - dVm/dx\,|_0}{r_a} \tag{7.2}$$

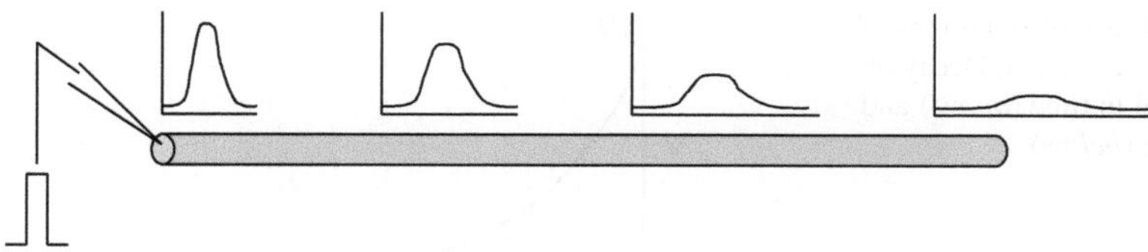

Fig. 7.3 Propagation of an electrical pulse down a cable

In the limit, their difference, over the length *Dx* of the compartment, is just the second derivative in space. We divide across by *Dx* to get

$$-\frac{1}{r_a}.\frac{\partial^2 Vm}{\partial x^2}+\frac{Vm}{r_m}+c_m\frac{\partial Vm}{\partial t}=0 \tag{7.3}$$

A little rearranging gives:

$$-\frac{r_m}{r_a}.\frac{\partial^2 Vm}{\partial x^2}+r_m c_m\frac{\partial Vm}{\partial t}+Vm=0 \tag{7.4}$$

Using

$$\lambda^2 = r_m / r_a \tag{7.5}$$

and

$$\tau = r_m.c_m \tag{7.6}$$

we get the familiar form of the cable equation:

$$\lambda^2\frac{\partial^2 Vm}{\partial x^2}-\tau\frac{\partial Vm}{\partial t}-Vm=0 \tag{7.7}$$

The general solutions of the cable equation are moderately complicated functions of time, space and boundary conditions (Rall 1959; Jack et al. 1975). The key intuitive results, however, are quite simple: a current pulse at one end of a cable undergoes three changes as it propagates through to the other end: decay, smoothing, and delay (Fig. 7.3).

To lend a slightly more quantitative view to these three effects, the cable equation has two specific solutions that are helpful in their simplicity.

1. If we eliminate space from the cable equation (reducing it to a single compartment), the time-course of charging or discharging is a simple exponential. This is a good approximation to what most cells do in the absence of active currents.

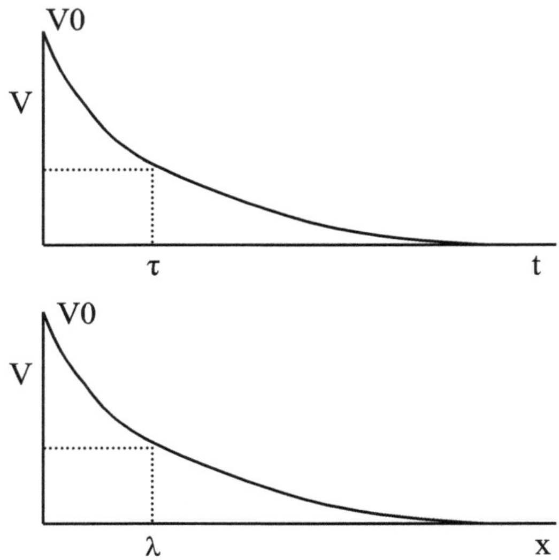

Fig. 7.4 Simple solutions of the cable equation. Decay of potential in time (*above*) and in space (*below*)

As we have seen from the cable equation, the charging time-course, τ, is related to the electrical properties as:

$$\tau = r_m.c_m \qquad (7.8)$$

2. Similarly, if we eliminate time from the cable equation (taking the solution at infinite time), and assume an infinitely long cable, the voltage profile along the cable is also a simple exponential (Fig. 7.4). This is a reasonable approximation to the decay of signals in a cell, again in the absence of active currents. The length constant of decay λ is given by:

$$\lambda = \sqrt{r_m / r_a} \qquad (7.9)$$

τ and λ characterize the temporal and spatial properties of neurons, respectively. For reference, τ is typically around $10 - 100$ *ms*, and λ for dendrites is in the range of $0.1 - 1$ *mm*. Axons are typically much thinner and myelination complicates their analysis, but they usually have shorter values of λ.

7.2.3 *Rall's Law and Electrotonic Length*

Wilfrid Rall showed that many neurons were electrically 'long' (Rall 1969). In terms of the steady-state solution in Fig. 7.4, this means that the geometrical length is one or more length constants. This ratio is called the electrotonic length, *L*.

$$L = geometrical\ length/\lambda \qquad (7.10)$$

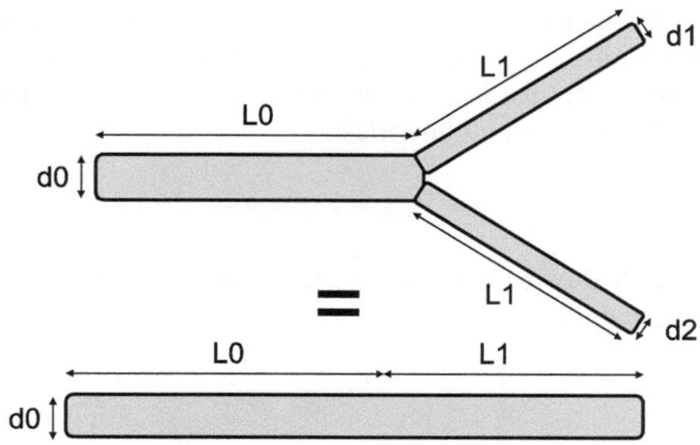

Fig. 7.5 Rall's Law. The branched structure is electrically equivalent to the single cylindrical cable below

This quantity is derived for a linear cable. As most neurons have extensive branching, the quantity *L* may seem of limited value. In one of his most important contributions, Rall showed that in fact most branching neurons can also be treated as simple cylinders for the purposes of calculating passive responses to electrical events at the soma (Rall 1959). Rall's Law states that, for any branch point, this equivalence holds provided the following four conditions are met:

1.

$$d_0^{3/2} = d_1^{3/2} + d_2^{3/2} \tag{7.11}$$

where d_0, d_1, and d_2 are the diameters of the parent and two child compartments.
2. The specific axial resistance *RA*, and specific membrane resistance *RM*, are uniform
3. The ends of the cylinders are similar (same boundary conditions)
4. The boundary conditions at matching values of *L* in the daughter cylinders are the same. In practice this typically means that the electrotonic lengths *L* of the daughter cylinders are identical (Fig. 7.5).

Note that this law is recursive. It applies to sub-branches of daughter branches, and converts each into an equivalent cylinder, so that an entire, multi-level branching tree can be treated as a single cylinder.

While these four conditions, especially the first, may seem very restrictive, many real neurons actually approximate them rather closely. A key point in applying Rall's Law is that it works for an input at the soma, but not at the dendrite. Remember condition 3: that the ends of the cylinders must have the same boundary conditions. Thus a signal at the soma propagates uniformly to all dendritic branches. However, an input to a single dendritic branch is not shared with all dendrites, so the condition fails. Thus a somatic signal (action potential or current injection) usually propagates well to all dendrites, but an input on a single dendrite is severely attenuated at the soma. Put another way, the electronic length from the soma to the dendrite is usually much smaller than the electronic length from the dendrite to the soma.

Another of Rall's useful results shows that the electrotonic length of a cable can be obtained from a close look at its charging time-course (Rall 1969). While a point cable, as discussed above, has a single charging/discharging time-course τ, extended cables have an infinite series:

$$V(t) = V_{rest} + V0\exp(-t/\tau 0) + V1\exp(-t/\tau 1) + \ldots \quad (7.12)$$

Rall showed that you can take any two of these time-courses and use them to compute L:

$$L = \frac{\pi}{\sqrt{\frac{\tau 0}{\tau 1} - 1}} \quad (7.13)$$

Most experiments are able to find $\tau 0$ accurately, $\tau 1$ less so, and higher-order time courses poorly if at all. So for all practical purposes, we can best estimate L using $\tau 0$ and $\tau 1$. This is a rather remarkable result. It shows that an estimate of the electrical size of the cell, L, can be obtained from a single recording at the soma of a neuron.

7.2.4 How Much Can You Learn About a Neuron by Recording from Its Soma?

At this point let us anchor some of these theoretical findings in a simple experiment. Supposing we have stuck an electrode into the soma of a passive neuron, that is, one without any ion channels to complicate matters. This criterion for passive behavior is usually well approximated if the charging curves are in the form of small current pulses, which will minimally activate active channels (Fig. 7.6). What can we learn about its passive properties?

1. *Vm*. The very first experiment one does on a neuron is to measure its resting potential, *Vm*. Here the electrode just records the intracellular potential with respect to ground.
2. Input resistance. Here one passes a small current I through the recording electrode. By monitoring the steady-state change in *Vm* and applying Ohm's Law, one obtains an estimate for the input resistance *Rin* of the whole cell.
3. Capacitance. By monitoring the time-course of the voltage in response to a step in applied current I, one can directly measure the charging time-course $\tau 0$. Plugging $\tau 0$ and the above input resistance into Eq. 7.8, one obtains the lumped cellular capacitance, *Cin*.
4. Electrotonic Length, L. By monitoring the same time-course as above, and using some curve-fitting to obtain $\tau 0$ and $\tau 1$, we can also estimate L using Eq. 7.13.

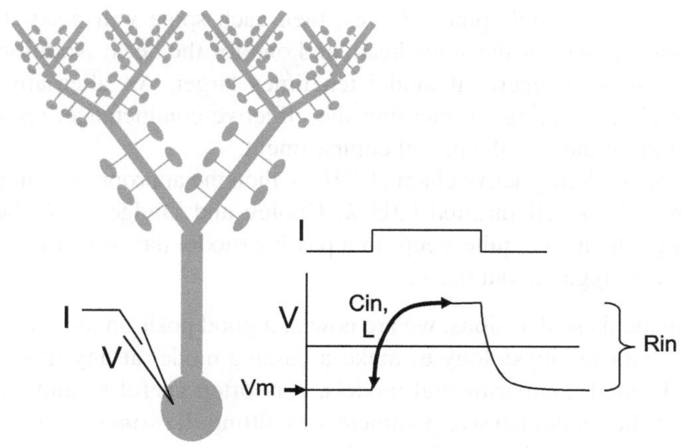

Fig. 7.6 Obtaining several passive parameters from a single somatic recording

This is quite a remarkable amount of information to obtain from a single electrode recording. For some purposes this is enough and more to make a neuronal model. For example, integrate-and-fire neurons can be mostly parameterized by points 1 – 3.

7.2.5 How Much Can You Learn by Looking in a Microscope?

Pictures of neurons are plentiful, and accurate if one accounts for staining distortions and depth. Detailed neuronal morphology images date back to the early staining techniques developed by Golgi, that were put to such effective use by Cajal. They provide the next line of data for compartmental modeling. As a first pass, we take a computerized neuronal morphology file (typically consisting of position of each dendritic segment and its diameter) and convert it into a compartmental model. This can partially be automated, but there are lots of cases where human judgement needs to come into the mix. Some of the key decisions in going from morphology data to a compartmental model are:

1. How much detail do you want in your model? This determines whether you need just an idealized spherical cell with just a somatic compartment, or a spatially faithful model with hundreds of compartments.
2. How small do you want your compartments? In part this is determined by the geometry: If the dendrites are highly non-uniform in diameter, or there are branches, then these have to be represented by smaller compartments. On the other hand, if you do not care about branches, you can bring Rall's law into play and collapse many dendrites into a single one.

3. Do you want to model spines? If yes, then each spine will need at least two compartments: one for the spine head, and one for the neck. Addition of spines can easily make a neuronal model ten times larger. An alternative is to do a surface area rescaling by merging the effective conductance, resistance and charge of the spine into the parent compartment.
4. Will you be modeling active channels? If so, then the appropriate compartmental length must be small (around 0.05 λ, Cooley and Dodge 1966) because the potential gradients are quite steep. In a passive model the compartmental length can be much bigger, about 0.2 λ.

Having made these decisions, we are now in a good position to combine the cell morphology with its physiology to make a passive model at any desired level of detail. For detailed compartmental models, it is often useful to undertake a final refinement of the model passive parameters by fitting the somatic charging curves to simulations, with the *Ra* and *Rm* as adjustable parameters. Estimates of *Cm* and *Vm* are usually quite accurate, so they can be left alone.

7.2.6 Beyond Cable Theory: Non-compartmental Analysis, Ephaptic Interactions

Cable theory provides a simple and general approximation to the passive properties of cells. Its resolution, however, is limited to a 1-dimensional analysis of a neuron and the other assumptions of compartmental models. So, for example, cable theory is not able to tell one what happens if there are field gradients around the cross-section of a dendrite, or to do a general analysis of non-cylindrical geometries. Such calculations require much finer subdivisions of the membrane, into small patches that tessellate the surface of the cell. Only a few simulators, such as STEPS, are currently capable of this (Wils and De Schutter 2009).

Another assumption of cable theory is that the outside of the cell is all at the same potential. This is obviously not the case, since all field and extracellular recordings rely on potentials recorded outside the cell. For intracellular potentials, it turns out that an explicit representation of extracellular resistances is mathematically equivalent to a small correction on the axial resistances (Ra) of the regular compartmental model. For extracellular potentials, one can compute the local potential of any point in space by summing up the contributions of each compartment as a little dipole (Holt and Koch 1999). This works even for potentials generated by many cells. Here, even though the potential is not assumed uniform, there is an assumption of homogeneity of the extracellular medium.

Finally, in a few cases there are interactions between sections of dendrite on the same or different cells, that arise from these extracellular potentials. For example, tightly packed non-myelinated nerve bundles may experiences ephaptic current flow between axons, due to local potential gradients. This is especially likely if a volley of impulses goes down several axons at once. There are currently no general simulators to solve such cases.

7.2.7 Summary of Cable Theory

Cable theory describes how electrical signals propagate through cells in the absence of active channels. This is important because neurons are typically long, thin cells. There are three main things that happen to signals as they go down a long section of dendrite or axon: They get smaller, smoother, and are delayed.

7.3 Ions and Ion Flow

We now turn to the cellular battery, which arises from concentration gradients of ions across the neuronal membrane. This battery is what maintains the cellular resting potential, which is around −65 to −70 mV. In other words, if you stick an electrode into a cell, the inside is about 65 mV more negative than the outside. The energy stored in these ion gradients is used to power the more interesting, active properties of neurons, that we will discuss in the next section. In terms of the equivalent circuit for a compartment, each ion channel is represented as a battery in series with a variable resistor. The potential on this battery is the Nernst potential.

7.3.1 The Nernst Equation

The Nernst equation is:

$$E_{rev} = \frac{RT}{zF} \ln\left(\frac{[out]}{[in]}\right) \approx 59\, mV. \log\left(\frac{[out]}{[in]}\right) \quad at\ T = 298K \tag{7.14}$$

What this means is that if there is an ion-selective channel across a membrane, then there will be no net current flow across the membrane when the cell is at *Erev*, the reversal potential. This situation arises because the free energy gained by an ion going down the concentration gradient is exactly balanced by the energy required for the ion to move up against the potential difference *Erev*. Note that this is an equilibrium situation: no energy is lost if the channel is left open at *Erev*, as there is no net flux of ions.

The Nernst potentials in (mV) for some common ions in vertebrate tissue is below (from Hille 1992):

Ion	Na^+	K^+	Ca^{2+}	Cl^-
Intracellular conc (mM)	12	155	0.0001	4.2
Extracellular conc (mM)	145	4	15,000	123
ENernst (mV)	60	−90	120	−85

One frequently used outcome of the Nernst equations is to depolarize a neuron by raising the extracellular potassium ion concentration. The resting potential of a neuron is mostly determined by the Nernst potential for potassium. By doubling [*Kext*] we increase its E_{Nernst} by approx. 20 mV. This is usually sufficient to take the cell above its action potential threshold.

Another important implication of this equation is to estimate how the reversal potentials of different ions change during neuronal activity. It turns out that the concentrations of most ions are simply too large to be much affected by the flux through ion channels over the time-course of a burst of action potentials. The major exception to this is calcium. The resting concentration of free calcium is about 0.1 μM, and channel openings routinely raise this to 1 μM. Calcium concentrations in microdomains may be still higher. These changes substantially alter the effective Nernst potential for calcium.

In a few cases a similar situation has been reported for extracellular potassium, which has been estimated to significantly rise during intense activity in some enclosed neural structures such as olfactory glomeruli. This rise of [*Kext*] leads to further cellular depolarization, and stimulates still further activity.

7.3.2 The Goldman-Hodgkin-Katz Equations

What happens when an ion channel lets multiple ions through? This is particularly common for ligand-gated channels, such as the nicotinic acetylcholine receptor or the NMDA receptor. Both of these have a reversal potential close to zero, because they allow many cations, including potassium, through. A simplistic view of this would simply be to compute some sort of weighted average of the Nernst potentials of the different ions. The actual calculation is more complicated, and gives rise to two equations called the Goldman-Hodgkin-Katz (GHK) equations.

$$I_s = P_s z_s^2 \frac{VmF^2}{RT} . \frac{[S]_i - [S]_o . \exp(-z_s VmF/RT)}{1 - \exp(-z_s VmF/RT)} \quad (7.15)$$

This is the GHK current equation. It states that the current carried by the ionic species S is expressed in terms of the permeability P_s and charge z_s for the ion, and a function of the membrane potential Vm. This equation is useful in simulations for calculating the fraction of current through a channel carried by a given ion, based on its relative permeability and concentrations.

$$E_{rev} = \frac{RT}{F} \ln \left(\frac{\sum P_s[S]_o}{\sum P_s[S]_i} \right) \tag{7.16}$$

The GHK voltage equation shown here computes the reversal potential of a channel where all ions S have a valence of $+1$. This reduces to the Nernst equation if only one ion is present. Univalent negative ions can also be placed in this equation, by putting the $[S]_i$ terms on top, and vice-versa. If the permeant ions differ in valence then the equation becomes still more complex (Hille 1992).

Why are the GHK equations so much more complex than the Nernst equation? This is because the Nernst equation is an equilibrium one: it defines the potential on purely energetic criteria. However, an ion channel passing multiple ions does not have an equilibrium even at its reversal potential. Instead, at this potential, the net charge passing through the channel is zero, but the flux of individual ions may still be substantial. This leads to a nasty singularity in the GHK (current) equation, because then the fraction of current carried by any ion is infinite. The usual way around this is to tabulate the potential dependence of current fraction, and to interpolate current around the singularity.

7.3.3 Summary of Ion Flow

Ionic concentration gradients provide the energy for neuronal electrical activity. The potentials generated by these ion gradients are described by the Nernst equation for single-ion channels, and by the more complex Goldman-Hodgkin-Katz equations for ion channels that carry multiple ions. In a few cases, specially involving calcium, changes in ion concentrations during activity significantly affect the potentials driving current through ion channels.

7.4 Voltage-Gated Channels

Passive cells sum current linearly. This is a fine thing for analysis, but their computational repertoire is limited. Furthermore, they cannot propagate information very far. Real neurons are adorned with a huge variety of voltage and ligand-gated ion channels, and these are what determine much of the electrical character of each cell type. In this section we discuss the modeling of voltage-gated ion channels. We will focus on the Hodgkin-Huxley description, which is the original and still the most commonly used description of how such channels act. To understand this, we need to take a look at the experimental voltage-clamp technique, which was used by Hodgkin and Huxley and remains an essential electrophysiological tool.

The Hodgkin-Huxley description is a little complex, so we will visit the Fitz-Hugh Nagumo equations that are more convenient for mathematical analysis. To wrap up the section, we will see how many channels are modeled in the same compartment, and deal with the further elaborations that are introduced by calcium dependence.

7.4.1 The Hodgkin-Huxley Formulation

The original and still by far the most commonly used description of voltage-gated ion channels is based on the work of Hodgkin and Huxley (1952a, b). Their formulation describes ion channel conductance in terms of a several 'gates', named m and h for sodium, and n for potassium:

$$G_{Na} = g_{\max} m^3 h \tag{7.17a}$$

$$G_K = g_{\max} n^4 \tag{7.17b}$$

The general form of this can have multiple gates, a, b, c, each with their own powers:

$$G = g_{\max} a^x \cdot b^y \cdot c^z \ldots \tag{7.17c}$$

These gating variables represent processes which all have to be in the open state for current to flow. Think of the each of these as a 'gate' in the channel, which can be in either of two states: closed or open. If the total fraction of the gate is 1, and the amount open is m, then the equation is:

$$(1 - m) \underset{\beta}{\overset{\alpha}{\longleftrightarrow}} m \tag{7.18a}$$

The crucial voltage dependence in these equations come from the α and β terms, which are functions of voltage. Although the α and β terms are the ones that define the dynamics of the system, from an experimental viewpoint, it is easier to transform these into steady-state and time-course parameters as follows:

$$m_\infty = \frac{\alpha_m}{\alpha_m + \beta_m} \tag{7.18b}$$

$$\tau_m = \frac{1}{\alpha_m + \beta_m} \tag{7.18c}$$

The form of the original Hodgkin-Huxley equations is illustrated for the m-gate of the Na channel in Fig. 7.7.

Even though these were expressed as analytic functions of voltage in the calculations of Hodgkin and Huxley, it is important to keep track of the experimental

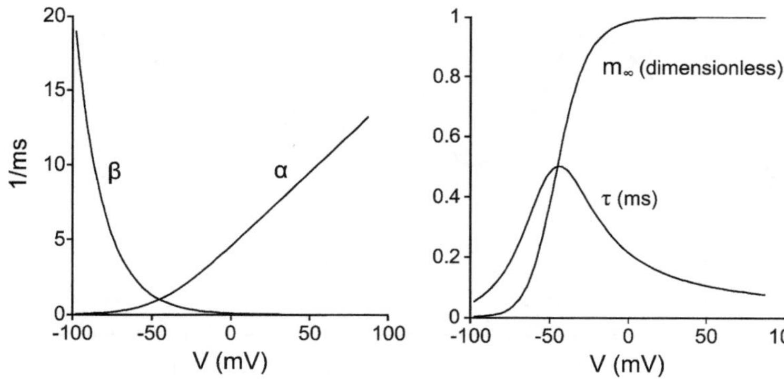

Fig. 7.7 Voltage dependence of the m-gate of the Hodgkin-Huxley sodium channel. On the left we have the α and β terms, and on the right the equivalent data plotted as m_{∞} and τ

roots of these functions. Specifically, these functions are curve-fits to tabulated experimental data. There is no particular reason for these fits to be simple functions of voltage. Fortunately, numerical computations do not need the analytic forms, and are typically done using tabulated lookups. The issues of specifying these channel parameters are discussed elsewhere in this volume.

7.4.2 FitzHugh-Nagumo Formulation

When analyzing neuronal dynamics, it may be useful to capture some of the channel dynamics at an intermediate level of description between integrate-and-fire neurons and the full Hodgkin-Huxley equations. Such a form was developed by FitzHugh (1961) and implemented using tunnel diodes by (Nagumo et al. 1962) (Fig. 7.8). The equation is

$$\frac{dV}{dt} = V - \frac{V^3}{3} - W + I \tag{7.19a}$$

$$\frac{dW}{dt} = 0.08\,(V + 0.7 - 0.8W) \tag{7.19b}$$

Where V is membrane potential and W is a recovery variable. This system is built around a cubic dependence of conductance on membrane potential. This third-order dependence gives it a region of negative-slope. The intersection of the curves $dV/dt = 0$ and $dW/dt = 0$ defines the resting state of the system, and when I is increased to 1 this intersection point moves into the negative-slope region, and the system oscillates. In brief, this equation generates membrane potential responses similar to those of the Hodgkin-Huxley equations, while remaining ana-

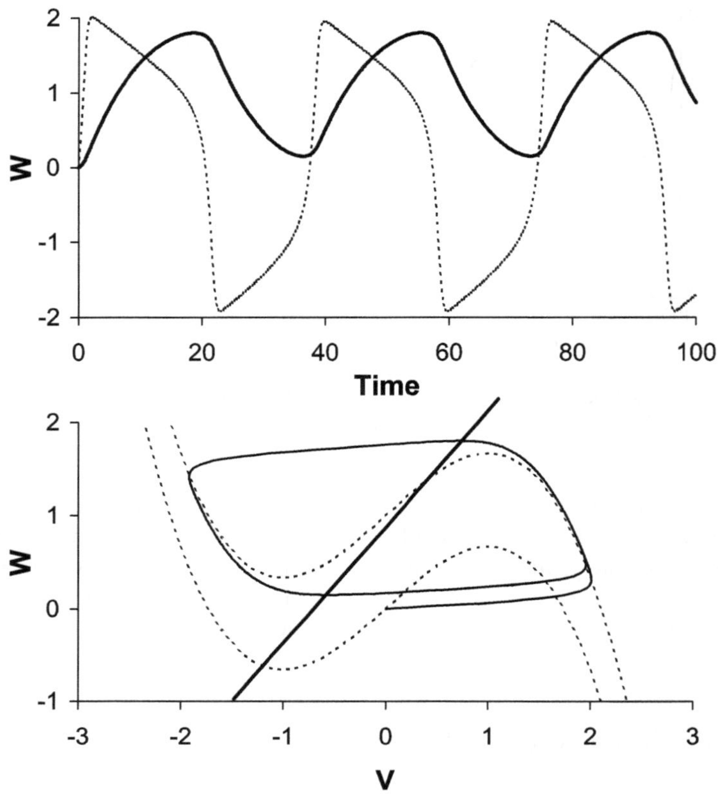

Fig. 7.8 FitzHugh-Nagumo equations. *Above*: the *dashed line* in the time-series represents the membrane potential, V, and the *solid line* represents W. *Below*: Phase diagram for oscillation of the system

lytically tractable. These equations are typically applied in mathematical treatments of single-compartment neurons. Despite its mathematical utility, the FitzHugh-Nagumo form is rarely used in compartmental modeling of the kind we discuss. The mathematical abstractions separate the firing activity from biological entities such as different channel classes. Thus, for doing detailed models, and with cheap computing power, it is more common to use standard channel simulations.

7.4.3 Experiments and the Voltage Clamp

Before going on to explore the properties of the Hodgkin-Huxley model, we need to make a small experimental excursion. We have already seen how we can monitor potential changes in the cell and apply current pulses. This is called current clamp:

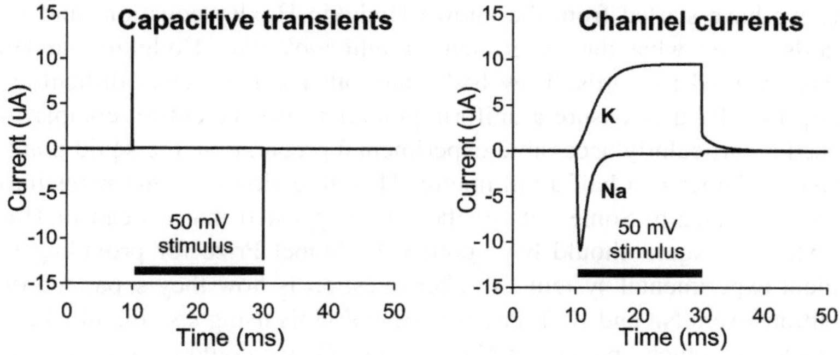

Fig. 7.9 Voltage-clamp of a membrane compartment. *Left*: Transient large currents are required to charge and discharge the membrane capacitance. *Right*: Na and K currents during the voltage step

we clamp the current to a desired value (often zero) and measure the voltage. The voltage clamp is the converse: we clamp the cellular potential to a desired value, and measure the current needed to keep it there. The voltage clamp is particularly valuable in understanding voltage-gated ion channels, because their properties are a dynamic function of voltage.

First, consider what happens when we have a passive compartment. Consider a voltage step: we clamp the voltage to resting potential (−65 *mV*) and then to −45 *mV* for 100 *ms* before stepping the voltage back to −65 *mV*. In order to raise the potential from −65 to −45 *mV*, we need to charge up the capacitor in the membrane. The voltage clamp circuit tries to do this very quickly, so we have a large but brief current pulse. The charge needed is $Q = C.\Delta V$, and this flows in a short time ΔT, so $I = C.\Delta V/\Delta T$. In order to bring the potential back down after the 100 *ms* are up, we need to discharge the capacitor. This results in another large but brief current pulse, with the opposite sign (Fig. 7.9).

Now, suppose we put in a potassium channel. From the Hodgkin-Huxley equations and Fig. 7.9, we see that it will slowly open when the potential is raised, reach a pleateau, and then slowly close again when the potential goes down again. The plateau value of the current is determined by the steady-state open fraction of the channel. This is precisely the parameter n_∞ from the equations above. The time-course of this opening is gives τ_n. This experiment is repeated for different values of voltage, in order to build up the parameters for the Hodgkin-Huxley model for the K channel.

The Na channel is more complicated. It has both an *m* and an *h*-gate. If we give the Na channel a voltage step, the current response looks like Fig. 7.9. Allowing for the fact that the sign of the current is the opposite to that for potassium, the first part of the curve is very similar. This is because the m-gate for the Na channel has a roughly similar form to the n-gate for potassium. However, after a small interval, the h (or inactivation) gate for Na kicks in. This begins to *close* at higher potentials. Thus the channel as a whole starts to close soon after opening, and is almost entirely shut (except for a small residual current) within 5 ms.

Here we have started from the known Hodgkin-Huxley equations and worked backwards to see what the experiments might look like. Hodgkin and Huxley, of course, worked forwards. They had many other experimental difficulties. For example, how did they ensure a uniform potential over the entire 'compartment'? They used a particularly accessible experimental preparation, the squid giant axon which has a diameter of half a millimetre. They threaded very fine wires into it to do the voltage clamp. Some wits of the time suggested that instead of Hodgkin and Huxley, the squid should have gotten the Nobel Prize for providing such a convenient experimental system. Another question is how they separated out the contributions from Na and K. Today we could do this using specific blockers such as tetrodotoxin, which knocks out Na channels. Or we could express one specific channel in the oocyte system. Hodgkin and Huxley had no such luxury. Instead they simply did the experiment in zero sodium, thus isolating the K channel currents. To find the Na channel currents they redid the experiments in regular solution, subtracting out the K contributions.

7.4.4 Properties of the Hodgkin-Huxley Model

The most striking property of the Hodgkin-Huxley model is that it fires and propagates action potentials. We start with the original Hodgkin-Huxley calculations, which eliminated space. When a depolarizing current pulse is applied to the compartment, the membrane potential rises. The sodium channels open, because the m-gate has an increasing opening probability with depolarization. This lets in more sodium ions, and the potential increases still further. This positive feedback loop is interrupted in two ways: first, the potassium channels begin to open as well, and then the inactivation (h-gate) process of the sodium channel starts to take effect. So, just as the sodium channels are shutting of their own accord, the potassium channels open wide. Many potassium ions leave the cell, and pull the potential down below resting levels. The net result of this cycle is that an action potential is generated and the cell is reset back down to resting potential or lower (Fig. 7.10).

There are many interesting features of the action potential and the Hodgkin-Huxley model, which are best explored with a simulator such as MOOSE or NEURON. Here is a list of properties to try to understand. Hint: many of them are outcomes of the inactivation of the Na channel.

- Why is the first action potential of a series usually the largest?
- Why is there a refractory period? That is, why is it harder to trigger a second action potential a short while after the first?
- Why is there only one action potential in a certain low current injection range?
- Why is there a little glitch on the upswing of the action potentials?
- What happens as the injection current is increased to very large values?
- How can you generate an action potential using a negative (hyperpolarizing) current?

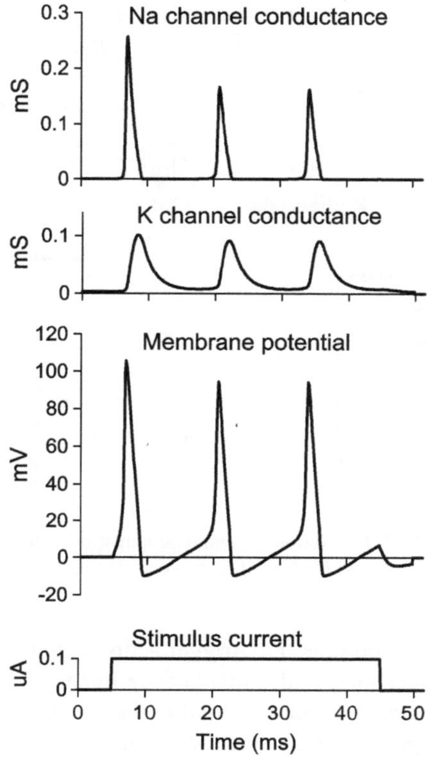

Fig. 7.10 Generation of action potentials by the Hodgkin-Huxley model in response to a current step

Hodgkin and Huxley painstakingly computed an action potential in a single compartment using a hand-cranked calculator to solve their differential equations. To model propagation in space, it is necessary to solve equations for lots of compartments connected up using their axial resistances. This was not an attractive prospect using manual calculations. Fortunately digital computers came to the rescue, and in 1966 Cooley and Dodge had done the first calculations of action potential propagation in a cable. Using our understanding of cable theory and the Hodgkin-Huxley equations, we can see the underlying mechanisms. Suppose an action potential is propagating along the cable from left to right. In the middle there is a region of strong depolarization, where sodium channels are open and potassium channels are not. Current flows in to this region, and through passive dendritic propagation it spreads to either side, leading to depolarization. At the leading edge, this triggers the same sequence of channel openings as we discussed for single-compartment action potentials. At the trailing edge, however, the passive current runs into the 'tail' where a spike has just been. Here the sodium channels are inactivated, and the potassium channel is still open and pulling the potential more negative. This confers direction to the action potential: it can only go forward. Note that 'forward' depends on the history of the axon, and it is quite possible

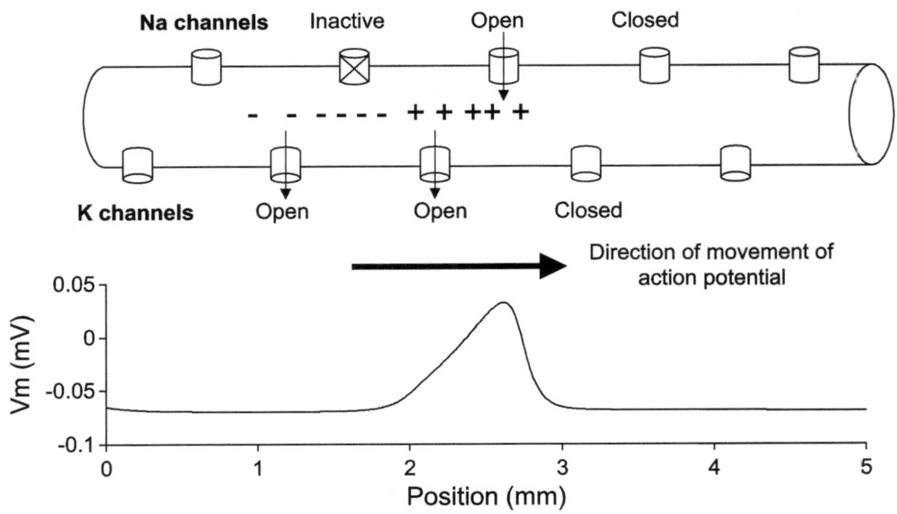

Fig. 7.11 Action potential propagation. The opening and closing of ion channels is indicated above, and the travelling wave of depolarization is shown below

to experimentally trigger action potentials that go from the end of the axon back to the soma. With suitable experimental (or model) manipulations it is also possible to start an action potential in the middle of the cable, or to have two action potentials approach each other on the same axon on a collision course. These are entertaining cases to figure out and simulate (Fig. 7.11).

From a physiological viewpoint, there are two related features of action potential propagation that are particularly interesting: speed and failure. The importance of propagation speed is apparent in any natural habitat: the faster-responding animal is more likely to survive. The simple and expensive way to speed up action potential propagation is to increase the axon diameter and the density of ion channels. While the detailed calculations are complex, cable theory tells us the general principle: with a larger diameter, current spreads further and faster. From empirical and theoretical studies, the following relationship seems to apply for non-myelinated fibers (Hursh 1939),

$$velocity \ \alpha \ \sqrt{diameter} \tag{7.20a}$$

This is why the squid so helpfully provided a giant axon for Hodgkin and Huxley to work with. The giant axon is used in the squid escape response, and the only way to speed it up using invertebrate technology was to make it bigger.

Vertebrates came up with a much better solution, which was to insulate the cable. This is done through myelination, in which an oligodendrocyte wraps an axon in layer after layer of insulating membrane. There are two good outcomes of this. First, the insulation (higher *Rm*) means that very little current leaks out of the membrane.

Second, the many layers of insulation reduce effective membrane capacitance (Cm) by increasing the distance between intracellular and extracellular potentials. Both of these mean that passive propagation is faster and longer-range. Myelinated fibres also conduct faster for a larger diameter (Rushton 1951).

$$velocity \propto diameter \tag{7.20b}$$

The active propagation of the action potential is now handled entirely at the nodes of Ranvier, which are the junctions between myelinated segments. Nodes of Ranvier act like booster stations along the axon, and they have very high Na and K channel densities to play this role. They are spaced 25 – 250 μm apart, and action potentials basically jump from one node to the next. This is called 'saltatory conduction'. The downside of this arrangement is that it is possible for propagation to fail if the nodes are too far apart, or equivalently, if there is too much electrical shunting of current between nodes. This is especially likely at branch points. In pathological cases it may also happen if the myelin degenerates. Experiments show that in normal cells the safety factor is quite large: the actual propagated potential is typically twice as high as needed to trigger the action potential at the next node of Ranvier.

7.4.5 Modeling Many Channel Types

While Hodgkin and Huxley centred their analysis around the squid ion channels, their formulation turns out to be general enough to work for many channel types. The key parameters needed to specify a Hodgkin-Huxley-type channel model are the channel conductance, number of gates, the power to use for each gate, and the voltage dependent terms (α and β, or equivalently τ and m_∞) for each gate. Typical curve-fits for the α and β terms for each gate involve five parameters each. This comes to 13 parameters for a typical channel. This is a lot of parameters. Most neurons express a few to dozens of varieties of ion channels, and one of the key jobs of in experiments and models is to decide which of these contribute most to the properties of the cell.

An experienced physiologist can just look at an intracellular recording and recognize a half-dozen subtypes of ion channel. In general, most of the variety comes from subtypes of potassium channels. In this figure we see some of the signatures of different ion channels (Fig. 7.12).

It is somewhat challenging to find parameters for ion channels, and there is no real substitute for the same painstaking kind of analysis that Hodgkin and Huxley did. The standard experimental sequence these days is:

- Identify channel, through whole-cell clamp data or blockers
- Obtain gene (RT-PCR or fly genetics)
- Express in frog oocytes
- Electrically characterize, using more or less the Hodgkin-Huxley protocol

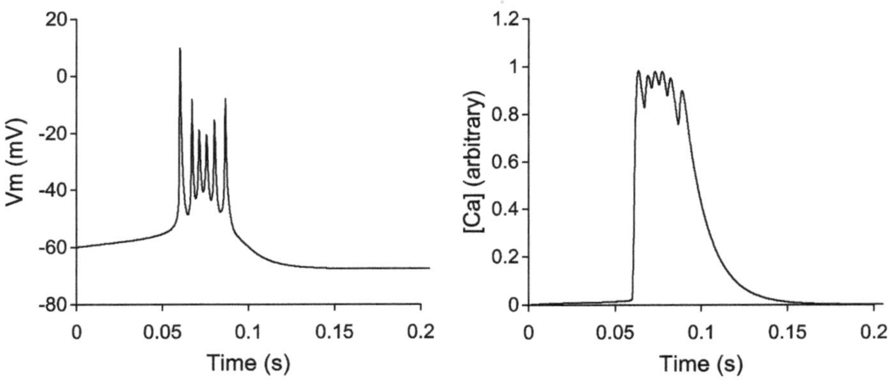

Fig. 7.12 Complex action potentials in a model with multiple ion channel types and calcium dynamics. Model based on Traub et al. 1991. *Left*: Membrane potential at soma. *Right*: Calcium buildup during burst of action potentials

– Determine calcium dependence
– Determine channel distribution in the cell.

Few labs currently go through this process systematically. There is now a slow accumulation of channel models in the databases, but many of these can trace their lineage to just a few original studies. One key point to note is that there is nothing sacred about the equations that Hodgkin and Huxley used to fit the alpha and beta terms. Simpler forms may do well, and most simulators are also quite happy to use the original data tables without any fitting at all.

7.4.6 Calcium Modeling

Calcium is a special case when it comes to modeling. The opening of many ion channels, especially potassium channels, depends on calcium as well as on voltage. The simplest approach is to define opening as a function of calcium, but this is not a good description in many cases. A fairly general approach which seems to work well in many cases is to treat calcium concentration itself as controlling the opening of a gate, using a similar form as for the voltage dependence of gating. This can get complicated, when for example the same gate depends on both voltage and calcium. In other words, the α and β terms are functions of both Vm and $[Ca]$. For this to work, we need to compute intracellular calcium as well as potential.

Unlike other major cellular ions, calcium is subject to very large concentration fluctuations. Resting levels of free calcium are around 100 *nM*, whereas stimulated calcium levels frequently exceed 1 μM. (Sabatini et al. 2002; Berridge 2006). Modeling calcium can become extraordinarily complicated, especially if we consider intracellular stores and the many signalling pathways involved in calcium dynamics.

However, there is a relatively simple biophysical approximation that is used in a very large fraction of compartmental neuronal models. Such models have a single calcium pool in each compartment, which sums ion fluxes from all the channels that are permeable to calcium. These models typically assume a standard resting concentration and a single extrusion (pump) process giving rise to exponential decay of calcium concentrations.

$$\frac{d[Ca^{2+}]}{dt} = k.\sum Ca^{2+}_{currents} - \left([Ca^{2+}] - [Ca^{2+}_{basal}]\right)/\tau \qquad (7.21)$$

Here k incorporates the Faraday constant and dendritic volume, and τ is the time-course of the pump.

Given the large concentration changes in intracellular levels of free calcium, it is important to model how ion fluxes change as the calcium gradients change. At the very least, this involves the use of the Nernst potential equation for the reversal potential for calcium-selective channels. Most voltage-gated calcium channels are quite calcium selective. However, ligand-gated channels frequently are permeable to several cations including calcium, and here we have to employ the GHK equations.

Many models demand more accurate estimation of calcium levels. For example, it is well known that there are steep gradients of calcium concentration in the cell, also known as microdomains (Berridge 2006). One way to handle this is to use an 'onion-shell' model for calcium levels. Each shell has a uniform concentration, and in addition to the fluxes from channels and pumps, also diffuses to the next inner and outer shells (Yamada et al. 1989). More complex spatial schemes are often necessary. In other chapters we discuss just how involved calcium signalling can become when spatial and biochemical details are added in. Few current neuronal models go beyond the shell model.

7.4.7 Beyond Hodgkin-Huxley

The Hodgkin-Huxley model has been extraordinarily successful as the basis for neuronal modeling, and to this day it is the standard representation in almost all modeling studies. The formulation is general enough to incorporate calcium dependencies, as we have seen above. The Hodgkin-Huxley description includes enough biological realism that single-channel stochastic responses can be modeled with it. This is done by considering the state transitions of individual 'gates' in the channel as if they were reactions, and using one of the standard stochastic methods (such as the Gillespie method (Gillespie, 1977) discussed elsewhere in this book) to compute trajectories of channel transitions.

Despite this remarkable endurance, the Hodgkin-Huxley model has a rather limited view of state transitions in a channel protein. It is now possible to resolve more detailed transitions using modern single-channel recording methods. From the viewpoint of modeling, the question is whether these additional details bring qualitative improvements in model fidelity.

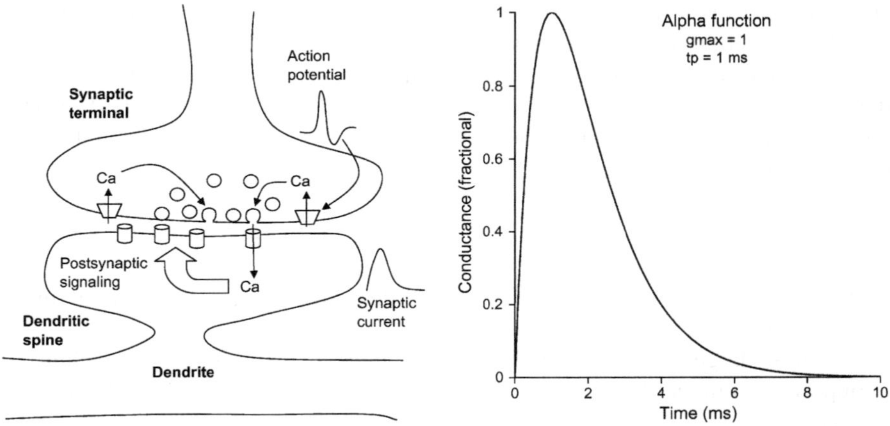

Fig. 7.13 Synaptic transmission. *Left*: Sequence of events during synaptic transmission. *Right*: alpha-function form for channel conduction following spike arrival

7.4.8 Summary of Hodgkin-Huxley Channels

Most voltage-gated ion channels, including calcium-dependent channels, are rather well described using the Hodgkin-Huxley formalism. In this, the channel consists of a number of gating entities (which have a loose correspondence to channel subunits), each of which switches between closed and open states in a voltage-dependent and sometimes calcium-dependent manner. The channel conducts when all gates are open.

7.5 Synaptic Channels

Synapses are enormously complex entities from a cell-biological viewpoint. They convert voltage signals (action potentials) into chemical signals (neurotransmitter release) and back again into electrical signals. They store information, and arguably are the major cellular basis for memory. Several of the chapters in this book analyze their signalling, their role in memory, and details of neurotransmission.

The biology underlying all this is packed into a $\sim 1\ \mu m$ structure including a presynaptic bouton, a very narrow gap between cells, and a postsynaptic site that in many cells sits on a mushroom-shaped bulge called a synaptic spine (Fig. 7.13). The presynaptic bouton contains elaborate machinery to put neurotransmitters into pools of tiny ($\sim 40\ nm$) vesicles. When an action potential reaches the presynaptic bouton, the resulting depolarization causes L-type calcium channels to open. The resulting calcium influx triggers the fusion of synaptic vesicles with the membrane, thus releasing neurotransmitter. This diffuses across the $\sim 20\ nm$ synaptic cleft. On the postsynaptic side, the neurotransmitter molecules bind to receptor channels and non-channel (metabotropic) receptors. The opening of the receptor channels leads to postsynaptic electrical effects, which is where the cell biology rejoins the electrical description of neuronal function. It is remarkable and very convenient for neuronal and network modeling, that much of this collapses into a very simple alpha-function description of their dynamics.

7.5.1 The Alpha Function and Its Derivatives

From a purely electrical viewpoint, a synapse is just a conductance which rises and falls in a stereotypical manner following synaptic input (Fig. 7.13). The shape of a typical single synaptic current is very well represented by an alpha function of the form

$$g(t) = g_{\max}.\frac{t}{tp}.\exp\left(1 - \frac{t}{tp}\right) \tag{7.22}$$

where t is the time since last spike, and tp is the time-to-peak of the synaptic current.

One can make this a little more general, and have different rise and fall times with a closing time parameter t_2 and a normalization factor A:

$$g(t) = \frac{Ag_{\max}}{\tau_1 - \tau_2}\left(\exp(-t/\tau_1) - \exp(-t/\tau_2)\right) \tag{7.23}$$

This is the dual alpha function. Most neuronal models stop here. The synaptic conductance term fits nicely into the existing representation of an electrical compartment, and the whole thing is numerically solved along with the conductance contributions of the Hodgkin-Huxley channels.

Using the alpha function as a core, there are a number of straightforward but important extensions that provide more detail to synaptic models.

- *Stochastic synaptic release*. The process of neurotransmitter release, as described above, is immensely complex, and in the end it typically comes down the fusion of an individual vesicle. This is stochastic. To make life interesting, it is also strongly history dependent. There are at least two processes acting, in opposite directions, to change the probability of release. One is the accumulation of calcium, which raises release probability. The other is the depletion of available vesicles, which reduces it. Each of these features is simple enough to model on its own, but few models take them all into account.
- *Plasticity*. The total synaptic weight is the effect one cell has on another through a synapse. This is the product of at least three terms: the number of receptor channels (n), the release probability (p), and the quantal size (q) (Stevens 1993; Lisman et al. 2007). The quantal size is itself a product of individual receptor channel conductances with the amount of neurotransmitter in each synaptic vesicle. Even leaving aside many known elaborations, there is plenty of scope

for model complexity in representing these three terms. Other chapters in this volume look at different models for plasticity. The most common forms ignore stochastic release, and modulate only the maximal conductance parameter gmax, of the synaptic alpha function.

- *Voltage dependence*. In addition to the presence of ligand, the opening of some synaptic receptor channels depends on membrane potential. This dual dependence results in the crucial property of associativity, in which activity in both the presynaptic and postsynaptic cell is required for synaptic plasticity. An important example is the NMDA subtype of glutamate receptors. The NMDA receptor is plugged by a Magnesium ion at low membrane potentials. Thus it opens only when the postsynaptic cell is depolarized *and* when there is ligand present. Since the NMDA receptor permits calcium entry into the postsynaptic cell, this acts as a key associative trigger for many aspects of synaptic plasticity. From a modeling viewpoint, the addition of the voltage dependence is a straightforward product term onto the channel conductance. However the NMDA receptor is permeable to several cations, so the GHK equations need to be added into the mix, resulting in a moderately complex channel model.

Like the venerable Hodgkin-Huxley model, the alpha function form arises from a particularly simple view of channel biophysics, in which we have a transition between open and closed states of a large enough population of individual channel proteins that a deterministic assumption is good. Like the Hodgkin-Huxley model, we now have the experimental resolution to construct multi-state models of many important synaptic channels.

7.5.2 Summary of Synaptic Channels

Synaptic events are rather well described by a simple alpha-function time-course of conductance on the postsynaptic receptor, following the arrival of an action potential. Many important physiological details, such as stochastic transmitter release, plasticity, and voltage dependence, can be added to this alpha function to give a more complete model of synaptic properties.

7.6 Practicalities of Modeling Neurons

In this chapter we have seen a number of equations: the cable equation, the Hodgkin-Huxley equations, the Nernst and Goldman-Hodkgin-Katz equations, equations for calcium dynamics, and finally the synaptic alpha function. How are these put together to numerically simulate neuronal dynamics?

7.6.1 Passive Model

The first step in doing a neuronal simulation is to solve the cable equation. This is a partial differential equation, as it depends both on space and time. It turns out that the approach used to simulate it is the same as that used to derive the equation in the first place: we subdivide the cell into small compartments, which we represent as electrical circuit elements. Recall that the derivation required us to take the limit as compartmental size approached zero. The fundamental requirement here is that all properties within a compartment should be uniform. It turns out that for passive models of a uniform cable one can get away with compartments up to about 0.2 length constants. However, if the cell is highly non-uniform with changes in diameter and dendritic spines, then the compartmental sizes should be dictated by the geometry. As in all modeling, the optimal compartment size is a matter of judgement, and depends on what information you want to find from your model.

Having chopped up the cell, we now have a branching circuit of little compartments coupled by their axial resistances. Sometimes there are branches. If we write down the current balance equations, we end up with a system of equations that is best expressed as a matrix. This system of equations has the unpleasant property of stiffness. This is because neurons have a wide range of time-scales for the electrical potential to settle. A small spine connected to a large dendrite will settle very quickly, but the charging time of the soma is orders of magnitude slower. Such systems need to be solved using a class of techniques called 'implicit' methods (defined elsewhere in the book). Fortunately, several people, notably Hines (1984) recognized that the branching structure of neurons allows the calculations to be done in order (N) time, where N is the number of compartments. Such algorithms are now the basis for almost all neuronal modeling programs.

7.6.2 Ion Channels

The next step in carrying out neuronal simulations is to solve for the channels, starting with the Hodgkin-Huxley terms. These are nasty nonlinear equations, but fortunately their coupling to the cable equation is through conductance terms, which are linear. This means that the gating variables of the channels are solved almost independently of the cable equation part, interacting only to take the voltage term from the cable equation. This gives us the values of the gating variables m, h, n and so on. Using these state variables we can compute the channel conductance using the three parts of Eq. 7.17. This term, with the channel reversal potential, now contributes back into the cable equation calculations for each compartment.

The channel calculations get a little more tedious, but not numerically difficult, if we have to add in elaborations for Nernst potential, calcium buildup, and other variations we have discussed above. Nernst potential calculations (Eq. 7.14) only need to be done if the ion concentration is likely to vary at run-time. This is usually only the case for calcium. Here too, we have to numerically integrate the concentration term, taking into account contributions from each of the ion channels, pumps and buffers, as in Eq. 7.21.

If some of the channels contributing to the calcium pool conduct multiple ions, it becomes necessary to evaluate calcium ion flux using the Goldman-Hodgkin-Katz equations (Eqs. 7.15 and 7.16).

Handling synapses is the next step. As far as the axon goes, this is trivially handled by taking the spiking threshold of the neuron and converting it into a digital event. The digital 'spike' event is specified by its time, and propagation delay to each of the target synapses. Once it reaches the target synapse, the system has to manage all incoming action potentials in a queue so that they are handled at the target synapse with the correct delay. Once the spike arrives at the synapse, there are several variations on the computations depending on the level of detail of the synaptic model. The simplest case is to always trigger an alpha-function for conductance change, of a fixed amplitude. As with the Hodgkin-Huxley channels, the coupling to the compartments is only through this conductance, so the calculations for synapses can be handled almost independently of each other and of the matrix solution for the cell as a whole. Elaborations to this, as discussed above, include probabilistic synaptic release, facilitation and depression, and various forms of synaptic plasticity which may in turn depend on calcium and other cellular context. Some of these elaborations are discussed in other chapters in this book.

7.6.3 Computational Costs

Modern (circa 2010) computers can solve a reasonably detailed 100-compartmental model in real time, on a single CPU core. Due to the linearity of the cable calculations, bigger models go proportionately slower. Various simulators have optimizations for parallel operation, or numerical integration using variable timesteps, which may improve on this somewhat. Although real-time may seem pretty good for any biological system, speed is important in a several ways. Network models may involve thousands of neurons. Parameter refinement for models may use huge numbers of runs (Van Geit et al. 2007). Additionally, real time itself is pretty slow for some of the most interesting neuronal phenomena which one would like to model, such as synaptic plasticity. These experiments take may take hours, so it would be nice to run these a lot faster than real time.

There are some surprises when it comes around to actually implementing neuronal compartmental calculations. It turns out that the cable equation calculations are actually the fastest to perform. In typical neuronal models, each compartment has half a dozen or so Hodgkin-Huxley-channels, and updating their state variables takes much longer than the cable calculations. Another surprise is that in large network models, managing the synaptic queues may take longest of all: the bookkeeping costs more than all the real numerical work put together.

This is because each neuron receives ~10,000 synaptic inputs, which are typically condensed onto a much smaller number of modeled synapses and compartments to make the network model manageable. Thus each synapse may have many incoming action potentials to sort in order of time of arrival. Plasticity rules make these calculations worse still. With such simulations one begins to appreciate that just like in the brain, the really heavy computation happens in the synapses.

Real simulations will, of course, emphasize one aspect or another of this overall picture. Typical detailed neuronal models have several hundred compartments and a dozen or more different channel types scattered over them. Network models typically use much simpler neuronal models, down to ten or fewer compartments. Many network models just employ integrate-and-fire models, which are essentially a single compartment with a reset mechanism to replace the Hodgkin-Huxley channels. Izhikevich has done an exhaustive survey of different neuronal firing patterns and the simplest way to model them (Izhikevich 2003). At the other extreme, there are now several models which look in exhaustive detail at one or a few synapses on a dendrite. These models include not only detailed electrical events, but also consider biochemical signalling and even molecular traffic as part of their purview.

7.6.4 Tools for Neuronal Modeling

There are a number of simulators that can handle all or some of the equation types discussed in this chapter. The most prominent of these are NEURON (Carnevale and Hines 2006) and GENESIS (Bower and Beeman 1995), both of which are now about 20 years old. NEURON is still actively developed primarily by Michael Hines. GENESIS in its original form is no longer being developed, but GENESIS 3 (Cornelis et al. 2012) and MOOSE (Ray and Bhalla 2008) are complete reimplementations that have varying degrees of backwards compatibility and many advanced forward capabilities. There are other simulators that focus on single neurons, good interfaces (Gleeson et al. 2007), or large networks (Eppler et al. 2008).

Databases and standards for neuronal modeling have been relatively slow to develop. The most established neuronal model database is ModelDB (Migliore et al. 2003; URL: http://senselab.med.yale.edu/senselab/modeldb). This is primarily a model repository in the original simulator formats, and models from NEURON are well represented.

Model exchange formats for neuronal models are nascent. The NeuroML project is the oldest of these, and is undergoing considerable revision at the time of writing (Crook et al. 2007). NeuroML started life as a common format that could be translated into either NEURON or GENESIS code, which is a subset of the role of a modeling standard. NeuroML models can be set up using the NeuroConstruct software tool (Gleeson et al. 2008). This situation is evolving, and both Neuron and MOOSE are now able to natively read at least some part of NeuroML. The International Neuroinformatics Coordinating Facility (INCF, www.incf.org) is pushing for such standards development.

7.6.5 The Practice of Neuronal Modeling

It has been 60 years since Hodgkin and Huxley made the first biophysically detailed neuronal model. Many decades have passed since Wilfrid Rall put together most of the rest of the analytical toolkit. Even the modeling tools themselves have had over 20 years to mature. Experimentalists have been churning out studies to analyze neurons and ion channels all this time. Where are the models?

The reason that even databases such as ModelDB have relatively few models is that it is hard to parameterize models. The outline of the process is straightforward enough: match geometry and passive properties, identify channels, parameterize them, and fine-tune. Like other modeling disciplines represented in this book, numbers are the problem. As a result, most existing neuronal models are an interesting chimera of faithful neuronal morphology, adequate passive properties, and ion channels adopted and mutated from assorted past models.

The passive model is relatively easy to fit. As discussed in the cable theory section, one starts with the neuronal geometry and decides how much detail is needed in the final model. Sub-threshold current clamp recordings provide sufficient information to work out most of the passive properties. The key requirement for the current-clamp data is that the current injection curves should involve small enough currents that the voltage-gated channels do not contribute. If we assume that passive properties are uniform over the whole cell, only four numbers need to be extracted from the recordings: *Vm*, *Rm*, *Cm*, and *Ra*. As discussed above, *Vm* is directly measured, and *Rm* and *Cm* can be estimated from the cell charging curves. The electrotonic length *L*, also available from these measures, can be used to estimate *Ra* if the cellular geometry is known. This estimate of *Ra* depends on how much detail you put into your cellular geometry. Assuming a cell model with complete morphology, it is relatively easy to do curve-fitting (manual or automated) to estimate *Ra* and to refine estimates for *Rm* and *Cm*.

Obtaining channel data is far more difficult. The process of experimentally characterizing a neuronal channel is outlined in the section on voltage-gated channels. Few groups have attempted to do the entire experiment-to-model building process on their own. Many published models use Hodgkin-Huxley channel parameters obtained in separate experiments (often from different species), and try to fit them to limited current- or voltage-clamp data by tweaking channel densities and other parameters. This process has been impeded further by the lack of standards for defining channel models. It is also clear that ion channels are distributed in a very non-uniform way over the surface of neurons. Many details of cellular responses depend on this distribution, which again is difficult to measure.

Synaptic channel data are not much better in terms of precision. However, the number of parameters involved is much smaller and there are fewer channel types. For example, the time courses for the alpha function are readily obtained from EPSC recordings. The big unknown in synaptic parameters is the weight of individual synapses. This is crucial for network properties, but is generally known only as a population mean and variance. Further synaptic data such as release probability are rarely modeled, and again, these are usually only known in statistical terms.

The final stage in developing a model is to fine-tune it so that the spiking output of the cell as a whole looks like experiment. This typically involves parameter sensitivity analysis and parameter optimization. Due to the highly nonlinear spiking responses of neurons, it is non-trivial to come up with a measure of 'goodness of fit' for models to experimental data. Nevertheless, this has been done in some cases (Van Geit et al. 2007). These tuning calculations typically require many hundreds of individual simulation runs. It is a good idea to try to fit cells using as many conditions as are experimentally available: often a cell optimised to match just one readout is very unphysiological on others. Typical experimental readouts include voltage traces for different current levels, time to first spike, responses in the presence of specific ion channel blockers, and voltage clamp recordings.

7.6.6 Summary of Practicalities

There are now well-established and efficient methods and simulators to handle the numerical aspects of neuronal modeling. Experimental data and actual models are not so easily forthcoming, although the process of going from experiment to model is well understood. The situation for model exchange is still at a very early stage, as modeling standards are only just taking shape and the database formats are consequently limited to single-simulator implementations.

7.7 Conclusion

Multicompartment neuronal models occupy an interesting position between different scales of brain function. They are sophisticated and highly refined models in their own right, with a long and distinguished history, excellent theoretical foundation, and good modeling tools. From a computational viewpoint, they do have interesting properties of their own, but they tend to be regarded as building blocks for larger neuronal networks.

A strikingly different view from recent years sees neurons themselves as enormous ecosystems of interacting networks, in this case molecular ones. There are many interesting cell-biology questions being tackled at the molecular level, including homeostasis, development, and structural elaboration. The topic that has attracted by far the most interest is that of synaptic plasticity: how are synaptic weights set, maintained, and altered?

The common thing about each of these viewpoints is that it is essential to have the detailed biophysical insights embodied in multicompartment models. One cannot understand real neural networks if the neurons in it do not behave properly. Nor can one accurately model subcellular processes in neurons without knowing about neurotransmitter input and calcium entry, and how in turn molecular events alter cellular excitability. Biophysically detailed neuronal models are the starting point to go up or down scales in exploring brain function.

References

Berridge MJ (2006) Calcium microdomains: organization and function. Cell Calcium 40(5–6):405–412

Bower JM, Beeman D (1995) The book of GENESIS. TELOS/Springer, Santa Clara

Carnevale NT, Hines HML (2006) The NEURON book. Cambridge University Press, Cambridge

Cooley JW, Dodge FA (1966) Digital computer solutions for excitation and propagation of the nerve impulse. Biophys J 6(5):583–599

Cornelis H, Rodriguez AL, Coop AD, Bower JM (2012) Python as a federation tool for GENESIS 3.0. PLoS One 7(1):e29018

Crook S, Gleeson P, Howell F, Svitak J, Silver RA (2007) MorphML: level 1 of the NeuroML standards for neuronal morphology data and model specification. Neuroinformatics 5(2): 96–104

Eppler JM, Helias M, Muller E, Diesmann M, Gewaltig M (2008) PyNEST: a convenient interface to the NEST simulator. Front Neuroinformatics 2:12

Fitzhugh R (1961) Impulses and physiological states in theoretical models of nerve membrane. Biophys J 1(6):445–466

Gillespie DT (1977) Exact stochastic simulation of coupled chemical reactions. J Phys Chem 25:2340–2361

Gleeson P, Steuber V, Silver RA (2007) NeuroConstruct: a tool for modeling networks of neurons in 3D space. Neuron 54(2):219–235

Hille B (1992) Ionic channels of excitable membranes. Sinauer Associates Inc., Sunderland

Hines M (1984) Efficient computation of branched nerve equations. Int J Biomed Comput 15(1):69–76

Hodgkin AL, Huxley AF (1952a) The components of membrane conductance in the giant axon of Loligo. J Physiol (Lond) 116(4):473–496

Hodgkin AL, Huxley AF (1952b) A quantitative description of membrane current and its application to conduction and excitation in nerve. J Physiol (Lond) 117(4):500–544

Holt GR, Koch C (1999) Electrical interactions via the extracellular potential near cell bodies. J Comput Neurosci 6(2):169–184

Hursh J (1939) Conduction velocity and diameter of nerve fibers. Am J Physiol 127:131–139

Izhikevich EM (2003) Simple model of spiking neurons. IEEE Trans Neural Netw 14: 1569–1572

Jack JJB, Noble D, Tsien RW (1975) Electric current flow in excitable cells. Calderon Press, Oxford

Lisman JE, Raghavachari S, Tsien RW (2007) The sequence of events that underlie quantal transmission at central glutamatergic synapses. Nat Rev Neurosci 8(8):597–609

Migliore M, Morse TM, Davison AP et al (2003) ModelDB: making models publicly accessible to support computational neuroscience. Neuroinformatics 1(1):135–139

Nagumo J, Arimoto S, Yoshizawa S (1962) An active pulse transmission line simulating nerve axon. Proc IRE 50:2061–2070

Rall W (1959) Branching dendritic trees and motoneuron membrane resistivity. Exp Neurol 1: 491–527

Rall W (1967) Distinguishing theoretical synaptic potentials computed for different soma-dendritic distributions of synaptic input. J Neurophysiol 30(5):1138–1168

Rall W (1969) Time constants and electrotonic length of membrane cylinders and neurons. Biophys J 9(12):1483–1508

Ray S, Bhalla US (2008) PyMOOSE: Interoperable scripting in Python for MOOSE. Frontiers Neuroinformatics 2(6):1–16

Rushton WAH (1951) A theory of the effects of fibre size in the medullated nerve. J Physiol 115(1):101–122

Sabatini BL, Oertner TG, Svoboda K (2002) The life cycle of Ca^{2+} ions in dendritic spines. Neuron 33(3):439–452

Segev I (2006) What do dendrites and their synapses tell the neuron? *J Neurophysiol*. 95(3):1295–1297

Seyfarth E (2006) Julius Bernstein (1839–1917): pioneer neurobiologist and biophysicist. Biol Cybern 94(1):2–8

Stevens CF (1993) Quantal release of neurotransmitter and long-term potentiation. Cell 72(Suppl):55–63

Traub RD, Wong RK, Miles R, Michelson H (1991) A model of a CA3 hippocampal pyramidal neuron incorporating voltage-clamp data on intrinsic conductances. J Neurophysiol 66(2): 635–650

Van Geit W, Achard P, De Schutter E (2007) Neurofitter: a parameter tuning package for a wide range of electrophysiological neuron models. Front Neuroinformatics 1:1

Wils S, De Schutter E (2009) STEPS: modeling and simulating complex reaction–diffusion systems with Python. Front Neuroinformatics 3:15

Yamada W, Koch C, Adams PR (1989) Multiple channels and calcium dynamics. In: Koch S (ed) Methods in neuronal modeling. MIT Press, Cambridge, pp 97–133

Chapter 8
Noise in Neurons and Other Constraints

A. Aldo Faisal

Abstract How do the properties of signalling molecules constrain the structure and function biological networks such as those of our brain? Here we focus on the action potential, the fundamental electrical signal of the brain, because malfunction of the action potential causes many neurological conditions. The action potential is mediated by the concerted action of voltagegated ion channels and relating the properties of these signalling molecules to the properties of neurons at the systems level is essential for biomedical brain research, as minor variations in properties of a neurons individual component, can have large, pathological effects on the physiology of the whole nervous system and the behaviour it generates. This approach is very complex and requires us to discuss computational methods that can span across many levels of biological organization, from single signalling proteins to the organization of the entire nervous system, and encompassing time scales from milliseconds to hours. Within this methodical framework, we will focus on how the properties of voltagegated ion channels relate to the functional and structural requirements of axonal signalling and the engineering design principles of neurons and their axons (nerve fibres). This is important, not only because axons are the essential wires that allow information transmission between neurons, but also because they play a crucial in neural computation itself.

Many properties at the molecular level of the nervous system display noise and variability, which in turn makes it difficult to understand neuronal design and behaviour at the systems level without incorporating the sources of this probabilistic behaviour. To this end we have developed computational methods, which will enable us to conduct stochastic simulations of neurons that account for the probabilistic behaviour of ion channels. This allows us to explore the relationship between individual ion channel properties, derived from high-resolution patch clamp data,

A.A. Faisal (✉)
Department of Bioengineering and Department of Computing, Imperial College London,
South Kensington Campus, London, UK
e-mail: a.faisal@imperial.ac.uk

N. Le Novère (ed.), *Computational Systems Neurobiology*,
DOI 10.1007/978-94-007-3858-4_8,

and the properties of axons. The computational techniques we introduce here will allow us to tackle problems that are (1) beyond the reach of experimental methods, because we can disambiguate the effects of variability and reliability of individual molecular components to whole cell behaviour, and (2) allow us to consider the many finer fibers in the central and peripheral system, which are experimentally difficulty to access and record from. We start with the well-established data that Ion channels behave with an element of randomness resulting in "channel noise". The impact of channel noise in determining axonal structure and function became apparent only very recently, because in the past findings were extrapolated from very large unmyelinated axons (squid giant axon), where channel noise had little impact due to the law of large numbers. However, the many axons in the central and peripheral nervous system are over 1,000 times thinner and the small number of ion channels involved in sustaining the action potential, imply that channel noise can affect signalling and constraint both the reliability of neural circuit function, but also sets limits to the anatomy of the brain as a whole.

8.1 A Systems Biology View of Neurons: Variability and Noise

Our brain processes information using electrical impulses, the action potentials, which are mediated by "protein transistors", so called voltage-gated ion channels. We understand exquisitley the mechanisms and molecular and cellular components involved in the action potential: such as the ion current basis of the action potential (Nobel prize for Hodgkin and Huxley in 1963), how action potentials are translated at the synapse (Nobel prize for Katz in 1963), how ion channels are the fundamental elements gating these currents, acting effectively like "protein transistors" (Nobel prize for Sakmann and Neher in 1991) and how the protein structure of ion channels allows for voltage-dependent conduction (Nobel prize for McKinnon in 2003). Yet, it remains still unclear how these characteristics determine the brain's design principles at the systems level. For example, our brain requires just 16 W of power – much less than any computer of equivalent computational power would need (and in fact less then current portable computers). This is, taking an engineering, view surprising. Brains use low quality components for electrical signals: fat as electrical insulator, protein transistors and salty water as conducting core.[1] How well can the brain process information being made up from such poor electrical components? In the following we will explore the fundamental implications of this question.

When Adrian began to record from neurons in the 1920s, he observed that neural responses were highly variable across identical stimulation trials and only the average response could be related to the stimulus (Adrian 1927, 1928). Biology viewed this variable nature of neuronal signalling as "variability", engineers called

[1]On the other side brains and organisms build themselves from themselves.

it "noise". The two terms are closely related, but as we shall see imply two very different approaches to think about the brain – one operating at the systems level, the other operating at the molecular level: On the one side the healthy brain functions efficiently and reliably, as we routinely experience ourselves.[2] Variability is a reflection of the complexity of the nervous sytem.

On the other side engineers viewed neurons as unreliable elements, subject to metabolic and other variations, but more importantly, perturbed by random disturbances of a more fundamental origin (von Neumann 1951, 1956). Taking this view the brain processes information in the presence of considerable random variability. Thus, individual neurons in the nervous system are highly variable because they are quite unreliable ("noisy"), and this poses a severe constraint on a neurons information processing power. Taking a molecular view of the nervous sytem, this is actually not suprising: we know that at the biochemical and biophysical level there are many stochastic processes operating in neurons: protein production and degradation, opening and closing of ion channels (i.e. confirmation changes of proteins), fusing of synaptic vesicles and diffusion and binding of signalling molecules.

In the classic of neurobiology it is implicitly assumed that averaging large numbers of such small stochastic elements effectively wipes out the randomness of individual elements at the level of neurons and neural circuits. This assumption, however requires careful consideration for two reasons:

1. First, neurons perform highly non-linear operations involving high gain amplification and positive feedback. Therefore, small biochemical and electrochemical fluctuations of a random nature can significantly change whole cell responses.
2. Second, many neuronal structures are very small. This implies that they are sensitive to (and require only) a relatively small number of discrete signalling molecules to affect the whole. These molecules, such as voltage-gated ion channels or neurotransmitters, are invariably subject to thermodynamic fluctuations and hence their behaviour will have a stochastic component which may affect whole cell behaviour.

This suggests that unpredictable random variability – *noise* – produced by thermodynamic mechanisms (e.g. diffusion of signalling molecules) or quantum mechanisms (e.g. photon absorption in vision) at the molecular level can have deep and lasting influence on variability present at the system level. In fact, as we shall we our deterministic experience of our own nervous system implies that the design principles of the brain must mitigate or even exploit the constraints set by noise and other constraints such as energetic demands.

It is worth considering what the implication of noise are for information processing: noise cannot be removed from a signal once it has been added to it.

[2]but, consider the following reflections of our nervous system's variability: the little random motions of a pointed finger, our uncertainty when try to understand a conversation in the presence of loud background noise, or when we seem not able to see our keys that were in plain view when we were searching for them.

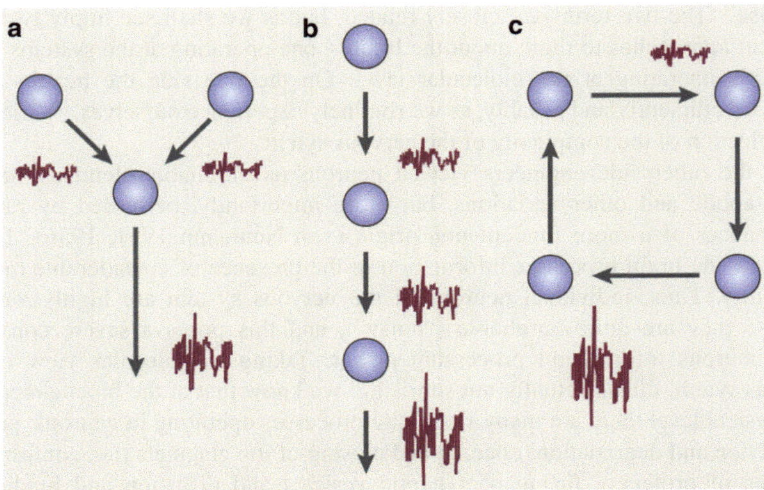

Fig. 8.1 How does network topology affect noise? (**a**) Convergence of signals onto a single neuron. (**b**) Serial propagation of a noisy signal through successive neurons. (**c**) Recurrent connections ("loops") in networks (Figure reproduced from Faisal et al. (2008))

Since signals can easily be lost, and noise easily added, this sets a one-sided limit on how well information can be represented – as measured by the signal-to-noise ratio. In many situations noise can be thought of as additive effect on a signal. It is reasonable to assume that a major feature of neurons is the summation-like integration of incoming signals, and so we can illustrate noise in three basic neural network topologies, using toy neurons that simply sum their synaptic inputs.

Figure 8.1a shows convergence of signals onto a single neuron. If the incoming pre-synpatic signals have independent noise then noise levels in the postsynaptic neuron will scale in proportion to the square root of the number of input signals (N), whereas the signal scales in proportion to N. Meaning that the more independent signals are integrated the more the integrated signal will stand out over the integrated noise (proportional to the square root of N). However, this signal-boosting effect underlies diminishing returns (especially if the cost scales proportionally to the number of inputs). If the noise in the signals is perfectly correlated, then the noise in the neuron will also scale in proportion to N. Also, consider that the integrating neuron itself will add it's own internal noise to the integrated signal. The only way to raise this signal-to-noise ratio, is by adding further relevant information. This usually implies information from the same source, but with noise that is independent (in some way) of the original message, e.g. by using another sensory neuron with overlapping receptive fields or using a second axon that carries the same information. However, decreasing the effects of noise by increasing the amount of redundant information comes at two costs: a fixed cost that results from maintaining redundant neurones, fibers, sensors and a dynamic cost, that results from signalling with these systems. Given the hight energetic efficiency of our brain and the low reliability of its individual molecular components, in contrast to the low energetic efficiency of computer chips and the very high reliability of their digital electronic components electronic circuits, mitigating the consequences of noise must have had a fundamental impact on the evolution and design of nervous systems: from the first "biochemical nervous system" in bacterial chemotaxis (where nutrient levels in the surrounding were used to control the degree of random swimming motion) to the mammalian cortex.

Figure 8.1b shows the passage of a noisy signal through successive neurons, and noise levels are incremented at each stage as internal noise is added to the signal. Note that parallel connections (not shown) do not augment noise through network interactions. In fact, it was suggested that the highly parallel and distributed yet compact structure of the CNS might help to limit the amount of noise that builds up from serial connections (Faisal et al. 2008).

Finally, Fig. 8.1c shows that recurrence ("loops") in neural circuits, can, if unchecked results in the build-up of correlated noise. Moreover, the whole nervous system operates in a continuous closed loop with the environment: from perception to action and back (see Fig. 8.2). Given this highly recurrent structure at all levels of biological organisation it is therefore important that noise is kept "private" to a neuron (De Weese and Zador 2004). Note, that the traditional view is that the all-or-nothing response of the action potential effectively thresholds out random fluctuations and provides an effective all-or- none response signal which is therefore thought to be noise free.

How can we know what constitutes noise when recording signals from the brain? For instance, neuronal membrane potential shows small variations even at rest, even if synaptic inputs are pharmacologically blocked. Synpatic or electrode inputs near action potential threshold, a neuron may or may not fire an action potentials, because of the all-or-nothing nature of the action potential (Skaugen and Wallœ 1979). It seems that some neurons can react in a very reliable and reproducible manner to fluctuating currents that are injected via intracellular electrodes. As long as the same time-course of the injected current is used the action potentials occur with precisely the same timing relative to the stimulation (Bryant and Segundo 1976; Mainen et al. 1995). A similar behaviour has been reported for neurons of the visual system in flies (de Ruyter van Steveninck et al. 1997) and monkeys (Bair and Koch 1996). On the other hand, neurons produce irregular spike trains in the absence of any highly temporally structured stimuli. Irregular spontaneous activity, i.e., activity that is not related in any obvious way to external stimulation, and trial-to-trial variations in neuronal responses are often considered as "noise" (Shadlen and Newsome 1995; Softky and Koch 1993).

The question whether this neuronal trial-to-trial variability is

- Indeed just noise (defined in the following as individually unpredictable, random events that corrupt signals)
- Results because the brain is to complex to control the conditions across trials (e.g. the organisms may become increasingly hungry or tired across trials)
- Or rather the reflection of a highly efficient way of coding information

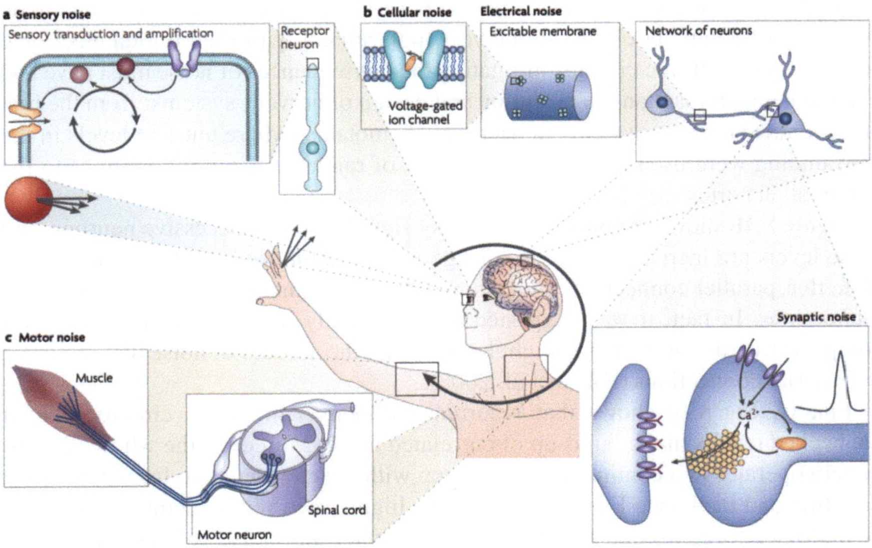

Fig. 8.2 Noise in the behavioural loop – the nervous system continuously perceives the world and executes actions (movements) in response to it. Noise is thereby encountered along all stations: from the conversion of physical stimuli into neuronal responses in sensory processing (**a**), neuronal information processing and signal transmission in axons and synapses, and in the motor system (**c**) when neuronal signals are converted into muscle forces and movements (Figure reproduced from Faisal et al. (2008))

cannot easily be answered. In fact, being able to decide whether we are measuring the neuronal activity that is underlying the logical reasoning and not just meaningless noise is a fundamental problem in neuroscience, with striking resemblance to finding the underlying message in cryptographic code breaking efforts (Rieke et al. 1997).

There are multiple sources contributing to neuronal trial-to-trial variability: deterministic ones, such as changes of internal states of neurons and networks, as well as stochastic ones, noise inside and across neurons (White et al. 2000; Faisal et al. 2008). To what extent each of these sources makes up the total observed trial-to-trial variability remains unclear. What has become clear is that to solve this question it is not sufficient to study neuronal behaviour only experimentally (as this measures only the total variability of the system): It requires taking a system biology view of neuronal information processing (Faisal et al. 2008). This is because noise is ultimately due to the thermodynamic and quantum nature of sensory signals, neuronal and muscular processes operating at the molecular level. Given that the molecular biology and biophysics of neurones is so well known, it allows us to use stochastic modelling of these molecular components to control and assess the impact of each source of (random) variability at the level of neurones, circuits and the whole organism. How we can link molecular noise to system level variability, and what new insights this offers us in understanding the design of the brain, is what we are going to explore in the remainder of the chapter.

8.2 Stochastic Versus Deterministic Views of Neurons: Small Means Noisy

Noise as a fundamental constraint to information processing and transmission, and variability is inherent in our brains and our behaviour. This variability however cannot captured by computational models that are deterministic in nature, such as the beautiful Hodgkin–Huxley model of the action potential. To account for variability we have to make use of stochastic modes.

Classically large neuronal structures, such as the squid giant axon have been key in understanding and explaining neural mechanisms such as the action potential. This is because, given their scale, they are experimentally easily accessible and appear to function deterministically. This is because random variability averages out quickly as size increases: the standard deviation of variability over the mean activity of a set of signalling molecules will go as the inverse square root of the number of involved molecules.

However, neurones and synapses in many pathways are tiny: In comparison to squid giant axon (0.5 mm diameter) neuronal connections in our cortex can be over 10,000 times smaller. Cerebellar parallel fibres have 0.2 μm average diameter, C-fibres involved in sensory and pain transmission range between 0.1 μm and 0.2 μm, while the unmyelinated pyramidal cell axon collaterals which form the vast majority of local cortico-cortical connections have an average diameter 0.3 μm (Faisal et al. 2005). Thus, only as few as a hundred ion channels will be involved in transmitting the action potential per given unit length of axon. This is in contrast to several millions of the same unit length in squid giant axon.

Similarly, the majority of central nervous system synapses (spiny- or bouton-type) are below a micrometer in size and biochemical processes and concentrations occur within volumes smaller than picoliters. For example, in the classic synaptic preparation of the frog neuromuscular junction several thousand post-synaptic receptors will be ‘listening’ to incoming neurotransmitters released by hundreds of vesicles. However, in the much smaller bouton-type synapses found in mammalian cortex as few as three post-synaptic receptors have to detect the release of a single vesicle (containing some 1,000–2,000 neurotransmitter molecules), triggered by a single action potential.

The action potential (AP) is the fundamental signal used for communication in the brain’s neural networks. Measuring the timing of APs in vivo and in vitro shows that neuronal activity displays considerable variability both within and across trials (Shadlen and Newsome 1998; Strong et al. 1998). This variability can have statistical characteristics that match those of simple random processes such as Poisson or Bernoulli processes. However, only because neuronal activity has some statistics of a random processes it does not necessarily follow that neuronal activity is generated by a random process itself. In fact, Shannon's theory of information (Shannon 1948) suggests that to maximize information transmission the optimal way to encode (neural) signals will make the stream of signals appear like a random (Cover and Thomas 1991; Rieke et al. 1997). Thus, to what extent neuronal variability is part of meaningful processing and meaningless noise remains a fundamental problem of neuroscience.

We will illustrate this approach now by looking at the initiation and propagation of the action potential.

8.3 Computational Neuroscience of Stochastic Neurons

A neurons AP is carried by the spread of membrane potential depolarisation along the membrane and is mediated by voltage-gated ionic conductances (Fig. 8.3a). The depolarisation of the membrane potential is (re)generated by the non-linear voltage-gated Na^+ conductances that open at low levels of membrane depolarisation, which depolarise the membrane further, thereby recruiting more Na^+ conductances. Thus, Na^+ conductactences act like positive feedback amplifiers. The resting membrane potential is then restored by the inactivation of the Na^+ conductance and is assisted by the (delayed) opening of K^+ and membrane leak conductances (and the two to three order of magnitude slower Na^+-K^+-pumps) that repolarise the membrane, thus K^+ channels and leak conductances provide negative feedback (Koch 1999; Hille 2001).

The patch-clamp technique showed that these ionic conductances resulted from populations of discrete and stochastic voltage-gated ion channels (Sakmann and Neher 1995; Katz 1971) (see Fig. 8.3b). These ion channels are transmembrane proteins that act as pores which can open and close in response to changes in the membrane potential ("channel gating") across the channel, thus acting like protein transistors. These voltage-gated ion channels operate with an element of randomness due to thermodynamic effects. This stochastic behaviour produces random electrical currents, called channel noise (White et al. 2000), which is by one to two orders of magnitude the most dominant source of intracellular noise in neurons[3] (Manwani and Koch 1999b; Faisal et al. 2005). What are the effects of channel noise on the action potential?

To answer this question we require first a stochastic model of the action potential. In most cases such stochastic models cannot be solved analytically, and require the need for computational approaches, through stochastic or "Monte-

[3]We ignore here synaptic input as a form of electrical "noise" and note that the common use of the term "synaptic background noise" denotes the (not necessarily random) variability produced by massive synaptic input in cortical neurons (Faisal et al. 2008).

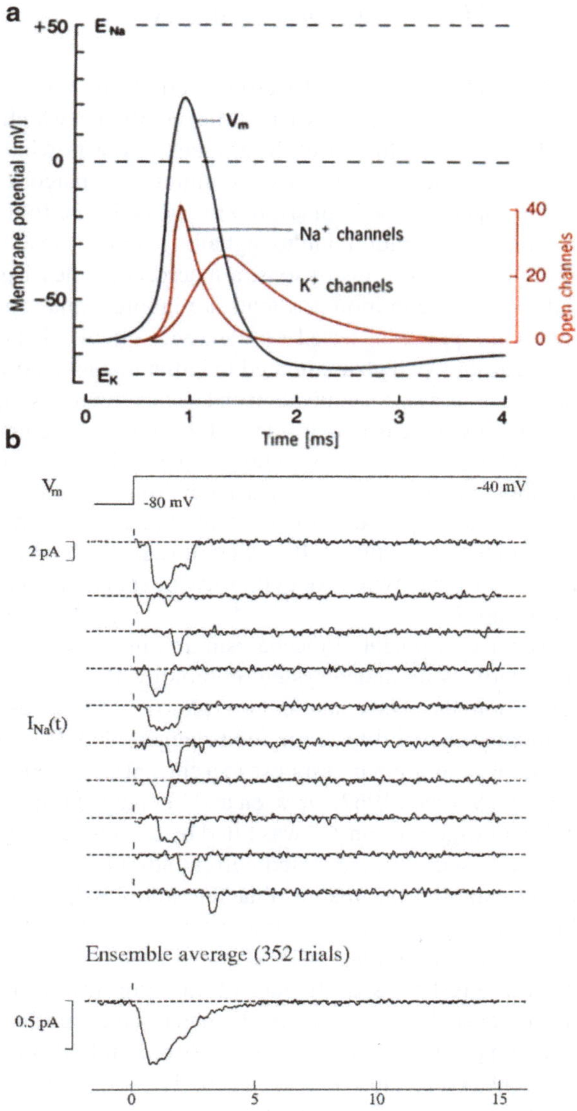

Fig. 8.3 (**a**) Schematic model of the action potential and the mediating Na^+ and K^+ currents. Note, that here the idealized deterministic ion channel behaviour is drawn, for small number of ion channels the more accurate picture would look like in (**b**). (**b**) Illustration of ion channel variability in repeated identical voltage-step trials. Patch-clamp recording of a few unitary Na^+ channels in mouse muscle during a 40 mV voltage step. The ensemble average – averaging over 352 repeated identical trials – approaches the idealized deterministic description (akin to the "Na^+ channels" curve in (**a**) (Figure reproduced from Faisal (2010) and generated using the Modigliani stochastic neuron simulator (Faisal et al. 2002, 2005)) (freely available from www.modigliani.co.uk)

Carlo" simulation. At the cellular level our stochastic simulations will use data on the responses of individual molecules – i.e. the properties of voltage-gated ion channels – to derive the responses of systems of interacting molecules – i.e. the response of a neuron.

8.3.1 Modelling Noisy Neurons

[OPTIONAL SECTION Electrical activity in neurons arises due to the selective movement of charged ions across the membrane, which we call membrane excitability. However, in most cases the amount of ions flowing through ion channels during episodes of electrical activity are minute compared to the number of ions present in the respective medium (Hodgkin 1964). In the following we ignore the changes in ionic concentration due to signalling, thus instead of a microscopic description of the neuron in terms of ions, a macroscopic description is used: individual ions and local concentration gradients are ignored, and replaced by a description of the membrane potential based on electrical circuit elements, including batteries and ionic currents (which are related to individual ions flows via Faraday's constant and the ionic charge). An equivalent electrical circuit description (see Fig. 8.4) is derived by equating the currents inside and through the membrane compartment according to Kirchhoff's current law. This method balances all currents flowing through the membrane and to other compartments (including branch points). Each transmembrane circuit describes an iso-potential patch of membrane and is represented by a membrane compartment. It is, therefore possible to mimic a neuron's morphology using tree-like networks of cylindrical or spherical compartments (Rall 1969a,b; see also Chap. 7)].

The action potential mechanism and its theory is arguably the most successful quantitatively modelled system in biology. Reliability and noise in action potential generation has been studied for almost as long as the ionic basis underlying membrane excitability (Blair and Erlanger 1933; Pecher 1939). Reliability of action potential generation in response to a current step input was measured at the Nodes of Ranvier (Verveen 1962; Derksen and Verveen 1966; Verveen et al. 1967). The probability of trigeering an AP was fitted by a Gaussian cumulative probability function, parameterised by the injected current stimulus amplitude. This phenomenological model captured the feature that the stimulus had to drive the membrane over a fluctuating threshold to generate an AP. Threshold fluctuations were postulated to result from an internal noise source of possibly ionic origins and it was concluded that the threshold's coefficient of variation (ratio of standard deviation over the mean) must depend on axon diameter. Later an analytical relationship between an assumed transmembrane noise source and AP threshold fluctuations (Verveen 1962; Derksen and Verveen 1966; Lecar and Nossal 1971). Transmembranen noise sources considered both thermal resistance noise produced by the neuronal membrane's resistance and noise resulting from discrete, stochastic ion channels (which at the time were not conclusively experimentally demonstrated). Noise that could result from ion channels was estimated to have an over 20 times larger effect on threshold fluctuations than thermal resistance noise (see also Hille 2001).

In stochastic simulation study by Skaugen and Wallœ (1979) well ahead of it's time, the impact of discrete, stochastic ion channels on AP initiation in squid-giant-axon type iso-potential membrane patches. They showed how current inputs below AP threshold in the deterministic Hodgkin–Huxley model could trigger APs. This was because channel noise linearised the highly non-linear input-current

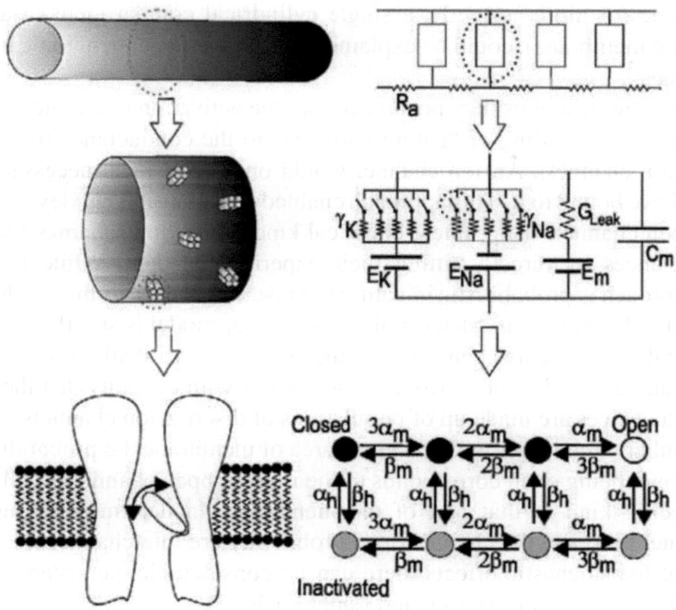

Fig. 8.4 Schematic view of how a compartmental model of an axon with stochastic ion channels. The *left* column shows the level of description and the *right* column the corresponding electrical and stochastic modelling abstraction. (*Top*) The axon is modelled as sequence of cylindrical compartments. (*Middle*) Each compartments of axon membrane contains two populations of ion channels. (*Bottom*) The voltagte-gated ion channel is described by a finite-state Markov random process. We depict here a Na channel model, which has a single open state, three closed, and four inactivated states (Figure adapted from Faisal et al. (2005))

versus firing-rate characteristic(when averaged over many trials). In other words, channel noise could increase and linearize the non-linear signalling range of neurons (Skaugen and Wallœ 1979). To introduce channel noise they replaced the deterministic linear kinetic equations representing the voltage-gated ion channels with stochastic Markov processes over populations of discrete ion channels (Fitzhugh 1965). This important modelling technique is worth understanding better, which we shall do in the following. As we will see this technique is ideal to model ion channels and is by now a well-established biophysical technique and confirmed by direct recordings of single ion channel behaviour (Sakmann and Neher 1995).

8.3.2 From Hodgkin–Huxley Conductances to Stochastic Ion Channels

First we have to establish the notation to model the action potential. Hodgkin and Huxley (1952) showed that the observed ionic currents within an iso-potential patch of membrane (or more correctly, a single cylindrical compartment of axon with iso-potential membrane), could be explained by voltage-dependent ionic membrane conductances.

To model the dynamics they postulated that the activation of a conductance was determined by the binding of "gating particles" to the conductance (or in modern terms the ion channel). An ion channel would only open if all necessary "gating particles" have bound to it. This approach enabled Hodgkin and Huxley to model the ratio of open channels using linear chemical kinetic reaction schemes for Na^+ and K^+ conductances by directly fitting their experimental data. While this gating-particle approach is probabilistic in nature the resulting model of the conductance is deterministic. Note, that this deterministic behaviour model is strictly speaking not always equal to the average behaviour of the system, as we shall see later.

Since Sakmann and Neher's work we now know with certainty that the voltgate-gated conductances are made up of populations of discrete ion channels. Assuming a large number of channels within a given area of membrane the probability p_i of a single channel being open corresponds to the ratio of open channels to all channels of the specific kind in that part of the membrane. In deterministic models the open channel ratio and the channel open probability are interchangeable, but when we account for stochastic effects there can be considerable, persistent deviations between the two quantities (Faisal and Laughlin 2007).

To simplify the notation we will use p_i as notation for both the open channel ratio and the channel open probability and highlight the differences later when necessary.

The ionic conductance per unit membrane area g_i is the product of the total ionic membrane conductance per unit membrane area $\bar{g}_i$ and the time and voltage dependent ratio of open channels p_i.

$$g_i(V,t) = \bar{g}_i\, p_i(V,t) \tag{8.1}$$

$\bar{g}_i$ is the product of the single channel conductance γ_i and the number of channels per unit membrane area (channel density) ρ_i – determining these parameters is important for stochastic simulations, as both the number of channels present and their individual conductance determine the level of noise

$$\bar{g}_i = \gamma_i \rho_i \tag{8.2}$$

Rewriting Eq. 8.2 into Eq. 8.1 and interpreting p_i as channel open probability yields a molecular level description of the ionic conductance per unit membrane area.

$$g_i(V,t) = \gamma_i \rho_i\, p_i(V,t) \tag{8.3}$$

In Hodgkin–Huxley's original gating particle model the probability that a channel will be open is given by the probability that all their gating particles are simultaneously bound. Implicitly, the gating-particle model assumes independence of the gating particles and the open channel probability p_i is therefore the product of the probabilities of each particle being bound. With q_j being the probability that a gating particle of type j is bound and $l(j)$ being the multiplicity of particles of type j that have to bind to support channel opening we can write

$$p_i(V,t) = \prod_j q_j(V,t)^{l(j)} \tag{8.4}$$

In the case of the standard squid axon Na^+channel (Hodgkin and Huxley 1952) we have $q_j \in \{m,h\}$ and $l(m) = 3$ and $l(h) = 1$, thus $p(V,t) = m^3h$. The q_j themselves are governed by linear chemical reactions:

$$\dot{q}_j(V,t) = \alpha_j(V)(1-q_j) - \beta_j(V)q_j \tag{8.5}$$

The reaction's kinetic rate functions $\alpha_j(V)$,$\beta_j(V)$ describe the rate of change of the probability q_j, Eq. 8.5. These rate functions are characteristic to the ion channel protein (and gene) studied (Hille 2001), such that they can be identified from whole cell behaviour (e.g. Faisal and Niven 2006; Faisal 2007, but see Prinz et al. 2004b). These rate functions are either sigmoidal or exponential functions empircially fitted to voltage-clamp data (e.g. Hodgkin and Huxley 1952) or Boltzmann functions derived from the ion channel's voltage sensor behaviour in a constant, uniform electrical field (Patlak 1991; Hille 2001).

We can define a voltage-dependent ion channel time constant τ_j characterising the time in which sudden changes in membrane potential will affect ion channel gating.

$$\tau_j(V) = \frac{1}{\alpha_j(V) + \beta_j(V)} \tag{8.6}$$

Directly related to the time constant is the voltage dependent steady-state value, $q_{j\infty}$, to which the q_j will converge for a constant membrane potential V.

$$q_{j\infty}(V) = \frac{\alpha_j(V)}{\alpha_j(V) + \beta_j(V)} = \frac{\alpha_j(V)}{\tau_j(V)} \tag{8.7}$$

The q_j are probabilities and have to satisfy $q_j \in [0,1]$. ntegrating.

Modelling each stochastic ion channel. Instead of describing the behaviour of lumped deterministic ionic conductance, we want to model (probabilistic) behaviour of individual ion channels.

One way to describe the gating behaviour of individual channels is the use of Markov processes (Markov 1906, 1971). A Markov process is a probabilistic system, that assumes that has set of discrete states. In each state there are a number of possible transitions that can cause a transition from the current state to another state. These transitions are probabilistic, i.e. have a specific probability of occurring. Like Hodgkin–Huxley's model these Markov models use linear chemical kinetic schemes to determine transition probabilities between discrete channel states, in fact these are the same transition rate functions as calculated by Hodgkin–Huxley model. The probability to transit from one state to another within a given time horizon is the transition rate.

Conveniently these Markov models can be recovered from gating-particle type models of conductances (Conti and Wanke 1975; Hille 2001). The deterministic gating-particle model is reformulated as a specific subclass of Markov model as described in the following: Every possible combination of bound and unbound gating particles corresponds to a discrete ion channel state. The deterministic kinetic functions that describe the binding and unbinding rate of gating particles are correspondingly used to describe the probability of transition per unit time between individual ion channel states. Each transition corresponds to the binding or unbinding of one gating particle. The Markov states and the associated transition probability rate functions together form a Markovian kinetic gating scheme.

A central assumption of the gating-particle models is that individual gating particles of the same type are indistinguishable and independent from each other. Multiple states, therefore, may have the same set of numbers of bound particles for each particle type. Without loss of generality these multiple, yet indistinguishable states are lumped together into one Markov state.

To account for lumped states the transition rates are multiplied by a factor k, which is determined as follows. A transition corresponding to the unbinding of a gating particle of type j has a factor k that equals the number of particles of type j bound to the state where the transition originated. A transition corresponding to the binding of a gating particle of type j has a k that equals the multiplicity $l(j)$ minus the j particles bound at the target state. This procedure allows one to transform any deterministic gating-particle model of conductances into a stochastic model of ion channel gating.

The Markovian kinetic scheme derived for the gating-particle model for Hudgkin–Huxley's squid axon $m^3\,h$-type gating-particle model of Na^+ channels is shown in Fig. 8.4 (bottom right): For example the transition rate from transition rate from the third closed state to the open state is $\alpha_m(V)$, the inverse transition has rate $3\beta_m(V)$ and the transition probability away from the open state is $3\beta_m(V) + \beta_h(V)$ (towards the two adjacent closed and inactivated states).

Having established how to model the stochastic nature of the action potential we now return to the question how we can link noise (resulting from ion channels) to variability in neuronal behaviour.

8.4 Neuronal Signalling Variability from Ion Channel Noise

The discrete and stochastic nature of individual Na ion channels (Sigworth and Neher 1980) was confirmed by the patch-clamp technique (Sakmann and Neher 1995). Experiments revealed that Na^+ channel fluctuations could be large enough in size to account for the observed threshold fluctuations in Nodes of Ranvier (several μm diameter, Na^+ channel densities $>1{,}000\,\mu m^{-2}$) (Sigworth 1980).

Moreover neuron simulations, in which stochastic models of Na^+ channels were the only source of variability, showed that the Na^+ channel noise alone produced AP threshold fluctuations which compared well with respect to experimental data (Clay and DeFelice 1983; Rubinstein 1995). Variability in the experimental context could be quantified as the coefficient of variation, defined as the standard deviation over the mean of a variable value. These studies suggested that the AP threshold's coefficient of variation was dependent on the square root of the number of Na^+ channel, N, present in the membrane. For large N this would imply that channel noise should have only little impact on spike-based information processing, as fluctuations in the number of open channels ΔN would have been small in most cells, because they are proportional to

$$\Delta N \propto \frac{\sqrt{N}}{N} = \sqrt{\frac{1}{N}}.$$

For Hodgkin–Huxley's squid giant axon this was certainly true as it measured several millimeters in diameter and possessed millions of ion channels per unit length. Similarly the myelinated nerves' Nodes of Ranvier considered in these studies, although about hundred-folds smaller in diameter then squid giant axon, featured 100-fold higher channel densities. Nodes of Ranvier were, thus, comparable to squid axon in terms of the expected variability. The general validity of this assumption required re-consideration for most neurons, as we shall see in the following.

Spike time variability measurements at the soma can be explained by channel noise. Random action potentials constitute form a major disruption of neuron-to-neuron communication. While the presence or absence of APs does carry information, it is known that also the precise timing of each AP carries information (e.g. Rieke et al. 1997). The trial-to-trial variability of AP timing in vivo and in vitro in many systems can be on the order of milliseconds (1–10 ms) and the timing of individual APs in the millisecond scale was shown to be behaviorally relevant in perception and movement of invertebrates (see Faisal et al. 2008 for review).

How large can the influence of channel noise be on neuronal firing variability? Schneidman et al. (1998) showed that in iso-potential membrane patches (of comparable area as pyramidal cell soma) with large numbers of ion channels channel noise can play a significant role for a neuron's spike time reliability, i.e. the timing precision at which an action potential is initiated (see also Fig. 8.5, top two plots). This is because during AP generation the instant when the membrane potential crosses AP threshold is determined by the small probability and thus small number of ion channels open around AP threshold $N^* = p_{\text{open}} N$ and not, as was often implicitly assumed, the much larger number N of ion channels present in total. The magnitude of the fluctuations of N^* could be considerable even for large N and were better described by a binomial random variable:

$$\Delta N^* \propto \frac{\sqrt{N p_{\text{open}}(1 - p_{\text{open}})}}{N p_{\text{open}}} = \sqrt{\frac{1 - p_{\text{open}}}{N p_{\text{open}}}}.$$

This implied that many neurons in the cerebral cortex, e.g. pyramidal cells, could be influenced by channel noise. The study was able to generate comparable spike time

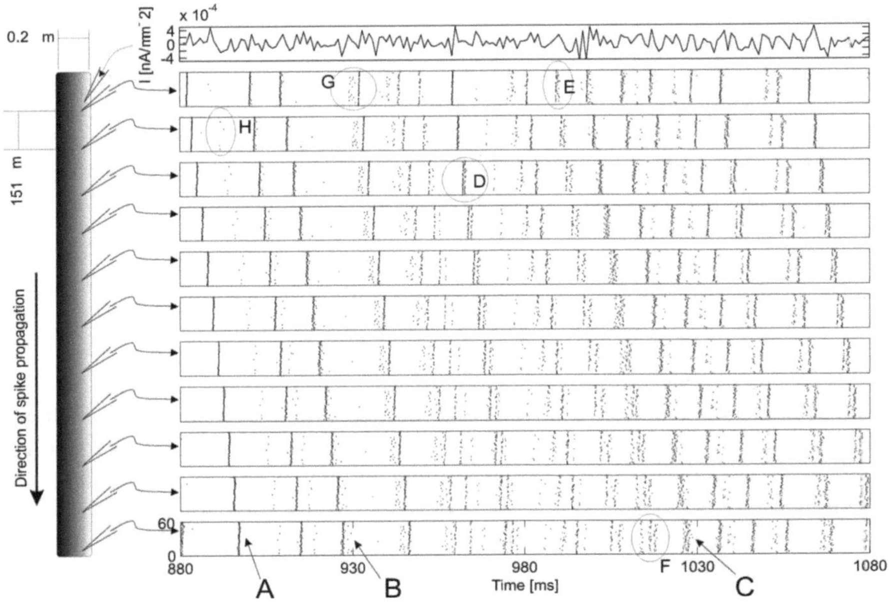

Fig. 8.5 The stacked raster plot visualizes traveling APs produced by identical repeated trials and is organized as follows. The top-most row shows the *white* noise current input. Below that, each row contains a spike raster plot recorded at equally spaced axonal positions (from the proximal stimulus site at the *top* to the distal end of the axon at the *bottom*). In each spike raster plot, the precise timing of a spike marked by a *dot* on an invisible time line. These time lines are stacked over each other for the N = 60 repeated trials. The linear shift visible in the overall spike pattern across rows reflects the APs traveling along the axon. The top-most raster plot reflects the spike initiation variability, all subsequent plots reflect variability produced during propagation which quickly exceed that of initiation. Data based on 10 s trial length, squid axon of 0.2 μm diameter (average diameter of cerebellar parallel fibers) with a frozen noise current stimulus (zero mean, 0.01 nA SD, band limited at 1 kHz) injected at the proximal end. See text for details (Figure adapted from Faisal and Laughlin (2007))

variability as found in cortical neurons in vitro. Furthermore, spike initiation had high temporal precision when the size of ionic current fluctuations near AP threshold were small compared to the injected stimulus current. Thus, weaker stimuli will produce more unreliable spiking in agreement with experimental data (Schneidman et al. 1998). These results were extrapolated to AP propagation in axons, where the current flowing ahead of the AP (re)generates the AP driving it forward. It was assumed that this axial current constituted a strong driving input and, hence, it was inferred that APs should propagate very reliably in axons, however as we shall discuss shortly conduction velocity will fluctuate significantly in thin axons due to channel noise (Faisal and Laughlin 2007).

Stochastic simulations showed that super-threshold inputs could fail to generate an AP and, more importantly, make random APs appear in the absence of any input (Strassberg and DeFelice 1993; Chow and White 1996; Faisal et al. 2005).

This deletion and addition of action potentials cannot be explained at all by deterministic models and signicantly contributes to neuronal variability. Moreover, Chow and White (1996) were able to derive analytically the rate at which these random action potentials were triggered in an iso-potential membrane patch due to the critical role of Na channel noise (and not K channel noise). Their analytical derivation of the random action potential rate was supported by detailed stochastic simulations that modelled both Na and K channels. Note, noise triggered action potentials were originally named 'spontaneous action potentials' (Chow and White 1996; Faisal et al. 2005). This term may be confounded with 'spontaneously activity' which describes neurons that are actively spiking in the absence of synaptic input. Such 'spontaneous activity' is often the result of purposeful instability of a neuron's resting state (e.g. when action potential threshold is below resting potential), and can thus appear also in deterministically modelled neurons. To disambiguate the term 'random action potentials' is used to refer to noise-initiated action potentials.

8.4.1 Variability of the Propagating Action Potential

While a neuron will contain many ion channels, typically these ion channels do not interact instantaneously (neurons are not iso-potential) and thus their fluctuations do not average out, as much smaller (and thus noisier) subsets of ion channels are responsible for driving activity locally in the neuron. However, little was known on how the stochasticity of ion channels influences spikes as they travel along the axon to the synapse and how much information arrives there. Experimentally axonal spike time jitter has previously been only measured in vitro at myelinated cat and frog axons of several μm diameter and was in the order of 0.01 ms (Lass and Abeles 1975a,b).

Biologically accurate stochastic simulations of axons showed (Faisal and Laughlin 2007) that the variability of action potential propagation (measured as spike time jitter) in unmyelinated axons between 0.1–0.5 μm diameter was in the order of 0.1–1 ms SD over distances of millimeters (cf. Fig. 8.5). Thus, axonal variability can grow several orders of magnitude larger then previously expected and having considerable impact on neural coding. Why can AP propagation become so variable? The spatial spread of membrane potential follows different input-response relationships than in point-like iso-potential membrane patches (Faisal et al. 2005): In fact, the current driving the AP ahead is one to two orders of magnitude smaller than the minimum stimulus current ("Rheobase") required to trigger an AP in a resting axon (Faisal and Laughlin 2007). Consequently the driving axial current is a weak input that is susceptible to channel noise. Channel noise acts in two ways that are implicit to the AP mechanism (see Fig. 8.6 for illustration): First, only a small number of Na^+ channels are involved in driving the AP, when the membrane is between resting potential and AP threshold, and these small Na currents are thus subject to large fluctuations. Second, the resting membrane ahead of the AP is far from being at rest, but fluctuates considerably.

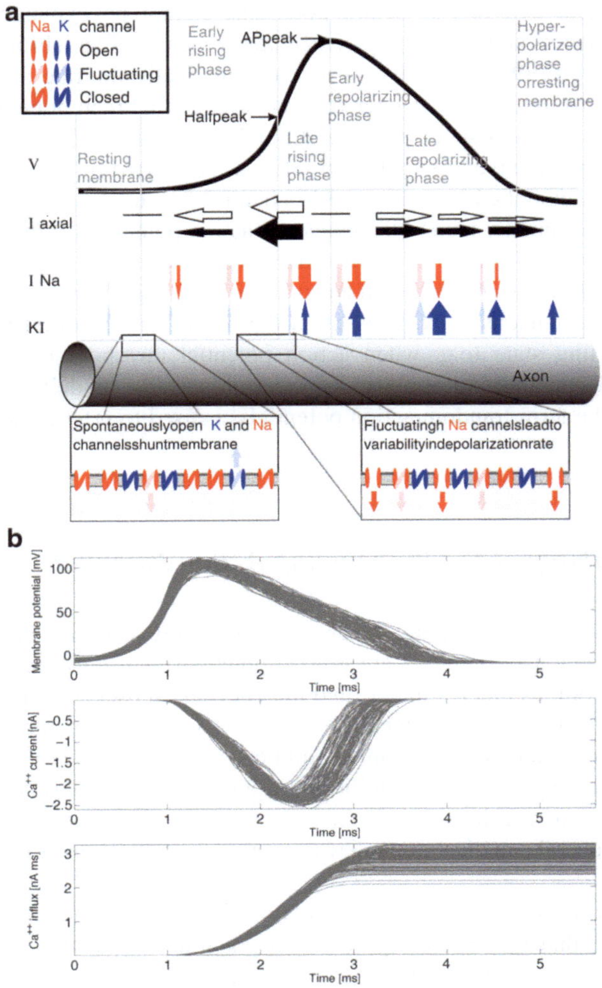

Fig. 8.6 (**a**) Diagrammatic representation of a traveling AP on the axon based on Faisal and Laughlin (2007). Stacked over each other, the leftward traveling membrane potential wave form of the AP (V) along the axon, axial currents flowing along the axon (I_{axial}), Na and K currents (I_{Na}, I_{K}) and the representation of the axon itself. Axial, Na^+ and K^+ currents are denoted by *black*, *red* and *blue arrows* scaled to represent the relative size of the current in the various phases. *Hollow* and *light shaded arrows* denote the size of the current fluctuations relative to the average currents. The AP wave form is subdivided into six phases, resting membrane, early rising phase, late rising phase, early repolarizing phase, late repolarizing phase and an optional hyperpolarized phase (Figure adapted from Faisal and Laughlin (2007)). (**b**) Synaptic variability from axonal variabiltiy. (*Top*) Wave forms of 713 consecutive APs arriving at the terminal end of a 1.6 mm long unmyelinated axon of 0.2 μm diameter. (*Middle* and *Bottom*) Ca^{++} current and total Ca^{++} influx resulting from the integration of the above AP wave forms into a model of a Calyx-of-Held type synapse (Figure adapted from Faisal and Laughlin (2004))

Inspection of Fig. 8.5 shows that APs generated by the same stimulus are not precisely aligned across trials, and the misalignment ("jitter") in this AP set grows considerably the further AP propagate. In general four distinct stochastic effects of channel noise on APs propagating in axons can be identified. To describe these effects the portion of the input stimulus which triggers an AP will be called a *stimulus event*. APs which were triggered across trials by the same stimulus event form an *AP set*.

The timing of APs in a set is jittered but remains unimodally distributed (Fig. 8.5, arrows A, B, C), or grow to be markedly multimodally distributed (Fig. 8.5, D, fourth row) – splitting into distinct groups of APs across trials. For a stimulus event we quantify the jitter at a given position on the axon as the standard deviation (SD) of spike timing in its corresponding AP set. For a 0.2 μm axon (shown in Fig. 8.5) AP generation at the proximal end of the axon had on average a SD of 0.38 ms, similar to spike generation in simulated membrane patches (Schneidman et al. 1998). However, spike time jitter increases over relatively short distance, such that at 2 mm the average jitter over all AP sets increased to $\approx$0.6 ms SD. This jitter implies that post-synaptic coincidence detection windows cannot be more precise than 2–3 m s at this short distance. Furthermore, at the site of spike generation the timings within each AP set are unimodally distributed (Fig. 8.5, top raster plot). However, during propagation the spike time distribution can become multimodal, with the different peaks several milliseconds apart. In other words the AP set splits into distinct groups (Fig. 8.5, D, fourth row). Thus, axonal channel noise sets limits to the precision at which neurons in the densely wired cortex can communicate with each other over a given distance. These effects become relevant in unmyelinated axons below 0.5 μm, as commonly found in the tightly packed circuits of the central nervous system.

Five unexpected features of noisy axons will provide more intuition how noise effects can have counterintuitive impacts on neuronal signalling and contribute to observable neuronal variability.

1. **Fewer action potential failures.** Although channel noise provides a means for APs to fail, stochastic simulations show (Faisal and Laughlin 2007) that conduction failures are rare – 1% in axons of the smallest known diameter and thus the noisiest axons – while empirically observed failure rates can be as high as 50%. Several neuronal mechanisms that purposely produce conduction failure are known, acting through membrane hyperpolarization, shunting effects, spike-frequency dependent block of APs at axonal branch points (Debanne 2004). In contrast channel noise has the opposite effect of promoting AP generation, because of the strong positive feedback of Na^+ channels. This suggests that channel noise cannot account for the AP failures observed in many systems and that other factors must be responsible. This suggests, that when propagation failures occur in the healthy nervous system, this is due to purposely designed mechanisms for pre-synaptic information processing, which allow the incorporation of local information not available at the site of spike initiation (Faisal and Laughlin 2007).

2. **Saltatory conduction in unmyelinated axons.** Some APs in thin axons travel faster than a continuously moving wave front, because the AP wave front suddenly jumps ahead. This mode of conduction results because of a collaborative channel effect, where random opening of nearby Na^+channels pre-depolarize the membrane by a few millivolts. The axial current from an incoming AP, then triggers AP threshold and the AP jumps several hundred micrometers ahead to the pre-depolarized region. Thus, the spike time at a given position of the axon appears to be shifted in the order of a millisecond. This stochastic micro-saltatory conduction effect resembles saltatory conduction between the morphologically specialized Nodes of Ranvier in myelinated nerve fibers. Here, however, it is produced by the stochastic behavior of individual channels embedded in an axon of uniform morphology and it occurs randomly. This adds considerable variability across trials and will enhance the effects of jitter and can initiate the following effect.
3. **Body temperature reduces the impact of noise.** Temperature is not only a key factor in determining the speed of biochemical reactions such as ion channel gating but also controls the amount of ion channel variability (Faisal et al. 2002, 2005). From a modelling perspective channel kinetics and simulations should be always accompanied by appropriate temperature-dependent scaling factors and base temperatures (or have to be assumed temperature invariant). Commonly, temperature-dependence is accounted for by scaling the transition rates $\alpha(V)$ and $\beta(V)$ by the factor $Q_{10}^{\frac{\theta - 6.3^\circ C}{10^\circ C}}$, where Q_{10} is an empirically determined, channel-specific parameter and θ is the temperature in Celsius.

   While commonly overlooked, temperature, and via it's effects on the kinetics of ion channels also the resulting channel noise, can vary greatly across the nervous system: cold-blooded insects can warm-up their body to over 40°C prior to taking flight, while human extremities and the sensory and motor neurons therein can be exposed to temperature differences of up to 10°C or more between their dendrites, cell bodies and axon terminals – as they span from the (cold) extremities to the (warmer) spinal chord.

   Once, one accounts for temperature-dependent stochasticity in computational models it can produce some counterintuitive effects – whereby increasing temperature can lower noise levels. Channel noise effects decrease with increasing temperature: As ion channel kinetics speed up with temperature, the duration of spontaneous depolarizing currents decreases and the membrane is less likely to reach AP threshold (this effect prevails over the increased rate of spontaneous channel openings). In other words increasing temperature shifts channel noise to higher frequencies where it is attenuated by the low pass characteristics of the axon (Steinmetz et al. 2000; Faisal et al. 2005). This may suggest that increasing temperature allowed polikliotherm animals, such as mammals, to develop more reliable, smaller, more densely connected and thus faster neural circuits.
4. **Synaptic transmission variability from axonal channel noise.** AP timing precision is bound to decrease the further the AP travels, thus long-range communication is in this respect noisier than short range communication, given

the same axon diameter. Axonal channel noise may have also an effect on information transmission in short range synaptic connections in unmyelinated axons of up to 1 μm diameter, because the shape of the AP wave form is perturbed by channel noise (Faisal and Laughlin 2004). The wave form of the presynaptic AP is of fundamental importance in determining the strength of synaptic transmission. It determines the calcium signal that controls synaptic transmitter vesicle release, by both controlling the opening of voltage-gated Ca^{++} channels and the driving force for Ca^{++} influx (Augustine 2001). Stochastic simulations showed that the traveling AP wave form fluctuated considerably (see Fig. 8.6b) (Faisal and Laughlin 2004) and the wave form of an AP mid-way down the axon and at the terminal end were little correlated, thus in thin axons below 1 μm somatically triggered AP are unlikely to carry much information in the AP wave form to the synapse, as has been measured in the soma of cortical neurons (de Polavieja et al. 2005).

Stochastic modelling from the soma to the synapse is essential, as synaptic reliability and variability has been in general attributed to mechanisms inside the cortical synapse, but knowledge is typically based on paired soma recordings or large synapses, while most of our brain's synapses are very small. Thus, it is difficult to dissociate synaptic and axonal stochastic effects in these preparations. Furthermore, most studies so far ignored, synaptic channel noise at presynaptic Ca^{++} channels, which may produce spontaneous postsynaptic potentials and further increase trial-to-trial transmission variability.

5. **Stochastic simulations for real neurons: The dynamic clamp technique.** While the discussion so far dealt with findings based on simulations, the difference between deterministic and stochastic channel behavior was recently investigated in living neurons using the dynamic clamp method. Dynamic clamp is an electrophysiological technique that uses a real-time interface between neurons and a computer that simulates dynamic processes of neurons (Sharp et al. 1993; Robinson and Kawai 1993; Prinz et al. 2004a). It reads the membrane potential of the neuron and calculates the transmembrane current produced by virtual, simulated voltage-gated or synaptic conductances. The simulated current is injected into the neuron, which therefore receives the same current as if it biologically contained the virtual conductances. Dorval and White used this technique to study the role of channel noise in cortical neurons in vitro, allowing them to investigate what happens if ion channels were to act deterministically in a real neuron (Dorval and White 2005). A Na^{+} channel population was blocked pharmacologically and using dynamic clamp replaced by an equivalent virtual population of Na^{+} channels. The virtual channels were simulated either deterministically or stochastically.[4] These neurons showed near threshold oscillations of the membrane potential, characteristic for their morphological class in vitro, and were able to phase lock their activity with other neurons, if and only if the virtual Na^{+} channels were simulated stochastically. Neurons lost these two

[4] using the Gillespie algorithm described.

properties with deterministically stimulated Na^+ channels. These experimental results provide the first direct demonstration that physiological levels of channel noise can produce qualitative changes in the integrative properties of neurons. This suggests, that channel noise could even have a more profound effect on the evolution and development of neurons.

8.5 Fundamental Biophysical Constraints on the Size of Axons

The impact of channel noise on membrane potential grows larger the smaller the membrane surface area. How small can neurons or axons be mode before channel noise effects disrupt action potential signaling? Hille (1970) suggested that in very fine axons the opening of a small number of Na^+ channels could generate an AP. This idea was subsequently used to highlight that channel noise could generate random action potentials (RAPs) posing limits to myelinated axons (Franciolini 1987). Based on a probabilistic argument predicting a RAP rate of 1 Hz in myelinated axons, it was suggested that RAPs would discourage the use of Nodes of Ranvier below 1 μm diameter. Unfortunately the calculation was flawed because ion channel state transition probabilities were confused with ion channel state transition rates. Furthermore, it was previously shown that in the mammalian nervous system myelinated Nodes of Ranvier exist with diameters as fine as 0.2 μm (Waxman and Bennett 1972). The first stochastic simulations of unmyelinated axons (Horikawa 1991), using simplified channel kinetics (discussed elsewhere in this chapter), showed that in fine axons more APs arrived at the distal end than were generated at the proximal end. Based on this single finding a lower limit to axon diameter of 0.2 μm was postulated. The relationship between diameter, biophysical parameters and RAP rate, however was not studied and the findings were not related to anatomical data. Anatomists had previously shown that axons as fine as 0.1 μm are commonly found in the central nervous system.

Detailed stochastic simulations (Faisal et al. 2005) showed that spontaneous opening of Na^+ channels can, in theory, trigger random action potentials below a critical axon diameter of 0.15–0.2 μm. Figure 8.7 shows this is because at these diameters the input resistance of a Na^+ channel is comparable to the input resistance of the axon. The single, persistent opening of a single Na^+ channel can therefore depolarize the axon membrane to threshold. Below this diameter, the rate at which randomly generated action potentials appear increases exponentially as diameter decreases. This will disrupt signaling in axons below a limiting diameter of about 0.1 μm, as random action potentials cannot be distinguished from signal carrying action potentials. This limit is robust with respect to parameter variation around two contrasting axon models, mammalian cortical axon collaterals and the invertebrate squid axon, show that the limit is mainly set by the order of magnitude of the properties of ubiquitous cellular components, conserved across neurons of different

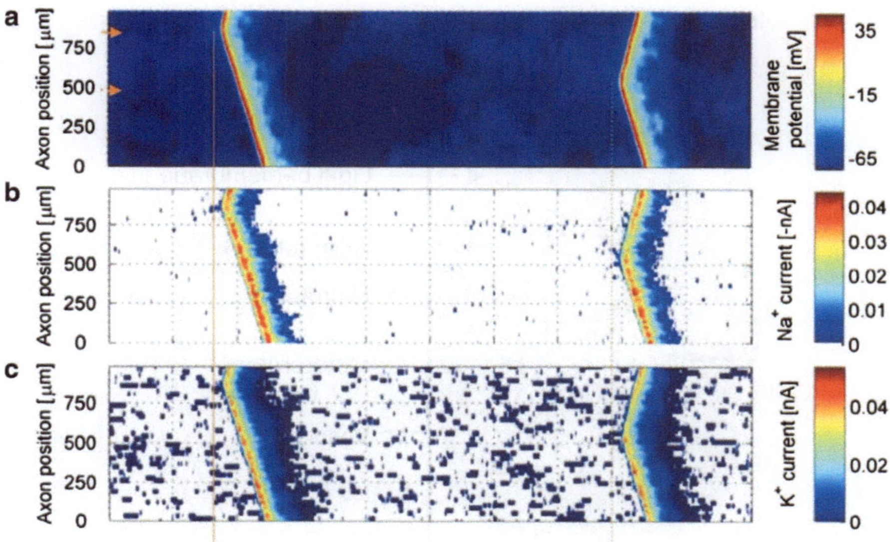

Fig. 8.7 The emergence and propagation of random action potentials (RAPs) in axons (in absence of any input). Space-time plots of membrane potential (**a**) and transmembrane Na and K currents (**b** and **c** respectively) in a simulation of a 1-mm- long pyramidal cell axon collateral ($d = 0.1\,\mu$m) at 23°C. In **b** and **c** the regions which show no transmembrane current was flowing are not color coded, making ionic currents from randomly opening ion channels at resting potential clearly visible (*dark blue dots*). The prolonged open time of single Na channels at t = 15 ms and t = 77 ms depolarizes the membrane to AP threshold, recruiting several nearby channels and resulting in spontaneous Aps, at t = 17 ms and t = 79 ms, that subsequently propagate along the axon. The horizonal time axis divisions has divisions 10 ms (Figure adapted from Faisal et al. (2005))

species. The occurrence of random action potentials and the exponential increase in RAP rate as diameter decreases is an inescapable consequence of the action potential mechanism. The stochasticity of the system becomes critical when its inherent randomness makes it operation unfeasible. The rate of RAPs triggered by channels noise counterintuitively decreases as temperature increases much unlike one would expect from electrical Johnston noise. Stochastic simulations (Faisal et al. 2005) showed that RAP rate is inversely temperature dependent in the cortical pyramidal cell and the squid axon which operated at 6.3° and 36°.

Other biophysical limits to axon size. How small can a functioning axon be constructed, given the finite size of it's individual components? Faisal et al. (2005) showed using. We use a volume exclusion argument to show that it is possible to construct axons much finer than 0.1 μm (Fig. 8.8). Neural membrane (5 nm thickness) can be bent to form axons of 30 nm diameter because it also forms spherical synaptic vesicles of that diameter. A few essential molecular components are required to fit inside the axon; this includes an actin felt work (7 nm thick) to support membrane shape, the supporting cytoskeleton (a microtubule of 23 nm

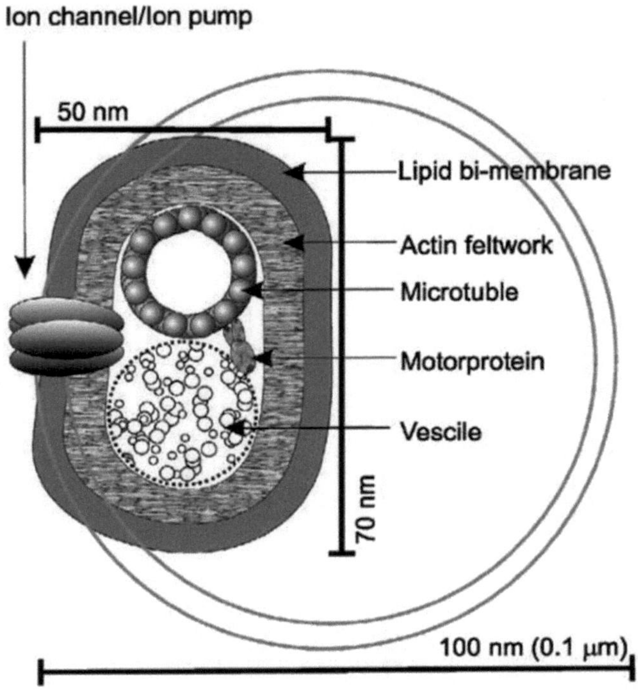

Fig. 8.8 Sterically minimal axon: this to scale drawing illustrating how components essential to a spiking axon can be packed into the cross-section of a fiber of 50 × 70 nm. The unfilled circle illustrates the finest known AP-conducting axons of diameter 100 nm (Figure adapted from Faisal et al. (2005))

diameter), the intracellular domains of ion channels and pumps (intruding 5–7 nm), and kinesin motor proteins (10 nm length) that transport vesicles (30 nm diameter) and essential materials (<30 nm diameter). Adding up the cross-sectional areas shows that it is possible to pack these components into axons as fine as 0.06 μm (60 nm). Indeed, the finest known neurites, those of amacrine cells in Drosophila lamina, are about 0.05 μm in diameter, contain microtubules, and connect to extensive dendritic arbors but do not transmit APs. The fact that the smallest known AP-conducting axons are about twice as large as the steric limit to axon diameter (0.1 μm cf. 0.06 μm) (Fig. 8.8), whereas electrically passive axons reach the physical limit, supports our argument that channel noise limits the diameter of AP- conducting axons to about 0.1 μm.

Electrically passive neurites of 0.05 μm diameter are known (which do not contain Na^+ channels or sustain AP conduction). A steric (volume-exclusion) argument shows that an axon of 0.05 μm diameter can be constructed, as the cross-sectional area of an axon can fit all required components to sustain AP conduction (i.e. Na^+ channels and Na–K pumps protruding into the axon, the thickness of

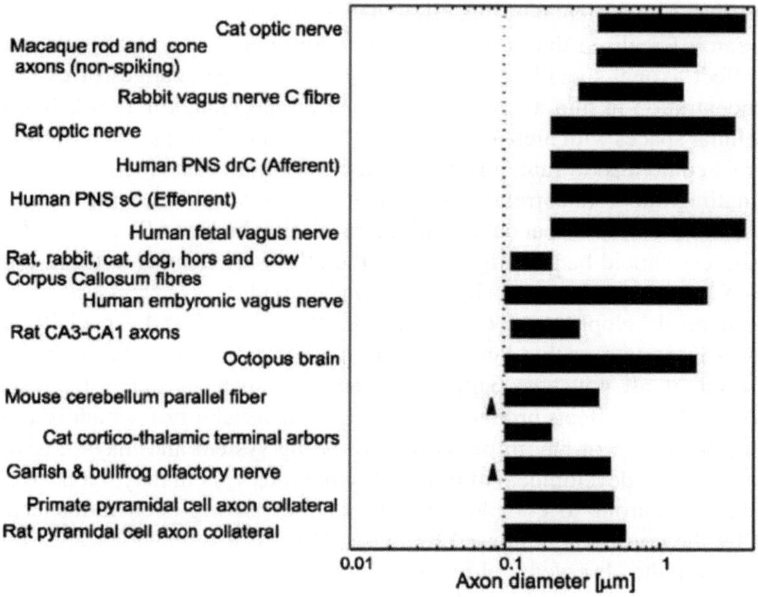

Fig. 8.9 Diameters of fine AP-conducting axons in a wide range of species and tissues, with the finest AP-conducting axons reaching a common lower limit diameter of 0.1 μm (*dotted line*). *Arrowheads* mark data points by developing fibers of 0.08 μm diameter (Figure adapted from Faisal et al. (2005))

the actin feltwork supporting cell shape, microtubules, motor proteins and vesicles supplying synapses). Furthermore, other molecular limits to axon diameter are well below the limiting diameter of 0.1 μm, thus AP conducting axons finer then 0.1 μm could, in theory, exist. Yet anatomical data across many species, invertebrate and vertebrate, extremely small insects and large mammals, shows an identical lower limit of diameter for action potential conducting axons of 0.1 μm. This suggests that channel noise limits axon diameter and thus the wiring density of the central nervous system and thus ultimately the size of the cortex.

Curiously the anatomical literature (see Fig. 8.9) demonstrated a common lower value for the diameter of axons for over 30 years, yet this was not noticed, till a systems biology view on the study on stochastic limits to cell size prompted to search for the smallest known axon diameters (Faisal et al. 2005).

What about other noise sources? The cable properties of thin axons isolate dendritic and somatic noise sources, such as synaptic background activity (Ho and Destexhe 2000). The presence of electrical resistances in the membrane introduces Johnson noise is magnitudes smaller than channel noise (Manwani and Koch 1999a; Lecar and Nossal 1971). Ephaptic coupling (Arvanitaki 1942; Katz and Schmitt 1940; Anastassiou et al. 2010), input through axo-axonic synapses (Tamas and Szabadics 2004) and axo-axonic gap-junctions (Schmitz et al. 2001) are ignored because little experimental or modelling data exists regarding their impact. During my research I realised that a previously not considered source of neuronal noise could affect axonal signalling. Signalling produces large changes of extracellular ion concentration in thin axons (Frankenhaeuser and Hodgkin 1956). In confined extracellular spaces with high tortuosity (Nicholson and Phillips 1981; McBain et al. 1990) this could lead to random changes in electromotive driving forces and affect AP signalling due to uncorrelated activity of other nearby axons (recall that rodent cortex features 3 of axon per cubic millimetre (Braitenberg and Schütz 1998). These noise sources should be investigated once the effects of channel noise as a general source of axonal variability has been fully understood.

In general developing a stochastic insight of biological and especially nervous systems is important, as this can further our understanding in at least three different ways. First of all novel properties may emerge, such as stochastic resonance. Second, stochastic effects produce noise – random variability – which is a limiting factor in the design of any information processing system and may be critical for the evolution and development of brains. The nervous system may not have evolved from simple neurons to complex circuits without having been influenced in its design by the constraints imposed by noise. Thirdly, it remains an open question to what extent the observed trial-to-trial variability in both the nervous systems and behaviour is caused by noise and to what part it may be due to "unfathomable" complex dynamics (reflecting chaotic dynamics and myriads of ever changing internal states). Noise will inescapably increase (neuronal) variability and we can thus compare the amount of variability produced by noise to the total experimentally observed variability. Stochastic models allow us to trace stochastic effects, from their sources at the molecular level to the level of neuronal function at the cellular and circuit level.

Implications of axon channel noise for the interpretation of experiments. The findings from stochastic simulations of the action potential prompt careful experimental consideration in thin axons, because typical experimental protocols are not geared to distinguish post-synaptic variability due to synaptic transmission from variability caused by axonal noise ("axonal variability"). Optical methods and extracellular recordings have limited stimulus control and signal resolution, thus intra-cellular paired-cell recordings, as used to study small synapses innervated by thin axons, would be more suitable. However, the impact of axonal channel noise in circuits may have gone so far unnoticed, because paired cell measurements which could reveal axonal variability are difficult to obtain. This is because typical cell-to-cell distances are limited by the microscopes field of view and the probability to find two mono-synaptically connected neurons with standard techniques. This limits the potential cell pairs to about about 500 μm distance, where conduction jitter standard deviation is small, in the order of about 0.1 ms. However, cortico-cortical axon collaterals and cerebellar parallel fibers, which are below 0.5 μm diameter, can extend up to a centimeter away from the soma, suggesting that jitter will limit the minimum width of reliable coincidence detection windows to 5–10 ms at their terminals.

The question to which extent channel noise generates variability in the brain and ultimately behaviour is will probably require to combine both experimental studies, where total variability can be assessed, and stochastic modelling, in which each source of variability can be controlled for. Noise levels set both hard limits on the central nervous system, such as the degree of miniaturization of the brain's circuits (Faisal et al. 2005) and soft constraints such as the achievable spike time reliability (Schneidman et al. 1998; Faisal and Laughlin 2007).

8.6 Towards Establishing Engineering Design Principles for Neurons

Stochastic simulations in neuroscience have contributed to develop our bottom-up understanding of how noise present at the molecular level, channel noise in neuronal membranes (as discussed here) and biochemical noise at synapses (Franks et al. 2003), affects information processing at macroscopic levels – whole neurons, neuronal circuits and behaviour (Faisal et al. 2008). The appropriate choice of algorithms for simulation is essential both in terms of resource efficiency and the accuracy of the results, as in many neurons of our brains modelling single ion channel behaviour counts. Here, we used stochastic methods to study the role of internal noise setting constraints on the function and structure of a typical cell signalling system, the action potential (Hartwell et al. 1999). Similar effects, constraints and limits due to stochasticity becoming critical will apply to other wave-based cell signalling systems, as they have to rely on inherently-noisy protein switches to generate and propagate signals.

Taking a telecommunications system viewpoint, there are several biophysical constraints on the performance in neurons using action potentials for transmitting information:

1. The finite response range of neurones – signals range over 100 mV in amplitude and less then 1 kHz in action potential frequency (Attwell and Gibb 2005).
2. The finite response speed of neural signalling mechanisms, as determined by the time delay and bandwidth of signals.
3. Any internal noise in the system's information transmission components will degrade and corrupt signals.

Note, that the first two factors determine how many different signals a neuron can communicate in a given time frame. The third factor, noise, will corrupt messages, resulting in signals that were formerly distinct now becoming confounded – this is often referred to as the representational capacity of communication system.

Studies in the earliest sensory processing stages have shown that neurons maximise the information (the number of bits) that can be retrieved from a single sensory sampling station, irrespective of its meaning (Rieke et al. 1997; Laughlin 1981). What happens at later stages, where neurons process this raw data to extract relevant parameters and features? Broadly speaking two changes take place as information passes to higher areas of the brain neural networks:

- Increasing convergence and divergence suggests that information is being represented by patterns of activity, distributed across populations or ensembles of neurons.
- The amount of information processing increases as the responses of neurons becomes more specific. A neuron ceases to respond to all aspects of incoming data and is driven by a particular set of parameters. The resulting response is equivalent to the extraction of important features and commonly involves non-linear operations that pick out particular combinations of input.

The information processed by higher order neurons is thus metabolically very valuable as many lower order neurones contributed to its computation and transmission. Thus, the more valuable each spike becomes, both because it carries more meaning for the actions of an organism and also because the organism invested it's resources in it the, the more the nervous system should make sure it is not corrupted by noise. Thus, taking this view, let's us appreciate in a new light why at the output of the brain's information processing pipeline – the muscles are driven by large, myelinated axons which should show virtually no spike time variability or random action potentials.

References

Adrian E (1928) Basis of sensation. Norton, New York

Adrian E, Matthews R (1927) The action of light on the eye. Part 1. the discharge of impulses in the optic never and its relation to the electric changes in the retina. J Physiol (London) 63:378–414

Anastassiou C, Montgomery S, Barahona M, Buzsáki G, Koch C (2010) The effect of spatially inhomogeneous extracellular electric fields on neurons. J Neurosci 30(5):1925–1936

Arvanitaki A (1942) Effects evoked in an axon by the activity of a contiguos one. J Neurophysiol 5:89–108

Augustine GJ (2001) How does calcium trigger neurotransmitter release? Curr Opin Neurobiol 11(3):320–326

Bair W, Koch C (1996) Temporal precision of spike trains in extrastriate cortex of the behaving macaque monkey. Neural Comput 8(6):1185–202

Blair E, Erlanger J (1933) A comparison of the characteristics of axons through their individual electric responses. Am J Physiol 106:524–564

Braitenberg V, Schütz A (1998) Cortex: statistics and geometry of neuronal connectivity, 2nd edn. Springer, Hamburg

Bryant H, Segundo J (1976) Spike initiation by transmembrane current: a white-noise analysis. J Physiol (London) 260:279–314

Chow C, White J (1996) Spontaneous action potentials due to channel fluctuations. Biophys J 71:3012–3021

Clay JR, DeFelice LJ (1983) Relationship between membrane excitability and single channel open–close kinetics. Biophys J 42(2):151–157

Conti F, Wanke E (1975) Channel noise in nerve membranes and lipid bilayers. Q Rev Biophys 8:451–506

Cover T, Thomas J (1991) Elements of information theory, 1st edn. Wiley series in telecommunications. Wiley-Interscience, New York

Debanne D (2004) Information processing in the axon. Nat Rev Neurosci 5(4):304–16

de Polavieja G, Harsch A, Kleppe I, Robinson H, Juusola M (2005) Stimulus history reliably shapes action potential waveforms of cortical neurons. J Neurosci 25(23):5657–5665

de Ruyter van Steveninck R, Laughlin S (1997) Light adaptation and reliability in blowfly photoreceptors. Intl J Neural Sys 7:437-477

Derksen H, Verveen A (1966) Fluctuations of resting neural membrane potential. Science 12:1388–1389

Deweese M, Zador A (2004) Shared and private variability in the auditory cortex. J neurophysiol 92(3):1840–1855

Dorval A Jr, White J (2005) Channel noise is essential for perithreshold oscillations in entorhinal stellate neurons. J Neurosci 25(43):10025–10028

Faisal AA (2007) Studying channelopathies at the functional level using a system identification approach. In: Siebes APJM, Berthold MR, Glen RC, Feelders AJ (eds) Computational lifesciences II, Lect Notes Comput Sci vol 940. AIP, Springer, pp 113–126

Faisal AA (2010) Stochastic methods in neuroscience. Chapter 11: Stochastic simulations of neurons, axons, and action potentials. Oxford University Press, Oxford, pp 297–343

Faisal A, Laughlin S (2004) Effect of channel noise on the propagating ap wave form and its potential impact on synaptic transmission. J Physiol 555P:492

Faisal A, Laughlin S (2007) Stochastic simulations on the reliability of action potential propagation in thin axons. PLoS Comput Biol 3(5):e79

Faisal AA, Niven JE (2006) A simple method to simultaneously track the numbers of expressed channel proteins in a neuron. Lect Notes Comput Sci 4216:257

Faisal A, Laughlin S, White J (2002) How reliable is the connectivity in cortical neural networks? In: Wunsch D (ed) Proceedings of the IEEE international joint conference on neural networks 2002, Honolulu. INNS, pp 1661–1667

Faisal A, White J, Laughlin S (2005) Ion-channel noise places limits on the miniaturization of the brain's wiring. Curr Biol 15(12):1143–1149

Faisal A, Selen L, Wolpert D (2008) Noise in the nervous system. Nat Rev Neurosci 9(4):292–303

Fitzhugh R (1965) A kinetic model of the conductance changes in nerve membrane. J Cell Comp Physiol 66:111–118

Franciolini F (1987) Spontaneous firing and myelination of very small axons. J theor Biol 128:127–134

Frankenhaeuser B, Hodgkin A (1956) The after-effects of impulses in the giant nerve fibers of loligo. J Physiol 131:341–376

Franks KM, Stevens CF, Sejnowski TJ (2003) Independent sources of quantal variability at single glutamatergic synapses. J Neurosci 23(8):3186–95

Hartwell L, Hopfield J, Leibler S, Murray A (1999) From molecular to modular cell biology. Nature 402 Supplement:C47–C42

Hille B (1970) Ionic channels in nerve membranes. Prog. Biophys. Mol. Biol 21:3–28

Hille B (2001) Ion channels of excitable membranes, 3rd edn. Sinauer Associates, Sunderland, p 814

Ho N, Destexhe A (2000) Synaptic background activity enhances the responsiveness of neocortical pyramidal neurons. J Neurophysiol 84(3):1488–96

Hodgkin A (1964) The ionic basis of nervous conduction. Science 145(3637):1148–1154

Hodgkin A, Huxley A (1952) Quantitative description of membrane current and its application to conduction and excitation in nerve. J Physiol (London) 117:500–544

Horikawa Y (1991) Noise effects on spike propagation in the stochastic Hodgkin–Huxley models. Biol Cybern 66:19–25

Katz B (1971) Quantal mechanism of neural transmitter release. Science 173(992):123–126

Katz B, Schmitt O (1940) Electric interaction between two adjacent nerve fibres. J Physiol 97:471–488

Koch C (1999) Biophysics of computation. Computational neuroscience. Oxford University Press, Oxford
Lass Y, Abeles M (1975a) Transmission of information by the axon: 1. Noise and memory in the myelinated nerve fiber of the frog. Biol Cybern 19:61–67
Lass Y, Abeles M (1975b) Transmission of information by the axon: 2. Information theoretic analysis(?). Biol Cybern 19:121–125
Laughlin S (1981) A simple coding procedure enhances a neuron's information capacity. Z Naturforsch 36c:910–912
Lecar H, Nossal R (1971) Theory of threshold fluctuations in nerves: 2. Analysis of various sources of membrane noise. Biophys J 11:1068–1084
Mainen ZF, Joerges J, Huguenard JR, Sejnowski TJ (1995) A model of spike initiation in neocortical pyramidal neurons. Neuron 15(6):1427–39
Manwani A, Koch C (1999a) Detecting and estimating signals in noisy cable structures, i: neuronal noise sources. Neural Comput 11:1797–1829
Manwani A, Koch C (1999b) Detecting and estimating signals in noisy cable structures, ii: information theoretic analysis. Neural Comput 11:11831–1873
Markov A (1906) Rasprostranenie zakona bol'shih chisel na velichiny, zavisyaschie drug ot druga. Izvestiya Fiziko-matematicheskogo obschestva pri Kazanskom universitete 15(2-ya seriya):135–156
Markov A (1971) Extension of the limit theorems of probability theory to a sum of variables connected in a chain (translation). In: Howard R (ed) Dynamic probabilistic systems, vol 1: Markov Chains. Wiley, New York. Reprinted in Appendix B
McBain C, Traynelis S, Dingledine R (1990) Regional variation of extracellular space in the hippocampus. Science 249:674–677
Nicholson C, Phillips JM (1981) Ion diffusion modified by tortuosity and volume fraction in the extracellular microenvironment of the rat cerebellum. J Physiol 321:225–57
Patlak J (1991) Molecular kinetics of voltage-dependent na+ channels. Physiol Rev 71(4): 1047–1080
Pecher C (1939) La fluctuation d'excitabilite de la fibre nerveuse. Arch Int Physiol Biochem 49:129–152
Prinz AA, Abbott LF, Marder E (2004a) The dynamic clamp comes of age. Trends Neurosci 27(4):218–224
Prinz AA, Bucher D, Marder E (2004b) Similar network activity from disparate circuit parameters. Nat Neurosci 7(12):1345–1352
Rall W (1969a) Distributions of potential in cylindrical coordinates and time constants for a membrane cylinder. Biophys J 9(12):1509–41
Rall W (1969b) Time constants and electrotonic length of membrane cylinders and neurons. Biophys J 9(12):1483–508
Rieke F, Warland D, de Ruyter van Steveninck RR, Bialek W (1997) Spikes : exploring the neural code. Computational neuroscience. MIT Press, Cambridge
Robinson H, Kawai N (1993) Injection of digitally synthesized synaptic conductance transients to measure the integrative properties of neurons. J Neurosci Methods 49(3):157–165
Rubinstein J.T (1995) Threshold fluctuations in an n sodium channel model of the node of ranvier. Biophys J 68(3):779–85
Sakmann B, Neher E (1995) Single-Channel recording, 2nd edn. Plenum Press, New York
Schmitz D et al (2001) Axo-axonal coupling: a novel mechanism for ultrafast neuronal communication. Neuron 31:831–840
Schneidman E, Freedman B, Segev I (1998) Ion channel stochasticity may be critical in determining the reliability and precision of spike timing. Neural Comput 10:1679–1703
Shadlen MN, Newsome WT (1995) Is there a signal in the noise? Curr Opin Neurobiol 5(2): 248–50
Shadlen MN, Newsome WT (1998) The variable discharge of cortical neurons: implications for connectivity, computation, and information coding. J Neurosci 18(10):3870–96

Shannon C (1948) A mathematical theory of communication. Bell Syst Tech J 27:373–423, 623–656
Sharp A, O'Neil M, Abbott L, Marder E (1993) Dynamic clamp: computer-generated conductances in real neurons. J Neurophysiol 69(3):992–995
Sigworth FJ (1980) The variance of sodium current fluctuations at the node of ranvier. J Physiol 307:97–129
Sigworth FJ, Neher E (1980) Single na+ channel currents observed in cultured rat muscle cells. Nature 287(5781):447–449
Skaugen E, Walløe L (1979) Firing behaviour in a stochastic nerve membrane model based upon the Hodgkin–Huxley equations. Acta Physiol Scand 107:343–363
Softky WR, Koch C (1993) The highly irregular firing of cortical cells is inconsistent with temporal integration of random epsps. J Neurosci 13(1):334–50
Steinmetz PN, Manwani A, Koch C, London M, Segev I (2000) Subthreshold voltage noise due to channel fluctuations in active neuronal membranes. J Comput Neurosci 9(2):133–48
Strassberg A, DeFelice L (1993) Limitation of the Hodgkin–Huxley formalism: effects of single channel kinetics on transmembrane voltage dynamics. Neural Comput 5:843–855
Strong S, Koberle R, de Ruyter van Steveninck R, Bialek W (1998) Entropy and information in neural spike trains. Phys Rev Lett 80(1):197–200
Tamas G, Szabadics J (2004) Summation of unitary ipsps elicited by identified axo-axonic interneurons. Cereb Cortex 14(8):823–826
Verveen AA (1962) Axon diameter and fluctuation in excitability. Acta Morphol Neerl Scand 5:79–85
Verveen AA, Derksen HE, Schick KL (1967) Voltage fluctuations of neural membrane. Nature 216(115):588–589
von Neumann J (1951) The general and logical theory of automata. Cerebral mechanisms in behavior, pp 1–41
von Neumann J (1956) Probabilistic logics and the synthesis of reliable organisms from unreliable components. Automata Studies 34:43–98
Waxman SG, Bennett MV (1972) Relative conduction velocities of small myelinated and non-myelinated fibres in the central nervous system. Nat New Biol 238(85):217–219
White J, Rubinstein J, Kay A (2000) Channel noise in neurons. Trends Neurosci 23(3):131–137

Chapter 9
Methodological Issues in Modelling at Multiple Levels of Description

Kevin Gurney and Mark Humphries

Abstract Computational neuroscience and Systems Biology strive to understand the nature of processing in the brain and its constituent parts. However, the comparative immaturity of the disciplines, and the complexity of their target systems, conspire to make the process of model building and model interpretation in these areas problematic. In this chapter we describe a general framework for thinking about the process of modelling, and illustrate it with examples including our own work on the basal ganglia and its dopaminergic modulation. We also advocate a multi-level approach in which low level features may be allowed to reveal their functionality in a wider system context.

9.1 Introduction

Computational neuroscience and Systems Biology are comparatively young, interdisciplinary areas in the life sciences, dealing with, arguably, the most complex systems we know of. All these factors conspire to make the status, and process, of building models in these areas problematic. Oftentimes modellers make tacit assumptions about their general approach, but we would argue that such assumptions should be explicit, and that establishing sound methodological principles is an important foundation stone for making progress. In this chapter we describe a general framework for thinking about modelling, and possible approaches to modelling within the framework. Much of the conceptual framework has already

K. Gurney (✉)
Department of Psychology, University of Sheffield, South Yorkshire, S10 2TP, UK
e-mail: k.gurney@shef.ac.uk

M. Humphries
Faculty of Life Sciences, University of Manchester, Manchester M60 1QD, UK
e-mail: m.d.humphries@shef.ac.uk

N. Le Novère (ed.), *Computational Systems Neurobiology*,
DOI 10.1007/978-94-007-3858-4_9,

been outlined in Gurney (2009a,b). Here, we present additional ideas, articulate existing ones in more detail, and focus more in our examples on the divide between computational neuroscience and systems biology.

9.2 Hierarchies and Computational Analysis

9.2.1 Levels of Structural Description in the Brain

It is usually acknowledged that the brain may be described at many structural levels of analysis (Fig. 9.1). At level 1 are intracellular signaling processes initiated by neuromodulators (like dopamine). Modelling at this level is the domain of *computational systems biology* (Kitano 2002) which forms the subject matter of Chaps. 1–5 and 15. Levels 2 and 3 deal with individual neurons. At level 2, patches of neural membrane or single neurons are modelled using, typically, the *Hodgkin Huxley formalism* dealt with in Chap. 7. In this scheme, the dynamics of the membrane is described in terms of the multiplicity of ionic currents it supports. At the next level, we deal only with whole neurons and are more interested in neural firing patterns. Models are often couched in a simplified or *reduced* form – using only two variables – without recourse to a detailed, multi-variable description of membrane biophysics (Izhikevich 2007, Chap. 6). Also included here are the extremely simplified leaky-integrate-and fire (LIF) neuron models which use only a single variable representing the membrane potential. At level 4 we consider microcircuits within a single brain nucleus – the example *par excellence* here, is the cortical microcircuit that extends over six neuronal layers. At level 5, microcircuits are agglomerated into brain nuclei and, beyond that, into entire functional modules (such as cortical sub-systems, hippocampus, basal ganglia etc). Models at these levels typically use rate coded neurons (leaky integrators) but may also use LIF spiking neurons. In practice, models may draw on features from adjacent levels and there is nothing special about the division into seven levels described here.

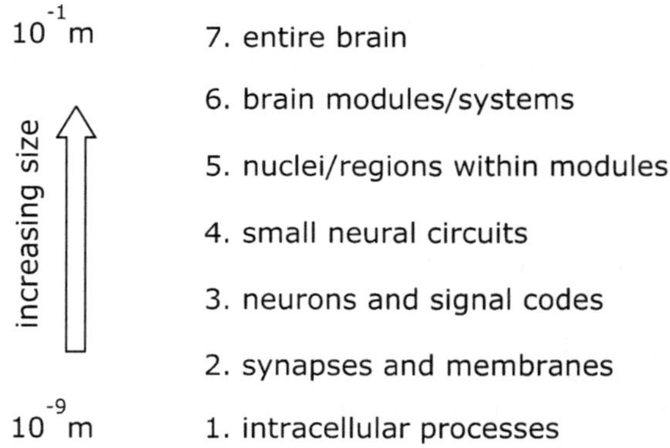

Fig. 9.1 Multiple structural levels of description in the brain

The hierarchy in Fig. 9.1 is based on empirical observation at different physical scales. While it provides a useful backdrop for considering what kind of model we are building, it tells us nothing about how to proceed; further guidance relies on more abstract, methodological principles.

9.2.2 A Computational Hierarchy

How should one go about developing a model of a particular brain system? Our answer is based on the proposal by Marr (1982) that brain functions address the solution of computational problems. Further, Marr suggested that these problems decompose (at least in the first instance) into three levels of analysis. At the top level is a description of 'What' is being computed and 'Why' – the computational task. This top level is sometimes referred to simply as the 'computation'. In this case the term 'computation' is used to mean 'function' rather than the act or process of computing. At the next level we describe 'How' the computation is carried out in terms of an algorithm and any associated representations. Finally we specify 'Where' the computation is carried out – which brain system implements the computation. This scheme, therefore, enjoins us to specify the neural computation as precisely as we can, before proceeding to detail an algorithm and implementation.

Marr's original example (Marr 1982) provides a very clear illustration, albeit outside the remit of brain modelling. Consider the computation of the bill in a supermarket with a cash register. In answer to the top level question of 'what' is being computed, it is the arithmetical operation of addition. As to 'why' this is being done, it is simply that the laws of addition reflect or model the way we should accumulate prices together when shopping; it is incorrect, for example, to multiply them together. Algorithmically, we use the normal procedure (add digits representing the same order of magnitude and 'carry' any overflow if needed). Further, in cash registers, this will be done in the decimal representation rather than binary (normally encountered in machine arithmetic) because rounding errors are incurred when converting between the everyday (decimal) representation of currency and binary. As for the implementation, this occurs using logic gates made out of silicon, silicon-oxide and metal.

Notice that choices at different levels are, in principle, independent of each other. For example, we could have chosen to use a binary representation, and alternative implementations might make use of mechanical machines or pencil and paper. The importance of discovering good representations for solving the problem is crucial. In the example, the use of a positional number system, with a number-base and sequentially increasing exponents (like decimal or binary numbers), is the key to the algorithm used here; algorithms for manipulating the number system of the ancient Romans are far more complex.

As a somewhat more realistic application to brain function, consider the problem of directing our visual gaze using ballistic eye-movements or saccades. While we will leave unanalysed several aspects of this problem, our treatment will highlight the need for expansion of the simple tri-level scheme described above. The specific *computational problem* we focus on is: how do we direct gaze to relevant, or 'salient', locations in a visually cluttered environment? We leave unanswered the related problem of finding how to compute what is 'salient' but assume that this is determined by a combination of bottom-up visual feature information (edges, corners and the like) and top-down task information or goal directed attention (Connor et al. 2004).

One *algorithm* for doing this is shown below.

Algorithm 1 An algorithm for directing visual gaze to salient points in space

divide visual space into a set of small regions, $\{R_i\}$ centred on $\mathbf{x}_i$
for each R_i **do**
  assign salience $s_i = S(\mathbf{x}_i)$
end for
find location of maximal salience $\mathbf{x}_{max} = \text{argmax}(S(\mathbf{x}_i))$
direct gaze to $\mathbf{x}_{max}$

The *representation* used in the algorithm is the set of the spatially localised saliences $S(\mathbf{x}_i)$). The *implementation* of the algorithm is presumably carried out in visuomotor circuits in cortical and subcortical structures throughout in the brain. However, there is, as it stands, a gap in the argument; we have not articulated how it may be possible to implement the mathematical operation of *argmax* in a neuronal network. This function makes no explicit reference to intra-neuronal integration or network operations as such, and so we propose that it is helpful to invoke another level of analysis which deals with the abstract *neural mechanisms* that are required to realise the algorithm. For example, the calculation of the maximally salient location $\mathbf{x}_{max}$ could be performed by a 'competitive network' of some kind. One such example takes the form of a single layer of neurons with lateral (within layer) recurrence arranged in an on-centre, off-surround pattern. We then suppose that each signal $S(\mathbf{x}_i)$ is topographically assigned as the input to an artificial neuron in such a network. The lateral connectivity will gradually suppress the output of neurons with small values of $S(\mathbf{x}_i)$ and enhance the output of those with larger inputs. In this way, with appropriately tuned weights, it is possible for the network to perform a 'winner-take-all' operation so that, at equilibrium, there is a single node with non-zero activity signaling the location of the maximal input.

While the relation between *argmax* and competitive networks may seem straightforward, the relation between the algorithm and neural mechanism is not always so transparent. Consider the proposition by Bar-Gad et al. (2003) that the striatum (part of the basal ganglia – see Sect. 9.3) performs information compression on its cortical input, and that the algorithm used is principal components analysis (PCA). It is indeed possible to perform PCA with a neural network in a variety of ways (Oja 1992) but these methods are not immediately intuitive. Hence, Bar-Gad et al's proposal that the striatum performs PCA was based on the similarity of the striatum's circuitry with one of the possible neural network implementation of PCA, rather than a direct claim that striatal circuitry can support PCA.

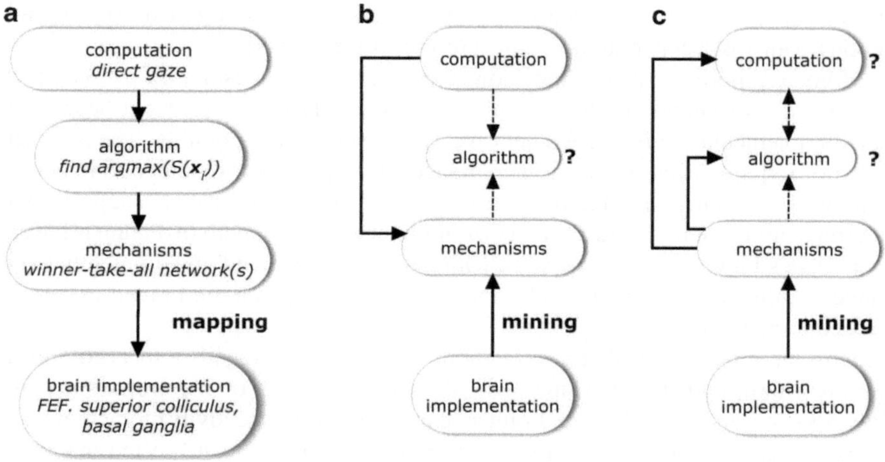

Fig. 9.2 Four-level scheme for analysing biological cognitive computation, and methods for using it. (**a**) *Top down*: mechanism mapping. (**b**), *Bottom-up*: mechanism mining. (**c**), *Bottom-up*: computation mining

This example highlights the utility of a separate, neuronal mechanistic level of analysis for conceiving of networks that can realise algorithms. However, the potential difficulty in identifying a *biological* network that can implement an *abstract* counterpart is also indicative that moving directly from algorithm to biological implementation is an explanatory gap too large to be bridged in one jump. The problem of identifying implementation is taken up again in Sect. 9.2.3.

To summarise, we suggest that Marr's tri-level scheme is augmented with an additional, mechanistic level of analysis yielding a 4-tier scheme of: computation, algorithm, neural mechanism, and implementation (Fig. 9.2).[1]

9.2.3 Modelling Routes Within the 4-tier Computational Framework

The existence of a hierarchical system of analysis does not, itself, indicate any specific *process* of model development. In practice, then, how does one take advantage of the analysis developed above? We illustrate this by returning to the toy example with visual salience. The application of the 4-tier analysis to this case is shown in Fig. 9.2a (Gurney et al. 2004b). We started with a proposed computation

[1]In Marr's original formulation of the computational framework, which appeared in an MIT technical report (Marr and Poggio 1976), a fourth level was described. However, this was dropped in the more popular account in Marr (1982). Independently, Gurney (1997) proposed a four level account which was subsequently developed further (Gurney et al. 2004b).

(how to direct gaze) proceeded to an algorithm (extracting argmax($S(\mathbf{x}_i)$)), and proposed an abstract neural mechanism (competitive networks).

However, it remains to show how this neural mechanism may be *mapped* onto biological neuronal circuits. This is a potentially thorny problem because a naive implementation of the winner-take-all network is not possible. The abstract network requires each neuron to simultaneously supply an excitatory 'on-centre' and inhibitory 'surround'. However, neuroscientific evidence strongly suggests that neuronal influence is either excitatory *or* inhibitory – but not both (a dictum referred to as Dale's Law). Further, a network with inhibitory 'surround' of sufficiently long range to influence all salience values $\mathbf{x}_i$ is not realistic given the limited extent of cortico-cortical inhibition.

It is far more likely that several brain circuits work together to accomplish the algorithm, and that they may use alternative variants of the winner-take-all networks (e.g using feedforward rather than recurrent connectivity) to achieve an approximation to the desired function. Figure 9.2a shows the suggested mapping involves frontal eye fields (FEF), superior colliculus and basal ganglia, all of which have been implicated in gaze control (Schall 2002; Hikosaka et al. 2006; Girard and Berthoz 2005).

In a more bio-inspired (rather than biologically constrained) approach, we may, indeed, be happy to subsume this function in to a single network which stands in for the gaze control circuit; but for the time being we assume we are striving for biological plausibility (at this level of description). Under these circumstances, the process of mechanism mapping – from abstract neural mechanisms onto brain circuits and systems – is the most challenging step in the top-down approach; will the abstract neural mechanism correspond in any simple way with a real biological substrate?

An alternative approach that avoids this problem is illustrated in Fig. 9.2b. Here, we are still working to understand a well-specified computational problem, but bypass, in the first instance, the algorithmic level altogether. Instead, we *mine* for mechanisms immediately available in the brain circuits supposed to serve the solution of the problem.

This bottom up approach is also not without its drawbacks because we are now bereft of an algorithm, and so there remains the problem of future *algorithm extraction*. However, depending on our position, we may be more or less indifferent to discovering algorithms. On the one hand, theoretical neuroscience demands that we establish a formal analytic (algorithm-like) framework for our model; on the other hand, pure simulation modelling does not. This is especially so for models working at lower levels of structural description, where the distinction between algorithm and abstract mechanism may, in any case, be subtle.

The situation with bottom-up simulation modelling is often not even constrained by the scheme shown in Fig. 9.2b. In this case, there may be no computational function posited *a priori* which can be used to guide the mining exercise. While we would argue that a functional hypothesis should be invoked wherever possible, many highly detailed, and well constrained neural models are constructed without any prior hypothesis about neural function. It is natural, given our emphasis on overall function, to ask what purpose these models serve.

To answer this, suppose we do indeed have a highly realistic model of an individual neuron (or neural microcircuit) but are unsure of its overall computation. To the extent that such a model has been extensively validated against its biological counterpart by ensuring similar behaviour, the model is a genuine surrogate for the biological system. As such, it may be subject to a slew of experimental manipulations in an attempt to unravel the mechanisms and, subsequently, the computations, that the circuit or neuron perform. These manipulations may be difficult and time consuming (taking many months) *in vivo* or *in vitro*, or (which is quite likely) they may be impossible, given the current state of experimental technology.

In contrast, *in silico*, the manipulations of the model and the harvesting of results may be quite rapid (hours or days) given readily available parallel and clustered computing. This leads to the notion that the model is more like an 'experimental preparation', having similar status to the *in vitro* preparations used by biologists. Such *in silico* preparations will allow us to perform *high throughput neuroscience* with the goal of discovering the computations the biological substrate performs.[2] This situation – which is quite radical within the functional/computational paradigm adopted here –is shown in Fig. 9.2c, and we dub it *computation mining*.

At the time of writing, the quintessential example of this approach is the Blue Brain project (Markram 2006) which is building biologically realistic models of cortical circuits.

9.2.4 Computational and Structural Hierarchies Combined

We are now enjoined to use two frameworks of analysis – the structural one described in Sect. 9.2.1, and the computational one of Sect. 9.2.2 – but how are they to mesh with each other? One possibility is to suppose that computations are defined at a high (or systems) level[3] dealing with brain modules, say, and that implementation occurs at the neural and molecular levels. Algorithms (and any other intermediate levels of analysis) then sit somewhere in between. This scheme interprets the two frameworks as running in parallel with each other, with each one somehow mirroring the other. However, we argue that the two frameworks deal primarily with different ontological categories and are therefore best thought of as 'orthogonal' to each other. Of the four levels of the computational decomposition, only the implementation refers to physical reality; the other three refer to concepts

[2] It is often argued that a 'divine gift' of a complete model of the brain would be useless. In the light of the above discussion, however, it would appear this is not true. It may be arduous to unravel the function of all aspects of the model-brain, but this task would certainly be easier than using biological experiments alone.

[3] We use the term 'system level' to denote a large scale ('low magnification') view of the brain, that incorporates at least one anatomically defined, functional set of nuclei. This is in contrast with the use of the term in 'systems biology' where it usually denotes the cellular level.

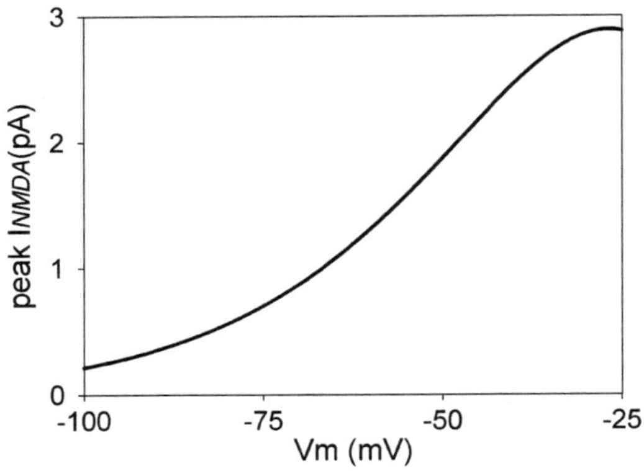

Fig. 9.3 Peak current due to NMDA receptors' activation as a function of membrane potential V_m in a typical sub-threshold regime. Function is plotted using a simple phenomenological fit of $I_{peak}(V_m)$ given in Eq. 9.1

of computation, algorithm, and abstract mechanism. In contrast, the structural hierarchy is rooted entirely in the physical and delineates objects typical of certain spatial scales. Clearly the common point of contact is implementation in the computational hierarchy, but there is no sense in which one hierarchy can fully account for, or 'mirror', the other.

We, and others (Churchland and Sejnowski 1992), therefore believe it makes more sense to allow the computational decomposition to become manifest at every structural level of description, a scheme which harmonises more naturally with an orthogonal relationship between the two frameworks. Thus, each level of structural description is a potential seat of computational function, and so it is just as valid to think of computation at the intracellular level – the remit of computational systems biology – as it does at the level of a brain nucleus.

To see how this works in more detail, we note that computations defined with respect to objects at a certain level in the structural hierarchy may invoke mechanisms implemented at a lower level of structural analysis. However, given the proposed 'structural-computational democracy', a mechanism may, in turn, be conceived of recursively as a computation at that lower level.

These ideas are illustrated using models of Mel et al. (1998) of neuronal receptive fields in the visual system. Here computation at the single neuron level is supported by dendritic and synaptic mechanisms. Mel showed that models of pyramidal neurons in visual cortex can have preferential response to oriented features in a translationally invariant way; the computation here is therefore visual feature detection. Mechanistically, this occurs through multiplicative-like operations between groups of synaptic inputs. This mechanism is, in turn, implemented, using the voltage dependent properties of NMDA receptors (NMDARs) shown in Fig. 9.3.

Notice that the response of NMDARs increases as the membrane become depolarised. Mel proposed that those inputs which need to interact in a multiplicative way are located near each other on the dendrites in 'synaptic clusters'. Then, depolarisation supplied by partially active glutamatergic synapses, supports the activity of their immediate (co-clustered) neighbours. In this way a synergy occurs between synapses in a cluster which facilitates an approximate 'AND' operation, or equivalently, multiplication, as required to perform the feature detection. Modelling this then requires morphologically extended, multi-compartment models of the form described in Chap. 7.

In the scheme as described so far, we have a neuronal level computation (feature detection) with an implementation at the membrane and synaptic levels of description which relies on synaptic current $I_{peak}(V_m)$ of the form in Fig. 9.3. It is known that NMDAR current depends on the concentration of Mg^{2+} ions external to the membrane according to (Jahr and Stevens 1990)

$$I_{peak} = \frac{-V_m g_0}{1 + \eta[\mathrm{Mg}^{2+}]e^{-\gamma V_m}} \tag{9.1}$$

where η, γ, g_o are constants (this was the function used to plot Fig. 9.3). At this level, $I_{peak}(V_m)$ is purely phenomenological and is simply fitted to corresponding I_{peak}, V_m data.

However, the *mechanism* that is constituted by the NMDAR current may be thought of as a *computation* more closely associated with a lower level of description. Specifically, the form $I_{NMDA}(V_m)$ in (9.1) – conceived of as a computation – arises through the interaction of Mg^{2+} ions with specific NMDAR sub-units (Qian et al. 2005). Such mechanistic accounts must rely on molecular level descriptions of the synapse, but we assume that their basic computational form maybe abstracted at a higher level, leading to (9.1).

One consequence of the recursion of mechanism and computation described here is that it is possible to track the way in which the low level (molecular) mechanisms described in systems biology may 'drill-upwards' through the structural hierarchy, and make themselves manifest at the top level via computations at the scale of entire brain systems. Thus, we suppose that, in general, low level mechanisms do not get 'smeared out' or lost as we model at higher levels. Nevertheless, we also anticipate that we do not have to incorporate every nuance of lower level mechanisms at higher levels; computational abstraction should serve well in many instances. This theme is explored further in the case study with dopaminergic modulation of cortico-striatal transmission (Sect. 9.3.4). It should now be apparent that we have access to a continuous line of explanation from the intracellular, all the way to behaviour, and that computational and systems neuroscience both have a vital role to play in such an exegesis. Ways in which this might play out in practice are taken up again Sect. 9.4.

9.3 A Case Study: Action Selection and the Basal Ganglia

We now illustrate the ideas outlined above in the light of our own work in modelling the basal ganglia at several levels of structural description. Several other methodological points will also emerge along the way, and the basal ganglia are dealt with....

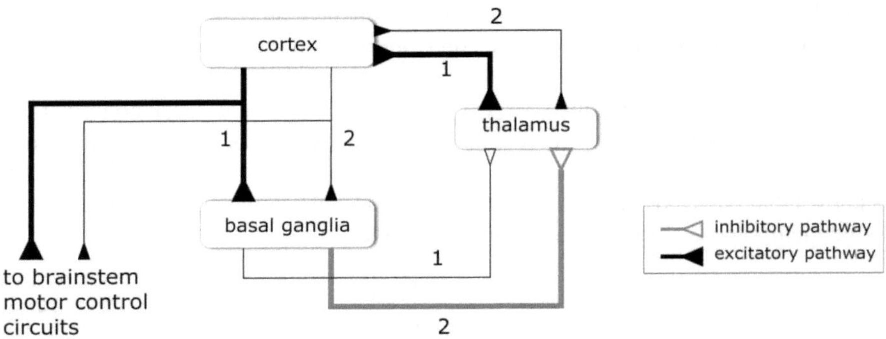

Fig. 9.4 Basic action selection mechanism for the basal ganglia. Two action channels (labelled '1' and '2' around the circuit) are shown in competition. *Thick/thin lines* indicate strong/weak signal strengths respectively)

9.3.1 *The Basal Ganglia and Action Selection*

The basal ganglia are the largest group of subcortical structures in the mammalian forebrain and have a critical influence over movement and cognition. The basal ganglia have been implicated in a wide range of processes, including perception and cognition (including working memory), and many aspects of motor function. However, one recurring theme (Mink and Thach 1993; Doya 1999) is that they are associated with some kind of selection processing. Our work has developed this idea of selection as a unifying computational theoretical framework for understanding basal ganglia function (Redgrave et al. 1999). Specifically, we proposed that the main computational function of the basal ganglia is to solve the problem of *action selection* – the resolution of conflicts between functional units within the brain that are in competition for behavioural (or cognitive) expression.

Fleshing out these ideas further, we suppose that functional command units send 'action requests' to the basal ganglia in the form of efferent copies of their encoding of action, and the basal ganglia acts as a central 'selector' or 'switch' mediating the competition for these action requests to be expressed (Fig. 9.4). Within the basal ganglia, these requests are sent through discrete information streams or *channels* which interact within selective or competitive processing mechanisms. Those requests with the largest overall activity or *salience* 'win' these competitions, resulting in depression of activity in the corresponding basal ganglia output channels. This, in turn, results in action selection as follows. Basal ganglia output is inhibitory and is normally active. The output channels form return loops, via thalamus, with the original functional units that made the action requests. On the winning channels therefore, there is *disinhibition* of the target thalamo-cortical circuits, allowing them to be active, thereby enabling their behavioural expression.

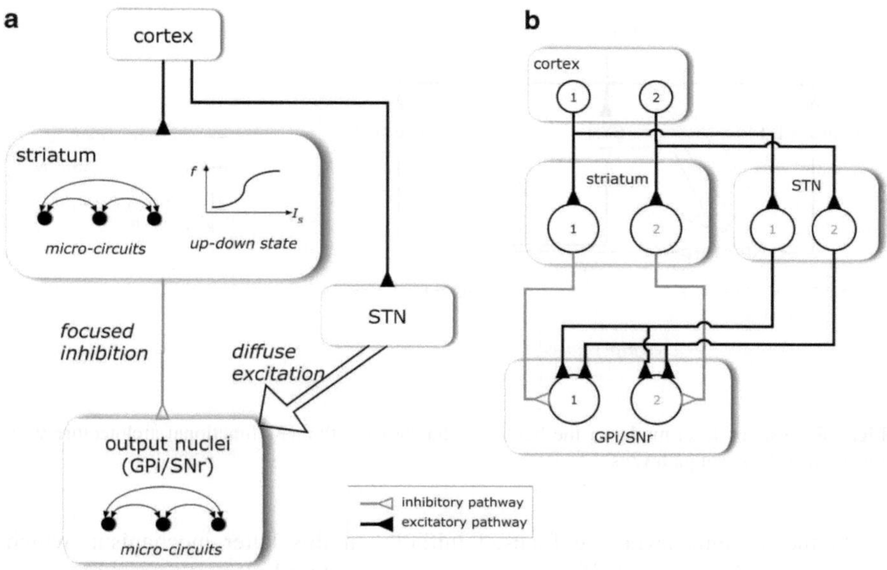

Fig. 9.5 Mining for selection mechanisms in the basal ganglia. (**a**), mechanisms at systems, circuit and neuronal level. (**b**), systems level mechanism constituted by feedforward, off-centre, on-surround network (two-channel instantiation)

9.3.2 Mechanism Mining in the Basal Ganglia at the Systems Level

Having proposed a computational function for the basal ganglia, we then proceeded to perform a bottom-up modelling study at the systems level of structural description. This, therefore, raises the question of what mechanisms can be mined from the anatomy that might support selection. Figure 9.5a shows some of these mechanisms in the context of a simplified, and partial view of basal ganglia anatomy in cartoon form (for a recent review of basal ganglia anatomy and physiology, see Redgrave 2007).

The main input nucleus in the basal ganglia is the *striatum*. This contains a complex microcircuit with several varieties of interneurons which may support competitive processing. In addition, the main population of projection neurons show a bimodal (up/down state) behaviour in which substantial, coherent cortical input is required to make these neurons fire. Such a mechanism may serve to prevent selection of weak action requests. The output nuclei in primates are the internal segment of the globus pallidus (GPi) and substantia nigra pars reticulata (SNr). These contain lateral inhibitory connections that may support competitive processing, and thereby, further selection processing. Finally, there is a system wide circuit formed from focused (‘intra-channel’) inhibition from striatum to the output nuclei, and diffuse (‘cross-channel’) excitation from another input station – the subthalamic nucleus (STN).

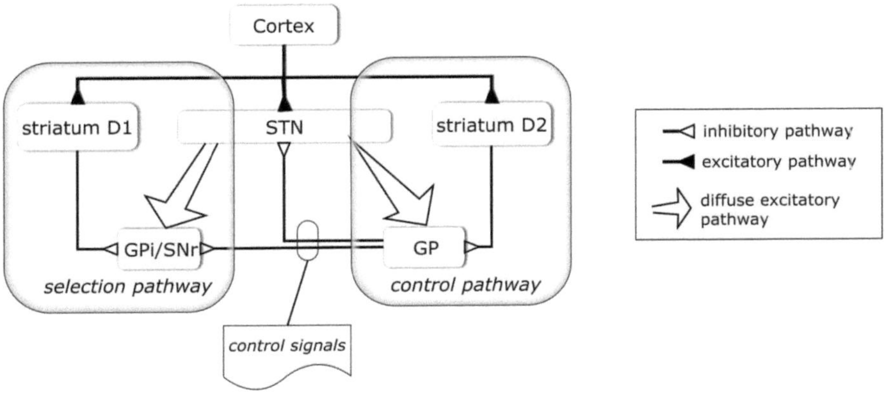

Fig. 9.6 Systems level model of the basal ganglia showing the new functional architecture with selection and control pathways

At the systems level, we focused initially on this latter mechanism, which constitutes a feedforward, off-centre on-surround network. It is illustrated in more detail for two channels in Fig. 9.5b. The polarity of the centre-surround scheme ensures a 'winner-lose-all' network, which is just what is required in the circuit of Fig. 9.4 in order to use release of inhibition as a means of gating actions.

While the circuit shown in Fig. 9.5b can, in principle, perform selection, it is not robust against widely varying signal levels (excitation can predominate without careful tuning of weights). However, our description of basal ganglia anatomy has, so far, been somewhat simplified. We now ask: what happens at the systems level, when the full basal ganglia circuit is used?

The basal ganglia also contain a nucleus – in primates, the external segment of the globus pallidus (GPe), in rats simply the globus pallidus (GP) – which receives input from striatum and STN, and which projects only internally to other basal ganglia nuclei. The striatum is also divided into two populations of projection neurons: one which projects primarily to the output nuclei (as shown in Fig. 9.5), and one projecting preferentially to the GPe. Further, the two populations are distinguished by their expression of dopaminergic receptor types, D1 or D2. We incorporated this additional circuitry into a model of the basal ganglia which used rate-coded neurons representing neural populations (Gurney et al. 2001a). The architecture is shown in Fig. 9.6.

This model was able to show robust selection and switching between actions consistent with the basal ganglia-action selection hypothesis (Gurney et al. 2001b). The model yielded a new functional architecture in which one role of the GPe is to supply control signals to the selection circuit to ensure its correct operation. Indeed, analysis and simulation confirmed that the closed loop formed by STN and GPe acted like an 'automatic gain control' on STN excitation to ensure the correct operation of the selection circuit over a wide signal range. Notice that this interpretation of the architecture is a natural imperative of the computational approach with selection in mind. System-wide selection is supported, in principle, by a subset of basal ganglia anatomy (D1-specific striatal neurons, GP/SNr and STN), and this forces us to seek an interpretation of the remaining circuitry that complements it.

The inclusion of this extra circuitry illustrates another general methodological point: that if we incorporate further biological constraints into a model and the resulting performance – in terms of the proposed system computation – is increased, then this is evidence in favour of the computational hypothesis. On the other hand, if performance declines, then this is evidence against the hypothesis. In subsequent models, (Gurney et al. 2004a; Humphries et al. 2003) incorporation of additional anatomical constraints (including the inhibitory collaterals in GPi and SNr noted above) did, indeed, show enhanced selection performance.

This mechanistic account has followed the mechanism mining scheme in Fig. 9.2b, and has so far ignored any mention of an algorithm. Recently, however, an algorithm has been proposed for the basal ganglia functional architecture described here (Bogacz and Gurney 2007). The algorithm assumes input to the basal ganglia is supplied by cortical integration of perceptual 'evidence' in a decision making paradigm. The basal ganglia may then be shown to be performing the multi-hypothesis sequential probability ratio test (Dragalin et al. 1999) which is an optimal method in that it guarantees minimal decision time for a given error rate.

9.3.3 Validation of Higher Level Models by Those at Lower Levels

In working at the systems level, many mechanistic abstractions are made. The most obvious example is that the basic neuronal signalling mechanism, namely discrete action potentials or 'spikes', is replaced by a population or ensemble firing rate. In addition, a plethora of neuronal features relating to spatio-temporal afferent signal integration are ignored. As a result, several possible relationships exist between a system level model and its lower level counterparts. First, the extra mechanisms of the biological neuron may work synergistically to produce an overall semilinear function of the kind found in the rate-coded 'neurons' (Koch 1999). This emergent linearity, means our systems level model is 'right for the right reasons'. Second, it is possible that, while the individual neural populations are highly nonlinear, their behaviour when embedded in the entire system is comparable to that of the systems level model. This global equivalence means we are 'right for the wrong reasons' – our computational hypothesis is still supported but its mechanistic explanation needs revising. Finally, it is possible that the neural nonlinearities lead to entirely different system behaviours, and our systems level model is simply 'wrong' – our computational hypothesis is refuted. Only by building the model at the lower level of description can we be sure which possibility applies.

Working at the level of neural spiking behaviour also enables a model to make contact with a much larger raft of physiological data – a situation that has two consequences. First, the additional data supplies further constraints to test any computational hypothesis for the system as a whole. Thus, confirmatory evidence of the hypothesis is obtained if the model can simultaneously implement the target function (here, action selection) *and* display the patterning of spikes observed *in vivo*. Conversely, if this is not possible, the hypothesis is brought into question. A second consequence of enriching the relevant constraining data is a corresponding burgeoning of possible opportunities for mechanism and computational mining.

To address these issues, we therefore constructed a model of the basal ganglia using spiking neurons (Humphries et al. 2006) which incorporates much of the known physiology of individual neuronal species including: synaptic currents with time constants for AMPA, NMDA and GABAergic synapses, spontaneous currents, dopamine in STN and GP as well as striatum, and inter-neuronal delays. In addition, while the neurons were based on simple LIF dynamics, they incorporated phenomenological (abstract) descriptions of shunting (multiplicative) inhibition at proximal dendrites and the soma (Koch 1999), and rebound bursting in STN neurons.

The model was pitted against a comprehensive data set related to low frequency oscillations (LFOs) under urethane anesthesia in the rat (Magill et al. 2001). In addition we made use of observations by Brown et al. (2002) on γ-oscillations (40–60 Hz). We found that the same model with the same parameters could simultaneously display selection, fit the LFO data, and that for γ-oscillations. This supported the hypothesis that the systems level model was 'right for the right reasons' but, of course, models of the basal ganglia with further mechanistic elaboration (in particular more realistic striatal microcircuits) may yet prove otherwise.

In addition to supporting the selection hypothesis, our model also highlighted new computational and mechanistic possibilities. First, we had explicitly hypothesised that certain features of GP-STN firing in the LFOs observed by Magill et al. (2001) were due to dopaminergic regulation of the strength of coupling between these nuclei; it did indeed transpire that this hypothesis was supported. Second, the source of γ-oscillations was attributed to the natural resonance between STN and GP. This dynamic is an inevitable consequence of the coupling between a source of excitation (STN) and one of inhibition (GP), with synaptic delay. This phenomenon was robust, and so two possibilities emerge; either the oscillation is an unwanted side-effect of the STN-GP circuit which uses firing *rates* as its main signal coding, or the oscillations serve a functional purpose. In order to 'mine' a possible computation here, we turn to theories of temporal synchrony in visual cortex (Engel and Singer 2001) in which γ-oscillations are believed to support a solution to the feature binding problem. In this account, neurons encoding features belonging to the same object are supposed to fire in synchrony at the resolution of γ-oscillation frequencies. This notions have been expanded recently by Fries (2009)

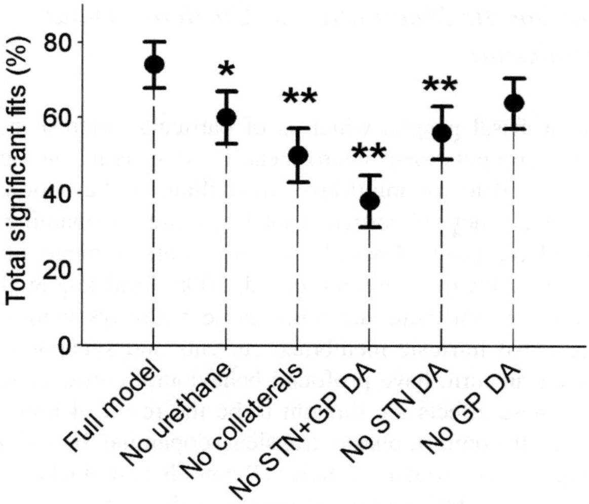

Fig. 9.7 Overall correlation between model simulation and animal data from (Magill et al. 2001) across variations of the model obtained by deleting certain features including: the model of urethane anesthesia, collaterals in GP and SNr, and dopamine in STN and/or GP. Fifty simulations were run for each condition, and a simulation was deemed a significant fit if it correlated with the data at the $p = 0.05$ level. The error bars are the standard deviations of the percentage of significant fits. A single asterisk denotes a significant difference between the expected value for the model variant and that of the full model with the one-tailed criterion $p < 0.05$; a double asterisk denotes a significant difference at $p < 0.01$

who describes a general role for γ-oscillations in visual cortical processing for segmentation, selection and gain modulation. These functions may also be relevant for action selection where 'action features' at the input to basal ganglia may be synchronised or segmented at the level of STN for further basal ganglia processing as integrated actions.

The model in (Humphries et al. 2006) incorporated several other mechanisms that were included to address specific features of the physiological data. However, while this mechanistic complexity is a source of explanatory power, it can also raise issues of its own. So, while each mechanism may have been introduced with a specific function in mind, their simultaneous operation may cause synergistic side effects that we had not anticipated. In addition, we should be wary of gratuitously including arbitrary mechanisms simply for the sake of 'extra realism'. We therefore tested for mechanistic redundancy by excluding each mechanism in turn, and examining the correlations with the data of Magill et al. (2001) over many simulations with small variations in the model parameters. All mechanisms contributed to the fit with the data (Fig. 9.7) supporting the conclusion that there was little or no redundancy in model mechanism.

9.3.4 Dopaminergic Modulation of Cortico-Striatal Transmission

One feature of the basal ganglia which is of particular interest in exploring the interface between computational neuroscience and systems biology, and which is an excellent candidate for multi-level modelling, is the action of dopamine in modulating the efficacy of cortico-striatal synapses. Dopamine receptors are G protein-coupled receptors (Chap. 2) which initiate a range of intra-cellular signal transduction pathways (Fernandez et al. 2006; Lindskog et al. 2006; Berke and Hyman 2000). At the neuronal level, these pathways manifest themselves in a complex array of intrinsic membrane currents and synaptic effects (Nicola et al. 2000) which, in turn, have profound behavioural consequences (Salamone et al. 2009). All these effects are thought to be the result of tonic (steady-state) levels of dopamine. In contrast, phasic (transient) dopamine 'bursts' also play a key role in mediating cortico-striatal plasticity (Reynolds and Wickens 2002) but we will not consider this dimension of dopaminergic action here.

At the cellular level in striatum, the phenomenology of dopamine modulation may be crudely approximated by stating that it facilitates cortico-striatal transmission when mediated by D1-type receptors (Akkal et al. 1996), and attenuates this process when activating D2 receptors (Gerfen et al. 1990; Harsing and Zigmond 1997). In our systems level model of basal ganglia (Gurney et al. 2001b), this was captured in the following way. Each cortio-striatal connection was assigned a nominal weighted value w. The tonic level of dopamine was represented by a parameter λ, and D1 and D2 effects were captured by forming new cortico-striatal weights

$$\begin{aligned} w_{D1} &= w(1+\lambda) \\ w_{D2} &= w(1-\lambda) \end{aligned} \tag{9.2}$$

for D1 and D2 striatal projection neurons respectively. At equilibrium, the output y of the rate-coded striatal projection neuron with input x, is then given by the piecewise linear form

$$y = \begin{cases} 0 & \text{if } x < \varepsilon/w_d \\ m(w_d x - \varepsilon) & \text{if } x \geq \varepsilon/w_d \text{ and } m(w_d x - \varepsilon) \leq 1 \\ 1 & \text{if } x \geq \varepsilon/w_d \text{ and } m(w_d x - \varepsilon) > 1 \end{cases} \tag{9.3}$$

where d is D1 or D2, m is the gradient of the increasing part of the function, and ε fixes a threshold in order that $y > 0$ (mimicking the up-state transition in these neurons). With this feature, the model was able to show dopamine-dependent selection regimes (Gurney et al. 2001b). Increased simulated dopamine (larger λ) led to easier selection (less salience required), and increased probability of multiple channel selection. In contrast, reduced dopamine made it harder to select and

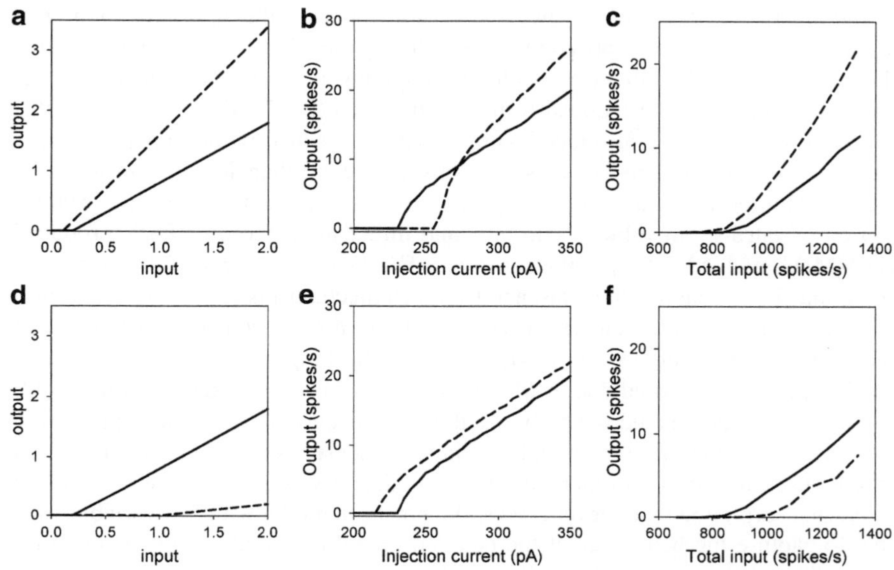

Fig. 9.8 Input–output relations for model striatal projection neurons under dopaminergic modulation. In all cases, the *solid line* is for no dopamine and the *dashed line* shows a level of simulated dopamine of 0.8. (**a**), rate-coded D1-type striatal neurons with functional form $y(x)$ from (9.3), but without saturation at 1. (**d**), similar to (**a**), but for D2-type neurons. (**b**, **c**), results for D1-type spiking neurons using the model in Humphries et al. (2009a). (**b**), shows the effect of dopamine on intrinsic membrane currents only (using current injection as input), **c** augments this with synaptic effects and shows output driven under synaptic barrage. (**e**, **f**) are the counterparts to (**b**), **c** for D2-type neurons)

prevented dual channel selection. In the limiting case, with $\lambda = 0$, the model was unable to select at all, irrespective of the salience of the action request. This is consistent with the model showing akinesia in Parkinson's disease (a pathology in which there is severe loss of dopaminergic innervation to the basal ganglia).

Aside from pathological extremes, it is plausible that the basal ganglia should be able to control the ease with which actions are selected. For example, it may be more or less useful for an animal to rapidly evaluate multiple options when exploring a new environment. We therefore elevate the above result to the status of a computational hypothesis – that the role of tonic dopamine in the basal ganglia is one of enabling *adaptive selection*.

In order to compare the dopaminergic modulation in (9.3) with that for the spiking model described subsequently, it is useful to ignore any saturation of output y at 1. Figure 9.8a shows this form of y for D1-type striatal projection neurons, with $\lambda = 0$ and 0.8 (solid and dashed line respectively). Figure 9.8d shows the corresponding result for D2-type projection neurons.

Given the underlying complexities of dopamine modulation in striatal projection neurons, we might anticipate that the simple functional relationships used thus-far would be rather inadequate; but exactly how poor is the approximation they represent? As part of our programme of building models of striatal microcircuits (Humphries et al. 2009b), we developed a spiking neuron model of the striatal projection neuron in the reduced form of Izhikevich (2007). However, within this general framework, we introduced a novel method for incorporating dopaminergic modulation that attempted to capture many of the mechanisms known to occur therein (Humphries et al. 2009a). Specifically, we included effects due to modulation of intrinsic membrane currents, and direct influence on EPSPs produced via NMDA and AMPA synaptic receptors. For D1-type neurons this gave three dopamine dependent mechanisms, and two such mechanisms for D2-type neurons.

Since there is insufficient experimental data to directly constrain all aspects of such a model, we validated our model by 'piggy-backing' it against a pre-existing, conductance-based model due to Moyer et al. (2007). This model has almost 200 compartments and includes nearly all the known currents in striatal projection neurons. It may therefore be thought of, in some sense, as a 'gold standard' which may be used to constrain other, simplified models, like our own. It was therefore somewhat surprising that our reduced spiking neuron model with few variables was able to capture the behaviour of the detailed model remarkably well (Humphries et al. 2009a). We argue that this approach of using low level models to help constrain those at higher levels, is a powerful one which could be deployed quite generally.

In the current context, the effects of dopamine on intrinsic membrane currents alone produced quite different results from the complete model that also incorporated synaptic influences (Fig. 9.8). For D1-type neurons, taking account of membrane properties alone, increasing levels of dopamine increases/decreases the response to large/small injection current inputs respectively (Fig. 9.8b). This process amounts to an increase in the signal-to-noise-ratio (snr) of the neural transfer function, an observation that has led to the well-known snr hypothesis for dopamine function (Servan-Schreiber et al. 1990). Other models of striatal projection neurons (Gruber et al. 2003) also predict similar relationships. However, these models do not take account of the action of dopamine at synapses. Incorporation of such effects in our model leads to the situation shown in Fig. 9.8c. Here, the trend, when measured against a barrage of synaptic input, is better described as one of increased gain at all input levels.

The story for D2-type neurons is also mixed. Dopamine acting only on intrinsic membrane currents is facilitatory (Fig. 9.8e) but this trend is reversed when synaptic effects are included (Fig 9.8f). Moreover, the D2-neuron relationship is best described as being a 'shift' in a constant gain transfer function.

The preservation of dopaminergic function in going from conductance-based models to reduced spiking form, is repeated in going up to rate coded neurons at the systems level. Comparing Fig. 9.8a, c, it is apparent that the D1-type behaviour in the rate coded model is almost perfectly rendered. It is remarkable that a single parameter, rate-coded model like that in the first part of (9.2) yields results comparable with a model which explicitly dissects three dopaminergic influences. In a previous attempt to make the rate-coded MSN models more realistic (Humphries 2007) we explicitly incorporated two effects of dopamine on intrinsic membrane currents for D1-type neurons, and captured the behaviour in Fig. 9.8b. However, this model did not explicitly incorporate synaptic effects, and therefore did not show the behaviour of Fig. 9.8c.

Comparison of Fig. 9.8d, f shows the D2-type behaviour for the rate coded model preserves the required function shift, but differs by introducing a spurious decrease in gain of the transfer function. Nevertheless, the overall pattern of behaviour is quite similar in both models. Therefore, since dopamine in the systems level model supports the adaptive selection hypothesis, we conclude that dopaminergic modulation in our spiking model also supports this hypothesis (more readily than one of control over signal-to-noise-ratio).

This section has shown how it is possible to capture the computational function of dopamine on non-plastic cortico-striatal transmission at several levels of description: from highly detailed compartmental models, through reduced spiking models, to rate-coded neurons. It is apparent that the function dopamine serves in this respect is well preserved at each level and may be modelled using progressively more abstract mechanisms. However, it remains to make the link between the complex web of intracellular signalling that mediates dopaminergic modulation of the neuron, and the phenomenological descriptions of dopamine that occur even at the level of compartmental modelling. This is clearly a key challenge for the interface between computational neuroscience and computational systems biology.

9.4 Multilevel Modelling

The previous section has emphasised the way in which molecular and membrane-based mechanisms can manifest themselves at higher levels, and may constitute crucial features of models at these levels of description. In the case of dopamine modulation, it transpired that its effects could be captured reasonably well in a phenomenological account using (9.2). However this may not always be the case and alternative approaches are required.

One possibility is to build a more detailed model at a level low enough to capture *all* mechanisms of interest at their native level of description, and which is sufficiently anatomically extensive to cover all large-scale (system-wide) interactions that would be included in a higher level model. This was the approach we took with our large-scale spiking model of the basal ganglia (Humphries et al. 2006). While not as vigorously bottom-up in design as full conductance-based, compartmental models, the resulting model is sufficiently complex to make it a virtual 'experimental preparation' (see Sect. 9.2.3) and it continues to be 'mined' for new insights (Humphries and Gurney 2012).

While such preparation-like models are valuable, mining them can be challenging because all mechanisms are universally present on a massive scale. It is as if we are confronted with a high magnification view of a very large piece of brain tissue without the ability to 'zoom out' to lower magnification to see the bigger picture. This metaphor is the inspiration for another approach we wish to advocate here. Thus, suppose, for example, we wish to explore the computations performed by a particular class of neurons using realistic, conductance-based (compartmental) models. In principle, it should be possible to study these functions using only a single model neuron (or at most a small microcircuit of only a few such neurons), so long as the *signal environment* of the target neuron model is similar to that which would occur in a homogeneous model of many thousands of such neurons. The signal environment has two aspects: the raw encoding scheme (point event spikes or firing rates) and the patterning and significance of the signals. In the case of single neuron models, an approximation to their veridical signal environment may be generated by a network of simplified spiking neurons, as long as they can supply spike trains with the right patterning (mean firing rate, bursting or tonic etc). We refer to the single, conductance-based target neuron as the *model core*. This is then embedded into a *model surround* comprising the simplified neuron network.

This *core-surround embedding* scheme may be replicated recursively ('Russian-doll' like) so that, the large-scale spiking neural network (previously considered a model surround) becomes, simultaneously, a model core by being embedded into a rate-coded, systems level model surround. The transition from spikes to rates (and vice versa) will, of course, require the construction of novel, *hybrid model neurons* but this exercise is a plausible one. Thus, a spikes-in to rate-out neuron might be comprised of a leaky-integrate-and-fire neuron in which the neural membrane (or some nonlinear function of it) is interpreted directly as a firing rate (rather than being used to induce spikes according to a threshold crossing). In contrast a rate-to-spikes neuron could be constructed by equipping a leaky integrator with a time-sampling event generator, with event probability based on the instantaneous firing rate.

The model core may also be defined, not by spatial localisation, but neural species (e.g. use a mix of compartmental and reduced neuron models in a micro-circuit) or, at higher levels, by a functional unit (e.g embed a single nucleus in a wider brain system).

We have successfully used this approach in our robotic level work, where biologically plausible basal ganglia models were embedded in surrounds that were largely engineered to produce realistic signal environment (Prescott et al. 2006; Gurney et al. 2004b). The basal ganglia models in (Gurney et al. 2001a) have also been embedded in abstract, connectionist neural network surrounds, in order to model high level cognitive phenomena (Stafford and Gurney 2007).

9.5 Summary

This chapter has outlined some principled approaches to building models in computational neuroscience. A recurring theme is that, ideally, we should always strive to specify a proposed top level function or computation for the target system (neuron, nucleus, brain system etc). If this is, indeed, possible, then we may or may not have access to abstract algorithmic/theoretical description of the function. This then determines a preference for a top down 'mapping' onto the biology or a bottom up ‘mining’ of potential mechanisms that support the computation. If no top level function is proposed, we should be using the model to seek one. We have emphasised that, low level features of the biology may have profound effect at higher levels, and that their influence may, in some cases, be captured in high level phenomenological descriptions. This will not in general be true, however, in which case we might resort to anatomically extensive model ‘preparations’, or seek a hybrid-level ‘embedding’ of the biological core in a higher level surround.

In any case, whichever approach is adopted, we argue that the *raison d’etre* for a particular low level mechanism may not be apparent until it is expressed, or embedded, in a wider functional domain. Our example here was the way that membrane dynamics and synaptic effects conspired to give a mathematical form for dopaminergic modulation of neural function which appeared to support a computation dealing with action selection – adaptive selection – which is expressed only at the level of the basal ganglia *in toto*.

This last observation supplies a compelling reason for a close relationship between computational systems biology and computational neuroscience. Thus, the consequences for neural behaviour of the intracellular signalling dealt with in systems biology will only be fully revealed in higher level models of brain function. Conversely, phenomenological models of membrane and synaptic function may be validated against reference (‘gold standard’) models of their molecular substrate. Finally, whatever opinion the reader has of the specific techniques espoused here, we would emphasise the importance of thinking about methodological issues in order to build appropriately motivated models, and extract the maximum benefit from them.

Acknowledgements This work was supported by UK EPSRC grant EP/C516303/1, and French grants: Marie Curie BIND, and an ANR Chaire d’Excellence.

References

Akkal D, Burbaud P, Audin J, Bioulac B (1996) Responses of substantia nigra pars reticulata neurons to intrastriatal d1 and d2 dopaminergic agonist injections in the rat. Neurosci Lett 213(1):66–70

Bar-Gad I, Morris G, Bergman H (2003) Information processing, dimensionality reduction and reinforcement learning in the basal ganglia. Prog Neurobiol 71(6):439–473

Berke JD, Hyman SE (2000) Addiction, dopamine, and the molecular mechanisms of memory. Neuron 25(3):515–532. doi:10.1016/S0896-6273(00)81056-9

Bogacz R, Gurney K (2007) The basal ganglia and cortex implement optimal decision making between alternative actions. Neural Comput 19(2):442–477. doi:10.1162/neco.2007.19.2.442

Brown P, Kupsch A, Magill PJ, Sharott A, Harnack D, Meissner W (2002) Oscillatory local field potentials recorded from the subthalamic nucleus of the alert rat. Exp Neurol 177(2):581–585

Churchland PS, Sejnowski TJ (1992) The computational brain. Computational neuroscience. The MIT Press, Cambridge

Connor CE, Egeth HE, Yantis S (2004) Visual attention: bottom-up versus top-down. Curr Biol 14(19):R850–R852. doi:10.1016/j.cub.2004.09.041

Doya K (1999) What are the computations of the cerebellum, the basal ganglia and the cerebral cortex? Neural Netw 12(7–8):961–974

Dragalin V, Tartakovsky A, Veeravalli V (1999) Multihypothesis sequential probability ratio tests – part I: asymptotic optimality. IEEE Trans Inf Theory 45(7):2448–2461

Engel AK, Singer W (2001) Temporal binding and the neural correlates of sensory awareness. Trends Cogn Sci 5(1):16–25

Fernandez r, Schiappa R, Girault J, Novre NL (2006) DARPP-32 is a robust integrator of dopamine and glutamate signals. PLoS Comput Biol 2(12):e176. doi:10.1371/journal.pcbi.0020176

Fries P (2009) Neuronal gamma-band synchronization as a fundamental process in cortical computation. Annu Rev Neurosci 32(1):209–224. doi:10.1146/annurev.neuro.051508.135603

Gerfen C, Engber T, Mahan L, Susel Z, Chase T, Monsma F, Sibley D (1990) D1 and d2 dopamine receptor-regulated gene expression of striatonigral and striatopallidal neurons. Science 250:1429–1432

Girard B, Berthoz A (2005) From brainstem to cortex: computational models of saccade generation circuitry. Prog Neurobiol 77(4):215–251. doi:10.1016/j.pneurobio.2005.11.001

Gruber A, Solla S, Surmeier D, Houk J (2003) Modulation of striatal single units by expected reward: a spiny neuron model displaying dopamine-induced bistability. J Neurophysiol 90(2):1095–1114

Gurney KN (1997) An introduction to neural networks. UCL Press (Taylor and Francis group), London

Gurney K (2009a) Computational models in neuroscience: from membranes to robots. In: Computational modelling in behavioural neuroscience: closing the gap between neurophysiology and behaviou, Advances in behavioural brain science. Psychology Press, East Sussex, p 107

Gurney KN (2009b) Reverse engineering the vertebrate brain: Methodological principles for a biologically grounded programme of cognitive modelling. Cogn Comput 1(1):29–41. doi:10.1007/s12559-009-9010-2

Gurney KN, Prescott TJ, Redgrave P (2001a) A computational model of action selection in the basal ganglia I: a new functional anatomy. Biol Cybern 84:401–410

Gurney KN, Prescott TJ, Redgrave P (2001b) A computational model of action selection in the basal ganglia II: analysis and simulation of behaviour. Biol Cybern 84:411–423

Gurney KN, Humphries M, Wood R, Prescott TJ, Redgrave P (2004a) Testing computational hypotheses of brain systems function: a case study with the basal ganglia. Network 15(4):263–290

Gurney KN, Prescott TJ, Wickens JR, Redgrave P (2004b) Computational models of the basal ganglia: from robots to membranes. Trends Neurosci 27(8):453–459

Harsing JLG, Zigmond MJ (1997) Influence of dopamine on gaba release in striatum: evidence for d1-d2 interactions and non-synaptic influences. Neuroscience 77(2):419–29

Hikosaka O, Nakamura K, Nakahara H (2006) Basal ganglia orient eyes to reward. J Neurophysiol 95(2):567–584. doi:10.1152/jn.00458.2005

Humphries MD (2007) High level modeling of tonic dopamine mechanisms in striatal neurons. arXivorg: q-bio/0701022, http://arxiv.org/abs/q-bio/0701022

Humphries MD, Gurney KN (2012) Network effects of subthalamic deep brain stimulation drive a unique mixture of responses in basal ganglia output. Eur J Neurosci, in press

Humphries M, Prescott TJ, Gurney KN (2003) The interaction of recurrent axon collateral networks in the basal ganglia. In: Kaynak O, et al. (eds) Joint International Conference ICANN/ICONIP. Lecture Notes in Computer Science, Istanbul, Turkey, Springer, pp 797–804

Humphries MD, Stewart RD, Gurney KN (2006) A physiologically plausible model of action selection and oscillatory activity in the basal ganglia. J Neurosci 26(50):12921–12942. doi:10.1523/JNEUROSCI.3486-06.2006

Humphries M, Lepora N, Wood R, Gurney K (2009a) Capturing dopaminergic modulation and bimodal membrane behaviour of striatal medium spiny neurons in accurate, reduced models. Front Comput Neurosci 3. doi:10.3389/neuro.10.026.2009

Humphries MD, Wood R, Gurney K (2009b) Dopamine-modulated dynamic cell assemblies generated by the GABAergic striatal microcircuit. Neural Netw 22(8):1174–1188. doi:10.1016/j.neunet.2009.07.018, PMID: 19646846
Izhikevich EM (2007) Dynamical systems in neuroscience: the geometry of excitability. MIT Press, Cambridge
Jahr CE, Stevens CF (1990) Voltage dependence of NMDA-activated macroscopic conductances predicted by single-channel kinetics. J Neurosci 10(9):3178
Kitano H (2002) Computational systems biology. Nature 420(6912):206–210. doi:10.1038/nature01254
Koch C (1999) The biophysics of computation: information processing in single neurons. Oxford University Press, New York
Lindskog M, Kim M, Wikstrm MA, Blackwell KT, Kotaleski JH (2006) Transient calcium and dopamine increase PKA activity and DARPP-32 phosphorylation. PLoS Computat Biol 2(9):e119. doi:10.1371/journal.pcbi.0020119, PMID: 16965177
Magill PJ, Bolam JP, Bevan MD (2001) Dopamine regulates the impact of the cerebral cortex on the subthalamic nucleus-globus pallidus network. Neuroscience 106:313–330
Markram H (2006) The blue brain project. Nat Rev Neurosci 7(2):153–160. doi:10.1038/nrn1848
Marr D (1982) Vision: a computational investigation into human representation and processing of visual information. Freeeman, New York
Marr D, Poggio T (1976) From understanding computation to understanding neural circuitry. Technical report AIM-357, MIT, Cambridge
Mel BW, Ruderman DL, Archie KA (1998) Translation-invariant orientation tuning in visual “complex” cells could derive from intradendritic computations. J Neurosci 18(11):4325–4334
Mink JW, Thach WT (1993) Basal ganglia intrinsic circuits and their role in behavior. Curr Opin Neurobiol 3(6):950–957
Moyer JT, Wolf JA, Finkel LH (2007) Effects of dopaminergic modulation on the integrative properties of the ventral striatal medium spiny neuron. J Neurophysiol 98(6):3731–3748. doi:10.1152/jn.00335.2007
Nicola SM, Surmeier DJ, Malenka RC (2000) Dopaminergic modulation of neuronal excitability in the striatum and nucleus accumbens. Annu Rev Neurosci 23(1):185–215. doi:10.1146/annurev.neuro.23.1.185
Oja E (1992) Principal components, minor components, and linear neural networks. Neural Netw 5:927–927
Prescott AJ, Gonzales FM, Gurney KN, Humphries M, Redgrave P (2006) A robot model of the basal ganglia: behavior and intrinsic processing. Neural Netw 19(1):31–61
Qian A, Buller AL, Johnson JW (2005) NR2 subunit-dependence of NMDA receptor channel block by external mg2+. J Physiol 562(2):319–331. doi:10.1113/jphysiol.2004.076737
Redgrave P (2007) Basal ganglia. Scholarpedia 2(6):1825
Redgrave P, Prescott TJ, Gurney KN (1999) The basal ganglia: a vertebrate solution to the selection problem? Neuroscience 89:1009–1023
Reynolds JNJ, Wickens JR (2002) Dopamine-dependent plasticity of corticostriatal synapses. Neural Netw 15(4–6):507–521, PMID: 12371508
Salamone J, Correa M, Farrar A, Nunes E, Pardo M (2009) Dopamine, behavioral economics, and effort. Front Behav Neurosci 3(13). doi:10.3389/neuro.08.013.2009
Schall JD (2002) The neural selection and control of saccades by the frontal eye field. Philos Trans R Soc Lond B Biol Sci 357(1424):1073–1082. doi:10.1098/rstb.2002.1098
Servan-Schreiber D, printz H, Cohen J (1990) A network model of catecholamine effects: gain, signal-to-noise ratio and behavior. Science 249:892–895
Stafford T, Gurney KN (2007) Biologically constrained action selection improves cognitive control in a model of the stroop task. Philos Trans R Soc Lond B Biol Sci 362(1485):1671–1684. doi:10.1098/rstb.2007.2060

Chapter 10
Virtues, Pitfalls, and Methodology of Neuronal Network Modeling and Simulations on Supercomputers

Anders Lansner and Markus Diesmann

Abstract The number of neurons and synapses in biological brains is very large, on the order of millions and billions respectively even in small animals like insects and mice. By comparison most neuronal network models developed and simulated up to now have been tiny, comprising many orders of magnitude less neurons than their real counterpart, with an even more dramatic difference when it comes to the number of synapses. In this chapter we discuss why and when it may be important to work with large-scale, if not full-scale, neuronal network and brain models and to run simulations on supercomputers. We describe the state-of-the-art in large-scale neural simulation technology and methodology as well as ways to analyze and visualize output from such simulations. Finally we discuss the challenges and future trends in this field.

10.1 Introduction

Biological neuronal networks vary in size from that of small and primitive animals like the roundworm *C Elegans* with some hundred neurons in its nervous system to man with some twenty billion neurons in the brain (Fig. 10.1). At least in the

A. Lansner (✉)
Departmentof Numerical Analysis and Computer Science, Stockholm University, Stockholm, Sweden

Department of Computational Biology, School of Computer Science and Communication, Royal Institute of Technology, Stockholm, Sweden
e-mail: ala@csc.kth.se

M. Diesmann
Institute of Neuroscience and Medicine (INM-6), Computational and Systems Neuroscience, Research Center Juelich, Juelich, Germany

Faculty of Medicine, RWTH Aachen University, Aachen, Germany

RIKEN Brain Science Institute, Wako City, Japan
e-mail: diesmann@fz-juelich.de

N. Le Novère (ed.), *Computational Systems Neurobiology*,
DOI 10.1007/978-94-007-3858-4_10,

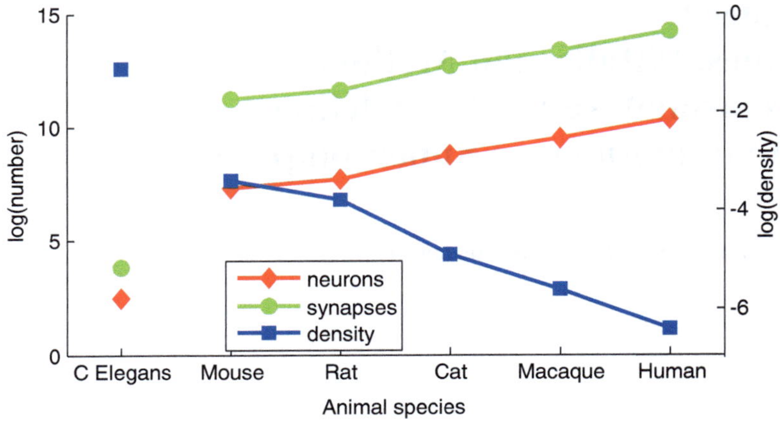

Fig. 10.1 Scaling of biological brains in terms of number of neurons, number of synapses, and connection density

mammalian series, from mouse to man, the number of synapses typically scale proportional to the number of neurons in the system, since the number of synapses made on average by each neuron, on the order of a few thousand, is practically the same from mouse to man. Already the mouse brain is quite sparsely connected – the actual number of existing synaptic connections relative to the pairs of neurons in the system is only about 0.4%. As we progress upwards in the mammalian series, connectivity density decreases further. This is in sharp contrast to most neural network models which have a dense connectivity and where the number of synapses is typically scaled like the square of the number of neurons.

Real vertebrate neuronal networks also have a complex and intricate structure, including several different types of neurons and a modular and layered layout, sometimes adequately represented by a microcircuit as the smallest repetitive unit. Even a microcircuit, e.g. the hypothetical cortical macrocolumn with a diameter of some 500 μm, contains tens of thousands of neurons and tens of millions of synaptic connections. A microcircuit is, by definition, a component in a much larger network and more than 50% of its synapses may originate from neurons located outside the local microcircuit. It can still sometimes be studied in isolation, especially when it can be related to a correspondingly reduced experimental *in vitro* preparation, e.g. a cortical slice. But much more often it is in reality embedded in a mosaic of similar modules and receives afferent input from several sources, and we need to model such a network of connected microcircuits.

Partly due to our limited computing resources, computational studies have so far often focused on the cellular and small network level, for instance, on signal transmission in single neurons and plasticity in single synapses or on information processing in local microcircuits, such as e.g. the emergence and dynamics of local receptive fields in the primary visual cortex.

Modeling at the network level poses specific challenges, but it is clear that taking into account global and dynamic network interactions is inevitable in order to better understand the functioning of a neuronal system operating *in vivo*. Closed loops within the brain and between brain and environment in sensory-motor control are abundant, but they only exist on the level of the entire nervous system and can thus only be simulated in brain-scale models.

It is thus overly optimistic to think that it would be possible to understand brain function from modeling a local network of some hundred cells, like a cortical minicolumn, or by dramatically sub-sampling the global network, e.g. by letting a few model neurons represent an entire cortical column or even area. One specific problem of using sub-sampled network models comprising model neurons tuned after their real counterparts, is how to provide sufficient synaptic input current to activate these model neurons. In a small network we have a very limited number of presynaptic neurons and synaptic terminals. Thus, either we need to exaggerate background noise level, connection probabilities or synaptic conductances, most of the time all of these. This tends to give a noise dominated network with few and strong signals circulating around, as opposed to the real cortical network where many weak signals interact. Such deviations from reality can distort significantly the dynamics of the networks. Thus, in many cases a large-scale model is a more accurate one than a smaller model.

The starting point for network modeling is the component cells and their pair-wise synaptic interactions, including transmission and conduction delays and synaptic plasticity. The level of complexity of model neurons and synapses now becomes an issue. Given limited computational resources, there is an obvious trade-off between complexity of the cell and synapse models and the size of the network. It is a painful fact that a standard PC of today gets unbearably slow as the model network size grows beyond some thousand cells. Furthermore, the amount of simulated time needed to study a certain phenomenon grows linearly with the time span of the phenomenon. This can vary widely, e.g. from a memory recall process lasting fractions of a second, to induction and expression of long-term memory which typically occurs over tens of minutes and even longer.

There is some light in the tunnel, however, since one reason for avoiding large-scale or full-scale simulations is gradually disappearing. The computational capacity of computers is still growing fast. Reduction of dimensions and growing clock speed of integrated circuits have contributed to this over many years. This development may be approaching its limits, but instead the number of processors operating in parallel is now increasing dramatically. Parallel computing has already reached the consumer market in the form of specialized graphics processing units (GPU) with a following drop in prices. Standard PC processors are now multi-core, and it is quite likely that we will have desktop boxes with thousands of parallel processors in the next five or so years. Fortunately, neuronal networks represent computationally quite homogenous computational structures where computation and memory are predominantly local, which make parallel simulation on cluster computers with distributed memory relatively straightforward. Thus, in the near future, computer power will no longer prevent us from putting together and study large and even full-scale models of global brain networks (Fig. 10.2).

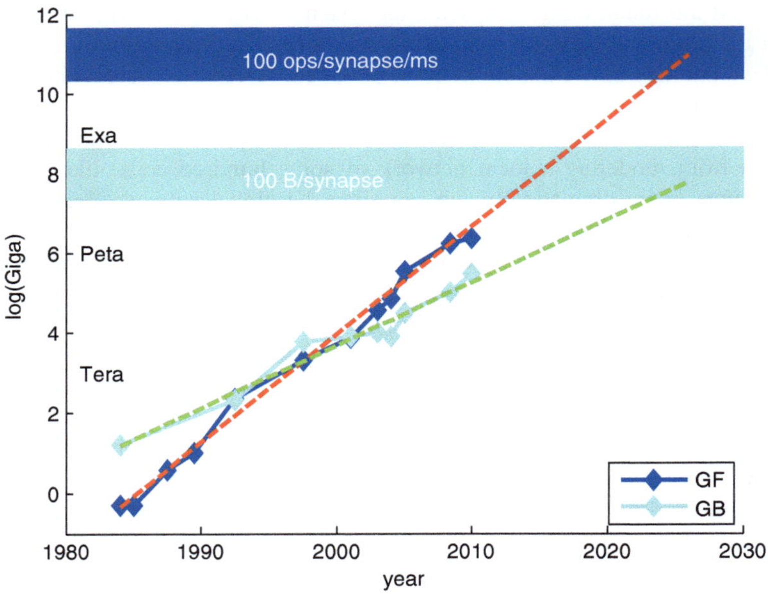

Fig. 10.2 Increase of computational power and memory of supercomputers over time. Data in *blue* and *cyan* respectively. Log-linear trends indicated as *dashed red* and *green lines*. The *blue* area at the *top* indicates the estimated computing capacity required to execute a human brain sized spiking neural network in real time. The *cyan* area *below* indicates the amount of memory needed to represent each synapse in the human brain with 100 byte

10.2 Terminology for Large-Scale Network Models

As the neuronal network models used diversify and increase in size and complexity, the need to characterize and categorize them with regard to different aspects becomes more important. The basic independent dimensions along which such models are often described include number of components, spatial extent, and level of biophysical and/or biochemical detail of network components. If the nature of the model components is specified, the number of dimension is well defined. With regard to the level of detail there is, however, a considerable range to consider. At the more fine-grained end, networks may be built from complex compartmental Hodgkin-Huxley type model neurons and conductance based synapses with ligand as well as voltage-gated properties, and with different forms of plasticity included. Even intracellular processes in the form of gene regulatory and biochemical signaling networks may be incorporated into a detailed neuron model. The model may comprise several different cell types, each with their typical blend of ion channels. At the more coarse-grained end, network models may be built from non-spiking units intended to represent, for instance, a local population of some hundred neurons, perhaps lumping together several neuron types in one unit. In even more coarse-grained models an entire cortical macrocolumn or cortical area may be represented by a single unit. At intermediate levels of detail we find different forms of single compartment "reduced" model neurons interacting via simple current of conductance based synapses. Though most network models today are homogenous with regard to granularity it is in principle possible and could be useful to design network models with a mixture of granularities, for instance, to embed a microcircuit modeled with a high degree of detail in a surrounding network of simplified or more coarse-grained units. This would then constitute a form of "multi-scale" model.

The spatial extent of a network model is also an important feature. It may reach from a local population or module, like a cortical column, to a network model representing several connected cortical areas or a whole brain. Since there are quite good estimates of the number of neurons and synapses in different brain structures, this measure is related to the number of components. In many cases, however, some form of sub-sampling is used, meaning that a brain structure is represented by fewer components than actually are there in reality, thus making these measures again somewhat independent.

As commonly used nowadays in the literature, however, the term "large-scale" does not refer to the absolute size of the system under investigation as for example specified in meters. The term rather expresses the relation between the smallest resolved unit in the system and its absolute size. A system with a high resolution (system size/unit size) is large if its absolute size is large because then it contains a large number of elements. Therefore, a model describing the activity of a whole brain would not be called a large-scale if it only consists of 100 units, i.e. is very coarse-grained. Technically one could define a model as large-scale if it taxes the main memory of a commodity workstation computer. With main memory being presently on the order of 1 GB this results in a system size of one billion units. A consequence of this definition is that the size of what is called a large-scale system changes over time and has been much smaller in the past than it is today.

Let us consider some typical parameters to see what the above notion means in terms of cortical neuronal networks. The probability that two neurons in the local cortical network have a synapse is about 0.1. Here, local refers to a diameter of about 1 mm and characterizes a volume in which a neuron can geometrically establish a local synapse with another neuron. For the sake of argument, we assume here that all synapses are local, which is not quite true. In a network of 10,000 neurons we then have a mean of 1,000 outgoing synapses per neuron and a total of ten million synapses. Such a model easily fits on workstations and laptop computers and would today not be called large-scale. However, the number of outgoing synapses of a neuron is not a free parameter. Anatomists have established this value to be in the order of 10,000 synapses. Fulfilling both constraints, connection probability and absolute number of synapses per neuron, simultaneously requires a minimal network size of up to 100,000 neurons. In this system the total number of synapses is already one billion, two orders of magnitudes larger than in the previous system. Here we have reached the region of large-scale models according to the above definition. As researchers in the past have not been able to investigate such systems, the number of synapses per neuron was usually scaled with the number of neurons N in the system like cN in order to conserve connection density.

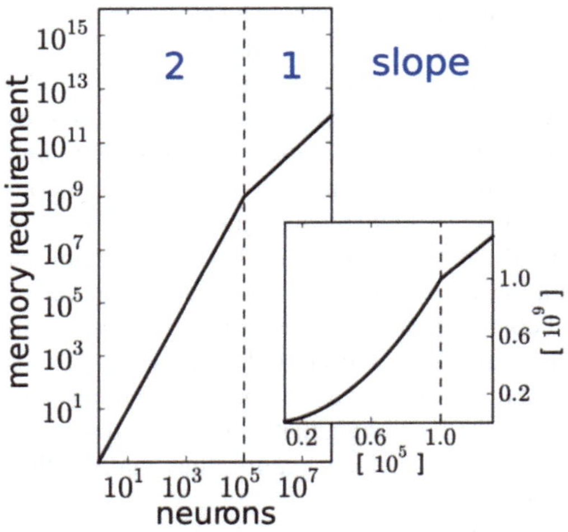

Fig. 10.3 Memory consumption as a function of network size in double-logarithmic representation. Below the threshold size of 100,000 neurons (*dashed vertical line*) the connection probability is constant at 0.1. Above the threshold the number of synapses per neuron is constant at 10,000. At threshold the slope drops from 2 to 1. The linear-linear representation in the inset shows the transition from the quadratic to the linear dependence of memory consumption on network size

Unfortunately in this scaling scheme, the total number of synapses increases quadratically with the number of neurons (see Fig. 10.3). Therefore even if the representation of a neuron takes up 1,000 times the computer memory of a synapse, at a network size of 10,000, the synapses in total take up the same amount of memory as the neurons. At a network size of 100,000 they already occupy 10 times more memory. This seems like a rather hopeless scenario. However, one has to realize that at a network size of 100,000 neurons each neuron is supplied with the biologically realistic number of synapses. Any larger network is less densely connected and from 100,000 neurons on memory consumption grows only linearly with network size. Already a few years ago researchers have broken through this barrier using distributed computing techniques and published the required algorithms and data structures (Morrison et al. 2005). Today such large-scale simulations are routinely done in several laboratories using computer clusters.

A network of 100,000 neurons is not only the smallest network in which a biologically realistic local connection probability of 0.1 can be combined with a realistic number of 10,000 synapses per neuron, it also cover the volume in which a cortical neuron establishes a large fraction of its synapses. Although, in reality a significant fraction of synapses originate outside this local module, in this anatomical sense we can regard a cubic millimeter of tissue as the elementary unit of cortex. We come back to the consequences of the relative memory consumption of neurons and synapses in Sect. 10.4.2.

10.3 Complexity of Large-Scale Brain Models

A reason sometimes given for staying away from large-scale brain models is that there would be no point to build a highly detailed model of a brain since it would be as complex as the system it represents and equally hard to understand. This is, however, an unimpressive statement, which severely underestimates the value of having easy access to every detail of a working complex system. Actually, one of the most severe obstacles to progress in brain science is the great difficulty to access the brain – especially under *in vivo* conditions. If we actually had a high fidelity model of the brain in the computer, the situation would be very different, since access to its internal mechanisms and operations would be much less of a problem. Consequently, having such a model would dramatically speed up progress in our understanding. Unfortunately, or perhaps fortunately, this is a complete utopia, and we have to bear with real brains and more inexact models for the foreseeable future.

A real and unavoidable but disturbing fact is that, the number of "free parameters" increases as you go from the single cell model to a network with a number of cell types and different types of interactions between them. Even rather simplistic conductance based multi-compartmental model neurons comprise tens of equations and hundreds of parameters. The number of synapses in a big network model may be counted in billions and they too require some state update equations and parameters each. Thus, any large-scale network model would contain on the order of billions of parameters. Even with advanced automated parameter search it would be hopeless to find a reasonably adequate combination of these massive amounts of parameters. This appears to put insurmountable demands on the availability of experimental data to constrain such a model. Fortunately, the situation is not quite that bad! Neuronal networks are typically described as comprising a quite limited number of cell types, with basically the same properties but some variation within the population. The parameter used for one neuron of a certain type is likely to do for the others as well, possibly with some compact description of a distribution around a mean. Though this view largely disregards, for example, possible interdependencies between parameter distributions it makes it possible to work with very large systems. This principle is also applied to the synaptic interactions as well, though now we need to consider pairs of cell types. Notably, synaptic conductances are not determined arbitrarily but are in many cases the result of the action of some kind of (so far possibly poorly known) learning rule coupling them to historical network activity. If we can represent this kind of synaptic plasticity rule and provide the system with a biologically plausible input, the synaptic matrix may be generated by means of training. An often used alternative is to set up a random connectivity and, not unexpectedly, the two ways may result in networks with dramatically different dynamical properties.

However, the good news is that if we adopt the above described strategy, the number of truly free parameters is more or less independent of the actual number of neurons and synapses used to instantiate the model. The same averages, distributions and learning rules hold for the huge network model as for the smaller one.

Knowledge about the distribution of cell and synaptic properties, of course, becomes important and is often lacking in available data. On the other hand, the minute details of e.g. dendritic arborisation of one individual cell is less important than in single cell modeling, since this is a typical thing that varies within the population. But what really increases with increasing numbers is the match between model and reality.

Large-scale network models shift the balance so that synaptic complexity takes over as the limiting factor. Somewhat unintuitive, for really large networks it comes with moderate extra cost to have complex cell models. Contrary to the case for single cells and small networks, the solution methods used for, for example, dendritic integration does not add significantly to the cost of computation, since synaptic computation dominates (see Sect. 10.4.2).

10.4 Technology for Large-Scale Simulations

10.4.1 Introduction and Background

Occasional publications on the technology for biological neuronal network simulations appear in the literature at least since the end of the 1980s (e.g MacGregor 1987). However, simulation technology is only seriously discussed as a problem in mainstream neuroscience journals since the middle of this decade. Prior to this time it was common practice to develop program code for one time use for the problem at hand or for in-house usage. There was no general awareness of the benefits of comparing and discussing algorithms, data structures and development strategies in a formal scientific setting and no forums and funding to do so. Scientists greatly underestimated the scope of the problem. For an early account of the technical, infrastructural, and sociological problems see Diesmann and Gewaltig (2002); an in-depth analysis is outside the scope of the present chapter. The situation began to change when large-scale neuroscience research projects involving hundreds of researchers required collaboration on the simulation of neuronal systems between many laboratories and individuals. It quickly became apparent that there is a serious lack of knowledge in three key areas: performance, reproducibility, and sustainability. One of the first reports of the arising discussion in the community is a review on tools and strategies for the simulation of spiking neurons administered by member of the European FACETS consortium (see acknowledgements). In this report (Brette et al. 2007) a group of 22 researchers assembled to develop a minimal set of network models which any simulation tool in the field should be able to represent and integrate. In fact it took more than a year before an agreement on the interpretation of the model definitions was reached and eight tools arrived at the correct results. In parallel the International Neuroinformatics Coordinating Facility (INCF) was founded in 2005 with the aim to support and coordinate work on computer science problems in neuroscience. A first result in the area of simulation science was the report by Djurfeldt and Lansner (2007) on the results of the 1st INCF Workshop on Large-Scale Modeling of the Nervous System. Since then the number of publications concerned with simulation technology in renowned journal like Neural Computation is increasing and the journal Frontiers in Neuroinformatics is a new forum for this discussion. In the past few years we have finally left the dark decades of computational neuroscience.

10.4.2 Neuron and Synapse Models for Large-Scale Network Simulations

In large-scale models of neuronal networks the degree of detail of the single neuron model has traditionally been low. When researchers started to investigate networks of spiking neurons they first focused on the dynamics of these systems. Therefore the models had static representations of the connectivity, or plasticity was introduced by offline changes to the synaptic weights. Aiming at networks with a realistically sparse connectivity, the typical number of neurons was between 1,000 and 100,000.

With an average spike rate of 1 Hz and 10,000 incoming synapses, a neuron receives about one input spike in a time interval of 0.1ms. This is the time-scale at which the neuronal dynamics needs to be updated to reflect the changes at its input independent of the details of the numerical solver in use. A time step of 0.1ms is also sufficient to resolve short fast components of neuronal dynamics like the rise time of post-synaptic potentials. Therefore a computation step size of $h = 0.1$ms is commonly used in simulations. If the state of a synapse is described by a single variable representing the, generally voltage dependent, current or conductance, the computational load of a neuron is dominated by the synapses. Another 10,000 state variables used for representing the time course of the potential along the dendrite, soma, and axon would only double the computational load. However, if we neglect all the spatial structure of a neuron and assume the system of equations governing a synapse as linear, we can collapse the state variables of all synapses with identical dynamics into the state variables of a single lumped synapse (Rotter and Diesmann (1999) provides a historical account of the idea). A neuron model without spatial extent is called a point neuron model. It still may have a set of state variables for each synapses or only a single set for the lumped synaptic state. In the most reduced version where a synaptic event is represented as a delta-kick of charge, no state variable is required at all for the synapses. In this view, where the point-neuron model is derived from the biophysics of a spatially extended model it is natural that the synaptic dynamics is governed by differential equations and that the state variables fully capture the complete history of all synaptic events. For researchers with a background in connectionism and artificial neural networks (ANN) turning to spiking neuron models in the 1990s this was less obvious. Here, the time course of a synaptic event was often represented by a finite table of discrete samples not constrained by a generating system of equations. Manuscripts using such techniques are still occasionally submitted to computational neuroscience journals.

The basic form of a point neuron model is the leaky integrate-and-fire model (Lapique 1907; reviewed by Brunel and van Rossum (2007)). Only a single state variable, the membrane potential, is required and important features of neuronal dynamics are already captured: decay of the potential to a resting level, generation of action potentials, refractoriness, and synaptic potentials of finite amplitude. Over the years a large variety of point-neuron models has been developed adding state variables for different purposes. Synaptic events are modeled by exponential, alpha, or beta functions; either representing the current or the conductance time course. The phenomenon of adaptation has been addressed by non-linear (Izhikevich 2004) or linear (Kobayashi et al. 2009) equations. Finally, the excitability of the neuron model has been modified by exponential (Fourcaud-Trocmé et al. 2003) and quadratic (Ermentrout and Kopell 1984) terms of the membrane potential near threshold. There are also single compartment versions of the Hodgkin-Huxley equations. The components can be combined like the ingredients on the menu of a pizzeria. Overwhelmed by the multitude of flavors some research consortia have agreed on a reference model. The FACETS project of the European Union, for example, has decided on the adaptive exponential integrate-and-fire model (Brette and Gerstner 2005). The modern simulation codes are, however, modular. When doubts occur whether the observations in a network simulation depend on the details of the single neuron model, the model can simply be replaced by another one and the simulation be repeated.

The reduction of biophysical detail to a point-neuron model with lumped synapses has reduced the computational costs by a factor of 10,000 and enabled us to study large-scale neuronal networks in manageable computation time. Contemporary network models require the synaptic weights to be heterogeneous or even plastic. Therefore, the usage of lumped synaptic dynamics does not lead to a reduction of memory consumption by a factor of 10,000; we still need a unit of memory for each synapse. Once we acquired the capability to simulate networks without downscaling the number of synapses at around 2005 (Morrison et al. 2005), the constraint to point-neuron models no longer existed. The same network with N instead of one electrical compartment per neuron runs roughly N times slower, but practically requires the same amount of memory. However, this argument neglects any cache effects. In present day simulation codes the perturbation of cache memory caused by the delivery of spikes to target neurons is so severe that the processing of the incoming spike may dominate the computation time for each neuron. In this case, the addition of a few compartments may not substantially alter computation time after all. After a decade of work on the simulation algorithms abstracted from a memory model, the proper layout and sequence of access of memory now comes into focus of research. The introduction of non-lumping synapses has a similar effect on the computational load as the use of multiple compartments. Consider a neuron with one differential equation for the membrane potential and K NMDA-synapses where the differential equation describing the conductance of a synapse depends on the value of the membrane potential. Now a system of $K + 1$ differential equations needs to be solved simultaneously. The difference is in the computational load the researcher needs to invest to achieve a better approximation of nature. While the segmentation of a neuron model into a few compartments may already capture significant aspects of the effect of dendritic processing on the spiking behavior of the cell, hundreds or thousands of synapses need to be equipped with NMDA dynamics before their effect becomes appreciable.

Mono-layered networks of point neuron models already explain fundamental features of cortical dynamics (Amit and Brunel 1997; Brunel 2000; van Vreeswijk and Sompolinsky 1996). These findings fostered the trust in the approximation and lead to its wide spread use. Recently, some characteristic features of the multi-layered structure of local cortical network were also explained using point-neuron models (Potjans et al. 2010). Nevertheless, any neuroscientist suspecting dendritic interaction as the cause of a particular phenomenon should be aware that today there is no principal limit in simulating such systems: using three instead of one compartment per neuron in the worst case triples the computation time, and a machine capable of storing the single compartment network can also store the other.

Early simulation codes did not distinguish between the representation of the dynamical equations and the solver used to integrate the equations. Often the computation step size was identified with a homogeneous small synaptic delay and implicitly coded in the parameters. The last 5 years have seen a rapid improvement of this situation, mainly triggered by the European FACETS project (Brette et al. 2007). State-of-the-art simulation codes now cleanly separate between the dynamics and the solver. At the end of the nineties it became apparent that many dynamical systems or sub-systems of relevance for neuroscience are "essentially linear"; they are linear up to non-linear state changes caused by point events. Between these point events the system state can be propagated exactly (Morrison et al. 2007b; Plesser and Diesmann 2009; Rotter and Diesmann 1999). The equations used to describe the experimental phenomena of spike-timing dependent plasticity (STDP) and short-term plasticity belong to this class (Morrison et al. 2008).

The first algorithms for the simulation of spiking neuronal networks on a temporal grid with a fixed spacing h required the spike times to be constrained to this grid. This was known to cause artifactual synchronization for a long time (Hansel et al. 1995). The discovery of STDP increased the awareness for this problem (Brette et al. 2007). Event-driven simulation schemes were proposed to overcome the problem because in these algorithms spikes are naturally represented in continuous time. Unfortunately, this lead to the perception that time-driven simulation schemes are approximate whereas event-driven schemes are precise. Morrison et al. (2007b) argued that a globally time-driven simulation scheme can be combined with the representation of spikes in continuous time and defined the efficiency of a simulation scheme as the computation time required to achieve a desired accuracy goal. Hanuschkin et al. (2010) further improved the new algorithm based on a detailed analysis of the computational costs of the various components of the solver. In addition the authors noticed that the algorithm extends to neuron models with arbitrarily non-linear subthreshold dynamics, in particular it enables the computation of the precise spike sequence of the adaptive exponential integrate-and-fire neuron model (Brette and Gerstner 2005). These advances are relevant because until today an efficient globally event-driven simulation scheme suitable for distributed computing has not been found although the class of neuron models accessible by event-driven simulation has been greatly expanded (Brette 2007; D'Haene and Schrauwen 2010; van Elburg and van Ooyen 2009). The solvers for point-neuron models are completed by algorithms for the integration of STDP (Morrison et al. 2007a), short term plasticity (Morrison et al. 2008) and neuromodulated plasticity (Potjans et al. 2010). The efficient implementation of gap-junctions, however, remains an open issue.

10.4.3 Communication in Large-Scale Network Models

At the end of the 1990s the motivation to employ distributed computing was to aggregate sufficient main memory for large-scale simulations. Nevertheless, once the technology was available scientists quickly recognized the benefits of a reduced wall-clock time. This enabled the study of large-scale plastic systems requiring observations extending over minutes and hours of biological time. However, the advent of distributed computing also qualitatively changed the work in the laboratory. Researchers are now investigating large-scale systems quasi-interactively, employing high-performance clusters as super-workstations, in the same way they used to study single equations on regular workstations before. Ideally increasing the number of processors by a factor of ten should reduce the required wall-clock time by an order of magnitude. In fact at intermediate levels of parallelization researchers even observed supra-linear speed-up. At first sight this seems to violate the fundamental limit of speed-up dictated by Amdahl's law. However, distributing a simulation over a larger number of processors also means that a larger amount of fast cache memory is available to the simulation. The resulting reduction in mean latency of memory access compensates for the run time of the components of the code not reducible by parallelization. One such serial component is the communication between processors. The best simulations codes available today require the exchange of data between the processors in steps of the minimal delay in the network (Hanuschkin et al. 2010; Morrison and Diesmann 2008). The costs of this component are increasing with the parallelization because for the neuronal network sizes studied today the neurons local to a processor have target neurons on an increasing number of processors. The serial components of the numerical integration of the neuronal dynamics can be sufficiently reduced such that the costs of communication are the dominating non-reducible contribution. Therefore, a simulation code well suited for supercomputing is communication bound. Here the numerical machinery has been parallelized to a degree that a further speed-up is only possible by an improved communication scheme or better communication hardware. Ironically this is not the limit prominently discussed in the literature because the regime of the software where the speed-up grows linearly with the number of processor looks more impressive. In addition the number of operations per second of a processor declines when communication starts to dominate. Nevertheless, we need to be aware that the linear speed-up can always trivially be rescued in this situation if the neuron model or the synapse model is replaced by a more complex one. We have seen above that models may differ in computational costs by a factor of 10,000. Alternatively, we can study larger networks to keep the processors from starving or study the scaled speed-up scenario, also called weak scaling, where the number of neurons is increased proportional to the number of processors thereby keeping the number of neurons per processor constant. We have seen in the last section that for simulations with simple neuron models, the memory consumption is dominated by the synapses. Consider a system with 1GB of main memory per processor. With 10,000 synapses per neuron and just 1,000 neurons per processor, this would allow 100 bytes per synapse. Thus, the representation of synapses should better be compact, otherwise the simulation may be communication bound just because we cannot supply the processors with a sufficient number of neurons. Note that this example specification would also mean that 100 processors are required to simulate a local cortical network of 100,000 neurons.

So is there a non-trivial sense in which a distributed simulation can be computation bound? One limit is reached when there is only a single neuron remaining on each processer and the neuron model is so complex that the time required by the local solver is still larger than communication time. There are projects operating in this regime and researchers have developed techniques to overcome this limit by distributing the equations of a neuron over several processors (Hines et al. 2008). Another limit is not so much computational but architectural. In the above considerations we have assumed that all the information on the synapses is stored on the machine where the postsynaptic neuron resides. This is useful for the computer systems used in the past decade where the number of processors did not exceed 10,000 and we can therefore expect a neuron to have a synapse on each processor. In this case the spikes generated on one processor are of interest for all the other processors. The 2005 paper by Morrison et al. (2005) uses the CPEX algorithm for this purpose, but it was later found by Hines (Migliore et al. 2006) that MPIAllgather() usually implements a much better algorithm. It is crucial for the parallelization that neurons interact by action potentials and therefore all target neurons only need to be informed with the same information: the point in time at which the action potential occurred. The sometimes called AER (Address Event Representation) based communication was probably invented several times at the beginning of 1990s in the areas of neuromorphic computing (see (Boahen 2000) and references therein) and computational neuroscience (e.g. (Bailey and Hammerstrom 1988; Hammarlund and Ekeberg 1998)).

If an all-to-all communication scheme is used a processor needs to find out which of the presynaptic spikes have local targets. Even if this test required only a single bit per neuron, this would limit the network size on our example system with 1GB per processor to eight billion neurons. Therefore, hashing methods are required to implement an effective data structure. In the context of CPEX Morrison et al. (2005) already proposed to store with the presynaptic neuron the information on which processors the neuron has targets. This enables schemes like CPEX to carry out only the required point-to-point communications, which will become increasingly important as the networks simulated on massively parallel architectures become as large and sparsely connected as the real mammalian brain and as the inter-processor communication becomes exceedingly sparse. The collective communication schemes used today synchronize larges parts of a supercomputer which becomes increasingly costly. For high-resolution brain-scale simulations we need to efficiently map the hierarchical and spatially organized architecture of the brain circuits to the architecture of the supercomputer. Domains of the brain coupled with longer delays require less frequent synchronization and uncoupled domains do not require direct synchronization at all. Future codes will need to rely on support by the operating system and the communication hardware for non-blocking communication routines to reduce synchronization to the required minimum for maintaining causality.

10.4.4 Other Technologies for Large-Scale Neuronal Network Simulations

The standard neural simulators use the MPI (Message-Passing Interface) and run on either distributed or shared memory architectures as well as on multi-core and many-core systems. Indeed, a multi-core PC is a very useful development environment for such parallel code. Though supercomputers provide extreme performance and a high flexibility and are enabling tools for simulation of the largest scale neuronal networks, they are also very expensive, power hungry, and demanding of space. They are obviously not suitable for embedded implementation and real-time applications, which is one of the technological end-points of brain-inspired neural network research and development. Many other technologies have been used for parallel implementation in this field, and we have surely not seen the end of the development of alternative solutions. The massively parallel, stochastic and robust computation performed by biological brains using low-precision and unreliable components provides challenging computational structures for implementation using future molecular scale computing devices (Misra and Saha 2010). Here we review briefly some currently available and developing technologies for large-scale simulation of neural networks, including neuromorphic hardware.

10.4.4.1 The SpiNNaker Project

A light-weight distributed computational structure based on conventional digital technology, i.e. the ARM processor, is being developed within the SpiNNaker project at Manchester university (Jin et al. 2010a, b). The core of this design is the ARM9 processor, commonly used in today's mobile phones. The project goal is to produce a low-power scalable system aimed at simulating neural networks comprising billions of neurons. Each SpiNNaker chip has 18 identical processing cores running at 200 MHz and 96K fast cache memory with and it uses off-chip memory for storage of connections. The general purpose ARM processor has been augmented with a multicast, packet-switched and self-timed innovative communication fabric with on-chip routers (Brown et al. 2010). Rather than every packet listing all destination addresses, routers make routing decisions based on the packet's source address (the identifier of the neuron that fired the spike). The network itself will deliver the packets to all chips containing neurons that have synaptic connections with the source neuron.

Being based on a well-established architecture, the design maintains programmability and generality, and different neuronal models and learning rules can be run. A notable feature of this construction is the asynchronous execution model. There is no requirement for explicit global synchronization in the computation. This leads to nondeterministic behavior, a property shared with the biological system being modeled. Thus far, the largest reported network simulation is of moderate size, i.e. 4,000 Izhikevich model neurons connected by about 100,000 synapses, using 16 bits fix-point arithmetic and with 1 ms time resolution (Jin et al. 2010a, b).

10.4.4.2 Graphic Processors (GPUs)

In recent years, it has become very popular to use the graphical processing units (GPU) of a personal computer for scientific computing. GPUs offer a high degree of parallelism but are restricted in terms of the amount of memory. Some reports indicate that a spiking neural network simulator on a GPU provides quite some flexibility, high-performance, and low-cost for large-scale simulation. For example, it is possible to run a network of 225K Izhikevich neurons with 225M synapses and synaptic plasticity (STDP) (Nageswarana et al. 2009). The simulator presented was only 1.5 times slower than real-time for a network of 100K neurons having 10M synaptic connections with an average firing rate of 9 Hz. The GPU implementation (on one NVIDIA GTX-280 with 1 GB of memory) was up to 24 times faster than a CPU version for simulation of 100K neurons with 50M synaptic connections, at an average rate of 7 Hz. Another study (Fidjeland and Shanahan 2010) reported that a network of 55K neurons and 55M plastic synapses could be simulated in real biological time. The performance of GPU solutions is often limited by the memory bandwidth supported by the GPU hardware rather than the number of scalar processors. For simulation of larger networks a cluster of GPUs can be employed, building upon the strategies outlined in this paper.

10.4.4.3 FPGA Implementations

An FPGA (Field-Programmable Gate Array) is a user programmable gate array chip that can be programmed using a hardware description language (VHDL, Verilog) to implement any logic function that an integrated circuit could perform. The component counts on contemporary FPGA chips today reach into the millions allowing for very complex designs. FPGAs are suitable for low-power neural simulation since they can take advantage of the massive parallelism inherent in the spiking neural network (Guerrero-Rivera et al. 2006; Maguire et al. 2007).

For large networks, the synaptic weight storage is the limiting factor, and external memory may have to be considered (Cheung et al. 2009). The largest simulations currently reported are typically of moderate size (like 1,000 fully connected neurons) and lack synaptic plasticity, but they run an order of magnitude faster than a standard dual core 3GHz CPU and also faster than some GPU implementations. The GPU performance is significantly affected by the density of spikes in the active network, which the FPGA implementation is not (Thomas and Luk 2009).

10.4.4.4 Neuromorphic Technology

Over the years, many efforts have been spent on developing dedicated hardware for simulation and emulation of neural networks. This is likely a necessary development for enabling the use of neural networks in embedded technical applications which require compact, power-efficient, and low-cost designs.

A recent and exciting development of this technology is analog or mixed signal VLSI implementation. This technology originates in the seminal work of Carver Mead at Caltech (Douglas et al. 1995) now represented e.g. by work at ETH in Zürich (Indiveri et al. 2009). Work aiming for large-scale network simulation is also in progress (Schemmel et al. 2010). In this case, the aim is to achieve a wafer scale integrated design in analog VLSI, featuring dynamic and plastic synapses (Schemmel et al. 2007). The network on a wafer will comprise on the order of 180K adaptive exponential integrate-and-fire model neurons and 40M plastic synapses with typically four bit precision weights. The speedup factor relative to biological real time is very high, in the range 10^3–10^5. This allows for simulation and training on massive amounts of input data over extended periods of time. The long-range and wafer-to-wafer connections are handled by a special purpose FPGA circuit via 1 to 10 Gbit Ethernet links.

The search for the best technology and algorithms to simulate and implement brain-inspired computational architectures is today intense. It remains an open question to what extent future designs will build on traditional digital or analog CMOS technology or on some currently more exotic type of computing substrate, e.g. based memristors (Jo et al. 2010) or CMOL (Likharev et al. 2003), or on some combination of these.

10.4.5 Visualization and Analysis of Output from Large-Scale Simulations

The number of state variables in a large-scale network simulation is typically very high, with several for each neuron and even each synapse. With a typical time step on the order of 100 microseconds the amount of generated new values in every update may be on the order of billions. This implies that logging even a fraction of the state variables on disk becomes extremely memory and time consuming. In some cases, like for the spike raster, the output is still quite compact and there are many advanced and well-known methods available to analyze this kind of data (for a review see e.g. Grün 2009). For other types of variables some strategy for sub-sampling over space and time will be required. The need for storing variables should be balanced against the opportunities of rerunning the same simulation for inspection of additional state variables. To cope with this situation we may also perform the required collection, data analysis, and data compression on-line by computing composite measures and variables within the simulation itself.

Advanced visualization of output from network simulations have been produced, for instance, within the Blue Brain Project using high end generic visualization software (Markram 2006). Yet, despite the apparent usefulness of 3D visualization of different network state variables of large-scale neural networks there are surprisingly few tools available to support this for large-scale network simulations. Static three-dimensional brain visualization is quite developed in the field of brain databases and an interesting opportunity would be *in situ* visualization of dynamic output from large-scale brain simulations.

Another quite useful strategy to visualize activity in network models with a significant spatial extent is to synthesize the kind of measures from the simulation that are comparable to mesoscopic and macroscopic experimental measurements such as, for instance, local field potentials (LFP), electro-encephalogram (EEG), and voltage sensitive dye (VSD) signals. This provides a number of new experimental constraints on the model, in addition to the ones on the cellular, synaptic and architectural level already used to set it up. In this way, the model helps to connect such macroscopic measurements with cellular and synaptic dynamics which helps to interpret and better understand such measurements.

The current toolbox for actually generating experimental measurements from a neural simulation, i.e for synthetic brain imaging, is today quite limited, but efforts are being made to expand it and increase the quality of the synthetic measurements generated. It is indeed a challenging problem in itself, since the biophysical mechanisms behind these measures are in many cases still unclear. Modelers are currently also addressing the question of integrating data from different imaging and measurement modalities, e.g. in the combined analysis of fMRI and EEG (Bojak et al. 2010; Sotero and Trujillo-Barreto 2008). Below follows a brief description of some different kinds of brain signals that may be synthesized from large-scale simulations.

Electrophysiological measures. The electrical activity generated by neural tissue can be measured in terms of multi-unit activity (MUA) local-field potentials (LFP), electro-encephalogram (EEG), and magneto-encephalogram (MEG). The source of such signals is the intracellular potential of neurons, generated by active membrane currents and synaptic input. The corresponding state variables are available in a more or less accurate form from most kinds of spiking neuron models. The extracellular field potential captures at least three different types of activity: single-unit activity representing action potentials of neurons close to the electrode tip, multi-unit activity generated by the spiking of small neural populations in a sphere of 100–300 mm radius, and perisynaptic activity of a neural population within 3 mm of the electrode tip, available in the low-frequency components of the extracellular field potential. Methods have been proposed to estimate the LFP from transmembrane currents of simulated multi-compartmental model neurons (Lindén et al. 2010) as well as from the local network spiking activity alone (Rasch et al. 2009). MEG and EEG signals derive from the net effect of ionic currents flowing in the dendrites of large number of neurons during synaptic transmission. Computational models for estimating EEG and MEG signals have been developed (Babajani-Feremi and Soltanian-Zadeh 2010; Dehghani et al. 2010)

VSD imaging. Voltage-sensitive-dye imaging offers a method to measure optically transmembrane potentials of neural tissue with a high spatial (20–50 μm) and millisecond temporal resolution. These measurements can be useful in studying function at the level of an individual cell, for determining how groups of neurons generate a behavior, and for studying the dynamic behavior of populations of neurons. The physics for generating the VSD signal is fairly straight-forward and it can be synthesized with relatively high accuracy (Chemla and Chavane 2010).

Ca-imaging. Changes in intracellular calcium (Ca) concentration following synaptic and supra-threshold activity are mediated by a wide range of processes and can be imaged using Ca-sensitive fluorescent dyes. Such signals can be synthesized from neuron models featuring Ca-channels and Ca-pools filled up, for instance, via voltage gated or NMDA receptor gated Ca-permeable channels. Ca-dyes may respond with high temporal resolution, but because they can act as Ca-buffers, the calcium signal in the presence of dye may substantially outlast the change in the cytosolic calcium concentration that would occur in the absence of the dye. In wide-field imaging the temporal resolution may be limited by the acquisition rate of the scanning. Ca-imaging can be used to visualize sub-cellular, single cell, and network activity (Göbel and Helmchen 2007; Homma et al. 2009; Stettler and Axel 2009).

BOLD and PET. Functional magnetic resonance imaging (fMRI) techniques are non-invasive and based on the assumption that changes in spike activity are accompanied by modulation in the blood oxygenation level-dependent (BOLD) signal. The exact relationship between neural activity and BOLD is, however, still quite unclear. This relative uncertainty about the relation between neural activity and BOLD prevents to device more precise synthetic BOLD signals from a simulation (Buxton et al. 2004; Sotero and Trujillo-Barreto 2008). Another non-invasive method to measure brain activity is positron-emission tomography (PET) and an attempt to synthesize this signal has been made as well (Arbib and Grethe 2001).

10.5 Two Examples of Large-Scale Simulations

In the early days of neural modeling, memory and computational capacity of computers allowed only simulation of tiny networks, i.e. with some tens of model neurons connected by some hundreds of synapses. It was a long time, before supercomputers became used for this application – one of the earliest published examples is the work by Roger Traub in the late 1990s (Traub et al. 1999). Later prominent examples of the use of parallel computing for brain simulation include the extreme size Beowulf cluster simulations by Eugene Izhikevich (2006), the biophysically detailed simulations of a cortical column performed within the Blue Brain Project headed by Henry Markram (King et al. 2009), and the cat cortex sized simulations by the Modha group at IBM (Ananthanarayanan et al. 2009).

As these examples demonstrate, research activities related to supercomputer simulations of the brain are developing rapidly and will likely bring forward new tools and knowledge about the brain in the next few years. In the remainder of this section, we describe somewhat more in depth such research performed in our own labs.

10.5.1 A Large-Scale Model of Cortical Layers 2/3

Associative memory is one of the fundamental operations of neocortex and this function has been modeled repeatedly in the neural network community (Lansner 2009). Most of the models studied were, however, based on abstract non-spiking units and one of the first cortical associative memory simulations of a network model composed of biophysically detailed network components was published in the early 1990s (Lansner and Fransén 1992).

In 2006, building on this early work, a very-large scale simulation of a spiking modular associative memory model of layer 2/3 neocortex, comprising 22 million neurons and 11 billion synapses was performed (Djurfeldt et al. 2008). The model integrates functional constraints suggested by a top-down view of the neocortex as an associative attractor memory network and neurobiological constraints provided by neuronal and cortical anatomy and physiology at multiple scales.

This cortex model is modular with smaller modules – functional columns, minicolumns – as the smallest local microcircuits bundled into larger complexes – hypercolumns or macrocolumns (Lundqvist et al. 2006). Minicolumns comprise on the order of 30 pyramidal cells in layer 2/3 recurrently connected with a density of 10–25% of excitatory synapses. Macrocolumns comprise some hundred minicolumns and a total of hundred basket cells provide lateral inhibition and competition between the local minicolumns, thus turning the hypercolumn into a soft winner-take-all structure.

Memory patterns stored in the model typically engaged one minicolumn unit in each hypercolumn and these units were connected via long-range excitatory synapses, thus forming a set of distributed cell assemblies. There was no long-term synaptic plasticity in the network model. Instead the memory matrix was formed by training an abstract non-spiking isomorphic network model and importing the resulting (long-range) connection matrix into the spiking network. A long-range connection between two distant minicolumns was formed stochastically if the two minicolumns belonged to the same pattern. Thus, each pyramidal cell received sparse long-range excitation from other cells in the memory pattern it was part of. Futher, units in different memory patterns were connected in an inhibitory fashion via dendritic targeting inhibitory interneurons, so called regular spiking non-pyramidal (RSNP) cells – corresponding to e.g. double bouquet and/or bipolar cells.

Neurons were modeled according to the Hodgkin-Huxley formalism and had between three and six compartments. The different types of neurons and synapses in the model were tuned to mimic their real counterparts. Pyramidal cells and dendritic targeting inhibitory interneurons featured calcium currents and calcium gated potassium channels to make them adapting. Glutamatergic transmission was mediated via AMPA gated as well as voltage dependent NMDA synapses and synapses between pyramidal cells were subject to fast synaptic depression.

This model was simulated using an MPI based simulator – SPLIT – developed in the mid-1990s with the aim of exploring how to efficiently use the resources of parallel computer architectures for large-scale biophysically detailed neuronal-network simulations (Hammarlund and Ekeberg 1998). The setup of this large size network model on the cluster made use of a novel abstraction, the connection-set algebra, which implements an efficient domain decomposition of the connectivity metadata (Djurfeldt 2012). It was implemented in a parallel fashion on the cluster to make the setup of the network efficient, but it still took up to an hour to perform the setup before starting the simulation itself.

The model was run on an IBM Blue Gene/L cluster at IBM Rochester, Minnesota, at the IBM Thomas J. Watson Research Center, Yorktown Heights, New York, and at The Royal Institute of Technology (KTH), Stockholm. Speedup was linear for models up to four million neurons and two billion synapses on up to 4K processors. The largest simulation using 8,192 processors of the BG/L system was run at the IBM Thomas J. Watson Research Center in Yorktown Heights. Its 22 million neurons and 11 billon synapses corresponds to a cortical surface area of about 16 cm^2, comparable in size to the cortex of a small mammal. The simulation, running in coprocessor mode, occupied 336 MB of memory at each node, giving a total of 2.8 TB. The setup time for this model was 6,927 s and it took 5,942 s to simulate one second of activity. While real pyramidal cells have 10,000 synapses, the average number of synapses per neuron was only 500 in this model because of the orthogonality of the memory patterns and the lack of connections to other cortical areas. The speedup dropped for the 8K processor simulation but still reached almost 80%.

In addition to pattern completion, the model had all the functionality usually ascribed to attractor networks, such as noise reduction and pattern rivalry. Recall times were comparable to psychophysical measurements, i.e. a couple of hundred simulated milliseconds. A stable ground state co-existed with the active memory states (Fig. 10.4). The ground state was characterized by population oscillations at a frequency of approximately 15 Hz, where the oscillations of individual minicolumns are phase-locked to other minicolumns in the hypercolumn. In this state, pyramidal cells fired at about 0.1 Hz, so the 15-Hz rhythm emerges as a collective network-level phenomenon.

In the active state, only the pyramidal cells of a single minicolumn were active in each hypercolumn. Pyramidal cells fired at 10–15 Hz, basket cells at 50 Hz, and RSNP cells at 25–35 Hz. One particularly interesting phenomenon, which consistently emerges in these simulations, is a rhythmic modulation of pyramidal

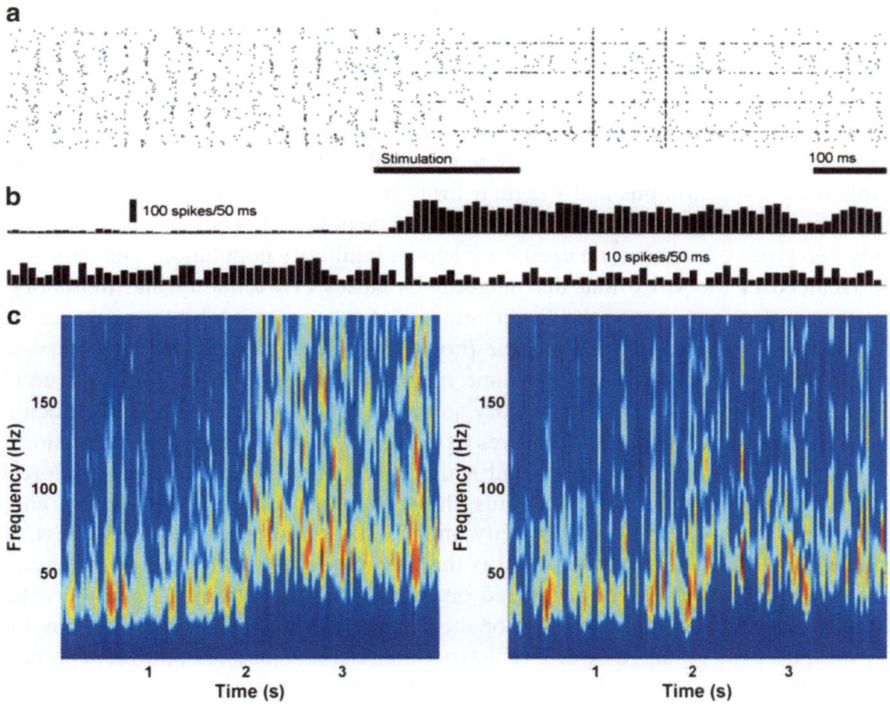

Fig. 10.4 (**a**): Spike raster of a subsample of 5,880 pyramidal cells is shown. In the beginning of the simulation the non-specific ground state was active. When a part of a pattern was stimulated it completed and was then persistently active, even after the stimulation terminated. (**b**): After stimulation, the foreground pyramidal cells elevated their firing rates and the background pyramidal cells lowered them as seen in the activity histograms (*upper* foreground, *lower* background). (**c**): Synthetic LFP spectrograms. The network started out in the ground state and entered an active state after 2 s due to stimulation. The signal was produced from 30 pyramidal cells entering foreground (*left*) and background (*right*) respectively

cell activity with a frequency of 25–40 Hz (Fig. 10.4c). It occurs only during active states and is reminiscent of the gamma-band activity observed during e.g. a working memory task.

Despite the regularities seen on a network level, the firing of individual pyramidal cells was Poisson distributed in the active state and showed exponentially distributed ISIs of pyramidal cells in the foreground. This result is consistent with physiology. One particularly attractive feature of the model is that it is robust to the perturbation of parameters, which is to be expected from a biological system. One answer we sought was whether such a large patch of cortex could maintain a stable attractor state despite the significant propagation delays caused by axonal conduction time. The simulation showed that this is indeed the case. This model is still under investigation with daily runs on the BG/L system at KTH and elsewhere. The networks are then typically smaller than described above, up to 100,000 neurons and some ten million synapses (Lundqvist et al. 2010a, b).

10.5.2 A Minimal Model of the Multi-layered Local Cortical Network

In the middle of the 1990s van Vreeswijk and Sompolinsky (1996) simultaneously with Amit and Brunel (1997) developed the balanced random network model to explain the asynchronous and irregular firing patterns of cortical neurons and the accompanying large fluctuations of the membrane potential. The model consists of only two types of neurons, an excitatory and an inhibitory population. The network is balanced in the sense that the outgoing synapses of excitatory and inhibitory neurons differ in sign and the lower number of incoming inhibitory synapses is compensated by a larger weight. The prominent effects can already be observed when each neuron establishes the same number of synapses with random source neurons and the dynamics of excitatory and inhibitory neurons is exactly identical. Up to the present day the model serves as a reference for theoretical investigations of the recurrent dynamics of cortical activity. A stream of publications explores the correlation structure of spike trains, the spectral properties of the activity, and the interaction of activity and plasticity on the basis of this model. Nevertheless, a fundamental limitation of this model is that it can not make any statements about the consequences of the multi-layered organization of the cortex. A considerable amount of literature is now available on the connectivity of the different cell types in the local network obtained by different experimental methods and in different preparations. However, theoreticians struggled to extract a consistent picture from these data and models based on individual data sets capture only part of the layers and lead to activity patterns hard to reconcile with the neuronal activity observed in vivo.

In this situation (Potjans and Diesmann 2011) decided to construct a minimal model of the multi-layered cortical network as an extension of the balanced random network model only taking into account the layered architecture. Accordingly the model has an excitatory and an inhibitory population of neurons in each layer. The single neuron dynamics of all neurons is identical, all excitatory postsynaptic potentials have the same amplitude, and the inhibitory postsynaptic potentials differ only by sign and a single scaling factor. The model comprises the layers 2/3, 4, 5, and 6. Thus, there are eight populations of neurons in four layers where the anatomical thickness of each layer and the neuron densities are taken from the literature. The model size in terms of cortical surface area determines the absolute size of the neuronal populations. The two most comprehensive connectivity data sets available are Binzegger et al. (2004) based on anatomical considerations and Thomson et al. (2002) based on electrophysiology. At first sight, the data sets do not seem to be in good agreement. The electrophysiological data set reports a probability of 0 for many connections, but for connections where synaptic contacts are found at all the probability is substantially higher than in the anatomical data set. This discrepancy is explained by the different experimental techniques. The data of Thomson et al. are collected by randomly inserting electrodes into two cells within a distance of not more than 100 μm in acute cortical slice preparations. A current pulse applied to one cell is adjusted to elicit an action potential. If the second neuron responds with a postsynaptic potential a connection is found. Statistics is collected by repeating the experiment many times. In contrast Binzegger et al. reconstruct the complete arborization of neurons previously labeled in vivo. For each target cell type the authors first determine the fraction of dendritic surface area the cell type contributes to a particular layer. Next they count the number of synaptic butons of a particular source cell type in the same layer. Assuming a random distribution of butons over the available dendritic surface Binzegger et al. compute the probability that neurons of the two cell types establish a synapse in this layer. The sum over all layers is the connection probability. As the authors attempt to reconstruct the full local dendrite, the range of measurement captures essentially all possible connection sites and spans a radius of up to several millimeters. Both measures assign a probability to the connection between two cell types irrespective of the lateral distance of the cell bodies. Therefore the probability needs to be interpreted as the average connection probability of the cells within the sample radius. Independent earlier studies demonstrate that the connection probability falls off with the distance of the cell bodies with a Gaussian profile.

Based on this additional literature Potjans and Diesmann argue that the discrepancies in connection probability between the data sets of Binzegger et al. and Thomson et al. are due to the different sample radii. The authors find that the two parameters of a hypothetical common Gaussian profile are uniquely determined by the average connection probabilities in the two data sets and the two sample radii. The resulting connection probability at the origin and the space constant are indeed in good agreement with the estimates of the earlier studies. Now the two data sets can be scaled to the same model size by deciding on a particular radius. Nevertheless the idea of the model is to construct a random network specified by the 64 average probabilities, one for each combination of cell types. With increasing models size the number of local synapses incorporated by the model and the recurrent circuit becomes more complete. Simultaneously though the average connection probability drops and the lateral structure of the network can eventually no longer be ignored. Potjans and Diesmann therefore settle at a model size where 90% of the local synapses are explained. This corresponds to a piece of tissue below a square millimeter of cortical surface and an average connection probability of 0.1.

Although the data sets of Binzegger et al. and Thomson et al. are overall in good agreement, some connections still exhibit large differences. Predominantly these are inter layer connections from excitatory to inhibitory cells where Thomson et al. report a higher connection probability than Binzegger et al. Potjans and Diesmann attribute this pattern to a systematic bias of the anatomical method to assign synapses to excitatory target cells. The reason is that excitatory neurons have larger dendrites and are therefore favored by a random distribution of synapses over the dendritic surface available in a layer. Nature, however, may not work in this way, in the cortical wiring process some cell types may specifically address inhibitory targets. In order to quantify such a preference Potjans and Diesmann introduce the measure of target specificity as the difference between the probability of neuron to target an excitatory cell and the probability to target an inhibitory cell in the same layer normalized by the sum of the probabilities. The measure is $+1$ if a neuron

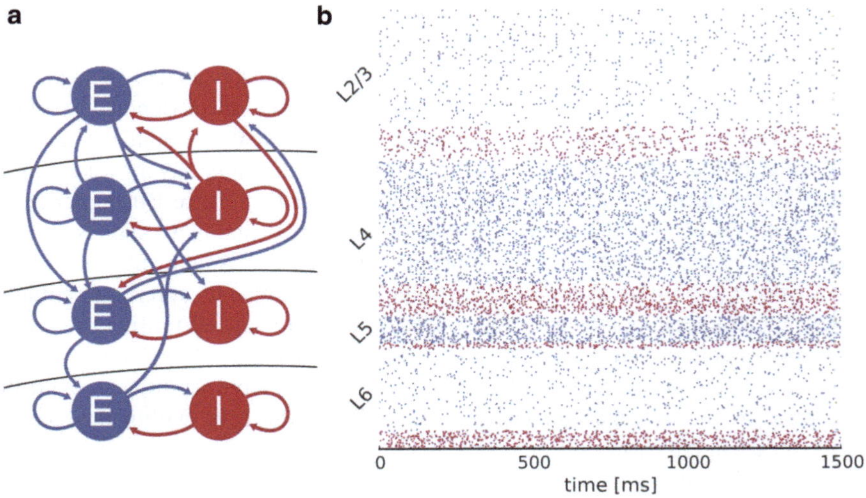

Fig. 10.5 Minimal multi-layered model of the local cortical network. (**a**) Model definition. Layers 2/3, 4, 5 and 6 are each represented by an excitatory (*blue filled circle*) and an inhibitory (*red filled circle*) population of model neurons. The number of neurons in a population is chosen according to Binzegger et al. (2004) based on the countings of Beaulieu and Colonnier (1983); Gabbott and Somogyi (1986). Input to the populations is represented by thalamo-cortical input targeting layers 4 and 6 and background input to all populations. Excitatory (*blue*) and inhibitory (*red*) connections with connection probabilities >0.04 are shown. The model size corresponds to the cortical network below 1 mm^2 surface. (**b**) Raster plot of spiking activity recorded for 400 ms of biological time of layers 2/3, 4, 5 and 6 (from *top* to *bottom*; *blue*: excitatory, *red*: inhibitory). Relative number of displayed spike trains corresponds to the relative number of neurons in the network (total of 1862 shown)

exclusively targets excitatory neuron, −1 if it exclusively target inhibitory neurons, and 0 if there is no preference in establishing a connection. The target specificities Binzegger et al. are all positive whereas they are more balance for Thomson et al. These insights are used to compile an integrated map from the two data sets reducing measurement error and bias where possible. For some connections also further data based on 2-photon imaging are taken into account. The resulting network structure is illustrated in Fig. 10.5a. There is strong recurrent connectivity within layers and the inter layer connectivity is mainly made up of excitatory projections. Potjans and Diesmann notice that some projections have a pronounced negative target specificity, especially the ones forming what the authors call the feedback projections. These run opposite to the classical feedforward pathway from layer 4 up to layer 2/3 down to layer 5 and then 6.

Simulations of this system where external synapses are replaced by a direct current result in an asynchronous irregular state of spiking activity accompanied by large fluctuations of the membrane potential without further fine tuning. Thus, characteristic properties of the mono-layered balanced random network model are conserved. The spike rates are different in the different layers and inhibitory neurons have a higher spike rate than the excitatory neurons (Fig. 10.5b). The distribution of spike rates across layers is in good agreement with experimental in vivo data. In particular the model reproduces the very low firing rates of layer 2/3 which are hard to achieve in a mono-layered model. Previously, the higher spike rate of inhibitory neurons has been attributed to single neuron properties like shorter time constants or a lower spike threshold. The study by Potjans and Diesmann shows that the experimentally determined structure of the network is sufficient to explain the elevated spike rates of inhibitory neurons. When the negative target specificity of the feedback projections is artificially removed the spike rates explode and a globally synchronized state emerges. If the increased excitation in the network is now compensated by increasing the scaling factor of the inhibitory postsynaptic potentials, the spike rates can be stabilized again at reasonable levels, the pattern of spikes rates across layers, however, is no longer compatible with the experimental data. Hence, the specific target type selection is required to achieve a stationary state compatible with in vivo recordings of spiking activity.

In a second simulation study Potjans and Diesmann stimulates the network by a train of spikes mimicking thalamic input arriving in layers 4 and 6. In response to the stimulus a pulse of activity propagates through the layers first activating layer 4, then simultaneously layers 2/3 and 5, and finally layer 6. The activation of the inhibitory neurons in a each layer follows the activation of the excitatory neurons and both end before the stimulation has even ended. Removing the specific target type selection destroys this crisp activation pattern. The activity shows a clear oscillatory structure and well outlasts the duration of the stimulus. Potjans and Diesmann analyze the interplay of excitation and inhibition in the multi-layered system and summarize the interaction as a handshaking. When a layer receives an excitatory message from a previous layer the receiving layer responds by exciting the inhibitory neurons of the sending layer as a handshake thereby terminating the activity of the sender.

This minimal multi-layered network model derived in a purely bottom-up fashion is already in use as building block in several other projects. The model may be a useful replacement of the mono-layered balanced random network model if a relation to mesoscopic measures like the local field potential needs to be made or the targets of functionally different inputs, for example top-down versus thalamic, need to be distinguished.

10.6 Challenges and Future of Large-Scale Simulations

Today large-scale models are either coarse-grained representations of brain-scale structures or attempting full-scale representations of local networks. The next generation of models, however, must combine both features: the models would be full-scale with respect to fundamental entities like the synapses and large enough to close functional circuits at the brain-scale. The capability to carry out simulations of such models marks the beginning of a new era of brain research. Now the behavior of a model can be studied in closed-loop interaction with the environment and at the same time the resulting activity and plasticity can be observed and compared to experimental data. This breakthrough is enabled by the roadmaps of supercomputer development for the next decade and the rapid experimental progress in uncovering the anatomy of the brain. At present data obtained with different techniques and at different spatial and temporal resolutions like fMRi, electrophysiology, 2-photon imaging, and electron microscopy cannot quantitatively be compared because no models spanning the scales and connecting structure with dynamics are available. Here supercomputers will work as data integration facilities. Inconsistencies between data sets can be discovered and resolved and as more data are coming in the models are refined. What are the challenges for simulation technology and more generally computational neuroscience to bring such models into existence?

The major challenges may be rather organizational and cultural than technical. In the past there has been no systematic build-up of knowledge on simulation technology for computational neuroscience in the literature. The reason for this is that generally senior researchers and journals did not consider the development of simulation technology as a relevant part of the science. Consequently, little of the knowledge created over the last 15 years was published, research groups had no chance to learn from each other and reinvented the wheel multiple times. As technologies were not subject to the proven scientific method of objective evaluation and comparison, progress was slow. Only in the last 2–3 years we have seen substantial progress. International committees have identified software as a bottleneck of neuroscience and slowly a culture of rigorous evaluation and publication is emerging. The foundation of the INCF and dedicated journals like Frontiers in Neuroinformatics are visible results, but also established journals like Neural Computation are publishing an increasing amount of high quality technological papers. Authors are still struggling to find good methods of communicating their insights; publications vary greatly in style and quality. It will be our task for the coming years to nurture this aspect of computational neuroscience and to teach not only the technologies but also the state-of-the-art of communicating them to the next generation.

The more advanced simulation systems we have available include formal test suites enabling the verification of their integrity. These enable the developer to make sure that the inclusion of a new feature does not break existing features. Furthermore, it also allows the researcher to verify that the software works as promised in the particular laboratory of choice. Having said this we need to realize that the present test suites mainly test fundamental properties of the simulation engines, like for example the ability to transmit a spike form neuron A to neuron B with a predefined delay. The tests rarely reach up from the technical level of communication and numerics to tests of invariants of neurobiological entities with little internal structure like the neurons in a simulator for point-neuron models. This is often due to the fact that for network models we just do not know what useful invariants are. Nevertheless, even for simple systems like an isolated neuron where the incoming synapses are governed by STDP and supplied by Poisson spike trains it is cumbersome to formulate the equilibrium distribution of synaptic weights as an invariant of a test script. In this situation we may have a good analytical approximation of the distribution at hand but the test script requires a considerable amount of code for model specification, simulation, data recording, and statistical analysis. The required effort rapidly increases for more complex models. The challenge is to develop more expressive simulation languages which enable the researcher to formulate simulation experiments and tests using concepts of the neurobiological problem domain with few lines of code. Progress in this area will also increase the reproducibility of simulation results in the sense that the description of a simulation experiment is not only understandable by the computer and the authors but also by their colleagues.

Correct simulation results still need to be communicated to the scientific community in order to advance research. It has been recognized (De Schutter 2008; Nordlie et al. 2009) that reproducibility is a problem in computational neuroscience. The chances to be able to reproduce the results of a published work without personal contact to the original authors are slim. This has been known in the community for several years – every practitioner privately tells a story of a personal failure to reproduce the work of others, but only recently researchers dared to speak up in public. The reason for this situation is not laziness or malpractice of the authors. We are just lacking the formal tools to describe complex models in a way that is understandable and at the same time captures all the details and is free of hidden assumptions. First important steps in addressing this issue are the development of NeuroML (Gleeson et al. 2010) and the introduction of the Nordlie tables (Nordlie et al. 2009). At present a task force of the INCF program on multi-scale modeling composed of the experts in the field is working towards a first version of a model description language suitable for complex large-scale networks.

In this chapter we have predominantly taken the bottom-up perspective of constructing the brain from its constituents. Implicitly we have assumed that the level of synapses and neurons or equipotential compartments of the latter is a useful lowest level. Obviously, for a given scientific question this is not guaranteed to be a reasonable choice – the dynamics of molecules and the transport inside the cell may be of importance as well. Strictly speaking, equipotential compartments are not even neurobiological entities but result from a specific choice of numerical accuracy. Furthermore, it is unlikely that following a pure bottom-up approach, brain function will suddenly emerge. We may never have enough data to sufficiently constrain the system. In practice scientists are also following top-down approaches where guided by observations of system behavior a network model is constructed exhibiting the desired function (Noble 2006). The danger here is that the model structure may be arbitrary because there are potentially many different structures implementing the same function. Therefore research programs need to combine aspects of both approaches. Multi-scale simulation software is required to identify at which level of description a certain phenomenon can be explained. Also individual simulations of brain-scale networks may need to be multi-scale. For the study of the molecular dynamics inside a nerve cell it may not be affordable to simulate the embedding network providing the required activity environment on the same level of description. When models of different components of the brain are combined to create a closed functional circuit, the models may not all be available on the same level of description: some may originate from bottom-up reconstruction others maybe the result of top-down arguments.

In present publications we seldom see that researcher use network models developed by other research groups to create more complex structure explaining higher-level features. This reuse of components seems to be mainly restricted to the level of neuron and synapse models. One reason may just be the complexity barrier of interfacing the various models very similar to the problems designers face in building large software systems (Winograd 1975). In order to overcome this situation we need to learn to build on the work of others and to learn to think in terms of modules which can be reused by other research groups. This requires changes in the mindset of researchers but software tool like MUSIC (Djurfeldt et al. 2010) and PyNN (Davison et al. 2009) are first software tools supporting the idea. The next generation of models is more likely to be created by international research consortia than individual groups.

An obstacle for progress in large-scale simulation is related to the traditional funding schemes of research. For the reasons discussed above there are still little resources directly available for the development of simulation technology. With the increased awareness of the software bottleneck in neuroscience this will hopefully change over the next few years. There is however a structural problem remaining. The software development today is to a large extent carried out by PhD students and postdocs. While the creation of new algorithms is now a valid target and leads to publishable results other aspects of the software life-cycle are not. The step of turning a new algorithm into a reliable and documented software product which can be used by others and maintained for a long term is standard practice of software engineering and not of scientific interest. Sustainment, not only theoretical sustainability, becomes the more important the more scientific progress depends on simulation. The critical and routine tasks cannot and should not be done by our precious students and postdocs but left to professional hands (Cannon and Gewaltig 2011). In the experimental branch of neuroscience it already has a long tradition that, for example, amplifiers and electrodes for electrophysiology are manufactured by dedicated companies and research funds can be used to purchase these products. This is an organizational challenge of simulation science we have to overcome. Whether commercial companies, dedicated facilities at research institutes equipped with long-term funding, or other mechanisms are a suitable needs to be discussed and tried.

The challenges ahead for large-scale brain modeling and simulation are truly exciting. We have shown that the ability to integrate experimental data from many sources and the capability to close the functional loop between the brain and the environment will qualitatively change the impact of simulation science on brain research. However, we also observe a second trend. So far computational neuroscientists have been busy with transferring established methods of supercomputing and software engineering into neuroscience. Eventually, we may be able to give back. The development of technology and reliable simulation environments coping with the multi-scale nature of the brain may stimulate the field of simulation science as a whole. Understanding how the hierarchical and recurrent communication pathways of the brain can be mapped to supercomputers and thereby learning more about brain function may inspire the design of brain-like and "intelligent" machines, implemented in efficient, compact and low-power neuromorphic hardware for use in many different and novel applications.

Acknowledgements Partially funded by EU Grant 15879 (FACETS), BMBF Grant 01GQ0420 to the Bernstein Center Freiburg, Next-Generation Supercomputer Project of MEXT, the Helmholtz Alliance on Systems Biology, the Swedish Science Council (VR-621-2004-3807), VINNOVA (Swedish Governmental Agency for Innovation Systems), the Swedish Foundation for Strategic Research (through the Stockholm Brain Institute), and the European Union grants 15879 (FACETS) and 269921 (BrainScaleS). Access to supercomputing facility through JUGENE-Grant JINB33. We also thank the DEISA Consortium (www.deisa.eu), co-funded through the EU FP6 project RI-031513 and the FP7 project RI-222919, for support within the DEISA Extreme Computing Initiative.

References

Amit DJ, Brunel N (1997) Model of global spontaneous activity and local structured activity during delay periods in the cerebral cortex. Cereb Cortex 7:237–252

Ananthanarayanan R, Esser SK, Simon HD, Modha DS (2009) The cat is out of the bag: cortical simulations with 10^9 neurons, 10^{13} synapses. In: Proceedings of the conference on high performance networking and computing (SC09), Portland, Oregon, USA: Article no. 63. ACM New York, NY, USA

Arbib MA, Grethe JS (2001) Computing the brain: a guide to neuroinformatics. Academic, San Diego

Babajani-Feremi A, Soltanian-Zadeh H (2010) Multi-area neural mass modeling of EEG and MEG signals. Neuroimage 52(3):793–811

Bailey J, Hammerstrom D (1988) Why VLSI implementations of associative VLCNs require connection multiplexing. International Conference on Neural Networks, San Diego, USA 2, pp 173–180

Beaulieu C, Colonnier M (1983) The number of neurons in the different laminae of the binocular and monocular regions of area 17 in the cat. J Comp Neurology 217:337–344

Binzegger T, Douglas RJ, Martin KAC (2004) A quantitative map of the circuit of cat primary visual cortex. J Neurosci 39(24):8441–8453

Boahen KA (2000) Point-to-point connectivity between neuromorphic chips using address events. IEEE Trans Circuits Syst II: Analog Digit Signal Process 47:416–434

Bojak I, Oostendorp TF, Reid AT, Kötter R (2010) Connecting mean field models of neural activity to EEG and fMRI data. Brain Topogr 23:139–149

Brette R (2007) Exact simulation of integrate-and-fire models with exponential currents. Neural Comput 19(10):2604–2609

Brette R, Gerstner W (2005) Adaptive exponential integrate-and-fire model as an effective description of neuronal activity. J Neurophysiol 94(5):3637–3642

Brette R, Rudolph M, Carnevale T, Hines M, Beeman D, Bower JM, Diesmann M, Goodman PH, Harris FC Jr, Zirpe M, Natschläger T, Pecevski D, Ermentrout B, Djurfeldt M, Lansner A, Rochel O, Vieville T, Muller E, Davison A, El Boustani S, Destexhe A (2007) Simulation of networks of spiking neurons: a review of tools and strategies. J Comput Neurosci 23:349–398

Brown A, Furber S, Reeve J, Wilson P, Zwolinski M, Chad J, Plana LA, Lester D (2010) A communication infrastructure for a million processor machine. In: Proceedings of ACM international conference on computing frontiers, Bertinoro, Italy, pp 75–76
Brunel N (2000) Dynamics of sparsely connected networks of excitatory and inhibitory spiking neurons. J Comput Neurosci 8(3):183–208
Brunel N, van Rossum MCW (2007) Lapicque's 1907 paper: from frogs to integrate-and-fire. Biol Cybern 97:337–339
Buxton RB, Uludag K, Dubowitz DJ, Liu TT (2004) Modeling the hemodynamic response to brain activation. Neuroimage 23(Supplement 1):S220–S233
Cannon R, Gewaltig M-O (2011) Fostering sustainability of software tools in neuroscience. arXiv:1205.3025v1 [q-bio.NC]
Chemla S, Chavane F (2010) Voltage-sensitive dye imaging: technique review and models. J Physiol Paris 104(1–2):40–50
Cheung K, Schultz SR, Leong PHW (2009) A parallel spiking neural network simulator. In: International conference on Field-Programmable Technology (FPT'09), Sydney, NSW, pp 247–254
Davison AP, Brüderle D, Eppler J, Kremkow J, Muller E, Pecevski D, Perrinet L, Yger P (2009) PyNN: a common interface for neuronal network simulators. Front Neuroinform 2(11). doi:10.3389/neuro.11.011.2008
De Schutter E (2008) Why are computational neuroscience and systems biology so separate? PLoS Comput Biol 4(5):e1000078
Dehghani N, Bédard C, Cash S, Halgren E, Destexhe A (2010) Comparative power spectral analysis of simultaneous electroencephalographic and magnetoencephalographic recordings in humans suggests non-resistive extracellular media: EEG and MEG power spectra. J Comput Neurosci 29(3):405–421
D'Haene M, Schrauwen B (2010) Fast and exact simulation methods applied on a broad range of neuron models. Neural Comput 22(6):1468–1472
Diesmann M, Gewaltig M-O (2002) NEST: An environment for neural systems simulations. *Forschung und wisschenschaftliches Rechnen, Beitraege zum Heinz-Billing-Preis 2001*. In: Plesser T, Macho V (eds). Göttingen, Ges fuer Wiss Datenverarbeitung: 43–70
Djurfeldt M (2012) The connection-set algebra–a novel formalism for the representation of connectivity structure in neuronal Network models. Neuroinformatics. doi: 10.1007/s12021-012-9146-1. Online first
Djurfeldt M, Lansner A (2007) Workshop report: 1st INCF workshop on large-scale modeling of the nervous system, Nature Precedings, Stockholm, Sweden
Djurfeldt M, Lundqvist M, Johansson C, Rehn M, Ekeberg Ö, Lansner A (2008) Brain-scale simulation of the neocortex on the IBM Blue Gene/L supercomputer. IBM J Res Dev 52:31–41
Djurfeldt M, Hjorth J, Eppler JM, Dudani N, Helias M, Potjans TC, Bhalla US, Diesmann M, Hellgren Kotaleski J, Ekeberg Ö (2010) Run-time interoperability between neuronal network simulators based on the MUSIC framework. Neuroinformatics 8:43–60
Douglas R, Mahovald M, Mead C (1995) Neuromorphic analogue VLSI. Ann Rev Neurosci 18(255–281)
Ermentrout GB, Kopell N (1984) Parabolic bursting in an excitable system coupled with a slow oscillation. SIAM J Appl Math 44:1133–1149
Fidjeland AK, Shanahan MP (2010) Accelerated simulation of spiking neural networks using GPUs. IEEE International Joint Conference on Neural Networks (IJCNN), Barcelona. IEEE, 345 E 47TH ST, New York, NY 10017
Fourcaud-Trocmé N, Hansel D, van Vreeswijk C, Brunel N (2003) How spike generation mechanisms determine the neuronal response to fluctuating inputs. J Neurosci 23:11628–11640
Gabbott PL, Somogyi P (1986) Quantitative distribution of GABA-immunoreactive neurons in the visual cortex (area 17) of the cat. Exp Brain Res 61(2):323–331
Gleeson P, Crook S, Cannon RC, Hines ML, Billings GO, Farinella M, Morse TM, Davison A, Ray S, Bhalla US, Barnes SR, Dimitrova YD, Silver RA (2010) NeuroML: a language for

describing data driven models of neurons and networks with a high degree of biological detail. PLoS Comput Biol 6:e1000815
Göbel W, Helmchen F (2007) In vivo calcium imaging of neural network function. Physiology 22:358–365
Grün S (2009) Data-driven significance estimation for precise spike correlation. J Neurophysiol 101:1126–1140
Guerrero-Rivera R, Morrison A, Diesmann M, Pearce TC (2006) Programmable logic construction kits for hyper-real-time neuronal modeling. Neural Comput 18(11):2651–2679
Hammarlund P, Ekeberg Ö (1998) Large neural network simulations on multiple hardware platforms. J Comp Neurosci 5:443–459
Hansel D, Mato G, Meunier C (1995) Synchrony in excitatory neural networks. Neural Comp 7(2):307–337
Hanuschkin A, Kunkel S, Helias M, Morrison A, Diesmann M (2010) A general and efficient method for incorporating precise spike times in globally time-driven simulations. Front Neuroinform 4:113
Hines ML, Markram H, Schürmann F (2008) Fully implicit parallel simulation of single neurons. J Comput Neurosci 25(3):439–448
Homma R, Baker B, Jin L, Garaschuk O, Konnerth A, Cohen L, Bleau C, Canepari M, Djurisic M, Zecevic D (2009) Wide-field and two-photon imaging of brain activity with voltage- and calcium-sensitive dyes. Methods Mol Biol 489:43–79
Indiveri G, Chicca E, Douglas RJ (2009) Artificial cognitive systems: from VLSI networks of spiking neurons to neuromorphic cognition. Cogn Comput 1:119–127
Izhikevich EM (2004) Which model to use for cortical spiking neurons? IEEE Trans Neural Netw 5(15):1063–1070
Izhikevich E (2006) 2011, http://www.izhikevich.org/human_brain_simulation/Blue_Brain.htm
Jin X, Galluppi F, Patterson C, Rast A, Davies S, Temple S, Furber S (2010a) Algorithm and software for simulation of spiking neural networks on the multi-chip SpiNNaker system. International Joint Conference on Neural Networks (IJCNN 2010), Barcelona, 18–23 July 2010, pp 649–656
Jin X, Lujan M, Plana LA, Davies S, Temple S, Furber S (2010b) Modeling spiking neural networks on SpiNNaker. Comput Sci Eng 12(5):91–97
Jo SH, Chang T, Ebong I, Bhadviya BB, Mazumder P, Lu W (2010) Nanoscale memristor device as synapse in neuromorphic systems. Nano Lett 10:1297–1301
King JG, Hines M, Hill S, Goodman PH, Markram H, Schürmann F (2009) A component-based extension framework for large-scale parallel simulations in NEURON. Front Neuroinform 3:1–11
Kobayashi R, Tsubo Y, Shinomoto S (2009) Made-to-order spiking neuron model equipped with a multi-timescale adaptive threshold. Front Comput Neurosci 3(9)
Lansner A (2009) Associative memory models – from cell assembly theory to biophysically detailed cortex simulations. Trends Neurosci 32:178–186
Lansner A, Fransén E (1992) Modeling Hebbian cell assemblies comprised of cortical neurons. Netw Comput Neural Syst 3:105–119
Likharev K, Mayr A, Muckra I, Turel O (2003) CrossNets high-performance neuromorphic architectures for CMOL circuits. Molecular electronics III. In: Reimers J, Picconatto C, Ellenbogen J, Shashidhar R (eds) New York, NY Acad Sci 1006: 146–163
Lindén H, Pettersen KH, Einevoll GT (2010) Intrinsic dendritic filtering gives low-pass power spectra of local field potentials. J Comput Neurosci 29(3):423–444
Lundqvist M, Rehn M, Djurfeldt M, Lansner A (2006) Attractor dynamics in a modular network model of the neocortex. Netw Comput Neural Syst 17:253–276
Lundqvist M, Compte A, Lansner A (2010a) Bistable, irregular firing and population oscillations in a modular attractor memory network. PLoS Comput Biol 6(6):1–12
Lundqvist M, Herman P, Lansner A (2010b) Theta and gamma power increases and alpha/-beta power decreases with memory load in an attractor network model. J Cogn Neurosci 23(10):3008–3020

MacGregor RJ (1987) Neural and brain modeling. Academic, San Diego
Maguire LP, McGinnity TM, Glackin B, Ghani A, Belatreche A, Harkin J (2007) Challenges for large-scale implementations of spiking neural networks on FPGAs. Neurocomputing 71:13–29
Markram H (2006) The blue brain project. Nat Rev Neurosci 7:153–160
Migliore M, Cannia C, Lytton WW, Markram H, Hines ML (2006) Parallel network simulations with NEURON. J Comput Neurosci 21(2):119–223
Misra J, Saha I (2010) Artificial neural networks in hardware: a survey of two decades of progress. Neurocomputing 74:239–255
Morrison A, Diesmann M (2008) Maintaining causality in discrete time neuronal network simulations. In: Beim Graben P, Zhou C, Thiel M, Kurths J (eds) Lectures in supercomputational neuroscience: dynamics in complex brain networks. Springer, Berlin, pp 267–278
Morrison A, Mehring C, Geisel T, Aertsen A, Diesmann M (2005) Advancing the boundaries of high connectivity network simulation with distributed computing. Neural Comput 17(8): 1776–1801
Morrison A, Aertsen A, Diesmann M (2007a) Spike-timing dependent plasticity in balanced random networks. Neural Comput 19:1437–1467
Morrison A, Straube S, Plesser HE, Diesmann M (2007b) Exact subthreshold integration with continuous spike times in discrete time neural network simulations. Neural Comput 19(1): 47–79
Morrison A, Diesmann M, Gerstner W (2008) Phenomenological models of synaptic plasticity based on spike-timing. Biol Cybern 98:459–478
Nageswarana JM, Dutt N, Krichmar JL, Nicolau A, Veidenbaum AV (2009) A configurable simulation environment for the efficient simulation of large-scale spiking neural networks on graphics processors. Neural Netw 22:791–800
Noble D (2006) The music of life: biology beyond genes. Oxford University Press, Oxford
Nordlie E, Gewaltig M-O, Plesser HE (2009) Towards reproducible descriptions of neuronal network models. PLoS Comput Biol 5(8):e1000456
Plesser HE, Diesmann M (2009) Simplicity and efficiency of integrate-and-fire neuron models. Neural Comput 21:353–359
Potjans T, Diesmann M (2011) The cell-type specific connectivity of the local cortical network explains prominent features of neuronal activity. arXiv:1106.5678 [q-bio.NC]
Potjans W, Morrison A, Diesmann M (2010) Enabling functional neural circuit simulations with distributed computing of neuromodulated plasticity. Front Comput Neurosci 4(141). doi:10.3389/fncom.2010.00141
Rasch M, Logothetis NK, Kreiman G (2009) From neurons to circuits: linear estimation of local field potentials. J Neurosci 29:13785–13796
Rotter S, Diesmann M (1999) Exact digital simulation of time-invariant linear systems with applications to neuronal modeling. Biol Cybern 81(5/6):381–402
Schemmel J, Bruderle D, Meier K, Ostendorf B (2007) Modeling synaptic plasticity within networks of highly accelerated I&F neurons. In: IEEE international symposium on circuits and systems art. no. 4253401, pp 3367–3370
Schemmel J, Brüderle D, Grübl A, Hock M, Meier K, Millner S (2010) A wafer-scale neuromorphic hardware system for large-scale neural modeling. In: ISCAS 2010 – IEEE International Symposium on Circuits and Systems: nano-bio circuit fabrics and systems, pp 1947–1950
Sotero RC, Trujillo-Barreto NJ (2008) Biophysical model for integrating neuronal activity, EEG, fMRI and metabolism. Neuroimage 39(1):290–309
Stettler DD, Axel R (2009) Representations of Odor in the Piriform Cortex. Neuron 63: 854–864
Thomas D, Luk W (2009) FPGA accelerated simulation of biologically plausible spiking neural networks. In: 17th IEEE symposium on field programmable custom computing machines, Napa, California, pp 45–52
Thomson AM, West DC, Wang Y, Bannister AP (2002) Synaptic connections and small circuits involving excitatory and inhibitory neurons in layer 2-5 of adult rat and cat neocortex: triple intracellular recordings and biocytin labelling in vitro. Cereb Cortex 12:936–953

Traub RD, Whittington MA, Buhl EH, Jefferys JGR, Faulkner HJ (1999) On the mechanism of the $\gamma - \beta$ frequency shift in neuronal oscillations induced in rat hippocampal slices by tetanic stimulation. J Neurosci 19(3):1088–1105

van Elburg RAJ, van Ooyen A (2009) Generalization of the event-based Carnevale-Hines integration scheme for integrate-and-fire models. Neural Comput 21(7):1913–1930. doi:10.1162/neco.2009.07-08-815

van Vreeswijk C, Sompolinsky H (1996) Chaos in neuronal networks with balanced excitatory and inhibitory activity. Science 274:1724–1726

Winograd T (1975) Breaking the complexity barrier (again). ACM SIGPLAN Notices 10:1 (Jan. 1975) (jointly issued as SIGIR FORUM 9:3 Winter 1974), pp 13–30. Reprinted. In: Barstow D, Shrobe H, Sandewall E (eds) Interactive programming environments. McGraw Hill, New York 1984 pp 3–18

Chapter 11
Co-operative Populations of Neurons: Mean Field Models of Mesoscopic Brain Activity

David T.J. Liley, Brett L. Foster, and Ingo Bojak

Abstract While the basic units of computation in the brain are the neuronal cells, their sheer number, complexity of structural organisation and widespread connectivity make it difficult, if not impossible, to perform realistic simulations of activity at millimetre range or beyond. Furthermore, it is becoming increasingly clear that a range of non-neuronal and stochastic factors influence neuronal excitability, and must be taken into account when developing models and theories of brain function. One answer to the these persistent difficulties is to model cortical tissue not as a network of spike-based enumerable neurons, but to take inspiration from statistical physics and model directly the bulk properties of the populations constituting the cortical tissue. Such an approach proves compatible with many experimental recording techniques and has led to a successful class of so-called "mean field theories" that, when constrained by meaningful physiological and anatomical parameterisations, reveal a rich repertoire of biologically plausible and predictive dynamics. The aim of this chapter is to outline the historical genesis of this important modelling framework, and to detail its many successes in accounting for the experimentally observed neuronal population activity in cortex.

D.T.J. Liley (✉)
Brain Sciences Institute, Swinburne University of Technology, P.O. Box 218, Hawthorn, VIC 3122, Australia
e-mail: dliley@swin.edu.au

B.L. Foster
Department of Neurology and Neurological Sciences, Stanford University, Stanford, CA 94305, USA
e-mail: blfoster@stanford.edu

I. Bojak
Donders Institute for Brain, Cognition and Behaviour, Centre for Neuroscience, Radboud University Nijmegen Medical Centre, Postbus 9101 // 126, 6500 HB Nijmegen, The Netherlands
e-mail: i.bojak@donders.run.nl

N. Le Novère (ed.), *Computational Systems Neurobiology*,
DOI 10.1007/978-94-007-3858-4_11,

> Not only is it not proven, but it is highly unlikely on general biological considerations, that a special sensory function is related to a cell type of a particular structure. The essential for the elaboration of any cortical function, even the most primitive sensory perception, is not the individual cell type but cell groupings.
>
> Korbinian Brodmann (1909)

> …the effective unit of operation in such a distributed system is not the single neuron and its axon, but groups of cells with similar functional properties and anatomical connections.
>
> Vernon B. Mountcastle (1997)

11.1 Introduction

Ever since the formulation of the neuron doctrine by Santiago Ramón y Cajal, Rudolf von Kolliker and others (López-Muñoz et al. 2006), neuroscience has strived to understand how consciousness and cognition arises out of the myriad and complex interactions between neurons of the central nervous system. Beginning with the work of McCulloch and Pitts (1943), in which single neurons were conceived as simple fixed threshold binary state devices organized into networks of great structural complexity, and culminating in the massively detailed single neuron models of the Blue Brain Project headed by Henry Markram (2006), brain function has been assumed to emerge out of the activity of networks of neurons. This has been an enormously successful paradigm and has led to models of great computational complexity and sophistication. However, it is becoming increasingly clear that a range of non-neuronal and stochastic factors and elements influence neuronal excitability and that these must be taken into account when developing models and theories of brain function, if we are to meaningfully simulate emergent neuronal activity. For example, it is now known that the supporting non-neuronal elements of cortical tissue, the glial cells (Perea et al. 2009; Perea and Araque 2010), interact synaptically with cortical neurons to influence the patterns of neuronal firing. Further, while the firing of individual neurons is regulated by deterministic factors their synaptic interactions may well not be – the reliability of synapses can be as low as 1%, i.e., only 1 in 100 pre-synaptic action potentials actually elicits a postsynaptic response (Branco and Staras 2009).

The problem is how to deal with this significant added complexity in the presence of often limited and non-specific empirical data. One possible solution has been to *not* consider cortical tissue as an network of enumerable neurons interacting by the transmission of spikes, and instead consider cortex in terms of a bulk or ensemble dynamics, such as the mean firing rate (and/or its moments) of a spatially circumscribed population of neurons (Freeman 1975; Nunez 1995; Deco et al. 2008; Coombes 2010). Such a rate-based reconfiguration has a number of advantages: (1) modelling populations of neurons corresponds more closely with the generally accepted contention that behaviour emerges out of the macroscopic manifestations of neuronal activity, (2) modelling the behaviour of populations of neurons implicitly deals with the unreliability of synaptic interconnections and incorporates the effects of non-neuronal elements, and (3) the spatial scale of modelled populations of neurons corresponds closely with the milli- to centimetre scales of spatial resolution of the non-invasive neuroimaging modalities typically utilised to interrogate brain function, like functional magnetic resonance imaging (fMRI), MEG (magnetoencephalogram) and EEG (electroencephalogram).

Describing cortical neuronal activity in terms of population averages gives rise to a class of models broadly known as mean field theories (Deco et al. 2008). Originally arising out of mathematical models of ferromagnetism in statistical physics, such models approximate the specific input a neuron receives from other neurons by the average activity in a neuronal surround defined by patterns of axonal and dendritic branching. In this way interactions between individual neurons are replaced by effective averages – the mean fields, i.e., cortical neurons can be viewed as *"sampling"* the activity of nearby populations of neurons based on the mean geometry of the axonal and dendritic arborisations. Thus the dynamics of populations of neurons are driven by mean fields, which are in turn determined by the activity of populations of neurons. The current mathematical approach for formulating equations of motion for the activity of neuronal populations or "masses" stems principally from the work of Wilson and Cowan (1972, 1973), Nunez (1974a), Freeman (1975) and Amari (1975, 1977). The resulting so-called mass action or neural field theories have formed a basis for the biomathematical exploration of macro- and mesoscopic neuronal dynamics. Mesoscopic neuronal activity is typically defined to be intermediate in scale between the activity of single neurons and the activity of large areas of cortex, i.e., at roughly millimetre scale.

The aim of this chapter is to outline in some detail the formulation of physiologically relevant mean field theories and how they can be utilised to account for a range of mesoscopic brain activity that includes the spatiotemporal dynamics of the resting EEG/electrocorticogram (ECoG), its perturbation during diseases such as epilepsy, and its modulation by a range of drugs that most importantly include anaesthetic and sedative agents. The chapter is organized into three main sections. The first describes the anatomical and physiological basis for modelling mesoscopic neuronal activity in mammalian cortex and the bulk and discrete approaches that have typically been employed to model it. It then focuses on the advantage of bulk approaches in the context of limited empirical knowledge and outlines the implicit microscopic constraints necessary in formulating the corresponding mean field theories. The second section outlines the existing mean field approaches by way of their historical development, firstly by describing the foundational models, and then their subsequent elaboration and development to include greater levels of physiological veracity. Finally, the third section details the patterns and types of mesoscopic brain activity that can provisionally be accounted for by the various mean field models.

11.2 Mesoscopic Neural Activity

Because the structure and function of the mammalian brain resists any simplistic representation or definition it has been difficult to conceive of generative theoretical frameworks to account for human behaviour on the basis of neural activity. The activity of human brains encompasses many aspects and spatial and temporal scales: from the millisecond flurry of the opening and closing of transmembrane ionic channels to socio-political machinations that can extend over many decades. Typically one wants to explain the behaviour observed at a higher, more meaningful level in terms of activity occurring at a lower, more mechanistically accessible level. In the case of neuroscience the long-term aim is to relate human intentional behaviour to the activity of neurons. However the gulf between the local actions of individual neurons (microscopic) and the intentional patterns of activity evinced by non-invasive neuroimaging modalities such as positron emission tomography (PET), single-photon emission computed tomography (SPECT) and fMRI (macroscopic) is too wide to bridge with current theories. An intermediate level of description is hence required. This mesoscopic level of the neuronal ensemble, mass or population is best justified on the basis of the anatomical structure of cortical tissue, which we proceed to outline, but can also be motivated using statistical mechanics (Deco et al. 2008).

11.2.1 Anatomical and Physiological Organization of the Cerebral Cortex at Different Scales

The thin outer rind of the mammalian brain, the neocortex, is generally thought to be the principle structure responsible for the generation and elaboration of purposeful activity. For a structure that is between 1 and 5 mm thick and has a surface area of only $\sim 0.19\ \mathrm{m}^2$ (Van Essen 2005), it has a truly staggering degree of structural complexity with about $\sim 2 \cdot 10^{10}$ neurons (Pakkenberg and Gundersen 1997) divided into six horizontal layers with at least a dozen major neuronal subtypes (Markram et al. 2004), each interacting via on average 6,900 synaptic connections with other neurons (Tang et al. 2001), synapses that utilise an array of chemical messengers and can be individually modified. Add in the non-neuronal glia known to influence cortical neuronal activity (Ben Achour and Pascual 2010; Araque and Navarrete 2010), astrocytes and microglia, which are equivalent in number and density to the neurons (Miguel-Hidalgo 2005; Azevedo et al. 2009), then the task of simulating cortical neuronal activity appears daunting, if not intractable. Fortunately, mammalian cerebral cortex is sufficiently well organized over a number of relatively distinct spatial scales to enable the construction of tractable models and theories of brain activity beyond that of the enumerable neural network. Furthermore, despite great variations in the size of the cortex among the various mammalian groups (Herculano-Houzel 2009), it nevertheless remains remarkably consistent in terms of its cellular elements, and its vertical and horizontal organization.

11.2.1.1 The Cellular Composition of Cortex

Cortex is comprised of neuronal and non-neuronal components with the neurons being broadly classified as belonging to two types: pyramidal and non-pyramidal cells. Pyramidal cells are the most numerous neuronal class making up somewhere between 60% and 85% (Braitenberg and Schüz 1998; Nieuwenhuys et al. 2008) of all cortical neurons. A typical pyramidal neuron is composed of a cell body from which a single axon descends and branches before exiting cortex, as well as a dendritic tree composed of two main branching structures (1) an apical dendritic tree composed of a trunk ascending in the direction of the pial surface and (2) a basal dendritic tree composed of multiple trunks giving rise to a cloud of local dendritic branches about the cell body. Both dendritic structures are typically extensively covered with small excrescences called spines where synapses form (Spruston 2008). Particularly in sensory cortices the apical dendrites of 50 or so pyramidal neurons distributed throughout the thickness of cortex can be clustered together into distinct and regularly spaced cylindrical groupings. These cylindrical groupings, referred to as dendrons by Eccles (1992), constitute a core component of the hypothesised *"minicolumn"*: a barrel shaped region representing the basic modular unit of neocortex (Rockland and Ichinohe 2004).

While pyramidal neurons generally show a fair degree of morphological variability the only atypical variant is the spiny stellate cell, an interneuron (see below) which lacks the characteristic ascending apical dendritic tree and descending axon. Pyramidal neurons constitute a functionally homogeneous group as they all exclusively release the excitatory monoamine glutamate from their axonal terminals. It is also worth noting that pyramidal neurons can be functionally subdivided based on their steady state firing pattern in response to step depolarising currents (Contreras 2004). As will be discussed later in Sects. 11.2.1.2 and 11.2.1.3, the branching pattern of pyramidal cell axons and the minicolumn form two possible characteristic scales for the spatial organization of cortex.

Despite their smaller numbers non-pyramidal cells are a morphologically much more differentiated class of cortical neuron that have a number of features in common (Nieuwenhuys et al. 2008): their dendrites are often spine free, their axons do not leave cortex (hence often called *local circuit* or *interneurons*), most release the inhibitory neurotransmitter $\gamma-$amino butyric acid (GABA) and a certain fraction (25–30%) also express one or more neuropeptides such as vasoactive intestinal polypeptide (VIP) or cholecystokinin, and various subpopulations show differential immuno-reactivity to one or more intracellular calcium binding proteins which can be used as subpopulation specific markers. It has been estimated that a dozen or so non-pyramidal cell sub-types can be identified morphologically (Markram et al.

2004), the most numerous of which are basket cells, whose axons form basket-like plexuses around pyramidal cells bodies, Martinotti cells, which project their axons to the superficial layers of cortex to interact with apical dendrites of pyramidal neurons, and bitufted cells, which have dendrites arising from upper and lower poles of the cell body.

Other notable non-pyramidal cells include the Chandelier, bipolar and double bouquet cells. Chandelier cells produce a profusely ramifying axonal tree with "candles", short vertical axonal segments containing rows of synaptic boutons, that are the pre-terminal components of axo-axonic synapses at initial segments of pyramidal neurons. Bipolar neurons, which are similar in morphology to bitufted neurons, represent the single known example in which a non-pyramidal neuron can be excitatory by releasing only VIP. They can also be inhibitory by releasing only GABA, while also expressing VIP (Markram et al. 2004). Double bouquet cells have a similar dendritic morphology to bitufted cells but produce radially (vertically) oriented dense axonal plexuses consisting of bundles of thin parallel axonal branches. Because the axonal system of a single double bouquet cell is closely associated with the apical dendrites of pyramidal neurons in a minicolumn and has a relatively well defined lateral extent of arborisation, their spacing (30–50 μm) provides a characteristic tangential (horizontal) scale for cortical organization. Like pyramidal neurons the non-pyramidal neurons can also be electrophysiologically classified.

The non-neuronal components of cortical tissue can be divided into the neuroglia and the cells of the perforating blood vessels. The neuroglia are comprised of astrocytes, microglia, oligodendrocytes and ependymal cells. Classically it used to be thought that the activity of these neuroglia did not contribute in any meaningful way to brain function: astrocytes, star shaped cells with multiple processes, provided biochemical, metabolic and structural support to the neurons and their interactions; microglia are the brain's macrophages; oligodendrocytes produced the myelin sheaths around axons to increase conduction velocities; and epithelial ependymal cells lining the ventricles produced the cerebral spinal fluid. However, beginning in the early 1990s research has revealed that astrocytes, like neurons, are excitable (with respect to intracellular Ca^{2+} levels) and respond to, and are influenced by, neuronal activity at the level of the synapse. To conceptualise this evidence the term "tripartite synapse" has been proposed (Perea et al. 2009), defined as consisting of one or more glial processes chemically interacting with the pre- and post-synaptic components of a synapse. Such 'synapses' seem to occur at the synapses of all neurons in cortex and have been shown to regulate interneuronal synaptic transmission and plasticity. Given these interactions and the fact that astrocyte–astrocyte interactions can be demonstrated (Dienel and Cruz 2003), it follows that functionally cortical tissue is more than just a network of neurons.

11.2.1.2 Vertical/Radial Organization of Neocortex

Beginning with Theodor Meynert and Vladamir Betz and culminating in the 1909 work of Korbinian Brodmann (Brodmann and Garey 2006), cerebral cortex was found to be divided into vertically stacked cellular laminae, the number, size and organization of which show substantial horizontal (regional) variation.[1] From an ontogenetic (developmental) perspective two broad structural forms of cerebral cortex can be identified based on the genesis of their laminar organization – homogenetic cortex and heterogenetic cortex. Homogenetic cortex, which is more commonly referred to as neocortex or isocortex, makes up the bulk of cerebral cortex and either consists of six reasonably well defined cellular laminae (homotypical cortex) or began as a six-layered cortex but during development addition or elimination of layers occurred (heterotypical cortex). In contrast, heterogenetic cortex is divided into primitive (or paleo-) cortex, in which there is no clear laminar cellular organization, and rudimentary (or archi-) cortex, in which there are only the crude beginnings of lamination. The olfactory bulb and amygdala are examples of paleocortex, whereas the hippocampus is an example of archicortex, in which there are only three identifiable cellular layerings.

The six neocortical layers labelled I–VI, see Fig. 11.1, are characterised by variations in cellular densities, types and morphologies as well as the patterns of termination and generation of cortical and subcortical afferents and efferents. Non-pyramidal cells occur in all layers and pyramidal cells in layers II–VI. Layer IV of sensory cortices is notable for the large numbers of tightly packed spiny stellate neurons, which are only found there, and the termination of sensory thalamocortical afferents on these neurons and the dendrites of other neurons passing through this layer (Thomson and Bannister 2003). In contrast it has been observed that associational and callosal cortico-cortical efferents arising from layer II and III pyramidal neurons preferentially terminate in layers IV, whereas layer V/VI pyramidal neuron long-range axons preferentially terminate in layers I and VI (Rockland and Pandya 1979).

While such cortical lamination suggests discrete horizontally arranged neuronal populations, such a distinction becomes less convincing when other radially organized cortical elements are included. Among the most (histologically) prominent of these are clusters or bundles of apical dendrites of layer V pyramidal neurons (Fleischhauer et al. 1972), bundles of descending myelinated axons of pyramidal cells generally referred to as the "radiations of Meynert" and column-like arrays of pyramidal cell bodies thought to be direct developmental descendants of organized clusters of cells in the embryonic precursor of the cerebrum. In addition, double bouquet interneurons (see Sect. 11.2.1.1), which are abundant in primate neocortex, give rise to tightly packed bundles of vertically oriented axonal collaterals called "horses tails" that span multiple laminae. Multiple radially organized cellular elements therefore bind pyramidal and non-pyramidal components across the various cortical laminae. Horizontal (or areal) periodicities in the radial organization of these neocortical cellular elements may provide a structural basis for defining the modular organization of neocortex.

[1] Here and in the following we mean by "vertically/radially" through the several millimetres thickness of cortex and by "horizontally/tangentially" parallel to its pial surface.

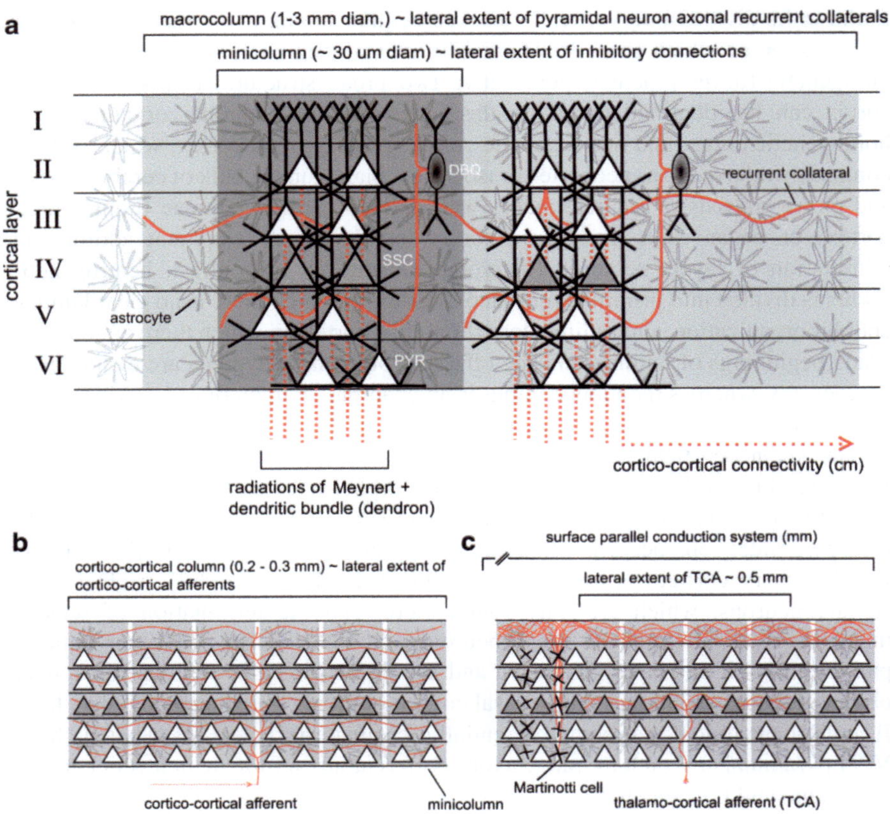

Fig. 11.1 Highly simplified sketch of the hypothetical modular organization of neocortical tissue. (**a**) The minicolumn was originally defined as a narrow chains of 100–200 neurons extending across layers II–IV and organized into repeating patterns (Mountcastle 1979). A variety of radially oriented units have a similar scale to the putative minicolumn. These include pyramidal (PYR) cell dendritic (dendron) and axonal (radiations of Meynert) bundles, as well as lateral arborisations of double bouquet neurons (DBQ). A macrocolumn binds together laterally about thousand minicolumns by recurrent axonal collaterals of an intracortical pyramidal axon. (**b**) The cortico-cortical column is defined by the lateral extent of intracortical terminations of afferent cortico-cortical fibres (associational and callosal). (**c**) Other hypothetical scales of modular organization include the lateral extent of thalamocortical afferents, shown here to principally synapse with spiny stellate cells (SSC), and the layer I plexiform arborisations of the axons of inhibitory Martinotti neurons

11.2.1.3 Horizontal/Tangential Organization of Neocortex

The idea that cortex is horizontally parcellated into anatomically well defined radially oriented columnar units has become virtual dogma. Commencing with the work of Lorente de Nó, who first proposed that a small radially oriented cylinder of cells extending through the full extent of cortex with a single thalamocortical axon as its axis defined an *"elementary unit"* of neocortical organization, the intervening years have seen a panoply of attempts to define a *"basic modular unit"* of cortical organization. Amongst the most well known are the micro-, mini-, macro- and cortico-cortical columns. Micro- and minicolumns typically refer to radially organized chords of $\approx$ 10–200 cells that span layers II–VI, arranged as horizontal mosaics with periodicities of the order of 15–80 μm (Jones 2000; Buxhoeveden and Casanova 2002; Rockland and Ichinohe 2004; Nieuwenhuys et al. 2008). In addition, there are a range of other elements repeating at this scale that could be said to define micro-/macrocolumnar organization. These include the periodically repeating bundles of radially oriented dendrites (pyramidal cell) and axons (double bouquet) mentioned above.

Columns were first hypothesised by Mountcastle (1957) on the basis of electrophysiological evidence in which radially co-localized neurons in cat somatosensory cortex shared receptive field properties in response to tactile stimulation. The lateral extent of this shared receptive field was estimated to be of the order of 0.5 mm. These columns, later designated macrocolumns (Mountcastle 1979), were subsequently considered by Mountcastle to be anatomically comprised of aggregations of several hundred minicolumns bound together by short range horizontal excitatory and inhibitory connections (Mountcastle 1997). In contrast to macrocolumns whose lateral extent is defined by the scale of short range horizontal connectivity, cortico-cortical columns (also referred to as neocortical columns) are typically defined in terms of the cylindrical aborisation volumes of either a single afferent cortico-cortical fibre or closely packed bundles of such fibres (Jones et al. 1975; Goldman and Nauta 1977; Szentágothai 1983). Estimates of the radial extent of such columns varies from 200 to 800 μm.

While some areas of cortex have a greater claim to displaying some form of columnar organization than others, visual and somatosensory in particular, cortical columns of any form or variety have not been substantiated by unequivocal anatomical evidence and therefore remain hypothetical. Neocortical columns (Markram 2008), intensely studied in barrel cortex (Petersen 2007; Lübke and Feldmeyer 2007), are perhaps closest to being established. What however is abundantly clear is that neocortex consists of populations of vertically well connected cellular elements interacting horizontally over a range of spatial scales. In general it is the lateral axonal ramifications of neocortical pyramidal neurons that define the spatial scales of horizontal connectivity within neocortex. The axons of all typical (i.e., not spiny stellate) pyramidal cells produce a number of horizontal branches (collaterals) in cortex before entering subcortical white matter where they form the long-range cortico-cortical fibres systems (see Fig. 11.1). Intracortical horizontal branches can either ramify in close proximity to the parent cell body or travel laterally for many millimetres depending on species, cortical area and layer (Nieuwenhuys et al. 2008). In general, it appears that the longer the branch the more likely it is to be myelinated.

Pyramidal axonal collaterals provide local input to GABAergic interneurons, which in turn form reciprocal synapses with pyramidal neurons (White 1989). In contrast, cortico-cortical axons can travel for many centimetres in subcortical white matter before re-emerging in cortex to form synaptic connections with all neuronal cell types, and in particular with the apical dendrites of typical pyramidal neurons. The cortico-cortical system of connectivity can be divided into commissural and associational fibre systems, depending on whether they respectively pass through the corpus callosum or remain ipsilateral. Somewhat arbitrarily, associational fibres can be divided into short- and long-range. The short-range system is believed to be fairly isotropic and homogenous in its distribution, while the long-range one is readily identified from gross dissection and non-invasive tractographic methods based on diffusion MRI (Johansen-Berg and Rushworth 2009). In humans the majority of commissural axons are myelinated (Aboitiz et al. 1992a,b) and based on measurements of fibre diameter are expected to have broadly distributed range of conduction velocities (Bojak and Liley 2010). Anatomical evidence suggests that the density of excitatory synapses arising from cortico-cortical afferents is similar to those made by recurrent axonal collaterals (Liley and Wright 1994; Braitenberg and Schüz 1998).

The lateral axonal ramifications of certain interneurons provide additional characteristic scales for the horizontal organization of cerebral cortex. We have encountered one such interneuron type previously – that of the layer II/III double bouquet cell whose descending bundles of axons have been shown to disperse horizontally in deeper layers. Peters and Sethares (1997) have estimated that the spacing of these so-called "horse tails", and by inference the extent of their terminal axonal arborisations, is 23 μm in rhesus monkey primary visual cortex. Another interneuron cell type that has been described as giving rise to extensive lateral axonal arborisations is the Martinotti cell. Martinotti cells, which occur in layers II–VI give rise to one or more ascending axons that project to laminae I, where they give rise to long horizontal branches that can run for several millimetres making synapses with the apical dendrites of pyramidal neurons. Szentágothai (1978) defined the *"surface parallel intracortical system"* to be comprised of these axons.

In addition, there have been attempts to topographically parcellate cerebral cortex according to shared anatomical, histological or histochemical features. The most consequential of these are those concerned with horizontal (areal) variations in the cellular architecture of the various neocortical laminae (cytoarchitectonics), in the organization of radially oriented bundles of myelinated fibres called *radii* or *radiations of Meynert* (myeloarchitectonics) and in the temporal order in which subgriseal white matter becomes myelinated during development. Of these architectonic parcellations the 1909 Brodmann map is still widely used (Zilles and Amunts 2010). By observing differences in the relative thickness and cell density of various layers and the size, shape and arrangement of neuronal cell bodies, Brodmann delineated 44 (paired) areas in the human neocortex (Nieuwenhuys et al. 2008; Brodmann and Garey 2006). However, because he and others only used a single stain (Nissl) and a limited number of brains, determined areal boundaries subjectively and ignored sulcal cortex, Nieuwenhuys et al. (2008) conclude that existing architectonic parcellations may substantially underestimate the number of juxtaposed structural areas. For example, modern approaches provide probabilistic maps of eight subdivisions of Brodmann areas 5 and 7 in the human superior parietal

Table 11.1 Major structural units of organization in mammalian neocortex and their approximate characteristics, see text

	Unit	Neurons per unit	No. of units[b]	Scale
Micro	Neuron (soma)	1	$2 \cdot 10^{10}$	15 μm
	Microcolumn[a]	20	10^9	15 μm
	Minicolumn	100	$2 \cdot 10^8$	35 μm
Meso	Dendritic tree	–	–	0.15 mm
	Cortico-cortical column	10^4	$2 \cdot 10^6$	0.35 mm
	Intracortical axonal tree	–	–	0.5 mm
	Macrocolumn	$2 \cdot 10^5$	10^5	1.5 mm
	Martinotti axonal tree	–	–	2 mm
Macro	Architectonic area	$2 \cdot 10^8$	100	5 cm[b]
	Cortico-cortical axon	–	–	10 cm[b]
	Brain region	$2 \cdot 10^9$	10	15 cm[b]
	Neocortex	$2 \cdot 10^{10}$	1	50 cm[b]

The table has been adapted from Nunez (1995), Jones (2000), Buxhoeveden and Casanova (2002) and Nieuwenhuys et al. (2008).
For neuronal groupings the diameter of a disk with equivalent cortical surface area is given as scale.
[a]The term "microcolumn" is sometimes used to refer to our "minicolumn".
[b]Figures are given for the human neocortex.

cortex (Scheperjans et al. 2008). The future of brain mapping efforts however likely belongs to comprehensive multimodal approaches, which for example integrate MRI data (Toga et al. 2006).

Table 11.1 summarises the various horizontal scales that have been identified or proposed on the basis of anatomical evidence. Based on this, and for the purposes of what follows, we choose to define the microscopic scale as commensurate with the level of the single neuron, micro- and minicolumn, whereas we establish the macroscopic scale as corresponding to the scale of the variously identified cyto/myelo-architectonic areas or larger. The mesoscopic level will thus represent an intermediate spatial scale including cortico-cortical and macrocolumns.

11.2.2 *Enumerable Network vs. Bulk Modelling Approaches*

The principle cellular substrate underlying brain function is without doubt the cortical neuron. On this basis, the most logical way forward to understanding the emergence of brain function is to simulate the activity of networks of neurons by modelling the properties and features of the individual neurons and the micro-circuitry of their connectivity. Yet such a research program faces a number of theoretical and practical problems: There are good reasons to believe that non-neuronal components of the brain, like the cortical astrocytes, play an important role in regulating interneuronal interactions, and thus neuronal activity.

Furthermore, cooperative neuronal activity dominates non-invasive measurements (e.g., EEG/MEG, fMRI BOLD) but often transcends the activity of the individual neuron. Practically, we have the problem of knowing how much detail to include – if the behaviour of individual neurons is believed crucial to understanding say the EEG, then the behaviour of individual ionic channels may be crucial to understanding the behaviour of individual neurons, and so forth. We soon find that trying to understand the behaviour of cortical tissue in terms of enumerating the functionally important components and their interactions leads to a combinatorial explosion in complexity. This is quite the opposite to what we want to achieve using modelling. Fortunately there is a way out of this epistemological bog.

Just as Waage and Guldberg in 1864 (Waage et al. 1986) sought to understand the kinetics of chemical reactions in bulk terms by defining the *principle of mass action* so a range of researchers, most notably Freeman (1975), have attempted to understand the dynamics of cortical neural activity in terms of the bulk interactions of functionally circumscribed masses or populations of neurons. The motivation for such masses in cortex depends upon two well established physiological principles: (1) firstly the vast majority of neurons in cortex chemically communicate using only a single neurotransmitter, and (2) the radial and horizontal organization of cortex (Sects. 11.2.1.2 and 11.2.1.3) defines domains of co-operative neural activity by virtue of synaptic interactions. Thus we can view cortex, at mesoscopic spatial scales, as networks of interacting populations or masses of excitatory (typical and atypical pyramidal) and inhibitory (non-pyramidal) neurons. In this way the cortical microcircuit is replaced by, and subsumed into, a *cortical mesocircuit*, see Fig. 11.2.

11.2.3 Population Densities, Mean Fields, and Continuum Approaches

One way forward to quantifying the dynamics of cortical mesocircuits is the ensemble density approach in which the time evolution of the probabilistic behaviour of large, potentially infinite, populations of neurons is quantified under the action of particular kinds of physiologically defined *"forces"* (Deco et al. 2008). Such an approach can include known stochastic fluctuations, such as variations in quantal transmitter release, as well as dynamically evolving the probability distributions associated with neuronal ensemble dynamics with all their moments and couplings. While providing a potentially rigorous approach to quantifying the dynamics of neuronal populations, in the context of empirical measurement problems arise: Firstly, probabilistic evolution, particularly in the presence of non-linearity, depends sensitively on initial states which in biological systems are in general unknowable. Secondly, actual measurements of the behaviour of populations of neurons will in general reflect only certain moments of the corresponding probability distributions, most prominently first moments (i.e., means). For example, a single EEG electrode records synaptically induced currents averaged over many thousands of neurons.

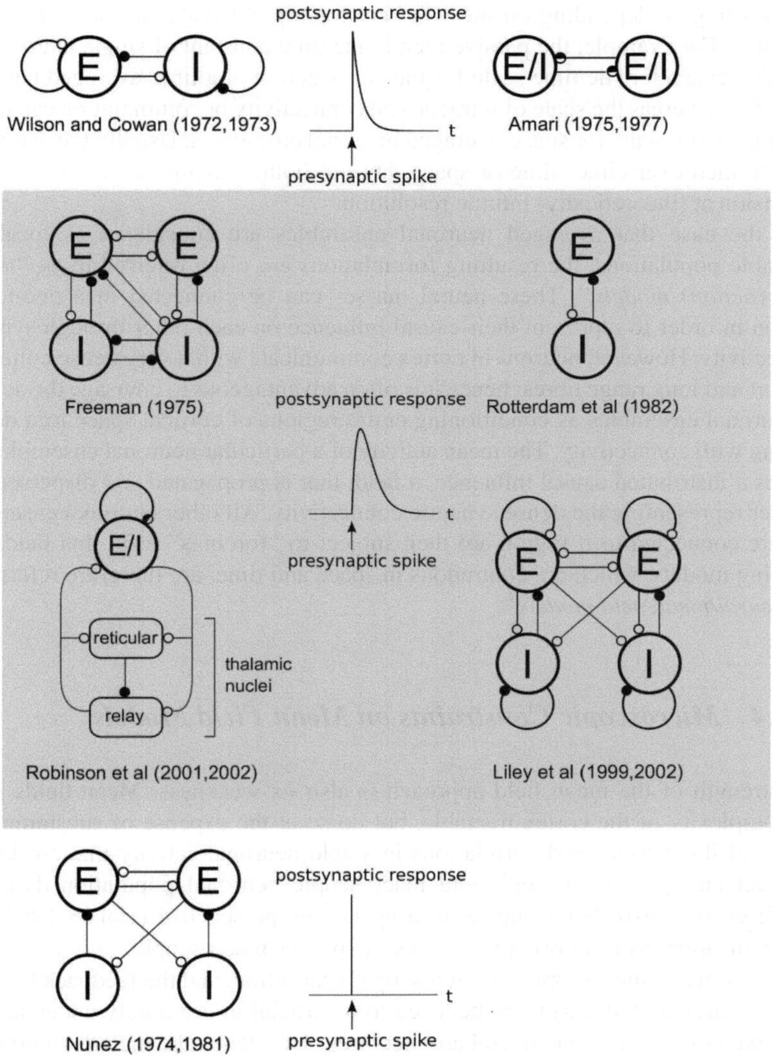

Fig. 11.2 Schematic representation of the circuit topologies of typical mean field models, segregated by their model for the postsynaptic response. All approaches consider two functionally differentiated neuronal populations: excitatory (E) and inhibitory (I) ones. *Open circles* represent excitatory connections, *filled circles* inhibitory ones, and *half-filled circles* both

For this and other reasons (Deco et al. 2008), approaches in which the dynamics of some appropriately defined first moment are tracked are often preferable. Such quantities typically include the *"mean soma membrane potential"* (Liley et al. 2002, 2003) and the *"mean firing rate"* (Wilson and Cowan 1972, 1973) of appropriately defined neuronal ensembles. These can be defined as either time- or space-averaged, depending on the what spatiotemporal scales are intended to be modelled. For example, the passive membrane time constant of single neurons can furnish a characteristic time scale for the construction of a time-averaged neuronal ensemble; whereas the scale of intracortical connectivity or columniation can define a characteristic scale for space-averaged neuronal ensembles. Usually the averaging is performed over either time or space, but not both, leaving the respective other dimension at (theoretically) infinite resolution.

In the case that averaged neuronal ensembles are considered as localized, separable populations, the resulting formulations are often referred to as *"neural mass (action) models"*. These neural masses can be connected in a one-to-one fashion in order to represent their causal influence on each other through synaptic connectivity. However, neurons in cortex communicate with a very dense collection of short and long range fibres; hence it is often advantageous to envisage the activity of neuronal ensembles as conditioning entire regions of cortical space to a degree varying with connectivity. The mean activity of a particular neuronal ensemble then defines a distributed causal influence, a field, that is propagated and dispersed in a manner representing the dense synaptic connectivity. All other neuronal ensembles that are connected to it region are then subject to "forcings" from this field. The resulting models, which are continuous in space and time, are therefore referred to as *"neural/mean field models"*.

11.2.4 Microscopic Constraints on Mean Field Models

The strength of the mean field approach is also its weakness. Mean fields make the complexity of the cortex tractable, but do so at the expense of subsuming the effects of fluctuations and correlations in single neuronal activity that are known to affect emergent mesoscopic and macroscopic neuronal population dynamics (Wolfe et al. 2010). For instance, in a synergetic perspective (Haken 1983), the effects of upwards (microscopic → mesoscopic → macroscopic) and downwards (macroscopic → mesoscopic → microscopic) causation, and the feedback between the two (circular causality), are believed to be crucial in accurately understanding the dynamics of multiscale neural architectures. This cannot be included into mean field models without making a range of additional assumptions that have, at present, only weak physiological support. Nevertheless, mean field models do provide a convenient starting point through the study of first and possibly second order moments (Faugeras et al. 2009; Buice et al. 2010).

11.3 History of Mean Field Innovations

The earliest models of neural mass action applied to the cortex that attempted to describe the spatial and temporal behaviour of these aggregate masses dealt mainly, if not exclusively, with excitatory interactions. Later models incorporated inhibitory interactions, paying more attention to the anatomical topology of connections between the neural masses, and took into account the conversion of efferent axonal activity into afferent dendritic activity (and the converse process), dendritic integration, axonal dispersion and synaptic delays. In what follows we will look at these developments in their historic context.

11.3.1 Foundations: Beurle, Griffith, Wilson and Cowan, Amari, Freeman

Perhaps the first approach to developing a mean field continuum theory of neural activity is that of Beurle (1956). In this theory, continuously distributed populations of excitatory neurons having a fixed firing threshold were considered in which the strength of interaction between individual neurons falls off exponentially with distance. By focusing on the fraction of excitatory neurons becoming active per unit time Beurle was able to show that this spatially continuous neural mass could produce propagating macroscopic waves of activity. While introducing a formalism that would later prove to be of great utility, the omission of inhibitory interactions meant its behaviour would be of no subsequent physiological significance. The later theory of Griffith (1963, 1965) suffered from the same problem, though he briefly discusses incorporating the influence of inhibition. However, his work is most notable for providing the first comprehensive derivation of a model for the spatiotemporal spreading of activation by using an equivalent partial differential equation (PDE). It was only through the later introduction by Jirsa and Haken (1996) of essentially the same idea that this approach became commonplace. We will discuss this in more detail below. The lack of an inhibitory component was subsequently rectified by the efforts of Wilson and Cowan (1972, 1973), Amari (1975, 1977) and Freeman (1975), who explicitly modelled inhibitory interactions.

Wilson and Cowan (1972, 1973) modelled cortical (and thalamic) neural tissue as comprised of two interacting, but functionally distinct, excitatory and inhibitory neuronal populations. The state of their bulk neuronal population model neural tissue was defined in terms of the time-averaged fraction of excitatory, $E(t)$, and inhibitory, $I(t)$, neurons firing per unit time, following the work of Beurle (1956). For point neural masses they were able to derive the following equations of motion

$$\tau_E \frac{dE}{dt} = -E + (1 - r_E E) S_E[c_{EE} E(t) - c_{IE} I(t) + P(t)] \tag{11.1}$$

$$\tau_I \frac{dI}{dt} = -I + (1 - r_I I) S_I[c_{EI} E(t) - c_{II} I(t) + Q(t)] \tag{11.2}$$

where $\tau_{E,I}$ are nominally the membrane time constants of the respective neural populations and determine their characteristic response times to incoming activity. The corresponding absolute refractory periods are denoted by $r_{E,I}$. The connectivity coefficients $c_{EE,IE,EI,II} \geq 0$ quantify the interactions, whereas the functions $S_{E,I}$ describe the relationship between neuronal population input and output in the absence of refractory effects. Because firing rates are bounded below by a zero and above by some physiological maximum, the $S_{E,I}$ are typically chosen as sigmoidal functions of their arguments e.g. $S_E \equiv 1/(1 + \exp[-a(E - \theta_E)]$. $P(t)$ and $Q(t)$ define the external input to the excitatory and inhibitory sub-populations.

While no analytical solutions exist for these equations they can, like most two-dimensional nonlinear systems, be analysed qualitatively in the phase plane. A considerable body of work has been devoted to such analyses of Eqs. 11.1 and 11.2, determining the number, type and properties of the equilibria of the system, bifurcations, and the behaviour of multiply coupled Wilson-Cowan type systems. For an in depth review of these results and related modelling approaches see Ermentrout (1998). The work of Wilson and Cowan (1972, 1973) introduced a number of conceptual innovations that virtually all subsequent mean field formulations have retained: the sigmoidal firing rate function and the cortical mesocircuit defined by all possible feedforward and feedback connections between spatially circumscribed populations of excitatory and inhibitory neurons. In the Wilson and Cowan model, equations of motion for time-averaged neuronal firing rates were derived. This and related models are therefore referred to as *activity based* models. However there also exists an alternative way of formulating mean field, or continuum, models, referred to as *voltage based* models. These are arguably more pertinent to modelling, and thereby understanding, the genesis of EEG dynamics. In this modelling approach the resulting equations of motion instead describe the spatiotemporal evolution of the average membrane potential of neurons.

One of the biomathematically most influential *voltage based* continuum models of cortical dynamics is that of Amari (1975, 1977). In its most general form, this model considers m distinct spatially distributed neuronal populations, in which the average membrane potential impulse response to incoming (axonal) input from other neuronal populations is $\exp[-t/\tau]$. The resulting field equations can then be written as

$$\tau_i \frac{\partial u_i}{\partial t} = s_i(\mathbf{x}, t) - u_i(\mathbf{x}, t) + \sum_{j=1}^{m} \int dt' \int d\mathbf{x}' \, w_{ij}(\mathbf{x}, \mathbf{x}'; t - t') \, f_j\left[u_j(\mathbf{x}', t')\right] \tag{11.3}$$

where $u_i(\mathbf{x}, t)$ is the average membrane potential of neurons of type i at time t and position $\mathbf{x}$, s_i represents extracortical input and f_i is a nonlinear function that describes the average firing rate (pulse emission rate) as a function of u_i. The functions $w_{ij}(\mathbf{x}, \mathbf{x}'; t - t')$ define the strength of connectivity between neuronal populations, i.e., they determine the input to neurons of type i at $\mathbf{x}$ from the pulse emission rate of neurons of type j at $\mathbf{x}'$, incorporating the effects of conduction and synaptic delays $t - t'$. As will be discussed below, the delay dependence is often factored out or simply ignored. The resulting function $w_{ij}(\mathbf{x}, \mathbf{x}')$ is then referred to as the *synaptic footprint*. A further common simplification is to consider the synaptic footprint as function of only the distance $r = |\mathbf{x} - \mathbf{x}'|$. The function $w(r)$

is then often defined to be excitatory $w(r) > 0$ (inhibitory $w(r) < 0$) for some defined neighbourhood $r < r_0$ and inhibitory $w(r) < 0$ (excitatory $w(r) > 0$) for more distant neurons $r > r_0$; this pattern of connectivity is typically referred to as *(inverted) Mexican hat connectivity*. For a detailed review of Amari-type models and their dynamics the reader is encouraged to consult Coombes (2005).

In contrast to the previous mathematical and constructive approaches, Freeman (1975) developed a schematic but empirically constrained mass action framework in order to explain the electrocortical dynamics of the mammalian olfactory bulb and pre-pyriform cortex. He developed a hierarchy of neural interaction – the well-known K-set hierarchy – in which functionally differentiated populations of neurons interact over progressively larger physical scales. The purpose of this hierarchy was to facilitate a more systems-oriented description. The simplest form of neural set that Freeman considered was the non-interactive or KO set. Members of this set have a common source of input and a common sign of output (excitatory or inhibitory), but do *not* interact synaptically or by any other means with co-members. At this level the characteristic form of the neuronal population response to incoming activity is specified. Unlike the first-order response of Wilson and Cowan, Eqs. 11.1 and 11.2, or Amari, Eq. 11.3, Freeman argued on the basis of detailed experiment that these population responses (or in his terminology *"pulse-to-wave"* conversion) could be described by third-order linear, time invariant, differential equations.

The K-set hierarchy was next extended to a non-zero level of functional interaction between members of the set. This defines the *KI* sets, broadly divided into mutually excitatory KI_e and inhibitory KI_i types. When there exists dense functional interaction between two *KI* sets, a *KII* set is formed. All possible interactions are in principle allowed to occur between the *KI* members of a *KII* set, which can be viewed as some local part of neocortex. The K-set hierarchy extends similarly to *KIII* and *KIV* sets, which nominally correspond to cortical areas and regions. The *KII* set of Freeman is equivalent physiologically and anatomically to the topology of cortex considered by Wilson and Cowan. Mathematically, the *KII* set is defined by four nonlinearly coupled sets of third order differential equations. The nonlinear couplings define how the induced population response to incoming synaptic activity is transduced into a neural population firing rate output. Freeman (1979) referred to the corresponding nonlinear function as the *"wave-to-pulse"* conversion function and has argued that such a function is an asymmetric sigmoid of the form $f(v) \propto \exp[-a \exp(-bv)]$.

11.3.2 Synaptic Dynamics: Lopes da Silva, Jansen and Rit, Wendling

While early models were successful in elaborating a cortical mesocircuit suitable for mean-field and mass action modelling, with the exception of Freeman (1975) they unrealistically assumed that the effects of synaptic activity are felt instantaneously at the neuronal soma. However, experiment suggests that the response of the neuronal membrane potential (and by inference the average membrane potential of a neuronal population) to incoming pre-synaptic spikes is at the very least second order: membrane potential rises to a peak and then decays away with characteristic time courses (Kandel et al. 2000). These "impulse" responses are referred respectively to as excitatory (EPSP) or inhibitory postsynaptic potentials (IPSP), depending on whether the spike arose from an excitatory or inhibitory neuron. Freeman (1975) calls this transduction of pre-synaptic activity into post-synaptic (soma) membrane variation *"pulse-to-wave"*. Such PSP delays are thought, on both empirical and theoretical grounds, to be important in defining the characteristic time scales of a range of electrocortical oscillatory phenomena that include alpha (8–13 Hz) and gamma (>30 Hz), and their modulation by, e.g., anaesthetic agents.

Probably the first to explicitly include PSPs in a mean field model were Lopes da Silva et al. (1974), cf. van Rotterdam et al. (1982), who constructed a bulk model of the EEG in which lumped or spatially distributed populations of excitatory and inhibitory neurons synaptically interacted via EPSPs and IPSPs having the form $\mathrm{PSP}(t) = V_{\mathrm{PSP}}\, t \exp[-t/\tau_{\mathrm{PSP}}]$, and where the mean neuronal population firing rate was a nonlinear (sigmoidal) function of the average membrane potential. The inclusion of such lumped postsynaptic dynamics was found sufficient to produce oscillatory activity in the alpha (8–13 Hz) electroencephalographic band. Jansen and Rit (1995), in a comprehensive extension of this model, investigated systematic variations of the model PSP parameters in order to account for observed changes in the visual evoked potential. Others have sought to better define the shape of the PSP in terms of a bi-exponential $\mathrm{PSP}(t) \propto \exp[-t/\tau_1] - \exp[-t/\tau_2]$ with $\tau_1 > \tau_2$, see for example Robinson et al. (2001) and Bojak and Liley (2005), or included IPSPs with different time scales in order to incorporate the effects of fast ($\mathrm{GABA_A}$) and slow ($\mathrm{GABA_B}$) inhibitory neurotransmitter kinetics (Wendling et al. 2005).

11.3.3 Activity Propagation: Nunez, Wright and Liley, Jirsa and Haken, Bojak and Liley

Consider a signal $S_j(\mathbf{x}', t')$, for example a brief "pulse" of excitatory ($j = e$) activity $S_e(\mathbf{x}', t') = \delta^{(2)}(\mathbf{x}' - \mathbf{x}_0)\delta(t' - t_0)$ which is generated at $t' = t_0$ and $\mathbf{x}' = \mathbf{x}_0$. How does this signal, and others generated in the brain, relate to the input received by a neural population k at position $\mathbf{x}$ and time t? A general expression is

$$\phi_{jk}(\mathbf{x}, t) = \int dt' \int d\mathbf{x}' \mathscr{G}_{jk}\left(\mathbf{x}, \mathbf{x}', t, t'\right) S_j\left(\mathbf{x}', t'\right), \tag{11.4}$$

i.e., an integration of signals from all times t' and places $\mathbf{x}'$ with a function $\mathscr{G}_{jk}$ weighting how much these signals contribute to the input. For a discretized model the integrals would be replaced by sums. The impact of this pulse on an inhibitory population $k = i$ is $\phi_{ei}(\mathbf{x}, t) = \mathscr{G}_{ei}(\mathbf{x}, \mathbf{x}_0, t, t_0)$, i.e., the response to the pulse is given by the corresponding $\mathscr{G}$ value. Such $\mathscr{G}$ functions are called "Green's functions".

Since the brain is a finite physical object, $\mathcal{G}_{jk}(\mathbf{x}, \mathbf{x}', t, t') \equiv 0$ for $t' \geq t$ and $\mathbf{x}' \notin C$, i.e., future brain activity does not influence earlier one and only connected sources contribute. One also often assumes continuous time invariance: $\mathcal{G}_{jk}(\mathbf{x}, \mathbf{x}', t, t') = \mathcal{G}_{jk}(\mathbf{x}, \mathbf{x}', t - t')$. This means activity propagation depends on conduction *delays* only. A typical model of this kind contains three factors

$$\mathcal{G}_{jk}\left(\mathbf{x}', \mathbf{x}, t - t'\right) = \int_0^\infty dv\, f_{jk}\left(v \,|\, \mathbf{x}, \mathbf{x}'\right) w_{jk}\left(\mathbf{x}, \mathbf{x}'\right) \delta\left(t - t' - \frac{|\mathbf{x} - \mathbf{x}'|}{v}\right). \tag{11.5}$$

The first factor gives the distribution of conduction velocities v of connecting fibres with $\int_0^\infty dv\, f_{jk}(v \,|\, \mathbf{x}, \mathbf{x}') = 1$. The second factor is the synaptic footprint, which models the strength and distribution of the these connections. The last factor calculates the delay by dividing (Euclidean) distance by the conduction velocity.[2] Only the synaptic footprint $w_{jk}(\mathbf{x}, \mathbf{x}')$ would be modified by synaptic plasticity or neuromodulation, whereas the other two factors express the arrangement and properties of the fibre tracts. If their speed of change is slow compared to propagation, then one can use a delay form and simply change the parameters of the synaptic footprint with time as needed.

Further simplifications come at the price of less biological fidelity. Continuous translation invariance $\mathcal{G}_{jk}(\mathbf{x}, \mathbf{x}', t - t') = \mathcal{G}_{jk}(\mathbf{x} - \mathbf{x}', t - t')$ implies homogeneity of the cortex, i.e., signal transmission then depends only on the *vector distance* between points, not on their location. Clearly this assumption does not hold true for specific connectivity between particular brain areas, yet it can be a reasonable approximation for the dense local "background" connectivity one finds all over the brain. One can further impose continuous rotation invariance: $\mathcal{G}_{jk}(\mathbf{x} - \mathbf{x}', t - t') = \mathcal{G}_{jk}(|\mathbf{x} - \mathbf{x}'|, t - t')$. This establishes isotropy, i.e., independence of fibre direction. Even for background connectivity this does not hold true everywhere in the brain, e.g., primary visual cortex can be modelled by homogeneous but anisotropic connectivity (Robinson 2006; Coombes et al. 2007; Bojak and Liley 2010). Models that are both homogeneous and isotropic are limited to describing qualitative features of brain activity, e.g., the existence of "brain waves" (Robinson et al. 1997; Wu et al. 2008) or drug effects on power spectra (Bojak and Liley 2005).

The "global" theory of Nunez (1974a,b, 1981, 1995), reviewed by Nunez and Srinivasan (2006), ignores the local neural dynamics and focuses on Eq. 11.4:

$$S_e(\mathbf{x}, t) = p(\mathbf{x}, t) + \sum_{j=e,i} q_j \phi_{je}(\mathbf{x}, t), \quad S_i(\mathbf{x}, t) = \sum_{j=e,i} q_j \phi_{ji}(\mathbf{x}, t), \tag{11.6}$$

$$\mathcal{G}_{je}\left(|\mathbf{x} - \mathbf{x}'|, t - t'\right) = \sum_{n=1}^{N} \int_0^\infty dv\, f_n(v) w_n\left(|\mathbf{x} - \mathbf{x}'|\right) \delta\left(t - t' - \frac{|\mathbf{x} - \mathbf{x}'|}{v}\right), \tag{11.7}$$

[2]For realistic cortical geometries Euclidean distance is not a good approximation to axonal fibre length. However, one can adjust the $f_{jk}(v \,|\, \mathbf{x}, \mathbf{x}')$ to compensate for geometry.

$$\mathcal{G}_{ji}\left(\left|\mathbf{x}-\mathbf{x}'\right|, t-t'\right) = w_i\left(\left|\mathbf{x}-\mathbf{x}'\right|\right)\delta\left(t-t'\right), \quad w_l(r) \equiv \frac{1}{2}\rho_l \lambda_l e^{-\lambda_l r} \tag{11.8}$$

We see that excitatory and inhibitory ($j, k = e, i$) neuronal populations are being considered. Their output S_k is directly determined by the inputs ϕ_{jk}, where excitatory contributions are weighted positively $q_e > 0$ and inhibitory ones negatively $q_i < 0$. In addition, the excitatory population receives excitatory extracortical innovations p as independent input. Propagation is in delay form, as well as homogeneous and isotropic. Inhibitory connectivity is taken as instantaneous due to a very short characteristic length $1/\lambda_i \simeq 30\ \mu$m. Excitatory connectivity consists of N distinct long range fibre systems.

For a (convoluted) strip of cortex of length $L \simeq 0.5$–1 m, functionally closed by excitatory fibre connections with a single conduction velocity $v_e \simeq 6$–9 m/s, one can estimate that the lowest "standing wave" mode oscillates at frequencies of $f_1 = v_e/L \simeq 6$–18 Hz consistent with awake EEG (Nunez 1995). An interesting consequence is that larger cortices (larger L) are predicted to oscillate at lower frequencies. It was hence suggested that people with larger heads have lower alpha rhythms (Nunez 1974b; Nunez et al. 1978). An experimental study by Valdés-Hernández et al. (2010) has shown recently that the size of the cortical surface does not correlate with the observed frequency of the alpha rhythm. However, Nunez' prediction can be rescued simply by assuming that axonal conduction velocity grows in tune with cortical size $v_e \sim L$. This prediction could be tested experimentally, and raises interesting questions about brain development.

In the following we will consider activity propagation with equivalent PDEs. A Fourier transform of Eq. 11.4 for a homogeneous delay form $\mathcal{G}_{jk}\left(\mathbf{x}-\mathbf{x}', t-t'\right)$ has convolution structure,[3] hence

$$\begin{aligned} \phi_{jk}(\mathbf{k},\omega) &= \int dt \int d\mathbf{x}\, e^{-i(\omega t + \mathbf{k}\cdot\mathbf{x})}\phi_{jk}(\mathbf{x},t) = \mathcal{G}_{jk}(\mathbf{k},\omega)S_j(\mathbf{k},\omega) \\ &\equiv \frac{Z(\mathbf{k},\omega)}{P(\mathbf{k},\omega)}S_j(\mathbf{k},\omega), \end{aligned} \tag{11.9}$$

For non-zero $P(\mathbf{k},\omega)$ we can then write

$$\begin{aligned} P\left(-i\nabla, -i\frac{\partial}{\partial t}\right) e^{i\omega t}e^{i\mathbf{k}\cdot\mathbf{x}}\phi_{jk}(\mathbf{k},\omega) &= Z\left(-i\nabla, -i\frac{\partial}{\partial t}\right) e^{i\omega t}e^{i\mathbf{k}\cdot\mathbf{x}}S_j(\mathbf{k},\omega) \\ \Rightarrow \quad P\left(-i\nabla, -i\frac{\partial}{\partial t}\right)\phi_{jk}(\mathbf{x},t) &= Z\left(-i\nabla, -i\frac{\partial}{\partial t}\right)S_j(\mathbf{x},t), \end{aligned} \tag{11.10}$$

where we have integrated over ω and $\mathbf{k}$ to perform the inverse Fourier transform. Equation 11.10 provides a "mathematically equivalent" PDE formulation wherein the structure of the differential operators P and Z reflects the chosen $\mathcal{G}$. Why is this

[3]For $\mathcal{G}_{jk}$ to be homogeneous, the connected region must be a closed (hyper)surface, e.g., a sphere.

rewriting helpful? The connected region C may well encompass the entire brain, and conduction delays can extend to several tens of milliseconds. This makes Eq. 11.4 difficult to evaluate, whereas the PDE evaluation is non-local only in proportion to the order of its differential operators. Using equivalent PDEs can hence greatly simplify analysis and speed up numerical computations.

As mentioned above, Griffith (1963, 1965) was the first to derive the commonly used kind of equivalent PDE, which we will briefly discuss below. However, the PDE approach became popular only through its reintroduction by Jirsa and Haken (1996), and was then investigated further by Robinson et al. (1997) and Liley et al. (2002, 2003). Consider the following

$$\begin{aligned}
\hat{\mathscr{G}}(r,\tau) = \frac{w_0}{2\pi\hat{\sigma}^2} e^{-r/\hat{\sigma}} \cdot \delta\left(\tau - \frac{r}{\hat{v}}\right) \quad &\overset{\mathscr{F}}{\Longrightarrow} \quad \hat{\mathscr{G}}(k,\omega) = \frac{w_0\left(\frac{\hat{\sigma}}{\hat{v}} i\omega + 1\right)}{\left[\left(\frac{\hat{\sigma}}{\hat{v}} i\omega + 1\right)^2 + \hat{\sigma}^2 k^2\right]^{3/2}} \\
\tilde{v} \equiv \sqrt{\frac{3}{2}}\hat{v}\,,\ \tilde{\sigma} \equiv \sqrt{\frac{3}{2}}\hat{\sigma} \qquad\qquad &\Downarrow\ k^2 \ll \frac{1}{\hat{\sigma}^2} + \frac{\omega^2}{\hat{v}^2} \\
\tilde{\mathscr{G}}(r,\tau) = \frac{w_0}{2\pi\tilde{\sigma}^2} e^{-\tilde{v}\tau/\tilde{\sigma}} \frac{\Theta\left(\tau - \frac{r}{\tilde{v}}\right)}{\sqrt{\tau^2 - \frac{r^2}{\tilde{v}^2}}} \quad &\overset{\mathscr{F}^{-1}}{\Longleftarrow} \quad \tilde{\mathscr{G}}(k,\omega) = \frac{w_0}{\left[\left(\frac{\tilde{\sigma}}{\tilde{v}} i\omega + 1\right)^2 + \tilde{\sigma}^2 k^2\right]}
\end{aligned} \tag{11.11}$$

where $r \equiv |\mathbf{x} - \mathbf{x}'|$, $\tau \equiv t - t'$, $k \equiv \sqrt{\mathbf{k}\cdot\mathbf{k}}$ and subscripts are left out for notational simplicity.

Firstly, the homogeneous and isotropic delay ansatz $\hat{\mathscr{G}}(r,\tau)$ propagates activity with a single velocity $\hat{v}$, and has an exponential decay with characteristic length $\hat{\sigma}$ as the synaptic footprint. Secondly, the Fourier transform $\hat{\mathscr{G}}(k,\omega)$ of this ansatz includes a fractional power $3/2$ that would translate into an infinite series of differential operators, negating any practical advantage of the PDE formulation. Hence thirdly, an expansion $\tilde{\mathscr{G}}(k,\omega)$ for small wavenumbers (i.e., long wavelengths $\lambda = 2\pi/k$) is performed, leading to the following equivalent PDE:

$$\left[\left(\frac{\tilde{\sigma}}{\tilde{v}}\frac{\partial}{\partial t} + 1\right)^2 - \tilde{\sigma}^2\nabla^2\right]\phi(\mathbf{x},t) = w_0 S(\mathbf{x},t)\,. \tag{11.12}$$

This "long-wavelength propagator" is an inhomogeneous telegraph equation well suited for analysis and numerics.

However, consider a gamma rhythm $\omega = 2\pi \cdot 38.2$ Hz $\simeq 240.0/\mathrm{s}$ with $\hat{v} = 600$ cm/s and $\hat{\sigma} = 3.33$ cm. The "long wavelength expansion" only holds for $k \ll 0.5/\mathrm{cm}$ or $\lambda \gg 13$ cm. Even taking cortical folding into account, coherent gamma activity at such scales seems unlikely. It is hence better to consider $\tilde{\mathscr{G}}(k,\omega)$ not as an expansion, but as a new ansatz in its own right, merely "inspired" by the original $\hat{\mathscr{G}}(r,\tau)$. Then fourthly, we can compute its $\tilde{\mathscr{G}}(r,\tau)$, which is easier to interpret in the form of Eq. 11.5:

$$\tilde{w}(r) = \frac{w^0}{2\pi\tilde{\sigma}^2}\mathrm{K}_0\left(\frac{r}{\tilde{\sigma}}\right), \quad \tilde{f}\,(v\,|\,r) = \frac{\tilde{v}}{v}\frac{\exp\left(-r/\tilde{\sigma}\cdot\tilde{v}/v\right)}{\mathrm{K}_0\left(r/\tilde{\sigma}\right)}\frac{\Theta\left(\tilde{v}-v\right)}{\sqrt{\tilde{v}^2-v^2}}, \tag{11.13}$$

with the Heaviside step function Θ. The distance-dependent velocity distribution $\tilde{f}\,(v\,|\,r)$ has a complicated distance dependence. However, we can see that the infinitely sharp $\hat{f}\,(v\,|\,r) = \delta\left(\tilde{v}-v\right)$ has been softened into a divergence $\sim 1/\sqrt{\tilde{v}-v}$ towards lower velocities $v < \tilde{v}$, whereas no $v > \tilde{v}$ are allowed. Thus most activity will arrive after a delay $\tau = r/\tilde{v}$, but some will arrive more slowly. The synaptic footprint has become a modified Bessel function. Since $\mathrm{K}_0(x) \sim e^{-x}/x$ for large x, this implies a more rapid decay of connectivity with distance.

Most spatially extended simulations use some variant of Eq. 11.12 for activity propagation, because the original ansatz $\hat{\mathcal{G}}(r,\tau)$ in Eq. 11.11 is intuitive and the equivalent PDE Eq. 11.12 is easy to use. Yet we have argued here that their connection is questionable due to the necessary long wavelength expansion. Furthermore, Bojak and Liley (2010) calculated the resulting *marginal* velocity distribution

$$\tilde{f}\,(v) = \int_0^\infty dr\,2\pi r\frac{\tilde{w}(r)\tilde{f}\,(v\,|\,r)}{w_0} = \frac{v}{\tilde{v}}\frac{\Theta\left(\tilde{v}-v\right)}{\sqrt{\tilde{v}^2-v^2}}, \tag{11.14}$$

and showed that it is severely incompatible with experimental data on axonal diameter distributions in both rat and human. They proposed new PDEs compatible with the data, in particular the so-called "dispersive propagator" of power $n > 0$:

$$\mathcal{G}_n(r,\tau) = \frac{w_0}{2\pi\sigma_n^2}e^{-r/\sigma_n}\cdot\frac{v_n}{4\sigma_n\Gamma(n)}\left(\frac{v_n\tau}{2\sigma_n}\right)^{n-2}e^{-\frac{1}{2}\frac{(\tau-r/v_n)^2}{\tau\sigma_n/v_n}}\Theta(t), \tag{11.15}$$

$$w_n(r) = \frac{w_0\left(r/\sigma_n\right)^{n-1}\mathrm{K}_{n-1}\left(r/\sigma_n\right)}{2^n\pi\Gamma(n)\sigma_n^2}, \quad f_n\,(v\,|\,r) = \frac{e^{-r/\sigma_n\cdot\left(v_n^2+v^2\right)/(2v_nv)}}{2v\left(v/v_n\right)^{n-1}\mathrm{K}_{n-1}\left(r/\sigma_n\right)}, \tag{11.16}$$

$$\left(\frac{2\sigma_n}{v_n}\frac{\partial}{\partial t}+1-\sigma_n^2\nabla^2\right)^n\phi\left(\mathbf{x},t\right) = w_0S\left(\mathbf{x},t\right), \quad f_n(v) = \frac{2nvv_n^{2n}}{\left(v_n^2+v^2\right)^{n+1}} \tag{11.17}$$

with the Gamma function Γ. Comparing $\mathcal{G}_n(r,\tau)$ of Eq. 11.15 with $\hat{\mathcal{G}}(r,\tau)$ in Eq. 11.11, we see that the first factor remains unchanged. The second factor of $\mathcal{G}_n(r,\tau)$ turns into a normal distribution of delays $n_\tau(\bar{\tau},\sigma_\tau^2)$ for $n = 1.5$, with mean $\bar{\tau} = r/v_n$ but *delay-dependent* standard deviation $\sigma_\tau = \sqrt{\tau\sigma_{1.5}/v_{1.5}}$. For longer delays, hence larger distances, the distribution of delays becomes broader. At other n, in particular integer ones providing convenient PDEs, this remains the case qualitatively.

We can furthermore see that for $n = 1$ and $\sigma_n = \tilde{\sigma}$, the synaptic footprints $\tilde{w}$ of Eq. 11.13 and w_n of Eq. 11.16 agree. Furthermore, in that case the equivalent PDEs Eqs. 11.12 and 11.17 agree but for the acceleration term, if $v_n = \tilde{v}$ as well. However, from fits to myelinated fibre data from human corpus callosum, Bojak and Liley (2010) rather suggest $n = 3$ with $v_3 = 14.91$ m/s, and as best comparable

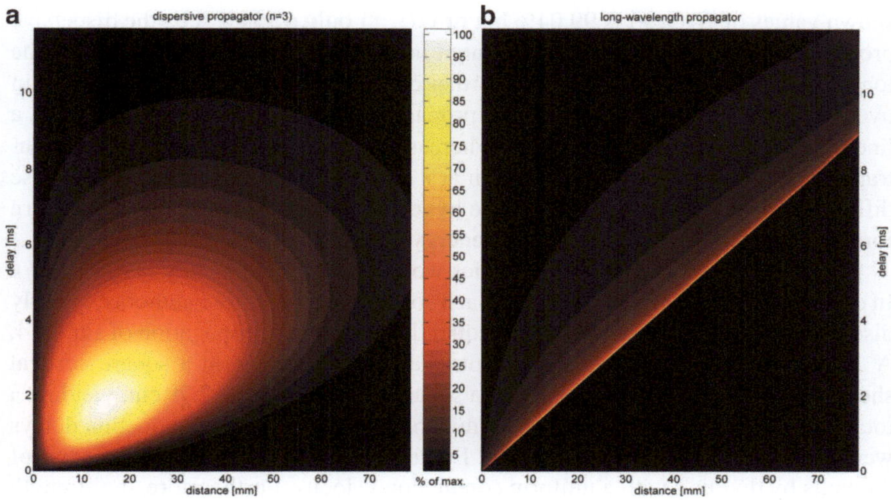

Fig. 11.3 Pulse spreading with the dispersive and long-wavelength propagators. (**a**) The activation delivery of the dispersive $n = 3$ propagator Eq. 11.17 is shown by integrating its Green's function, cf. Eq. 11.18. (**b**) Likewise for the long-wavelength propagator Eq. 11.12. Synaptic footprints of both propagators approximate an exponential decay with characteristic length 18 mm. The median conduction velocity of both propagators is 7.601 mm/ms. The colour scale is normed to the respective maximum values as indicated by the central colourbar

values for the long-wavelength propagator $\tilde{v} = 8.782\,\text{m/s}$ and $\tilde{\sigma} = 4.930\sigma_3$. The former leads to the same median velocity 7.601 m/s for both propagators. The latter means that the synaptic footprint of both approximates the same exponential decay, e.g., $\sigma_3 = 0.871\,\text{cm}$ and $\tilde{\sigma} = 4.29\,\text{cm}$ both approximate an exponential decay with a characteristic length of 1.8 cm. Such an effective length scale can be motivated functionally by noting that the coherence of subdural electrode recordings falls to 0.25 at about 2–3 cm (Bullock et al. 1995). Note also the significant difference between the long-wavelength $\tilde{\sigma}$ and the exponential scale, which had been ignored in the literature prior to Bojak and Liley (2010).

In Fig. 11.3 on the left we show the function

$$G_3(i,j) = \int_0^{\Delta\tau} dt \int_0^{\Delta r} dx \, \frac{1}{w_0} \mathscr{G}_3(i \cdot \Delta r + x, j \cdot \Delta\tau + t) \tag{11.18}$$

with $\mathscr{G}_3$ of Eq. 11.15 and $i, j = 0, 1, 2, \ldots$ specify the discretized values. Note that $\sum_{i=0}^{\infty} \sum_{j=0}^{\infty} G_3(i,j) = 1$, i.e., this shows the spatiotemporal delivery of activity from a single pulse in a properly normed fashion. For comparison we show on the right of Fig. 11.3 the function $\tilde{G}(i,j)$ defined in a like manner using the $\tilde{\mathscr{G}}$ of Eq. 11.11. This method integrates out the discontinuity of $\tilde{\mathscr{G}}$ at $r = v\tau$ and hence facilitates a meaningful comparison of the two propagators. We have used the values mentioned above for comparable median velocity 7.601 mm/ms and characteristic length 18 mm, as well as $\Delta r = 157.1\,\mu\text{m}$ and $\Delta\tau = 23.41\,\mu\text{s}$. The sum of the shown values of $G_3(i, j)$ is 99.04% but of $\tilde{G}(i, j)$ only 62.51%, i.e., the dispersive propagator has delivered most of the input in the shown spatiotemporal range, the long-wavelength propagator extends further out. We also see the clear difference in overall structure: the long-wavelength propagator is highly concentrated around a line of constant conduction velocity, whereas the dispersive propagator is concentrated in a "blob" around the maximum at $r = 11.6\,\text{mm}$ and $\tau = 2.22\,\text{ms}$. These different characteristics for example are hypothesised to allow spontaneous pattern formation with the expenditure of less energy (Bojak and Liley 2010).

Finally we briefly consider the earlier work of Wright and Liley (1995, 1996). In contrast to the equivalent PDE approach so far discussed, these authors explicitly discretized the cortical sheet and consequently the activity propagation of Eq. 11.4. A 20 by 20 matrix of neural mass units was used to represent a square cortical sheet, where every unit corresponds to a square with side length 2.79 cm, yielding a total area equivalent to roughly one human hemisphere. Axonal conduction delays were then simply calculated from the Euclidian distance between the centers of the units by dividing with a uniform conduction velocity. Furthermore, the strength of connectivity was determined by a normal distribution with this distance. These assumptions specify two matrices (the strength of connectivity between any two units, as well as their assumed conduction delay), but could easily be replaced with other matrices implying inhomogeneity and anisotropy of the connectivity and complicated conduction velocity profiles with positional dependence.

While this approach is very flexible, it suffers from two fundamental drawbacks: First, the number of possible connections grows with the square of the number of units. Hence increasing the spatial resolution typically comes at a significant computational costs. Second, one needs to keep track of past output from every unit as far back as the maximum conduction delay. If the conduction delays are sizable, then a lot of past values must be kept in memory. It is hence not surprising that in 1995 the chosen grid size was fairly small. While equivalent PDEs numerics employs spatial grids as well, their computation is much less costly. To evaluate Eq. 11.12, minimally one needs to keep track of only two past values of ϕ at every grid point for the time derivatives and consider the four nearest neighbours of every grid point for the Laplacian. Nevertheless, this localized PDE computation instantiates large scale, dense connectivity. However, with ever increasing computational power the greater flexibility of the discretized approach is becoming more important, hence as discussed below this approach is making a comeback in cortical mesh computations.

11.3.4 Realistic Geometry and Connectivity: Robinson et al., Kötter et al., Sotero et al., Bojak et al.

Prior to Robinson et al. (2001), all mean field formulations of cortical activity had posited that any emergent dynamics arises through reverberant interactions between at least two spatially distributed, functionally differentiated, cortical or thalamic, neuronal populations. However, there exists significant reciprocal connectivity between cortex and the subcortical structure that determines and controls its input, the thalamus. Robinson et al. (2001) have argued that the inclusion of such cortico-thalamic feedback in a bulk or mean-field theory is crucial in order to plausibly model the essential dynamical features of normal (e.g., alpha rhythm) and abnormal (e.g., spike wave epilepsy) EEG activity (Rodrigues et al. 2006; Breakspear et al. 2006). van Albada and Robinson (2009) have extended the subcortical extent of this model by including interactions between the nuclei of the basal ganglia and cortex and thalamus. We will discuss the influence of subcortical structures on cortical activity further below.

The description of cortical connectivity itself is also far from optimal. While "background connectivity" in cortex (Liley and Wright 1994; Hellwig 2000; Kaiser et al. 2009) is roughly compatible with assumptions of homogeneity and isotropy, a functionally significant part of the connectivity in the brain is more specific (Biswal et al. 2010): it connects only particular brain regions to each other and then often over long distances (Hagmann et al. 2008). While it is possible to simulate anisotropy with PDE approaches (Robinson 2006; Coombes et al. 2007; Bojak and Liley 2010), this requires simple periodicity to limit computational expense, since the number of necessary PDEs grows with the complexity of the angular profile. Furthermore, the requirement of homogeneity cannot be relaxed, hence this method is only applicable where the anisotropic pattern of connectivity (roughly) repeats across a patch of cortex, like for example in primary visual cortex. The long-range specific connectivity in the brain however does not repeat in this manner.

In order to localize specific cortical connectivity appropriately, one needs anatomically accurate representations of cortex. A first crucial step is hence the extraction of cortical geometry. The imaging technique of choice is structural MRI, which can distinguish different brain tissues on a voxel basis. However, cortical grey matter is basically a layered sheet (2–5 mm thick with 0.2 m^2 total area), and often a representation as two-dimensional surface is more suitable. Several software packages are available for extracting triangular mesh surfaces from voxels identified with structural MRI, for example, the Civet pipeline (Kim et al. 2005), see Fig. 11.4a. Often these meshes need some post-processing, e.g., in order to reduce the number of vertices to limit computational cost, see Fig. 11.4b. It should be noted that proper modelling of EEG/MEG requires *oriented* local current dipoles as sources. Their orientation is ultimately dependent on dendritic arborization growing roughly perpendicular to the pial surface. Hence a pure voxel-based approach is insufficient for EEG/MEG signal prediction, and must anyhow be augmented with some estimate of surface normals.

Once one has obtained a cortical surface, the question becomes how to employ it for anatomical mapping. Two methods have been explored: The first is to deform the cortical surface into simpler geometries for computation, e.g., representing each brain hemisphere by a sphere (Jirsa et al. 2002). The two hemispheres are deformed separately to avoid strong distortions from inflating the compact callosal pathway. However, consider a source point on the simulation sphere, and two target

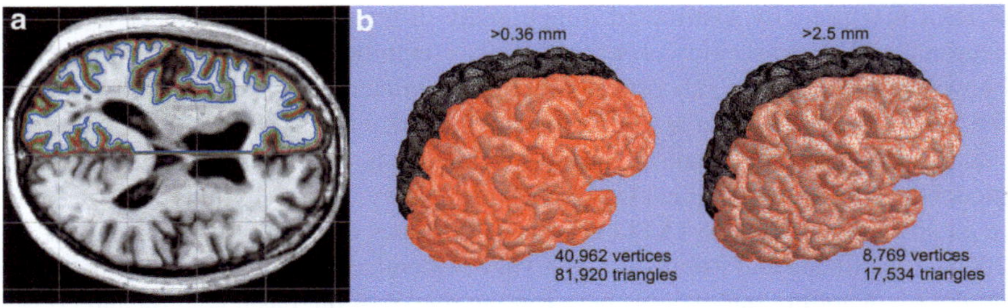

Fig. 11.4 Extraction of a cortical triangular mesh. (**a**) Cross-section of a Civet (Kim et al. 2005) extraction of grey matter interfaces with white matter (*blue*) and cerebrospinal fluid (*red*), respectively, in the *left* hemisphere. Layers can be introduced along vectors connecting corresponding vertices; here a halfway one (*green*) is shown. (**b**) Custom decimation algorithm (Bojak et al. 2011) working on the right hemisphere (*red*) of a Civet cortical mesh. The number of vertices is reduced fivefold while the minimum edge length is increased sevenfold, yet the surface loss is only about 10%

points equidistant along the spherical surface. Given equal conduction velocities, a signal from the source points will reach the target points at the same time. The deformation to the real cortical surface is non-conformal, thus in general the target points will be at different distances from the source point. However, the signal will still arrive at the same time in both, hence the deformation has implicitly made the conduction velocities differ. Such uncontrolled changes to the connectivity are unacceptable beyond qualitative studies. One could ameliorate the situation by introducing compensatory tuning of the conduction velocity. But then one would need to consider conduction between points individually, in effect turning the first method into the second.

The second method works directly on the discretized cortical surface. Then connectivity is instantiated by transferring activity signals from any vertex to each connected vertex, where conduction delays determine the time when the signal is actually released to the receiving vertices. Exactly the same method can be used for connecting grey matter voxels, if one does not extract a cortical surface first (Sotero et al. 2007). This second method is basically the anatomically realistic version of the method introduced by Wright and Liley (1995, 1996), which we have discussed previously. However, as mentioned this method scales badly if connectivity is not sparse. For example, Bojak et al. (2010, 2011) were forced to include about 30 million connections for 17,000 vertices just in order to approximate the dense local connectivity implicit in typical PDE approaches (Bojak and Liley 2010). Furthermore, unlike for networks of spiking neurons, for neural populations typically firing rate information and hence functions continuous in time need to be transferred. This means an event-based approach is impossible, and hence that the necessary parallelization on compute clusters is more difficult.

A series of works involving the late Rolf Kötter (Honey et al. 2007; Ghosh et al. 2008; Deco et al. 2009) established the idea of using tracer connectivity data from macaque monkey collected in the CoCoMac database (Stephan et al. 2001), by using a "regional map" of brain areas (Kötter and Wanke 2005) across species. See also the extensive review of Deco et al. (2011) focusing on these papers. This may appear odd, since diffusion MRI tractography is readily available (Mori et al. 2005) and there are some interspecies differences that clearly cannot be accounted for in such a mapping, for example concerning areas involved in speech production. However, diffusion MRI is incapable of determining in which direction information flows along reconstructed fibre tracts. This is a major drawback, since it generally makes a significant difference whether a connection is $A \rightarrow B$, $A \leftarrow B$ or $A \leftrightarrow B$. Furthermore, diffusion MRI tends to be biased to short distance connections. Tracer data does not have these drawbacks, which explains its continued popularity in spite of the uncertainties of cross-species mapping (Bojak et al. 2010, 2011).

While modelling is still rapidly improving concerning anatomical fidelity, the basic building blocks are now in place: The geometry of cortical and potentially subcortical structures – as well as skull and scalp for volume conduction for the EEG signal expression – are typically extracted from structural MRI. Gray matter activity is then predicted by assigning neural masses at the chosen resolution level to voxels or mesh vertices. The long-range connectivity of these masses is estimated from diffusion MRI tractography or tracer data. Finally, activity propagation is then performed directly in terms of this discretization, rather than by an equivalent PDE formalism. Nevertheless, the simplicity of the equivalent PDE approach means that we should expect it to remain superior for gaining qualitative understanding or modelling gross changes to the overall brain state, e.g., due to potent drugs.

11.3.5 Bayesian Inversion: Friston et al., David et al., Moran et al., Daunizeau et al.

A particular problem encountered in all modelling efforts, including the mean field approach, is to fit the parameters of the model to available experimental data. This is difficult enough in the case of a single model, if it contains a large number of parameters – as is invariably the case for attempts at "biological realism". The anatomical and physiological constraints on the parameters are then typically weak. Furthermore, as argued in more detail by Liley et al. (2011), projecting parameter space from an "ideal" model to the actual one under investigation generally results in a complicated distribution of the resulting dynamics in parameter space. This may foil straightforward attempts at fitting parameters, potentially requiring novel methods like the "metabifurcation analysis" of Frascoli et al. (2011) to make any further progress.

Hence it may seem hopeless then to fit entire networks of neural masses to experimental data. However, progress has been made in this direction thanks to the invention of Dynamic Causal Modelling (DCM) by Friston et al. (2003). This original work was intentionally abstract in its assumptions about local dynamics, positing a bilinear form that connected local neural dynamics and external stimuli. Furthermore, it was targeted at fMRI BOLD rather than EEG or MEG. Yet a crucial novelty was the ability to estimate both parameters of the local bilinear model and of the effective connectivity between several "regions of interest" (ROIs) based on data. To this end advanced Bayesian parameter and model estimation was used, cf. Friston (2002) and Friston et al. (2002, 2007). Briefly, these methods consist in making reasonable assumptions expressed via probability distributions for model parameters, the so-called "prior densities", followed by an estimate of how likely the available evidence (e.g., experimental data) is given these assumption, the so-called "likelihood". The probability distributions are then updated according to Bayes' rule to take the evidence into account, resulting in the so-called "posterior densities".

Soon after DCM was invented, David and Friston (2003) studied a modification of the mean field model of Jansen and Rit (1995) with two separated ROIs, and later extended it to model event-related responses in EEG and MEG (David et al. 2005). It was then only natural to combine these efforts with DCM (David et al. 2006; Kiebel et al. 2006). The fused model included state equations based on the mean field model and observer equations based on an electromagnetic forward model, in effect providing a neurobiologically constrained source reconstruction scheme for the EEG/MEG inverse solution problem. Using this approach, David et al. (2006) were for example able to demonstrate learning-related changes in connectivity for an auditory oddball paradigm. Next Moran et al. (2007) constructed a frequency-domain version of the Jansen and Rit (1995) model, in order to investigate steady-state spectral responses. Again Bayesian parameter estimation (Moran et al. 2008) and integration with DCM (Moran et al. 2009) followed. They used this DCM to analyze multi-channel local field potentials from mice. A final crucial step was then taken by Daunizeau et al. (2009): instead of considering sources as point processes, i.e., as equivalent current dipoles, they were distributed over the cortical surface by using a standing wave approximation of the long-wavelength equivalent PDE we have discussed above. Thus one can now say that the entire range of mean field descriptions has been given a DCM-style counterpart suitable for Bayesian inversion.

However, some limitations must be mentioned. Firstly, DCM-style approaches will remain limited to a handful of ROIs for the foreseeable future. The combinatorial explosion of possible connectivity and the per se difficult computation of Bayesian estimates mean that DCMs rapidly become unwieldy when more ROIs are introduced. Secondly, the dependence on prior densities for a specified model structure means that the posterior estimates should not be considered as the ground truth. Thirdly, it seems likely that the present Bayesian inversions gloss over the complexity of the underlying mean field model parameter space to some extent. For the Jansen and Rit (1995) model used throughout the DCM work discussed above, Spiegler et al. (2010) have demonstrated the expected complicated dependence of dynamics on parameter values in the physiological range (Liley et al. 2011). It hence remains to be seen how comprehensively Bayesian inversion can probe the full dynamical repertoire of mean field models. Nevertheless, the mentioned works represent a pioneering effort in matching mean field models to experimental data.

11.4 Understanding Mesoscopic Brain Activity

Much of what we know about human brain function is derived from non-invasive recording methods that are able to sensitively measure changes in electromagnetic and hemodynamic cortical activity that attend behaviour and cognition. The EEG and MEG measure the spontaneous and evoked electromagnetic activity of large populations of cortical neurons whereas fMRI, based on blood oxygen-level dependent (BOLD) contrast, and near infrared spectroscopy (NIRS) quantify local variations in cerebral blood flow in response to such population activity. Thus the mean field modelling approach, having roughly the same spatiotemporal scales as these recording techniques, is ideally placed to provide physiological explanations of a range of empirical phenomena, some examples of which we discuss below. We believe however that this is just the beginning, agreeing with Freeman (1975) that the field of neurodynamics, of which mean field models are an integral part, *"still can be regarded as a giant sleeping in infancy"*.

11.4.1 EEG Alpha Rhythm: Stochastic, Non-linear, or Both?

The alpha rhythm is arguably the most ubiquitous rhythm seen in scalp recorded EEG. First discovered by Berger (1929, 1930), see also Gloor (1969), and later confirmed by Adrian and Matthews (1934), it has played a central role in phenomenological descriptions of brain electrical activity in cognition and behaviour ever since. While the definition of classical alpha is restricted to 8–13 Hz oscillatory activity recorded over the occiput, which is reactive to eyes opening and closing, activity in the same frequency range can be recorded from multiple cortical areas. However, despite decades of detailed empirical research involving the relationship of this rhythm to cognition, we remain essentially ignorant regarding the mechanisms underlying its genesis and its relevance to brain information processing and function (Steriade 2005).

To date two broad approaches have emerged for explaining the origin of the alpha rhythm and similar activity. The first approach conceives of alpha as arising from cortical neurons being paced or driven at alpha frequencies: either through the intrinsic oscillatory properties of other cortical neurons (Llinás 1988; Silva et al. 1991), or through the oscillatory activity of a feed-forward subcortical structure such as the thalamus (Hughes and Crunelli 2005, 2007). In contrast, the second approach assumes that alpha emerges through the reverberant activity generated by reciprocal interactions of synaptically connected neuronal populations in cortex, and/or through such reciprocal interactions between cortex and thalamus (Nunez et al. 2001). Two principle lines of evidence have arisen in support of the latter view. Firstly, multichannel MEG (Williamson and Kaufman 1989; Ciulla et al. 1999) and high density EEG (Nunez et al. 2001) have revealed that scalp recorded alpha activity arises from a large number or continuum of equivalent current sources

(ECD) in cortex. Secondly, a raft of physiologically plausible computational (Liley et al. 1999a) and theoretical models (Nunez 1981; van Rotterdam et al. 1982; Robinson et al. 1997, 2001; Liley et al. 2002, 2003), reveal that electroencephalographically realistic oscillatory activity can arise from the synaptic interactions between distributed populations of excitatory and inhibitory neurons.

The work of Robinson et al. (2001) suggests that characteristic loop delays in cortico-thalamic and thalamocortical connectivity are responsible for the generation of alpha, and beta (13–30 Hz) electroencephalographic activity. In contrast the work of Liley et al. (2002, 2003) suggests that alpha emerges principally through the reciprocal interactions of populations of inhibitory neurons. In the this case the alpha rhythm can either arise as noise induced perturbations of a cortical system at steady state having an intrinsic alpha resonance or as deterministic (limit cycle or chaotic) oscillation of a far from equilibrium system. Noise driving is more in line with empirical evidence that finds, except for short bursts of subdural and scalp-recorded EEG, that the alpha rhythm is indistinguishable from linearly filtered white noise (Stam et al. 1999; Stam 2005). Other more recent work (Liley et al. 2010) suggests that resting alpha represents a combination of noise-driven and deterministic dynamics and that transitions between the two are not only possible, but are a regular occurrence in resting EEG.

11.4.2 Drug Effects: Anaesthesia, Sedatives, Stimulants

Efforts to maintain biological plausibility in mean field models have enabled the exploration of physiological dynamics during a range of specific brain states. Researchers have therefore focused on how key dynamical attributes are changed during parametric perturbations which attempt to simulate genuine physiological events. As is often customary in neurophysiology, the core of these efforts have focused on changes in excitability through selective modification of inhibitory and excitatory processes. Shifts in brain excitability can occur both through endogenous mechanisms, such as sleep, or pathological conditions, such as seizures (both discussed below) – however, drug induced changes in excitability are another typical starting point. Particular attention has been placed on the modelling of depressant agents like anaesthetics and conversely stimulants such as pro-convulsants (Foster et al. 2008; Liley et al. 2011).

Anaesthesia induced changes are a preferable domain for mean field simulation for several reasons. Firstly, the cellular and molecular targets of anaesthetics are selective and known in some detail. This allows not only meaningful model parameterization, but furthermore enables the use of empirical data to explore the effects of model perturbation within meaningful physiological ranges. Secondly, the effects of anaesthetics at the cellular level produce clear changes in macroscopic scale activity, such as the EEG, in a dose-dependent manner, allowing fairly straightforward experimental validation of the simulations. Finally, the mechanisms by which the cellular or microscopic targets of anaesthetic drugs relates to the observed macroscopic effects recorded by the EEG are far less clear, promoting the use of mesoscopic methods to help link these physical scales and elucidate the intriguing processes occurring during the transition to unconsciousness.

Generally, anaesthetics, as well as sedatives (e.g. benzodiazepines), produce their depressant effects by targeting inhibitory neurotransmission through enhancement of $GABA_A$ receptor function. By potentiating $GABA_A$ receptor based IPSPs, anaesthetics can progressively suppress neural firing. Less obviously, one also finds a slowing of the EEG that appears in the time domain as a shift from low amplitude desynchronised EEG to large amplitude slow oscillations, and in the frequency domain as a shift from peaked alpha power spectra to ones dominated by the delta band (0–4 Hz). This overall cortical slowing of activity has been successfully modelled by a number of investigators, however explanations for the spectral changes which occur during the *transition* between wakeful rest and deep anaesthesia are somewhat more controversial. During the induction of anaesthesia, the EEG in humans (Kuizenga et al. 1998) and other species (Dutta et al. 1997) show a rise in power prior to the onset of slow wave activity, which approximately coincides with the loss of consciousness (LOC). Typically referred to as the *biphasic response*, studies have shown that a variety of anaesthetic agents produce transient increases in EEG activity in the frequency ranges more associated with alertness and arousal during both induction and emergence (Kuizenga et al. 1998, 2001; Feshchenko et al. 2004). Interestingly, because such effects may be more pronounced during emergence than induction, a number of modelling approaches have taken this as an explanatory predicate in simulations of the biphasic response and its hypothesised hysteresis (system dependence on prior history).

Steyn-Ross et al. (1999, 2004) modelled the biphasic response as a first-order phase transition using the mean field cortical model of Liley et al. (1999b). By selectively modifying the inhibitory neurotransmitter decay rate constant, they found three stages of analytic equilibrium for changes in the level of anaesthesia defined by λ, a scaling-factor representing the fractional change in IPSP duration from the non-anaesthetized state. These stages were termed (1) "coma" (high anaesthesia $\lambda > 1.53$); (2) "awake" (low/no anaesthesia $\lambda = 1$), and (3) "seizure" (retracted inhibition $\lambda < 0.3$). During the transitions between these states estimated EEG spectra were shown to display a clear biphasic response. It was therefore posited that the biphasic response was coincident with the transition between stable states, from (2) to (1) or (3), via a saddle node bifurcation, reflecting a phase transition in the mean soma membrane potential. While this phase transition theory was an innovative and appealing integration of the biphasic response and the loss of consciousness during anaesthesia, the model depended upon neural firing rate parameters outside physiological ranges with "awake" power spectra devoid of any resting alpha activity.

Subsequent work by Bojak and Liley (2005) showed that biphasic phenomena can still be simulated without the occurrence of a phase transition and within plausible firing rate ranges. They used mean field model parameter sets restricted by physiological limits with proper resting state behaviour, and realistically modelled the effects of the anaesthetic agent isoflurane by allowing independent modifications of IPSP amplitude, rise and decay times based on empirical data. The simulated biphasic response was then not coincident with rapid shifts in firing rates, but rather consistent with empirical observations of progressive dose dependent spike suppression. Bojak and Liley (2005) were also pioneering in the mean field context by using a large number of base parameter sets to gain an understanding of the robustness of the predictions under conditions of typical biological variability. This approach has become popular in the field of individual neuron models (Marder and Taylor 2011). Molaee-Ardekani et al. (2007) later showed that the application of a slow adaptive function for the modelling of firing rates was capable of both simulating biphasic spectral activity and relating the slowing of firing rates to the well know occurrence of up/down state firing modes during anaesthesia and slow wave sleep. More recently Hutt and Longtin (2010) have sought to connect both descriptions of biphasic activity. For a simplified model they were able to derive conditions under which the anaesthetic agent propofol causes the biphasic effect either by a phase transition or without.

The possibility of linking the biphasic response of cortical populations with the LOC is a fascinating area for future computational research. Yet it remains unclear if the biphasic changes in EEG spectral power are even causally linked to LOC or whether these two events arise independently in response to pharmacological action. For example, empirical and theoretical work has shown that benzodiazepines such as diazepam and alprazolam, which also selectively target $GABA_A$ receptors, can produce arousal responses in EEG spectral power producing acceleration of peak alpha frequency into the low beta range (Liley et al. 2003b). This "beta buzz" is not greatly different from that produced by alcohol, a $GABA_A$ agonist also. Therefore the occurrence of this form of biphasic response is interpreted as being reflective of low dose anaesthesia and not as a marker of a specific dynamical transition, for this effect may extended over a period far greater than the interval associated with the loss of consciousness. Such effects can be informed by the observed sensitivity of inhibitory synapses and the time delays associated with their reverberant activity (Liley et al. 2003b): enhancement of inhibitory–inhibitory coupling through GABA agonists will strengthen, and therefore shorten, the interactions of this population shifting alpha power to higher frequencies (Liley et al. 2002, 2003).

There has been little application of mean field modelling to understanding the macroscopic action of stimulants, because the involved pharmacological and physiological mechanisms are less straightforward to include in the established formulations. Interestingly, the study by Rowe et al. (2005) concerning the action of stimulants in attention deficit hyperactivity disorder points to yet another *inhibitory* effect, namely decreased firing activity in the locus coeruleus. The authors speculate that more generally stimulants could act to suppress the activity of the thalamic reticular nucleus, thereby increasing thalamocortical and synaptic activity. If true, then once more one would find that inhibitory control is a key locus for the control of brain dynamics.

11.4.3 Cognitive States: Sleep, Rest, Memory, Attention

Following progress in the mean field modelling of anaesthesia, a number of researchers have gone on to theoretically account for human sleep cycles and a range of other cognitive phenomena. Some investigators have extended existing mean field models to incorporate further differentiated neural populations (typically subcortical nuclei), whereas others have focused on purely parametric perturbation of a putative (noise driven) cortex alone. Regarding the modelling of sleep cycles, Steyn-Ross et al. (2005a) have explored the latter approach by modelling the role somnogenic (sleep promoting) factors have in modulating cortical arousal. Conversely Phillips and Robinson (2007) have explored the former approach by applying mean field theory to simulate the modulatory effects of ascending brain stem projection systems. We will discuss each of these approaches in turn.

Steyn-Ross et al. (2005a) model sleep states with particular attention to the transition between slow wave sleep (SWS) and rapid eye movement (REM) phases. In this approach the authors employ a macrocolumn whose characteristic time scales have been greatly increased from their standard parameter values in order to better incorporate the time courses associated with the ultradian human sleep cycle ($\approx$90 min). Subsequently, the transition between SWS and REM states are parameterized as arising from the differential modulation of the basal K^+ leak currents by the somnogenic adenosine ($\uparrow K^+$ leak) and activating acetylcholine (ACh; $\downarrow K^+$ leak). In combination, these effects were best captured through modification of the coupling strength at excitatory synapses with a factor λ_{ACh} scaling the excitatory synaptic gain. Similar to their work in anaesthesia, this parameterization produced a first-order phase transition, typified by a rapid shift from coherent slow firing (SWS) to a classical desynchronized high firing state. Approaching the point of transition between these two states was characterized by increasing coherent frequency power in progressively slower frequencies. This system also supports the occurrence of K-complexes (transient spike-like waves), which typically occur during sub-arousal stimulation (e.g., auditory) during REM sleep. Wilson et al. (2006), explored the response of cortex to such transient *kicks* in input. It was found that depending on system state (SWS or REM), transient shifts in membrane potential cause different spatiotemporal oscillations.

Finally, these authors have also explored the capacity of putative Hebbian learning in such systems in relation to memory consolidation during sleep. Steyn-Ross et al. (2005b) postulate a mechanism for memory erasure or unlearning during the suppressive dynamics occurring during the approach to the SWS-REM transition. Then SWS oscillations become more coherent and expanded in phase-space, producing a state preferentially enhancing inhibitory synaptic weights, after which the REM sleep will reset synaptic weights back to a state capable of learning. Wilson et al. (2007) extended their model to allow estimation of synaptic weight distribution and other statistical attributes, which help to link Hebbian network descriptions with mesoscopic columnar field models.

In contrast to the previous approach, Phillips and Robinson (2007) have included the cortically ascending projections of the various brain stem nuclei which are involved in mediating arousal and therefore the sleep-wake cycle. This ascending arousal system has cortical projections which terminate diffusely across the cerebrum. Phillips and Robinson (2007) model this group of nuclei by defining a monoamine group (MA) and a ACh group, both of which receive driving (e.g., circadian) inputs from the ventrolateral preoptic area (VLPO; hypothalamus). Since the interaction of the defined MA and ACh groups reflect nuclei whose reciprocal action is thought to control SWS-REM switching, Phillips and Robinson (2007) focus only on interactions between the VLPO and MA groups, while setting the ACh group constant. Thus they focus on the transition between waking and sleeping states and not the ultradian sleep cycle (i.e., sleep stages). The mutual inhibition between the MA group (wake promoting) and the VLPO (sleep promoting) are parametrically explored as putative sleep-wake cycling. By modulating the driving input, they observed clear changes in the mean soma membrane potential of the MA group, such that wake-sleep transitions occur as a saddle node bifurcation at several driving values which produces a hysteresis in the return from sleep to waking states. This hysteresis produces a zone wherein the transition between awake and sleep states is more easily achieved and more sensitive to noise. This unstable transition is suggested by the authors to dynamically underpin the phenomenon of narcolepsy.

Attempts to model learning process through the modification of synaptic properties represents one example of how mean field theories may begin to make contact with cognition and cognitive systems. However, it remains difficult to meaningfully capture the dynamics of cognitive activity with these simple biological models. Currently the focus is on explaining activity patterns and to use data-driven approaches to help identify physiological parameters that sensitively control neuronal population dynamics. For example, Steyn-Ross et al. (2009) have explored the utility of mean field models to capture differing dynamical properties associated with activity patterns during resting and cognitive states. In brief, this work suggests that the ongoing activity of the brain may be the result of spatiotemporal instabilities, with idling activity slowly oscillating (1 Hz) around a rest state that can be perturbed into an active cognitive state defined by faster (40 Hz) activity. Such slow oscillations may relate to those observed during resting state neuroimaging (Steyn-Ross et al. 2009, 2011) – the so called *"default mode"* (Raichle et al. 2001). Concrete studies of cognitive action have also been performed, in particular of (visual) attention (Deco and Rolls 2005; Mavritsaki et al. 2011). Interestingly in these cases, hybrid descriptions in terms of both spiking neurons and mean field models were employed, which interacted consistently with each other by construction (Brunel and Wang 2001). It is likely that such hybrid approaches will become more common in the future.

11.4.4 Special Dynamics: Metastability, Gamma Synchrony, Epilepsy

Although early attempts to dynamically describe brain function sought to prescribe explicit attractor dynamics to neural activity, more recent thinking focuses on transitory non-equilibrium behaviour (Rabinovich et al. 2008a). Neuronal population dynamics are conceived as evolving transiently, rarely reaching stability. On this basis a number of authors have opted to describe this type of dynamical regime as metastability (Kelso 1995; Friston 1997; Bressler and Kelso 2001; Freeman and Holmes 2005; Rabinovich et al. 2008b). Common to many of these descriptions is an ongoing occurrence of transitory neural events, or state transitions, which define the flexibility of cognitive and sensori-motor function. Some dynamical examples include the chaotic itinerancy of Tsuda (2001), in which neural dynamics transit in a chaotic motion through unique Milnor attractors, or the liquid-state machine of Rabinovich et al. (2008a), where a more global stable heteroclinic channel is comprised of successive local saddle states. More specific neurodynamical approaches include the work of Kelso (1995), Friston (2000) and Freeman and Holmes (2005).

The idea that the brain activity switches between qualitatively different states is reflected in the dynamical behaviour of many mean field model formulations. For this reason it has been conjectured that mean field models may offer a suitable explanatory framework in which to understand the genesis and evolution of epileptic activity (Wendling et al. 2000; Robinson et al. 2002; Lopes da Silva et al. 2003; Suffczyński et al. 2005; Kramer et al. 2005; Liley and Bojak 2005; Breakspear et al. 2006; Marten et al. 2009; Molaee-Ardekani et al. 2010) as well as the formation of coherent mesoscopic gamma band activity (Wright 1997; Rennie et al. 2000; Robinson 2006; Bojak and Liley 2007). What relates these two topics is that such dynamics are typically conceived as limit cycle (or chaotic) behaviour to which the brain transits from a regular state under certain conditions, e.g., the change of some physiological parameter. The difference between epilepsy and gamma synchrony is then mainly related to the dominant frequency of these "special" dynamics. That the dominant frequency typically can be adjusted with some model parameter suggests that these phenomena may indeed be physiologically related.

However, a general caveat applies to such studies: it is often comparatively easy to obtain activity that resembles epilepsy or coherent gamma oscillations from complicated (mean field) models. That per se should hence not count as a success of the model. Rather, it is precisely the mechanistic explanation of the transit from regular to special dynamics which then becomes a hallmark of biological fidelity. For example, Liley and Bojak (2005) showed that seizure-inducing properties of some general anaesthetic agents could be reproduced with the Liley mean field model, rather than just that the model supported seizure-like limit cycles. At the very least the switching mechanism should be post hoc interpreted, like the emergence of large scale gamma synchrony in terms of changes in presynaptic thalamocortical input (Bojak and Liley 2007).

11.4.5 Beyond Electrical Activity: fMRI BOLD and Multimodal Integration

The EEG's continued popularity in research and clinical practice (Niedermeyer and Lopes da Silva 2005) is largely due to its excellent time resolution, which allows for example the tracking of evoked potentials (Regan 1989; Rennie et al. 2002). EEG taken with a large number of electrodes and MEG can provide data of brain activity with comparatively high spatial resolution in the centimetre range. However, due to its millimetre resolution fMRI BOLD has taken over as the de facto standard for researching cognition (Norris 2006; Matthews et al. 2006), in spite of weak temporal resolution and uncertainties concerning its genesis (Logothetis 2008). The rapidly growing field of multimodal imaging (Stufflebeam and Rosen 2007; Shibasaki 2008; Mulert et al. 2008; Blinowska et al. 2009; Freeman et al. 2009) also promises to combine the excellent temporal resolution of EEG with the good spatial resolution of fMRI BOLD. This combination is particularly attractive because it is possible to record EEG and fMRI BOLD simultaneously (Laufs et al. 2008), which avoids all the issues of brain state dependence that arise in combining data sets recorded at different times (or even from different people). Furthermore, the relation of fMRI resting state networks to EEG cortical microstates (Britz et al. 2010; Musso et al. 2010) suggests close dynamical links between data recorded with these modalities.

Therefore it is of obvious interest to extend the mean field formalism first to the description of fMRI BOLD, and then to the predicition of simultaneous EEG/fMRI. The first issue to confront is the addition of appropriate hemodynamics. Almost all mean field modellers have adopted the so-called "Balloon-Windkessel" hemodynamic model in the form introduced by Friston et al. (2000, 2003) based on prior work by Buxton and Frank (1997), Buxton et al. (1998) and Mandeville et al. (1999). The main reason is that this model merely adds four ordinary differential equations (ODEs) to the equations, with specified (prior) parameter values. The Balloon-Windkessel model will predict the BOLD contrast based on neural activity innovations, making it most suited for task paradigms. It is somewhat less suited for resting state activity (Bojak et al. 2010, 2011), a point that will require scrutiny in future. Note that these equations are local, hence in a large scale model comprising many neural masses or in a continuous formulation discretized for numerical computation, the ODE systems must be evaluated at every unit. Furthermore, the question arises what neural activity to use as input for the Balloon-Windkessel hemodynamics, or in other terms, how the neurovascular coupling is modelled. In the works we will describe below, a bewildering variety of models have been used, though conceptually they reduce to either total synaptic activity, glutamate release or energy consumption. We believe that this confusion reflects our limited knowledge about the genesis of the BOLD signal and refrain here from commenting further on this issue. Suffice to say that all current modelling efforts must be considered speculative concerning the fMRI signal generation.

We have already mentioned the works using CoCoMac connectivity (Honey et al. 2007; Ghosh et al. 2008; Deco et al. 2009), recently reviewed by Deco et al.

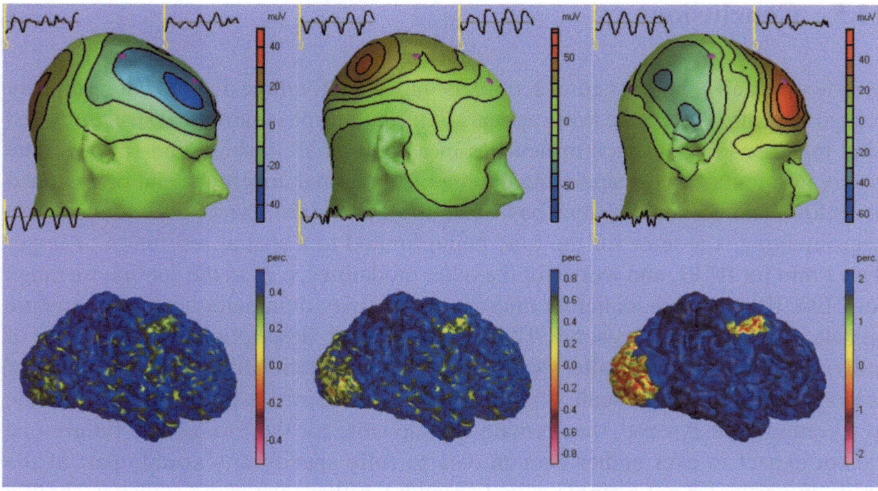

Fig. 11.5 Simultaneous EEG/fMRI BOLD prediction. The *top* row shows snapshots of EEG scalp potentials, as well as EEG time traces from three electrodes (*purple*). The *bottom* row shows corresponding fMRI BOLD predictions (Bojak et al. 2010, 2011). Columns from *left* to *right* show results for specific connectivity at 60%, 85%, and 90% strength relative to that of local ("background") connectivity. Scales vary between panels and are indicated by colourbars. A clear jump in fMRI BOLD contrast occurs for increasing the strength from 85% to 90%

(2011). They all provide multimodal EEG and fMRI predictions, whereas the related paper by Honey et al. (2009) used diffusion MR tractography and predicted only fMRI signals. The focus of these works is on resting state oscillations and the so-called "default mode network" (Raichle et al. 2001). General experimentally observed features were reproduced; however, Honey et al. (2009) raises questions about the dependence of such dynamics on the cortical parcellation. Babajani et al. (2005), Babajani and Soltaninan-Zadeh (2006) and Babajani-Feremi et al. (2008) convincingly fit both MEG and fMRI BOLD data arising from an auditory task with a single underlying model. Recently Babajani-Feremi and Soltaninan-Zadeh (2010) also extended the electrocortical part of this model to multiple areas. Riera et al. (2005, 2006, 2007) have successfully fit simultaneous EEG/fMRI data in visual and auditory tasks, though their focus was more on the technical development of inversion and neurovascular coupling models. Sotero et al. (2007), Sotero and Trujillo-Barreto (2008) and Valdes-Sosa et al. (2009) pioneered highly resolved descriptions of the brain geometry and obtained intriguing results on "alpha blocking" and BOLD-alpha rhythm (anti-)correlation, respectively. Deneux and Faugeras (2010) studied in-depth the extent to which simultaneous EEG/fMRI can extract information beyond the use of single modalities. To this end they developed Kalman filters and smoothers for the inverse problem. Finally, Bojak et al. (2010, 2011) show that minor changes in the implemented effective connectivity can lead to drastic changes in the simulated dynamics, see Fig. 11.5.

11.5 Conclusions

Mesoscopic and coherent neural activity is important for three distinct but intimately related reasons. Firstly, the brain appears to have multiple anatomical structures that bind together larger groups of neurons in a "columnar" fashion and so parcellate cortex into units. A corresponding degree of functional coherence can be assumed to follow, and to some extent has been shown to exist. Secondly, non-invasive neuroimaging is at least for the time being limited to a spatial resolution of larger than 1 mm for fMRI, and worse for the other modalities, e.g., in the centimetre range for EEG. Furthermore, coherent microscopic activity dominates such macroscopic signals for statistical reasons. Thirdly, theoretical descriptions at the level of individual neurons (or smaller) scale badly in computational costs for a variety of reasons. Yet even if large scale simulations of this kind increasingly become possible in a computational sense, they remain questionable for the foreseeable future: one cannot expect to ever gather enough data to fully specify any sizable part of the brain, in particular not a single actual structure, rather than some statistics about a large number of similar structures.

Bulk models of neural activity try to describe mesoscopic activity directly, eschewing a description in terms of enumerable microscopic units. Yet they typically retain selected "microscopic features" that are salient for the description of the phenomena of interest. Thus for example if it is known that some drugs affects the PSPs of individual neurons, then it is advantageous if the bulk approach employed contains a PSP description that can be matched to this "microscopic" knowledge. The simplest bulk models consider only the mean (i.e., first moment) properties of neuronal ensembles. Here the connection to the microscopic world is generally direct, replacing a microscopic quantity by its mesoscopic mean. We have described in great detail how these models developed historically, and what sort of brain activity they are able to describe successfully. We will not repeat our observations here. Suffice to say that mathematical and computational progress has been good and seems to be accelerating still, and a great number of applications have been found already.

To this clear success story we have told in the preceding pages we now wish to add a more critical perspective and some caveats. Firstly, the microscopic-mesoscopic connection remains fairly ill-defined. It is unclear to what extent a population description can and should be made "consistent" with existing single neurons models: the brain consists of much more than just neurons and bulk approaches should subsume these non-neural contributions in their "effective equations". The necessity to maintain salient microscopic features however means that the bulk descriptions often closely resemble single neuron models after all, to allow direct mapping of known effects. Yet it is unclear to what extent this similarity of form remains justified for an effective description that includes other elements. Indeed, as comparing Newton's equations with the ideal gas law suggests, it is far from certain that the best mesoscopic description of even a purely neuronal and homogeneous population would resemble the individual neuron model at all.

A certain amount of "hand-waving" seems to be required at the moment to motivate all bulk models. We suggest that much future work is required to clarify the situation.

Secondly, applications of bulk models have yet to deliver an insight of such obvious neuroscientific or medical impact that their future importance is established beyond reasonable doubt. In other words, this field still requires an universally accepted breakthrough. An important limiting factor in this regard is the lack of information intrinsically associated with the predicted activity. While the single neuron modeller can dream of cracking the "spike code" of information processing in the brain, it is currently not obvious what bulk activity *means*. We need to learn better how to associate computation with bulk activity, how to assign meaning to the energtic flows that are presently being modelled. There seem to be two ways forward here: On one hand one can try to elucidate further the effect of coherent group activity on single neurons, and thus establish a meaning for the mesoscopic activity in terms of whatever meaning single neuron activity may have. On the other hand one can try to connect to the more abstract models of psychology and psychophysics. Their natural "neural basis" may very well be found in bulk approaches. It seems to us that both paths to greater impact must now be pursued vigorously.

Thirdly, at the cutting edge of technical development, bulk approaches are meeting challenges that are rather reminiscent of the problems single neuron modellers face. If one attempts to describe the entire brain at mesoscopic resolution in the millimetre range, one ends up with about the same number of mesoscopic units (many tens of thousands) as large scale single neuron simulations. Thus the same issues of computational cost, combinatorial explosion of connectivity and experimental indeterminacy arise. We cannot, and likely will not ever, simulate *this* brain here and now even at a mesoscopic level, just as attempts to simulate *this* macrocolumn here and now are doomed to failure. What we need to work out is hence the appropriate level of detail: where can and should we introduce individual features, where should we stick to stereotypical and qualitative descriptions. Furthermore, the issue of fitting to experimental data the many parameters of a large number of functional units in a complicated network remains unresolved, and limits progress at both the mesoscopic and microscopic level. We will likely need the help from experts in the statistical field to break this gridlock.

However, such technical challenge also provide chances for future growth, and the need for bulk approaches becomes increasingly clear. We would like to particularly encourage our experimental colleagues to consider the mesoscopic level of investigation as a field with enormous growth potential. Mesoscopic theory is now getting to a level where it can speak to experimental recordings from multi-electrode arrays recording LFPs in stimulated slice preparations all the way to complex task paradigms tracked with fMRI BOLD. Modern experimental technologies can go in many ways beyond the patch-clamping of a single neuron, and most of those involve the consideration of the collective activity of neurons. The marriage of experiment and theory has not always been easy in the neurosciences, but there is a clear convergence of interest here concerning the activity of many neurons working together. As neuroscientists we should respond to this opportunity by working together as well.

References

Aboitiz F, Scheibel AB, Fisher RS, Zaidel E (1992a) Fiber composition of the human corpus callosum. Brain Res 598:143–153

Aboitiz F, Scheibel AB, Fisher RS, Zaidel E (1992b) Individual differences in brain asymmetries and fiber composition in the human corpus callosum. Brain Res 598:154–161

Adrian ED, Matthews BHC (1934) The Berger rhythm, potential changes from the occipital lobe in man. Brain 57:355–385

Amari SI (1975) Homogeneous nets of neuron-like elements. Biol Cybern 17:211–220

Amari SI (1977) Dynamics of pattern formation in lateral-inhibition type neural fields. Biol Cybern 27:77–87

Araque A, Navarrete M (2010) Glial cells in neuronal network function. Philos Trans R Soc B 365:2375–2381

Azevedo FAC, Carvalho LRB, Grinberg LT, Farfel JM, Ferretti REL, Leite REP, Jacob Filho W, Lent R, Herculano-Houzel S (2009) Equal numbers of neuronal and nonneuronal cells make the human brain an isometrically scaled-up primate brain. J Comp Neurol 513:532–541

Babajani A, Soltaninan-Zadeh H (2006) Integrated MEG/EEG and fMRI model based on neural masses. IEEE Trans Biomed Eng 53:1794–1801

Babajani A, Nekooei MH, Soltaninan-Zadeh H (2005) Integrated MEG and fMRI model: synthesis and analysis. Brain Topogr 18:101–113

Babajani-Feremi A, Soltaninan-Zadeh H (2010) Multi-area neural mass modeling of EEG and MEG signals. NeuroImage 52:793–811

Babajani-Feremi A, Soltaninan-Zadeh H, Moran JE (2008) Integrated MEG/fMRI model validated using real auditory data. Brain Topogr 21:61–74

Ben Achour S, Pascual O (2010) Glia: the many ways to modulate synaptic plasticity. Neurochem Int 57:440–445

Berger H (1929) Über das Elektrenkephalogramm des Menschen. Arch Psychiatr Nervenkr 87:527–570

Berger H (1930) Über das Elektrenkephalogramm des Menschen. Zweite Mitteilung. J Psychol Neurol 40:160–179

Beurle RL (1956) Properties of a mass of cells capable of regenerating pulses. Philos Trans R Soc B 240:55–94

Biswal BB, Mennes M, Zuo XN, Gohel S, Kelly C, Smith SM, Beckmann CF, Adelstein JS, Buckner RL, Colcombe S, Dogonowski AM, Ernst M, Fair D, Hampson M, Hoptman MJ, Hyde JS, Kiviniemi VJ, Kötter R, Li SJ, Lin CP, Lowe MJ, Mackay C, Madden DJ, Madsen KH, Margulies DS, Mayberg HS, McMahon K, Monk CS, Mostofsky SH, Nagel BJ, Pekar JJ, Peltier SJ, Petersen SE, Riedl V, Rombouts SARB, Rypma B, Schlaggar BL, Schmidt S, Seidler RD, Siegle GJ, Sorg C, Teng GJ, Veijola J, Villringer A, Walter M, Wang L, Weng XC, Whitfield-Gabrieli S, Williamson P, Windischberger C, Zang YF, Zhang HY, Castellanos FX, Milham MP (2010) Toward discovery science of human brain function. Proc Natl Acad Sci USA 107:4734–4739

Blinowska K, Müller-Putz G, Kaiser V, Astolfi L, Vanderperren K, Van Huffel S, Lemieux L (2009) Multimodal imaging of human brain activity: rational, biophysical aspects and modes of integration. Comput Intell Neurosci 2009:813607

Bojak I, Liley DTJ (2005) Modeling the effects of anesthesia on the electroencephalogram. Phys Rev E 71:041902

Bojak I, Liley DTJ (2007) Self-organized 40 hz synchronization in a physiological theory of EEG. Neurocomputing 70:2085–2090

Bojak I, Liley DTJ (2010) Axonal velocity distributions in neural field equations. PLoS Comput Biol 6:e1000653

Bojak I, Oostendorp TF, Reid AT, Kötter R (2010) Connecting mean field models of neural activity to EEG and fMRI data. Brain Topogr 23:139–149

Bojak I, Oostendorp TF, Reid AT, Kötter R (2011) Towards a model-based integration of co-registered EEG/fMRI data with realistic neural population meshes. Philos Trans R Soc A 369:3785–3801

Braitenberg V, Schüz A (1998) Cortex: statistics and geometry of neuronal connectivity, 2nd edn. Springer, Berlin

Branco TP, Staras K (2009) The probability of neurotransmitter release: variability and feedback control at single synapses. Nat Rev Neurosci 10:373–383

Breakspear M, Roberts JAG, Terry JR, Rodrigues S, Mahant N, Robinson PA (2006) A unifying explanation of primary generalized seizures through nonlinear brain modeling and bifurcation analysis. Cereb Cortex 16:1296–1313

Bressler SL, Kelso JAS (2001) Cortical coordination dynamics and cognition. Trends Cogn Sci 5:26–36

Britz J, Van De Ville D, Michel CM (2010) BOLD correlates of EEG topography reveal rapid resting-state network dynamics. NeuroImage 52:1162–1170

Brodmann K, Garey LJ (2006) Brodmann's localisation in the cerebral cortex: the principles of comparative localisation in the cerebral cortex based on cytoarchitectonics – translated with editorial notes and an introduction, 3rd edn. Springer, New York

Brunel N, Wang XJ (2001) Effects of neuromodulation in a cortical network model of object working memory dominated by recurrent inhibition. J Comput Neurosci 11:63–85

Buice MA, Cowan JD, Chow CC (2010) Systematic fluctuation expansion for neural network activity equations. Neural Comput 22:377–426

Bullock TH, McClune MC, Achimowicz JZ, Iragui-Madoz VJ, Duckrow RB, Spencer SS (1995) EEG coherence has structure in the millimeter domain: subdural and hippocampal recordings from epileptic patients. Electroencephalogr Clin Neurophysiol 95:161–177

Buxhoeveden DP, Casanova MF (2002) The minicolumn and evolution of the brain. Brain Behav Evol 60:125–151

Buxton RB, Frank LR (1997) A model for the coupling between cerebral blood flow and oxygen metabolism during neural stimulation. J Cereb Blood Flow Metab 17:64–72

Buxton RB, Wong ECC, Frank LR (1998) Dynamics of blood flow and oxygenation changes during brain activation: the balloon model. Magn Reson Med 39:855–864

Ciulla C, Takeda T, Endo H (1999) MEG characterization of spontaneous alpha rhythm in the human brain. Brain Topogr 11:211–222

Contreras D (2004) Electrophysiological classes of neocortical neurons. Neural Netw 17:633–646

Coombes S (2005) Waves, bumps, and patterns in neural field theories. Biol Cybern 93:91–108

Coombes S (2010) Large-scale neural dynamics: simple and complex. NeuroImage 52:731–739

Coombes S, Venkov NA, Shiau LJ, Bojak I, Liley DTJ, Laing CR (2007) Modeling electrocortical activity through improved local approximations of integral neural field equations. Phys Rev E 76:051901

Daunizeau J, Kiebel SJ, Friston KJ (2009) Dynamic causal modelling of distributed electromagnetic responses. NeuroImage 47:590–601

David O, Friston KJ (2003) A neural mass model for MEG/EEG: coupling and neuronal dynamics. NeuroImage 20:1743–1755

David O, Harrison LM, Friston KJ (2005) Modelling event-related responses in the brain. NeuroImage 25:756–770

David O, Kiebel SJ, Harrison LM, Mattout J, Kilner JM, Friston KJ (2006) Dynamic causal modeling of evoked responses in EEG and MEG. NeuroImage 30:1255–1272

Deco GR, Rolls ET (2005) Neurodynamics of biased competition and cooperation for attention: a model with spiking neurons. J Neurophysiol 94:295–313

Deco GR, Jirsa VK, Robinson PA, Breakspear M, Friston KJ (2008) The dynamic brain: from spiking neurons to neural masses and cortical fields. PLoS Comput Biol 4:e1000092
Deco GR, Jirsa VK, McIntosh AR, Sporns O, Kötter R (2009) Key role of coupling, delay, and noise in resting brain fluctuations. Proc Natl Acad Sci USA 106:10302–10307
Deco GR, Jirsa VK, McIntosh AR (2011) Emerging concepts for the dynamical organization of resting-state activity in the brain. Nat Rev Neurosci 12:43–56
Deneux T, Faugeras O (2010) EEG-fMRI fusion of paradigm-free activity using Kalman filtering. Neural Comput 22:906–948
Dienel GA, Cruz NF (2003) Neighborly interactions of metabolically-activated astrocytes in vivo. Neurochem Int 43:339–354
Dutta S, Matsumoto Y, Gothgen NU, Ebling WF (1997) Concentration-EEG effect relationship of propofol in rats. J Pharm Sci 86:37–43
Eccles JC (1992) Evolution of consciousness. Proc Natl Acad Sci USA 89:7320–7324
Ermentrout BG (1998) Neural networks as spatio-temporal pattern-forming systems. Rep Prog Phys 61:353–430
Faugeras O, Touboul J, Cessac B (2009) A constructive mean-field analysis of multi-population neural networks with random synaptic weights and stochastic inputs. Front Comput Neurosci 3:1
Feshchenko VA, Veselis RA, Reinsel RA (2004) Propofol-induced alpha rhythm. Neuropsychobiology 50:257–266
Fleischhauer K, Petsche H, Wittkowski W (1972) Vertical bundles of dendrites in the neocortex. Z Anat Entwicklungsgesch 136:213–223
Foster BL, Bojak I, Liley DTJ (2008) Population based models of cortical drug response: insights form anaesthesia. Cogn Neurodyn 2:283–296
Frascoli F, van Veen L, Bojak I, Liley DTJ (2011) Metabifurcation analysis of a mean field model of the cortex. Physica D 240:949–962. doi:10.1016/j.physd.2011.02.002
Freeman WJ (1975) Mass action in the nervous system: examination of the neurophysiological basis of adaptive behavior through the EEG, 1st edn. Academic Press, New York, also electronic edn.: http://sulcus.berkeley.edu/MANSWWW/MANSWWW.html,2004
Freeman WJ (1979) Nonlinear gain mediating cortical stimulus-response relations. Biol Cybern 33:237–247
Freeman WJ, Holmes MD (2005) Metastability, instability, and state transition in neocortex. Neural Netw 18:497–504
Freeman WJ, Ahlfors SP, Menon V (2009) Combining fMRI with EEG and MEG in order to relate patterns of brain activity to cognition. Int J Psychophysiol 73:43–52
Friston KJ (1997) Transients, metastability, and neuronal dynamics. NeuroImage 5:164–171
Friston KJ (2000) The labile brain. I. Neuronal transients and nonlinear coupling. Philos Trans R Soc B 355:215–236
Friston KJ (2002) Bayesian estimation of dynamical systems: an application to fMRI. NeuroImage 16:513–530
Friston KJ, Mechelli A, Turner R, Price CJ (2000) Nonlinear responses in fMRI: The Balloon model, Volterra kernels, and other hemodynamics. NeuroImage 12:466–477
Friston KJ, Penny WD, Phillips C, Kiebel SJ, Hinton GE, Ashburner J (2002) Classical and Bayesian inference in neuroimaging: theory. NeuroImage 16:465–483
Friston KJ, Harrison LM, Penny WD (2003) Dynamic causal modelling. NeuroImage 19: 1273–1302
Friston KJ, Mattout J, Trujillo-Barreto NJ, Ashburner J, Penny WD (2007) Variational free energy and the Laplace approximation. NeuroImage 34:220–234
Ghosh A, Rho YA, McIntosh AR, Kötter R, Jirsa VK (2008) Noise during rest enables the exploration of the brain's dynamic repertoire. PLoS Comput Biol 4:e1000196
Gloor P (1969) Hans Berger on the electroencephalogram of man. Electroencephalogr Clin Neurophysiol S28:350

Goldman PS, Nauta WJH (1977) Columnar distribution of cortico-cortical fibers in the frontal association, limbic, and motor cortex of the developing rhesus monkey. Brain Res 122: 393–413

Griffith JS (1963) A field theory of neural nets: I: derivation of field equations. Bull Math Biol 25:111–120

Griffith JS (1965) A field theory of neural nets: II: properties of the field equations. Bull Math Biol 27:187–195

Hagmann P, Cammoun L, Gigandet X, Meuli RA, Wedeen VJ, Sporns O (2008) Mapping the structural core of human cerebral cortex. PLoS Biol 6:e159

Haken H (1983) Synergetics: an introduction. Nonequilibrium phase transitions and self-organization in physics, chemistry, and biology, 3rd edn. Springer, Berlin

Hellwig B (2000) A quantitative analysis of the local connectivity between pyramidal neurons in layers 2/3 of the rat visual cortex. Biol Cybern 82:111–121

Herculano-Houzel S (2009) The human brain in numbers: a linearly scaled-up primate brain. Front Hum Neurosci 3:31

Honey CJ, Kötter R, Breakspear M, Sporns O (2007) Network structure of cerebral cortex shapes functional connectivity on multiple time scales. Proc Natl Acad Sci USA 104:10240–10245

Honey CJ, Sporns O, Cammoun L, Gigandet X, Thiran JP, Meuli RA, Hagmann P (2009) Predicting human resting-state functional connectivity from structural connectivity. Proc Natl Acad Sci USA 106:2035–2040

Hughes SW, Crunelli V (2005) Thalamic mechanisms of EEG alpha rhythms and their pathological implications. Neuroscientist 11:357–372

Hughes SW, Crunelli V (2007) Just a phase they're going through: the complex interaction of intrinsic high-threshold bursting and gap junctions in the generation of thalamic alpha and theta rhythms. Int J Psychophysiol 64:3–17

Hutt A, Longtin A (2010) Effects of the anesthetic agent propofol on neural populations. Cogn Neurodyn 4:37–59

Jansen BH, Rit VG (1995) Electroencephalogram and visual evoked potential generation in a mathematical model of coupled cortical columns. Biol Cybern 73:357–366

Jirsa VK, Haken H (1996) Field theory of electromagnetic brain activity. Phys Rev Lett 77: 960–963

Jirsa VK, Jantzen KJ, Fuchs A, Kelso JAS (2002) Spatiotemporal forward solution of the EEG and MEG using network modeling. IEEE Trans Med Imaging 21:493–504

Johansen-Berg H, Rushworth MFS (2009) Using diffusion imaging to study human connectional anatomy. Annu Rev Neurosci 32:75–94

Jones EG (2000) Microcolumns in the cerebral cortex. Proc Natl Acad Sci USA 97:5019–5021

Jones EG, Burton H, Porter R (1975) Commissural and cortico-cortical "columns" in the somatic sensory cortex of primates. Science 190:572–574

Kaiser M, Hilgetag CC, van Ooyen A (2009) A simple rule for axon outgrowth and synaptic competition generates realistic connection lengths and filling fractions. Cereb Cortex 19: 3001–3010

Kandel ER, Schwartz JH, Jessell TM (2000) Principles of neural science, 4th edn. McGraw-Hill, New York

Kelso JAS (1995) Dynamic patterns: the self-organization of brain and behavior. The MIT Press, Cambridge

Kiebel SJ, David O, Friston KJ (2006) Dynamic causal modelling of evoked responses in EEG/MEG with lead field parameterization. NeuroImage 30:1273–1284

Kim JS, Singh V, Lee JK, Lerch J, Ad-Dab'bagh Y, MacDonald DJ, Lee JM, Kim SI, Evans AC (2005) Automated 3-D extraction and evaluation of the inner and outer cortical surfaces using a Laplacian map and partial volume effect classification. NeuroImage 27:210–221

Kötter R, Wanke E (2005) Mapping brains without coordinates. Philos Trans R Soc B 360:751–766

Kramer MA, Kirsch HE, Szeri AJ (2005) Pathological pattern formation and cortical propagation of epileptic seizures. J R Soc Interface 2:113–127

Kuizenga K, Kalkman CJ, Hennis PJ (1998) Quantitative electroencephalographic analysis of the biphasic concentration-effect relationship of propofol in surgical patients during extradural analgesia. Br J Anaesth 80:725–732

Kuizenga K, Wierda JMKH, Kalkman CJ (2001) Biphasic EEG changes in relation to loss of consciousness during induction with thiopental, propofol, etomidate, midazolam or sevoflurane. Br J Anaesth 86:354–360

Laufs H, Daunizeau J, Carmichael DW, Kleinschmidt AK (2008) Recent advances in recording electrophysiological data simultaneously with magnetic resonance imaging. NeuroImage 40:515–528

Liley DTJ, Bojak I (2005) Understanding the transition to seizure by modeling the epileptiform activity of general anesthetic agents. J Clin Neurophysiol 22:300–313

Liley DTJ, Wright JJ (1994) Intracortical connectivity of pyramidal and stellate cells: estimates of synaptic densities and coupling symmetry. Netw Comput Neural Syst 5:175–189

Liley DTJ, Alexander DM, Wright JJ, Aldous MD (1999a) Alpha rhythm emerges from large-scale networks of realistically coupled multicompartmental model cortical neurons. Netw Comput Neural Syst 10:79–92

Liley DTJ, Cadusch PJ, Wright JJ (1999b) A continuum theory of electro-cortical activity. Neurocomputing 26-27:795–800

Liley DTJ, Cadusch PJ, Dafilis MP (2002) A spatially continuous mean field theory of electrocortical activity. Netw Comput Neural Syst 13:67–113

Liley DTJ, Cadusch PJ, Dafilis MP (2003) Corrigendum. Netw Comput Neural Syst 14:369

Liley DTJ, Cadusch PJ, Gray M, Nathan PJ (2003b) Drug-induced modification of the system properties associated with spontaneous human electroencephalographic activity. Phys Rev E 68:05190

Liley DTJ, Bojak I, Dafilis MP, van Veen L, Frascoli F, Foster BL (2010) Bifurcations and state changes in the human alpha rhythm: theory and experiment. In: Steyn-Ross DA, Steyn-Ross ML (eds) Modeling phase transitions in the brain. Springer series in computational neuroscience, vol 4. Springer, New York, pp 117–145

Liley DTJ, Foster BL, Bojak I (2011) A mesoscopic modelling approach to anaesthetic action on brain electrical activity. In: Hutt A (ed) Sleep and anesthesia: neural correlates in theory and experiment. Springer series in computational neuroscience, vol 15. Springer, New York, pp 139–166

Llinás RR (1988) The intrinsic electrophysiological properties of mammalian neurons: insights into central nervous system function. Science 242:1654–1664

Logothetis NK (2008) What we can do and what we cannot do with fMRI. Nature 453:869–878

Lopes da Silva FH, Hoeks A, Smits H, Zetterberg LH (1974) Model of brain rhythmic activity: the alpha-rhythm of the thalamus. Kybernetik 15:27–37

Lopes da Silva FH, Blanes W, Kalitzin SN, Parra J, Suffczyński P, Velis DN (2003) Dynamical diseases of brain systems: different routes to epileptic seizures. IEEE Trans Biomed Eng 50:540–548

López-Muñoz F, Boya J, Alamo C (2006) Neuron theory, the cornerstone of neuroscience, on the centenary of the Nobel Prize award to Santiago Ramón y Cajal. Brain Res Bull 70:391–405

Lübke J, Feldmeyer D (2007) Excitatory signal flow and connectivity in a cortical column: focus on barrel cortex. Brain Struct Funct 212:3–17

Mandeville JB, Marota JJA, Ayata C, Zaharchuk G, Moskowitz MA, Rosen BR, Weisskoff RM (1999) Evidence of a cerebrovascular postarteriole windkessel with delayed compliance. J Cereb Blood Flow Metab 19:679–689

Marder E, Taylor AL (2011) Multiple models to capture the variability in biological neurons and networks. Nat Neurosci 14:133–138

Markram H (2006) The blue brain project. Nat Rev Neurosci 7:153–160

Markram H (2008) Fixing the location and dimensions of functional neocortical columns. HFSP J 2:132–135

Markram H, Toledo-Rodriguez M, Wang Y, Gupta A, Silberberg G, Wu C (2004) Interneurons of the neocortical inhibitory system. Nat Rev Neurosci 5:793–807

Marten F, Rodrigues S, Benjamin O, Richardson MP, Terry JR (2009) Onset of polyspike complexes in a mean-field model of human electroencephalography and its application to absence epilepsy. Philos Trans R Soc A 367:1145–1161
Matthews PM, Honey GD, Bullmore ET (2006) Applications of fMRI in translational medicine and clinical practice. Nat Rev Neurosci 7:732–744
Mavritsaki E, Heinke D, Allen H, Deco GR, Humphreys GW (2011) Bridging the gap between physiology and behavior: Evidence from the sSoTS model of human visual attention. Psychol Rev 118:3–41
McCulloch WS, Pitts W (1943) A logical calculus of the ideas immanent in nervous activity. Bull Math Biophys 5:115–133, reprinted 1990 in Bull Math Biol 52: 99–115
Miguel-Hidalgo JJ (2005) Lower packing density of glial fibrillary acidic protein-immunoreactive astrocytes in the prelimbic cortex of alcohol-naive and alcohol-drinking alcohol-preferring rats as compared with alcohol-nonpreferring and Wistar rats. Alcohol Clin Exp Res 29:766–772
Molaee-Ardekani B, Senhadji L, Shamsollahi MB, Vosoughi-Vahdat B, Wodey E (2007) Brain activity modeling in general anesthesia: enhancing local mean-field models using a slow adaptive firing rate. Phys Rev E 76:041911
Molaee-Ardekani B, Benquet P, Bartolomei F, Wendling F (2010) Computational modeling of high-frequency oscillations at the onset of neocortical partial seizures: from 'altered structure' to 'dysfunction'. NeuroImage 52:1109–1122
Moran RJ, Kiebel SJ, Stephan KE, Reilly RB, Daunizeau J, Friston KJ (2007) A neural mass model of spectral responses in electrophysiology. NeuroImage 37:706–720
Moran RJ, Stephan KE, Kiebel SJ, Rombach N, OConnor WT, Murphy KJ, Reilly RB, Friston KJ (2008) Bayesian estimation of synaptic physiology from the spectral responses of neural masses. NeuroImage 42:272–284
Moran RJ, Stephan KE, Seidenbecher T, Pape HC, Dolan RJ, Friston KJ (2009) Dynamic causal models of steady-state responses. NeuroImage 44:796–811
Mori S, Wakana S, van Zijl PCM, Nagae-Poetscher LM (2005) MRI atlas of human white matter. Elsevier, Amsterdam
Mountcastle VB (1957) Modality and topographic properties of single neurons of cat's somatic sensory cortex. J Neurophysiol 20:408–434
Mountcastle VB (1979) An organizing principle for cerebral function: the unit module and the distributed system. In: Schmitt FO, Worden FG (eds) The neurosciences: fourth study program. The MIT Press, Cambridge, pp 21–42
Mountcastle VB (1997) The columnar organization of the neocortex. Brain 120:701–722
Mulert C, Pogarell O, Hegerl U (2008) Simultaneous EEG-fMRI: perspectives in psychiatry. Clin EEG Neurosci 39:61–64
Musso F, Brinkmeyer J, Mobascher A, Warbrick T, Winterer G (2010) Spontaneous brain activity and EEG microstates. A novel EEG/fMRI analysis approach to explore resting-state networks. NeuroImage 52:1149–1161
Niedermeyer E, Lopes da Silva FH (eds) (2005) Electroencephalography: Basic principles, clinical applications, and related fields, 5th edn. Lippincott Williams & Wilkins, Philadelphia
Nieuwenhuys R, Voogd J, van Huijzen C (2008) The human central nervous system, 4th edn. Springer, Berlin, pp 491–679
Norris DG (2006) Principles of magnetic resonance assessment of brain function. J Magn Reson Imaging 23:794–807
Nunez PL (1974a) The brain wave equation: a model for the EEG. Math Biosci 21:279–297
Nunez PL (1974b) Wave-like properties of the alpha rhythm. IEEE Trans Biomed Eng 21:473–482
Nunez PL (1981) Electric fields of the brain: the neurophysics of EEG, 1st edn. Oxford University Press, New York
Nunez PL (1995) Neocortical dynamics and human EEG rhythms. Oxford University Press, New York
Nunez PL, Srinivasan R (2006) A theoretical basis for standing and traveling brain waves measured with human EEG with implications for an integrated consciousness. Clin Neurophysiol 117:2424–2435

Nunez PL, Reid L, Bickford RG (1978) The relationship of head size to alpha frequency with implications to a brain wave model. Electroencephalogr Clin Neurophysiol 44:344–352
Nunez PL, Wingeier BM, Silberstein RB (2001) Spatial-temporal structures of human alpha rhythms: theory, microcurrent sources, multiscale measurements, and global binding of local networks. Hum Brain Mapp 13:125–164
Pakkenberg B, Gundersen HJG (1997) Neocortical neuron number in humans: effect of sex and age. J Comp Neurol 384:312–320
Perea G, Araque A (2010) GLIA modulates synaptic transmission. Brain Res Rev 63:93–102
Perea G, Navarrete M, Araque A (2009) Tripartite synapses: astrocytes process and control synaptic information. Trends Neurosci 32:421–431
Peters A, Sethares C (1997) The organization of double bouquet cells in monkey striate cortex. J Neurocytol 26:779–797
Petersen CCH (2007) The functional organization of the barrel cortex. Neuron 56:339–355
Phillips AJK, Robinson PA (2007) A quantitative model of sleep-wake dynamics based on the physiology of the brainstem ascending arousal system. J Biol Rhythms 22:167–179
Rabinovich MI, Huerta R, Laurent G (2008a) Neuroscience. Transient dynamics for neural processing. Science 321:48–50
Rabinovich MI, Huerta R, Varona P, Afraimovich VS (2008b) Transient cognitive dynamics, metastability, and decision making. PLoS Comput Biol 4:e1000072
Raichle ME, MacLeod AM, Snyder AZ, Powers WJ, Gusnard DA, Shulman GL (2001) A default mode of brain function. Proc Natl Acad Sci USA 98:676–682
Regan D (1989) Human brain electrophysiology: evoked potentials and evoked magnetic fields in science and medicine. Elsevier, New York
Rennie CJ, Wright JJ, Robinson PA (2000) Mechanisms of cortical electrical activity and emergence of gamma rhythm. J Theor Biol 205:17–35
Rennie CJ, Robinson PA, Wright JJ (2002) Unified neurophysical model of EEG spectra and evoked potentials. Biol Cybern 86:457–471
Riera JJ, Aubert E, Iwata K, Kawashima R, Wan X, Ozaki T (2005) Fusing EEG and fMRI based on a bottom-up model: inferring activation and effective connectivity in neural masses. Philos Trans R Soc B 360:1025–1041
Riera JJ, Wan X, Jimenez JC, Kawashima R (2006) Nonlinear local electrovascular coupling. I: a theoretical model. Hum Brain Mapp 27:896–914
Riera JJ, Jimenez JC, Wan X, Kawashima R, Ozaki T (2007) Nonlinear local electrovascular coupling. II: from data to neuronal masses. Hum Brain Mapp 28:335–354
Robinson PA (2006) Patchy propagators, brain dynamics, and the generation of spatially structured gamma oscillations. Phys Rev E 73:041904
Robinson PA, Rennie CJ, Wright JJ (1997) Propagation and stability of waves of electrical activity in the cerebral cortex. Phys Rev E 56:826–840
Robinson PA, Rennie CJ, Wright JJ, Bahramali H, Gordon E, Rowe DL (2001) Prediction of electroencephalographic spectra from neurophysiology. Phys Rev E 63:021903
Robinson PA, Rennie CJ, Rowe DL (2002) Dynamics of large-scale brain activity in normal arousal states and epileptic seizures. Phys Rev E 65:041924
Rockland KS, Ichinohe N (2004) Some thoughts on cortical minicolumns. Exp Brain Res 158: 265–277
Rockland KS, Pandya DN (1979) Laminar origins and terminations of cortical connections of the occipital lobe in the rhesus monkey. Brain Res 179:3–20
Rodrigues S, Terry JR, Breakspear M (2006) On the genesis of spike-wave oscillations in a mean-field model of human thalamic and corticothalamic dynamics. Phys Lett A 355:352–357
Rowe DL, Robinson PA, Gordon E (2005) Stimulant drug action in attention deficit hyperactivity disorder (ADHD): inference of neurophysiological mechanisms via quantitative modelling. Clin Neurophysiol 116:324–335
Scheperjans F, Eickhoff SB, Hömke L, Mohlberg H, Hermann K, Amunts K, Zilles K (2008) Probabilistic maps, morphometry, and variability of cytoarchitectonic areas in the human superior parietal cortex. Cereb Cortex 18:2141–2157

Shibasaki H (2008) Human brain mapping: hemodynamic response and electrophysiology. Clin Neurophysiol 119:731–743
Silva LR, Amitai Y, Connors BW (1991) Intrinsic oscillations of neocortex generated by layer 5 pyramidal neurons. Science 251:432–435
Sotero RC, Trujillo-Barreto NJ (2008) Biophysical model for integrating neuronal activity, EEG, fMRI and metabolism. NeuroImage 39:290–309
Sotero RC, Trujillo-Barreto NJ, Iturria-Medina Y, Carbonell F, Jimenez JC (2007) Realistically coupled neural mass models can generate EEG rhythms. Neural Comput 19:478–512
Spiegler A, Kiebel SJ, Atay FM, Knösche TR (2010) Bifurcation analysis of neural mass models: impact of extrinsic inputs and dendritic time constants. NeuroImage 52:1041–1058
Spruston N (2008) Pyramidal neurons: dendritic structure and synaptic integration. Nat Rev Neurosci 9:206–221
Stam CJ (2005) Nonlinear dynamical analysis of EEG and MEG: review of an emerging field. Clin Neurophysiol 116:2266–2301
Stam CJ, Pijn JPM, Suffczyński P, Lopes da Silva FH (1999) Dynamics of the human alpha rhythm: evidence for non-linearity? Clin Neurophysiol 110:1801–1813
Stephan KE, Kamper L, Bozkurt A, Burns GAPC, Young MP, Kötter R (2001) Advanced database methodology for the collation of connectivity data on the macaque brain (CoCoMac). Philos Trans R Soc B 356:1159–1186
Steriade M (2005) Cellular substrates of brain rhythms. In: Niedermeyer E, Lopes da Silva FH (eds) Electroencephalography: basic principles, clinical applications, and related fields, 5th edn. Lippincott Williams & Wilkins, Philadelphia, pp 31–83
Steyn-Ross ML, Steyn-Ross DA, Sleigh JW, Liley DTJ (1999) Theoretical electroencephalogram stationary spectrum for a white-noise-driven cortex: evidence for a general anesthetic-induced phase transition. Phys Rev E 60:7299–7311
Steyn-Ross ML, Steyn-Ross DA, Sleigh JW (2004) Modelling general anaesthesia as a first-order phase transition in the cortex. Prog Biophys Mol Biol 85:369–385
Steyn-Ross DA, Steyn-Ross ML, Sleigh JW, Wilson MT, Gillies IP, Wright JJ (2005a) The sleep cycle modelled as a cortical phase transition. J Biol Phys 31:547–569
Steyn-Ross ML, Steyn-Ross DA, Sleigh JW, Wilson MT, Wilcocks LC (2005b) Proposed mechanism for learning and memory erasure in a white-noise-driven sleeping cortex. Phys Rev E 72:061910
Steyn-Ross ML, Steyn-Ross DA, Wilson MT, Sleigh JW (2009) Modeling brain activation patterns for the default and cognitive states. NeuroImage 45:298–311
Steyn-Ross ML, Steyn-Ross DA, Sleigh JW, Wilson MT (2011) A mechanism for ultra-slow oscillations in the cortical default network. Bull Math Biol (in press). doi:10.1007/s11538-010-9565-9
Stufflebeam SM, Rosen BR (2007) Mapping cognitive function. Neuroimaging Clin N Am 17: 469–484
Suffczyński P, Lopes da Silva FH, Parra J, Velis DN, Kalitzin SN (2005) Epileptic transitions: model predictions and experimental validation. J Clin Neurophysiol 22:288–299
Szentágothai J (1978) The Ferrier lecture, 1977 – the neuron network of the cerebral cortex: a functional interpretation. Proc R Soc Lond B 201:219–248
Szentágothai J (1983) The modular architectonic principle of neural centers. Rev Physiol Biochem Pharmacol 98:11–61
Tang Y, Nyengaard JR, De Groot DM, Gundersen HJG (2001) Total regional and global number of synapses in the human brain neocortex. Synapse 41:258–273
Thomson AM, Bannister AP (2003) Interlaminar connections in the neocortex. Cereb Cortex 13: 5–14
Toga AW, Thompson PM, Mori S, Amunts K, Zilles K (2006) Towards multimodal atlases of the human brain. Nat Rev Neurosci 7:952–966
Tsuda I (2001) Toward an interpretation of dynamic neural activity in terms of chaotic dynamical systems. Behav Brain Sci 24:793–810

Valdés-Hernández PA, Ojeda-Gonzéalez A, Martínez-Montes E, Lage-Castellanos A, Virués-Alba T, Valdés-Urrutia L, Valdes-Sosa PA (2010) White matter architecture rather than cortical surface area correlates with the EEG alpha rhythm. NeuroImage 49:2328–2339

Valdes-Sosa PA, Sánchez-Bornot JM, Sotero RC, Iturria-Medina Y, Alemán-Gómez Y, Bosch-Bayard J, Carbonell F, Ozaki T (2009) Model driven EEG/fMRI fusion of brain oscillations. Hum Brain Mapp 30:2701–2721

van Albada SJ, Robinson PA (2009) Mean-field modeling of the basal ganglia-thalamocortical system. I. Firing rates in healthy and Parkinsonian states. J Theor Biol 257:642–663

Van Essen DC (2005) A Population-Average, Landmark- and Surface-based (PALS) atlas of human cerebral cortex. NeuroImage 28:635–662

van Rotterdam A, Lopes da Silva FH, van den Ende J, Viergever MA, Hermans AJ (1982) A model of the spatial-temporal characteristics of the alpha rhythm. Bull Math Biol 44:283–305

Waage P, Guldberg CM, Abrash HI (1986) Studies concerning affinity (English translation). J Chem Educ 63:1044–1047

Wendling F, Bellanger JJ, Bartolomei F, Chauvel PY (2000) Relevance of nonlinear lumped-parameter models in the analysis of depth-EEG epileptic signals. Biol Cybern 83:367–378

Wendling F, Hernández AI, Bellanger JJ, Chauvel PY, Bartolomei F (2005) Interictal to ictal transition in human temporal lobe epilepsy: insights from a computational model of intracerebral EEG. J Clin Neurophysiol 22:343–356

White EL (1989) Cortical circuits. Synaptic organization of the cerebral cortex: structure, function and theory. Birkhäuser, Boston

Williamson SJ, Kaufman L (1989) Advances in neuromagnetic instrumentation and studies of spontaneous brain activity. Brain Topogr 2:129–139

Wilson HR, Cowan JD (1972) Excitatory and inhibitory interactions in localized populations of model neuron. Biophys J 12:1–24

Wilson HR, Cowan JD (1973) A mathematical theory of the functional dynamics of cortical and thalamic nervous tissue. Kybernetik 13:55–80

Wilson MT, Steyn-Ross DA, Sleigh JW, Steyn-Ross ML, Wilcocks LC, Gillies IP (2006) The K-complex and slow oscillation in terms of a mean-field cortical model. J Comput Neurosci 21:243–257

Wilson MT, Steyn-Ross ML, Steyn-Ross DA, Sleigh JW (2007) Going beyond a mean-field model for the learning cortex: second-order statistics. J Biol Phys 33:213–246

Wolfe J, Houweling AR, Brecht M (2010) Sparse and powerful cortical spikes. Curr Opin Neurobiol 20:306–312

Wright JJ (1997) EEG simulation: variation of spectral envelope, pulse synchrony and ≈40 hz oscillation. Biol Cybern 76:181–194

Wright JJ, Liley DTJ (1995) Simulation of electrocortical waves. Biol Cybern 72:347–356

Wright JJ, Liley DTJ (1996) Dynamics of the brain at global and microscopic scales: neural networks and the EEG. Behav Brain Sci 19:285–320

Wu JY, Huang XY, Zhang C (2008) Propagating waves of activity in the neocortex: what they are, what they do. Neuroscientist 14:487–502

Zilles K, Amunts K (2010) Centenary of Brodmanns map – conception and fate. Nat Rev Neurosci 11:139–145

Chapter 12
Cellular Spacing: Analysis and Modelling of Retinal Mosaics

Stephen J. Eglen

Abstract A key step in nervous system development is the spatial positioning of neurons within a structure. In this chapter I review the mechanisms by which the cellular spacing of neuronal networks emerges. In particular, I focus on the spatial distribution of neurons within the retina. The retina is ideal for studying such developmental mechanisms because of its multilayered structure and specific neurochemical markers can reliably label all neurons of a given type. This chapter describes the quantitative methods used for assessing spatial regularity of neuronal distributions and computational methods for simulating these distributions.

12.1 Introduction

The retina is a relatively small neural structure located at the back of the eye. It is a multilayered structure (Fig. 12.1): photoreceptors toward the back of the eye convert light into neural activity which then propagates through several layers where it is modulated by lateral connections (via horizontal and amacrine cells) until reaching the ganglion cell layer. Retinal ganglion cells (RGCs) encode the visual scene into spike trains which then leave along the optic nerve and into the brain for further processing. For a general overview of retinal processing, see Wässle (2004).

The layered organisation of the retina makes for relatively easy identification of cell type, as each cell type tends to occur in only one layer of the retina. Taking a cross section through one of the layers, such as the ganglion cell layer, reveals a further aspect of structural organisation within the retina. Cells of a given type are positioned semi-regularly through a layer, forming what is commonly termed a retinal mosaic, due to the way that the cell bodies and their dendrites tile the surface (Fig. 12.2). (For the rest of this article, when I refer to retinal mosaics, it

S.J. Eglen (✉)
Cambridge Computational Biology Institute, University of Cambridge, Cambridge, UK
e-mail: sje30@cam.ac.uk

N. Le Novère (ed.), *Computational Systems Neurobiology*,
DOI 10.1007/978-94-007-3858-4_12,

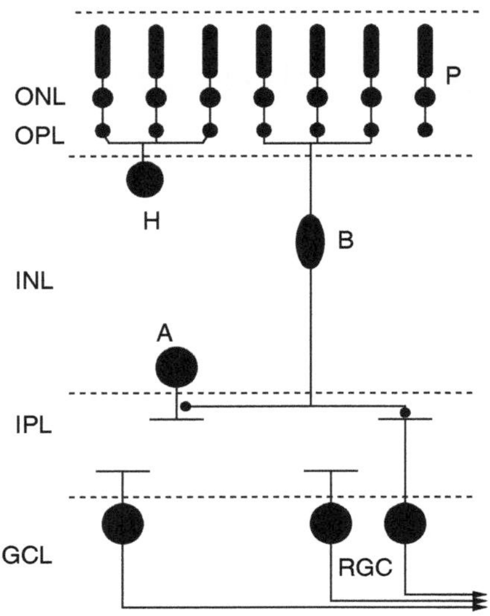

Fig. 12.1 Cross-section of a vertebrate retina. The retina is organised into three cell-dense layers (*ONL* outer nuclear layer, *INL* inner nuclear layer, *GCL* ganglion cell layer). Connections between layers are predominantly made in the *OPL* (outer plexiform layer) and *IPL* (inner plexiform layer). Each of these layers can be regarded as a two-dimensional sheet. There are five major classes of retinal neuron: photoreceptors (*P*), horizontal cells (*H*), bipolar cells (*B*), amacrine cells (*A*) and retinal ganglion cells (*RGC*). Photoreceptors transduce light into neural activity, which then propagates to the RGCs via the bipolar cells. Both the horizontal and amacrine cells have widespread lateral interactions that modulate neural activity. Finally, axons of the RGCs form the optic nerve, carrying the neural signal from the eye to the brain for further processing. Cells of a particular class are usually restricted to a given layer within the retina, allowing for easier identification of individual cell classes. The spatial location of all neurons within a layer can then be revealed using various staining techniques (see Figs. 12.2 and 12.3)

will mostly refer to the positioning of the cell body, assuming that the surrounding dendritic arbor is also tiled. For further details on modelling dendritic growth, see Chap. 13.) Most cell types form independent mosaics (see later for a rigorous definition of independence), such that the presence of a retinal mosaic is often used to determine whether a given population forms an independent type (Cook 1998). Together with reliable biochemical markers for reliably staining individual cell types, this has meant the catalogue of cell types (five classes, divided into about 60 types, depending on species) within the retina is nearing completion (Masland 2004).

What function might such retinal mosaics perform? For photoreceptors, having a regular spacing of neurons is presumably necessary to sample the entire visual field, avoiding any 'blind spots'. However, the spatial distribution of photoreceptors is slightly different to that of other cell types as they tend to be tightly packed against each other (Fig. 12.3). Short wavelength cones ('blue cones') tend to be regularly spaced, as they are relatively sparse compared to the medium/long wavelength cones, which are randomly arranged (Roorda et al. 2001). However, in other layers

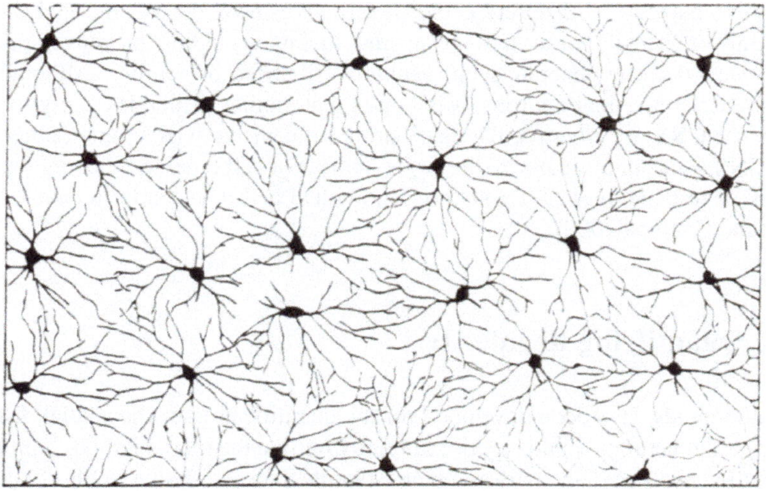

Fig. 12.2 Regular arrangement of on-centre alpha retinal ganglion cells from cat retina. The area shown is approximately 1.7 × 1.2 mm. The dendrites around each cell body tile the retinal surface, and the cell bodies seem roughly equally-spaced from each other. In this article, 'field' means the area of tissue within which the neurons are observed. Other neurons (e.g. off-centre alpha retinal ganglion cells) within the same layer are not shown (Reproduced by permission from Macmillan Publishers Ltd.: Nature **292:**344–345, copyright 1981)

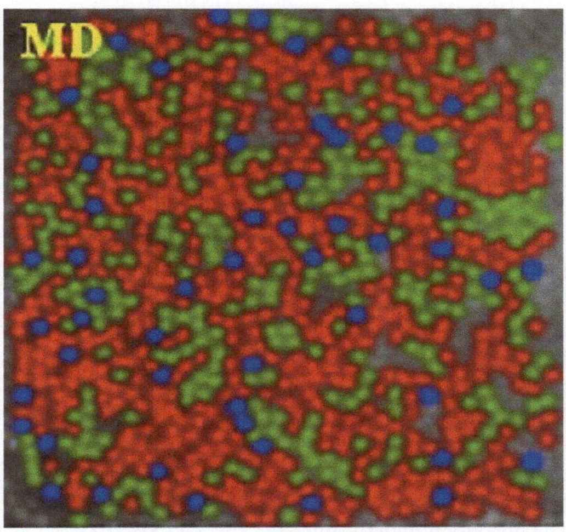

Fig. 12.3 Close packing of cone photoreceptors in a human retina (subject named MD). The three different classes of photoreceptor are coloured *blue* (short wavelength), *green* (medium wavelength) and *red* (long wavelength). Approximate width of view: 20 arc min (Reproduced from Hofer et al. 2005 with permission of the Society for Neuroscience)

of the retina, after sampling of the visual world, the advantages of a regular mosaic are not so obvious. One hypothesis is that regular arrangements of individual cell types in different layers may aid in the developmental wiring of connections between cell types (Galli-Resta 2002). (Chap. 14 discusses the wiring of connections between neurons.) However, this wiring hypothesis has yet to be explored.

In this chapter I will describe the quantitative methods for the analysis and modelling of retinal mosaics, with an aim to understanding the developmental mechanisms that can generate such regular distributions of neurons. Although this work focuses on retinal neurons, it is hoped that similar principles apply to other parts of the CNS. Whether regular distributions of neurons exist or not in other parts of the CNS is still unclear, due to the larger number of cell types in other regions, and the lack of reliable markers for staining individual cell types early in development (Cook and Chalupa 2000).

12.2 Quantifying Regularity

Several methods have been developed for the quantification of the spatial distribution of retinal neurons. In this section, I briefly outline the main methods employed. The methods will be demonstrated on an example data set, shown in Fig. 12.4.

12.2.1 Regularity Index

The most popular method for quantifying mosaic measures is the regularity index (Wässle and Riemann 1978). For each neuron in the field, the distance to the

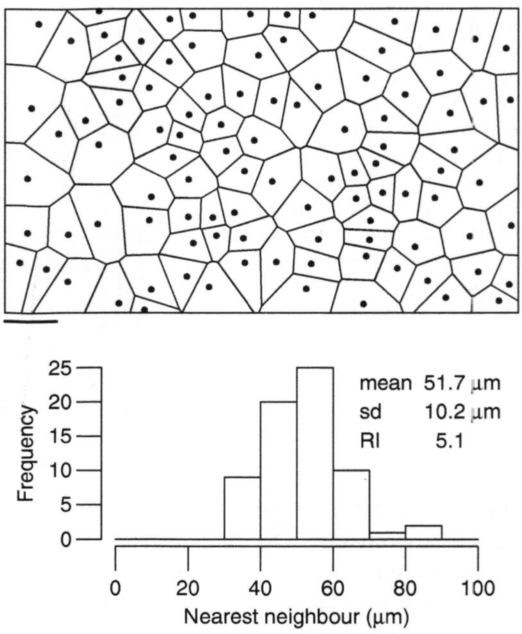

Fig. 12.4 Example mosaic (synthetic data set). Neurons are drawn as *circles* with 10 µm diameter representing typical soma size; scale bar: 100 µm. Each neuron is surrounded by its Voronoi polygon, showing the region of space closest to that point. The histogram underneath shows the distribution of nearest-neighbour distances, along with the regularity index (*RI*) of 5.1. This RI is typical of regular mosaics, such as cholinergic amacrine neurons

nearest-neighbouring neuron is measured and plotted in a histogram. The regularity index (RI) is simply the mean of this distribution divided by its standard deviation. For the example in Fig. 12.4, the RI of 5.1 indicates a highly regular mosaic.

Calculating a measure such as this immediately raises the question of how to interpret this number. Cook (1996) first investigated the properties of the RI (termed the conformity ratio in his article). The baseline to compare against is when the neurons are placed at random throughout the field—this is termed complete spatial randomness (CSR). The RI for neurons arranged randomly is 1.9, and the more regular the arrangement, the higher the RI. For retinal mosaics observed to date, the RI is typically 3–8. However, the exact threshold for determining whether the mosaic is non-randomly arranged depends on the number of neurons and the geometry of the field (Cook 1996). Furthermore, the physical size of the soma may introduce lower limits onto the size of the nearest-neighbour distances. However, all of these can be handled appropriately by using Monte-Carlo techniques, see later.

12.2.2 Autocorrelation Methods

A key limitation of the RI measure is that it is based only on the distribution of distances to nearest-neighbours. Autocorrelation-based methods are more powerful as they include the relative distance of all points, not just the nearest-neighbour. Such autocorrelation methods were made popular in the retinal mosaic literature by Rodieck (1991), although these methods were introduced much earlier in the spatial statistics literature (Ripley 1976). An autocorrelation plot is created by taking one cell as the reference neuron, and plotting the relative position of all other neurons in a plot. This is repeated using each neuron as a reference neuron to build up the autocorrelation plot shown in Fig. 12.5a. Annuli are drawn 10 μm apart, and clearly show a 'exclusion zone' effect: no two neurons are closer than about 40 μm apart, but beyond this distance, there is no further structure to the plot. This indicates that neurons perhaps are operating under the rule that they simply should avoid becoming 'too close' to each other, but there are no further constraints imposed.

Within each annulus of the autocorrelation plot, there is rarely little spatial variation. Each annulus can thus concisely be described by one number, the density of points in that annulus. The density of each annulus then forms the density recovery profile (DRP; Fig. 12.5b) which is the usual way of summarising the autocorrelation. Again this clearly shows the exclusion zone principle acting up to around 40 μm, and beyond that the density of each annulus fluctuates around the mean density (horizontal line). Further quantification of the DRP is possible; for example, the size of the exclusion zone can be quantified by the effective radius shown in the figure as a vertical line; see (Rodieck 1991) for further details. The DRP and its associated statistics are useful complements to the nearest-neighbour methods.

One limitation of the DRP approach is that the results may be dependent on the size of each annulus: smaller annuli should lead to more sensitive estimation of DRP

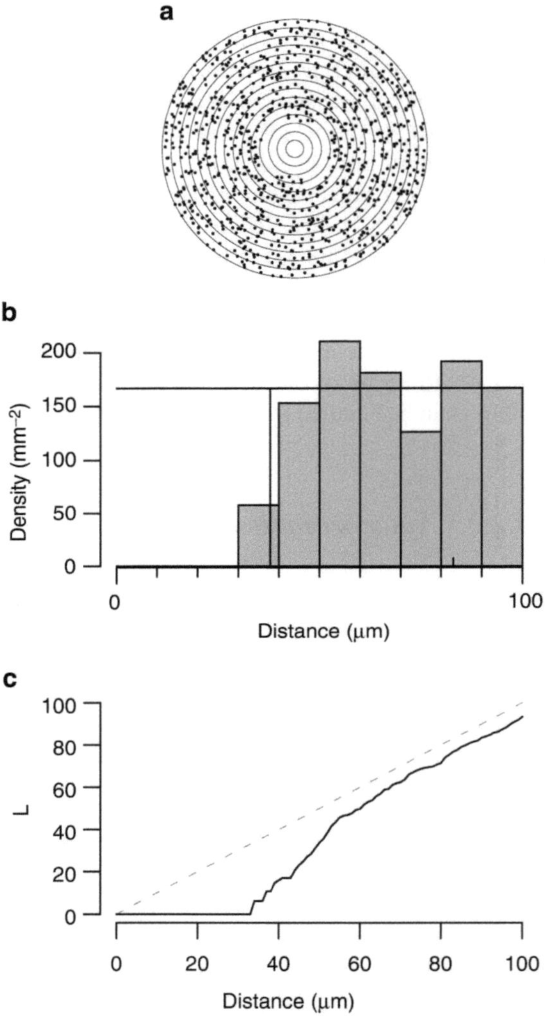

Fig. 12.5 Autocorrelation-based analysis of the mosaic shown in Fig. 12.4. (**a**) Autocorrelation plot. Each *dot* represents the position of a cell relative to another neuron in the field. Annuli are spaced 10 μm apart. The lack of cells in the first *four annuli* indicate the presence of an exclusion zone. (**b**) Density recovery profile (*DRP*). Each bar in the histogram shows the density of points in the corresponding annulus of the autocorrelation plot. The *horizontal line* indicates the mean density of points (166 cells/mm^2) and the *vertical line* (at 38 μm) shows the effective radius (see text). (**c**) The *L* function is the scaled integral of the DRP. *Solid line* indicates the *L* function for the mosaic; *dotted line* indicates the curve that would be expected if the points were arranged randomly

parameters such as the effective radius. However with smaller annuli, the neuronal counts within each annulus can be quite small and thus the density estimates may vary significantly. Cumulative histograms of counts are more robust, by avoiding bin sizes, and have been proposed in spatial statistics (Ripley 1976; Diggle 2002). In particular, Ripley's *K* function is effectively the integral of the DRP. The *K* function is defined as:

$$K(t) = \frac{|A|}{n(n-1)} \sum_{i=1}^{n} \sum_{j \neq i} w(i, j)^{-1} I(||\mathbf{x}_i - \mathbf{x}_j|| \leq t) \tag{12.1}$$

In this function, $\mathbf{x}_i$ is a 2-d vector representing the position of neuron i; hence $I(\cdot)$ counts how many pairs of neuron are less than or equal to some distance t apart. $|A|$ is the field area, and n is the number of neurons. The term $w(i, j)$ is a weighting factor to correct for border effects, as described in the next section (Diggle 2002). Under the null hypothesis of CSR, the theoretical K function is $K(t) = \pi t^2$. Finally, $L(t) = (K(t)/\pi)^{0.5}$ is used for plotting purposes.

Figure 12.5c shows the L function corresponding to the DRP in panel B. The null hypothesis of CSR is given by $L(t) = t$, shown in the dotted line, and deviations of $L(t)$ below that line indicate regularity, as is the case here. ($L(t) > t$ would indicate that the neurons are clustered, rather than spaced-apart.)

Many other statistics are also available, including Voronoi-based measures, as well as other cumulative distance functions from the spatial statistics literature (notably the F and G functions); for further details, see (Diggle 2002). It is an open question as to which of these functions are most useful for discriminating patterns, hence it is good practice to compare the effectiveness of several functions.

12.2.3 Boundary Effects

Figure 12.6 demonstrates the problems associated with boundary effects when quantifying retinal mosaics. For example, when finding the nearest neighbours, for cells in the centre of the region (e.g. cell 1) it is clear which cell is the nearest neighbour (cell 2). However, for a cell close to the boundary, such as cell a, although cell b is the closest within the field, there might have been another cell just to the right of the field that was closer to cell a (e.g. at point c or anywhere within the circle outside the field). Hence the estimate of the nearest neighbour for cells at the border is unreliable. To determine which cells are located at the border, we can add a ‘safety zone’, marked by the dotted line, and consider only the nearest-neighbour distances for neurons within the safety zone (filled symbols). However, what size of safety zone should be imposed? The larger the safety zone around the edge of the field, the smaller the impact of boundary cells. With larger safety zones however, fewer neurons are left within the safety region, and hence fewer samples to estimate the RI.

Imposing a safety zone is therefore simple, but requires another parameter (the width of the safety zone) and often discards a lot of data. Another technique for identifying border cells is to use the Voronoi tessellation, and label neurons as being at the border if their Voronoi polygon intersects with the field boundary. However, a subtler approach to handling boundary effects is to use weighting factors such that a contribution of e.g. each nearest-neighbour distance is measured, but the distances are weighted according to how close a neuron is to the border. One such edge-correction technique is to measure the fraction of the circumference of a circle (e.g. shown for point a in Fig. 12.6) that lies within the field (Ripley 1976; Diggle 2002). This edge-correction term accounts for the $w(i, j)$ term in Eq. 12.1.

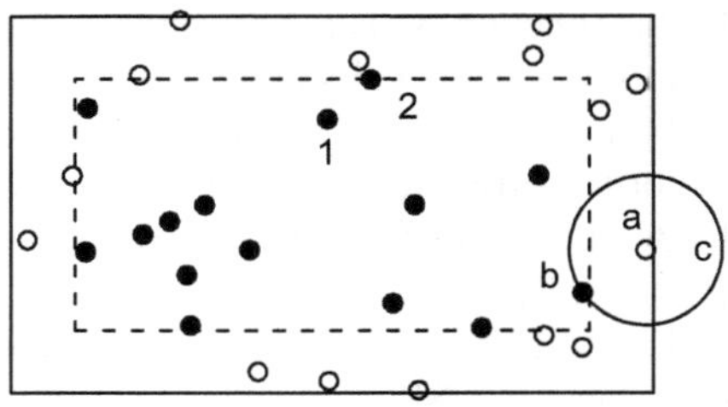

Fig. 12.6 Demonstration of boundary effects. The *solid rectangle* indicates the boundary region under study, with a central safety zone shown as *dotted lines*. Individual points (1, 2, a, b and c) are referred to in the text

Another concern with boundary procedures that is often overlooked is the size of the field itself. Often, retinal mosaics are described simply by the x,y locations of each neuron—the coordinates of the (usually rectangular) field which determine which neurons are recorded are often not kept. As seen above, the position of the boundary is important, and affects the reliability of the measures taken from the mosaic. In the absence of a reported boundary region, one can be estimated by using the extreme x and y coordinates of all the neurons. This is the smallest possible field, and although it is the maximum likelihood estimate (Ripley and Rasson 1977), it is obviously an underestimate.

Ideally therefore, the field is decided in advance, placed onto the retinal tissue, and the positions of all neurons within that field should be recorded. What size should the field be? For practical purposes, most software assumes rectangular region (although some, such as SPLANCS (Rowlingson and Diggle 1993) can handle arbitrary closed polygons). It should also be large enough to contain enough cells (e.g. at least 50), but small enough so that long-range spatial variations in density can be ignored. Again, some methodologies exist for handling non-homogeneities in spatial density across the field (Baddeley and Turner 2005). However, often the long-range density variations observed across the retinal surface mean that investigators do not use very large fields, typically smaller than 1×1 mm.

12.3 Phenomenological Approaches to Modelling

What are the mechanisms underlying the development of these retinal mosaics? Progenitors of retinal neurons divide at the location of the photoreceptor layer, and once the neurons become postmitotic (i.e. stop dividing), they migrate through the retina to the appropriate layer for a given cell type. Certain cell types then migrate laterally within a layer to reach their final position (Reese and Galli-Resta 2002). As well as these migratory processes, many other developmental mechanisms are thought to be involved, including lateral inhibition of cell fate and cell death (Reese and Galli-Resta 2002). For a general review of the developmental mechanisms, see Cook and Chalupa (2000).

In addition to experimental approaches to understanding mosaic formation, theoretical modelling can help us evaluate the potential of different developmental mechanisms for generating such regular patterns. In this chapter I compare two styles of modelling:

Phenomenological the focus is on generating model output that looks similar to observed data, using mechanisms that may or may not be biologically plausible.

Mechanistic the primary concern is on modelling the cellular processes thought to be involved, rather than focusing on model output.

These two approaches are common in areas of biological modelling (e.g. see Nathan and Muller-Landau 2000). In this section, I describe the phenomenological approaches; mechanistic approaches are discussed in the following section.

12.3.1 Exclusion Zone Models

The exclusion zone model is fairly straightforward and simply embodies the local rule that no two neurons should come closer to each other than some minimal distance. This local exclusion zone should then be able to recreate the hole seen in autocorrelation plots. This style of model was first applied to retinal mosaics by Shapiro et al. (1985), who examined the spatial distribution of blue cone photoreceptors in macaque retinas. However, the exclusion zone model has been popularised by the more recent work of Galli-Resta and colleagues (Galli-Resta et al. 1997), where the model is termed the d_{min} model, where d_{min} is the main parameter of the model, representing the diameter of the exclusion zone. The value of d_{min} is normally not fixed, but drawn from a normal distribution with a given mean and standard deviation. The other parameters of the model (the field size and the number of cells) are taken from the observed mosaic being modelled.

The d_{min} model is an example of a serial model, where neurons are positioned one-by-one into the field (Fig. 12.7). The starting point therefore is an empty field, the same size as the mosaic being modelled. A trial point is selected at random within the field, and a value for d_{min} is sampled from the normal distribution. If the nearest-neighbouring neuron in the field is closer than d_{min}, the trial cell is rejected, otherwise the trial cell is added into the field. This process continues until either the desired number of neurons have been added into the field or until it is no longer possible to fit any more neurons.

Once a field has been simulated using the d_{min} model, it can be compared against the observed mosaic (Fig. 12.8). Visual comparisons are often inadequate, and so we use the quantitative methods (outlined in Sect. 12.2) to compare observed with simulated mosaics. We take advantage of the fact that we can generate many instances of simulated mosaics to estimate the goodness of fit. For example, if we use the RI as the metric to quantify regularity, we calculate the RI of the observed mosaic and the RI of each of 99 simulated mosaics from the d_{min} model (fixing the parameter values, and just varying the random number generator for positioning neurons). Informally, for a good fit, the RI of the observed mosaic should fall within the range of RIs generated by the d_{min} model.

This assessment of goodness of fit can then be quantified by calculating an empirical p value. If the RI of the observed mosaic is x_1 and the RI of $n - 1$

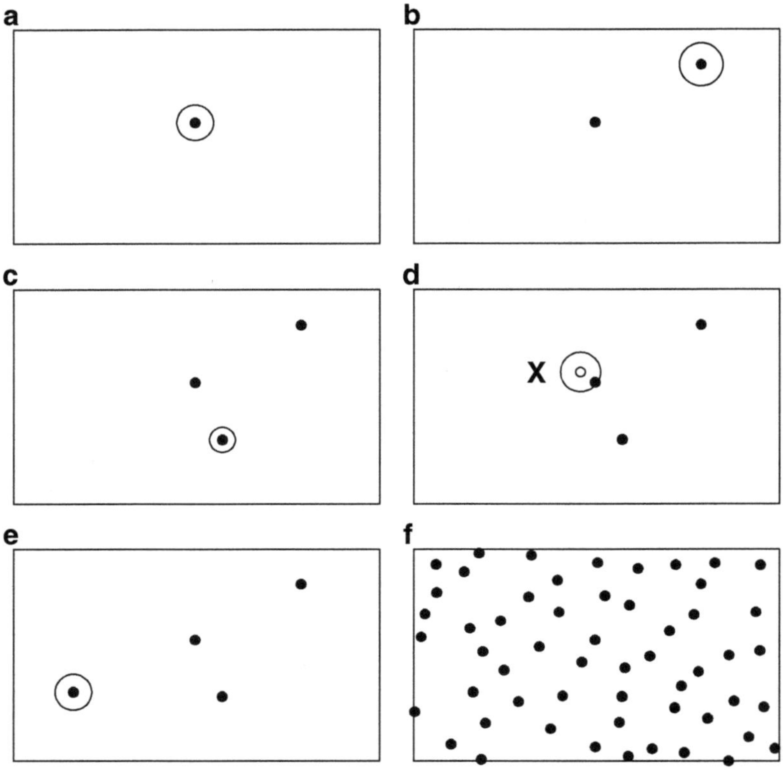

Fig. 12.7 Generation of a simulated mosaic using the d_{min} exclusion zone. In each panel, the rectangle shows the field with the trial cell surrounded by a circular exclusion zone. If no other cell is positioned within the exclusion zone, the trial cell is accepted into the field (**panel d** shows a trial cell being rejected). **Panel a** shows the starting condition: since there are no previous cells, the first trial cell is always accepted. **Panel f** shows the final mosaic after the desired number of cells have been added

simulated mosaics are $x_2 \ldots x_n$, then for each mosaic i we calculate a u_i value which determines the difference between the RI for mosaic i and the average RI of all other mosaics:

$$u_i = \text{abs}\left(x_i - \frac{1}{n-1}\sum_{j \neq i} x_j\right)$$

The expectation then is that if the model is a good fit to the data, u_1 should be of similar magnitude to all other u scores. A p value can then be calculated by sorting the values of u, largest first, and then counting the position of u_1 and dividing by n. For example, if u_1 was the ninth largest value of u out of 100, the p value would be 0.09. In this context, small p values indicates a poor fit of the model to the data, and the better the fit, the larger the p value. (Hence this test is one-tailed,

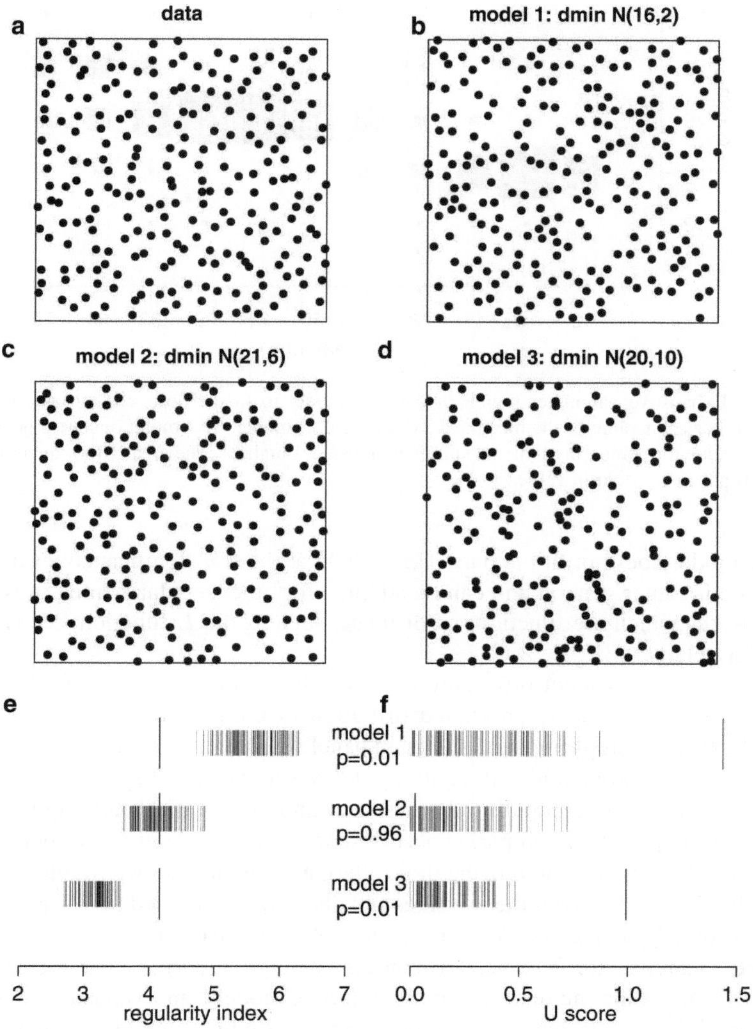

Fig. 12.8 Fitting the d_{min} model to an example mosaic. (**a**) Example observed mosaic (cholinergic amacrine cells in rat). The field of view is $400 \times 400\ \mu m^2$. (**b**)–(**d**) Example simulations using three different values for d_{min} parameters; in each case the d_{min} value is drawn from a normal distribution with given mean and s.d. (**e**)–(**f**) Assessing the fit of each model to the data. In (**e**), each row shows the regularity index from 99 simulations of each model; the larger vertical line in each case is the regularity of the observed data in (**a**) (4.16). Informally, the model fits the data if the observed RI falls within the range of the RIs generated by the model. **Panel f** shows the *u* score for each mosaic (real or simulated), with the score for the real mosaic drawn with a larger line. Whereas model 1 and 3 can be rejected, model 2 fits the data

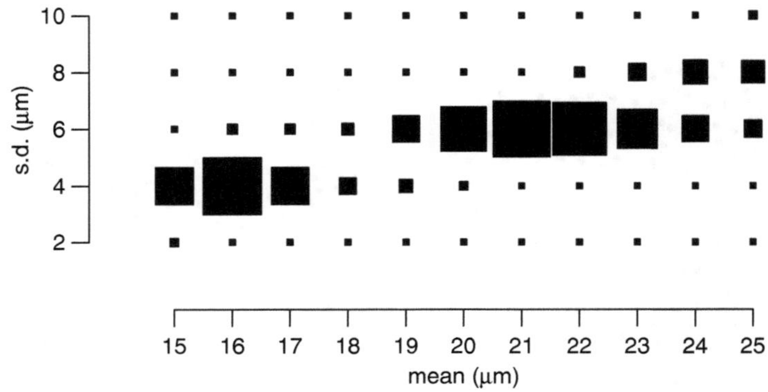

Fig. 12.9 Exhaustive parameter search of the d_{min} model to fit the observed mosaic shown in Fig. 12.8a. For each value of the mean and s.d. of the d_{min} model, 99 simulations were generated, and the p value for comparing real and simulated mosaics obtained. The area of the square in each case is proportional to the p value

and the model does not fit the data if $p \leq 0.05$ at the 5% significance level.) This procedure is rather general and can be adopted for other regularity measures, even for measures which are functions of distance, such as the L function; see (Diggle 1986) for details.

This model assessment procedure is demonstrated in Fig. 12.8. Three different parameter values for the d_{min} normal distribution are compared to see how well they can replicate an example mosaic (shown in panel A). Out of the three models, model 2 shows the best fit of simulated regularity indexes to the observed regularity index. This is confirmed by computing the u scores and p values. Clearly, quantitative methods are required for comparing observed data and model output, as there are no strong visual differences among the three alternative models shown in Fig. 12.8b–d.

Finally, Fig. 12.9 shows the p values obtained by this procedure for a range of different model parameters. As the d_{min} model is relatively fast, such exhaustive parameter searches are feasible, and can easily pinpoint parts of parameter space where the model fits the data. For more complex models, an exhaustive approach is not feasible, and instead a heuristic search procedure should be used. Some other phenomenological models from the spatial statistics literature have specialised fitting procedures—for example see the R package SPATSTAT for details (Baddeley and Turner 2005).

12.3.2 Evaluation of d_{min} Model

The d_{min} model has been used to fit a wide range of mosaics of different cell types and different species (Galli-Resta et al. 1997, 1999; Cellerino et al. 2000; Raven et al. 2003). This strongly suggests that a homotypic exclusion zone is *sufficient* to generate a retinal mosaic. (In this context, *homotypic* means that interactions are restricted to cells of the same type; *heterotypic* interactions involve cells of different types.) This means it is unlikely that long-range interactions are required between cells of the same type, nor are interactions needed between cells of different types, confirming results from cross-correlation analysis (Rockhill et al. 2000; Mack 2007). However, the d_{min} model does not say anything about the biological mechanisms underlying the generation of such local exclusion zones. I return to this topic in Sect. 12.4.

12.3.3 Other Phenomenological Models

The d_{min} model is one instantiation of a whole class of phenomenological models whereby spatial points exhibit mutual exclusion (Diggle 2002). A generalisation of this style of model is the pairwise interaction point process whereby a non-negative function $h(t)$ influences the probability of any two cells being a distance t apart. The shape of $h(t)$ can then determine both excitatory and inhibitory interactions between pairs of points, as demonstrated by Diggle (2002). These models also allow for a 'birth-and-death' style of cell positioning: cells are initially positioned randomly within the field, and then individual cells are killed and move to new positions. Such birth-and-death algorithms need several iterations to converge, but are preferable to the serial methods which may introduce order artifacts (cells added later into the field are more difficult to position than earlier-born cells). For further details see (Diggle 2002; Eglen et al. 2005).

Finally, in contrast to the models whereby local order emerges from random initial conditions, another class of model has been proposed for modelling retinal mosaics whereby an initially regular hexagonal mosaic is distorted to match the observed pattern. This 'distorted lattice' approach has been used to model the distribution of horizontal cells (Ammermüller et al. 1993) and retinal ganglion cells (Zhan and Troy 2000). Although these models can recreate the spatial properties of observed retinal mosaics, they are of limited utility in informing us about the developmental mechanisms underlying mosaic generation as they require hexagonal mosaics to be first created and then distorted.

12.4 Mechanistic Models

In this section I briefly discuss mechanistic models that have been proposed for generation of retinal mosaics. The key focus of these models is to help further understand the developmental mechanisms underlying pattern formation, as opposed to observing a good statistical fit between model and data. For further details of these models, the reader is referred to the original references and (Eglen 2006).

12.4.1 Lateral Migration

Once retinal neurons become postmitotic, they migrate radially through the layers of the retina until they arrive at the layer that is appropriate for their cell type. Whilst they migrate through the layers, it is thought that cells of the same type do not respect any minimal spacing rule (Galli-Resta et al. 1997). They therefore arrive randomly spaced over a period of several days. However, once they arrive in the destination layer, they appear to move laterally within the layer. The amount of lateral movement observed varies by cell type (Reese et al. 1999), and those that move more tend to have more regular mosaics.

What causes the lateral movement of neurons within their destination layer? Early evidence suggested a correlation between the time of movement and the first emergence of neurites in horizontal cells (Reese et al. 1999). This suggested that dendritic interactions might underlie the lateral migration, a hypothesis that was investigated using modelling techniques (Eglen et al. 2000), described in the next paragraph. Subsequently, further evidence for the role of dendritic interactions in mosaic formation came from work by Galli-Resta showing that temporary disruption of microtubules in dendrites caused mosaics to collapse; once microtubule function restored, mosaic organization returned (Galli-Resta et al. 2002). Most recently, mosaics are disrupted in mice lacking the cell adhesion molecule DSCAM, possibly as a consequence of altered dendritic fasciculation among homotypic neurons (Fuerst et al. 2008).

In the lateral migration model (Eglen et al. 2000), neurons initially have small circular dendritic arbors. Each cell receives input from its neighbours in proportion to the amount of dendritic overlap, and arbor size varies to maintain a fixed amount of input from neighbouring cells (van Ooyen and van Pelt 1994). In addition, cells repel each other in proportion to their dendritic overlap. In this manner, as dendritic arbors develop, cells gradually begin to repel each other; once arbor sizes have stabilised, the cells then gradually settle into a regular hexagonal-like mosaic layout. The amounts that each cell moves is small, in line with the lateral distances observed experimentally (Reese et al. 1999). One limitation of the model is that it usually generates mosaics with regularity indexes that are much higher than those observed experimentally. This is because the model dendrites are perfectly circular and the amount of overlap between arbors is calculated exactly. Reducing the precision with which the amount of overlap is detected produces more realistic mosaics (Eglen et al. 2000). Subsequent modelling work has also examined in detail the mechanical forces that might compose the dendritic interactions, thus moving towards more realistic description of the developing dendrites (Ruggiero et al. 2004).

12.4.2 Lateral Inhibition of Cell Fate

The eventual identity of any given neuron in the retina is not predetermined early in development but is influenced by many intrinsic and environmental factors during development. Many cell fate mechanisms influence the identity of a given cell. One of the most common is lateral inhibition: neighbouring neurons compete to inhibit each other from acquiring a particular fate. There are many molecular pathways by which this lateral inhibition is mediated, but most notable is that of Delta-Notch signalling (Frankfort and Mardon 2002). Cell fate mechanisms can therefore naturally impose minimal distance constraints as they prevent neighbours from being the same type of neuron.

The effect of cell fate interactions upon the relative numbers of primary and secondary fate neurons was studied by Honda et al. (1990). This early modelling study showed that lateral inhibitory mechanisms are sufficient to generate the correct relative numbers of primary and secondary fate neurons in developing grasshopper neuroblasts. We have subsequently shown that lateral inhibition can generate regular primary fate mosaics from an initial irregular distribution of undifferentiated neurons (Eglen and Willshaw 2002). However, if the initial population of undifferentiated neurons is already regular, the subsequent mosaic of primary fate neurons is not more regular than the initial population. Stochastic cell fate processes have also been shown theoretically to be sufficient to account for the generation of regular mosaics in zebrafish photoreceptors (Tohya et al. 1999). Further work by this group showed that these zebrafish mosaics could equivalently be generated by cell rearrangement processes (Mochizuki 2002; Tohya et al. 2003).

12.4.3 Cell Death

Many more neurons are produced in development than survive to adulthood. For example, estimates suggest that 50–90% of RGCS that are born will die before adulthood (Finlay and Pallas 1989). This programmed cell death may have many roles in development, including the refinement of retinal projections to their targets (O'Leary et al. 1986). Cell death might be an active process in forming retinal mosaics, by removing those inappropriately-positioned neurons that are too close to their neighbours (Jeyarasasingam et al. 1998; Cook and Chalupa 2000). The mechanisms by which neurons detect that they are too close too each other are however unknown. Furthermore, computer modelling of this process suggests that the cell death would need to be highly selective or the level of cell death would need to be very high to transform an irregular mosaic into a regular mosaic (Eglen and Willshaw 2002). These modelling studies would therefore suggest that cell death alone does not account for the emergence of RGC mosaics (Jeyarasasingam et al. 1998). Cell death could however account for the generation of other mosaics, e.g. dopaminergic amacrine neurons (Raven et al. 2003), as the level of naturally-occurring cell death is very high and the final mosaics are only mildly regular.

12.4.4 Interactions Between Developmental Mechanisms

Although cell death alone could not account for the emergence of RGC mosaics, it is likely that many mechanisms can co-operate to generate regular mosaics. Indeed, combining lateral inhibition of cell fate with cell death is sufficient to generate highly regular RGC-like mosaics (Eglen and Willshaw 2002). The effects of interactions between several developmental mechanisms has been studied within the context of cellular patterns in the chick inner ear, the basilar papilla (Goodyear and Richardson 1997), where primary fate cells are regularly distributed across the surface (Podgorski et al. 2007). Three different mechanisms were studied: lateral inhibition of cell fate, cell death, and differential adhesion. Individually, no single mechanism could account for the generation of the primary fate mosaics. However, iteratively coupling these mechanisms robustly generated regular patterns over a wide range of initial conditions. These results suggest that modelling the interactions between developmental mechanisms is clearly important before one can fully understand the relative role of individual processes, such as cell death.

12.5 Exclusion Zone Modelling: Application to Two Types of Neuron

This previous section has outlined several mechanisms that could underlie the generation of retinal mosaics, and in particular how an exclusion zone might be generated. If we assume that exclusion zones can somehow be generated, then it is natural to return to the d_{min} model and see how else it can be used to investigate mosaic formation. In particular, in this section we consider whether the d_{min} model can account for the generation of cellular patterns involving two related cell types.

Out of the 60+ cell types in the retina, there are several types of cell that come in complementary pairs (Cook and Chalupa 2000). For example, the most prominent example of complementary pairing is the classification of alpha and beta RGCs into two types: on-centre or off-centre, depending on their response to light (Wässle et al. 1981a,b). Likewise, in both cat and macaque, horizontal cells are divided into two types, each regularly arranged (Wässle et al. 1978, 2000). In this section I show how exclusion zone modelling can test whether heterotypic developmental interactions are required to generate these mosaics.

Figure 12.10a shows the regular arrangement of two types of horizontal cell in macaque (Wässle et al. 2000). There are roughly twice as many type 1 neurons as type 2 neurons. The regularity index for all neurons (irrespective of type) is just under 4.0 (Fig. 12.11), which is relatively high and thus lead to the suggestion that the two types of neuron might interact to create this high regularity (Wässle et al. 2000). To test this hypothesis, we extended the exclusion zone model to include two types of neurons (Eglen and Wong 2008). Each neuron respected the exclusion

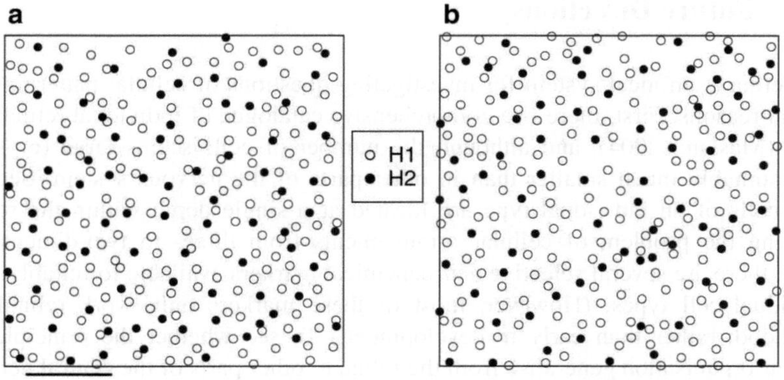

Fig. 12.10 Regular arrangement of two types of horizontal cells. (**a**) Observed distribution from macaque retina. Type 1 neurons are drawn as *open circles*, type 2 cells are filled. (**b**) Example output from the extended d_{min} model, assuming no interactions between cell types except for preventing somal overlap

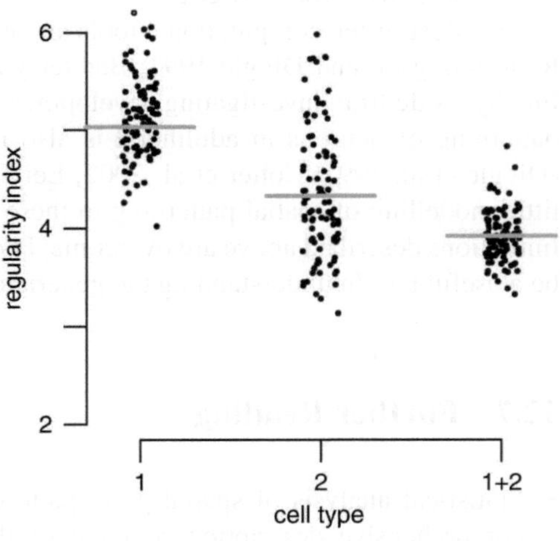

Fig. 12.11 Quantitative comparison of the extended d_{min} model with the macaque horizontal cells. The horizontal *grey line* shows the observed regularity index for either type 1 neurons, type 2 neurons, or all neurons, irrespective of type. *Black dots* indicate the regularity index from 99 simulations. The observed regularity index falls within the range of the 99 simulations, indicating a good fit between model and data

zone only of cells of the same type; the only interaction between cells of different type was that they could not come closer than about 12 μm, the average soma diameter, to prevent somal overlap. This model generated retinal mosaics that were both visually (Fig. 12.10b) and quantitatively similar, as assessed by distribution of regularity indexes (Fig. 12.11) and *L* functions (Eglen and Wong 2008). Thus, the exclusion zone model predicts that horizontal cell mosaics can emerge without heterotypic interactions. A similar conclusion was reached for the generation of two types of beta RGCs in cat, using a more flexible exclusion zone technique (Eglen et al. 2005).

12.6 Future Directions

The retina is an ideal system for investigating questions of cellular patterning for several reasons. First, there is a comprehensive catalogue of individual retinal cell types (Masland 2004), and although the number of cells seems large (60+), it is presumably much smaller than in other parts of the nervous system. Second, most cells of an individual type are located at a single depth within the retina, reducing the problem of cellular arrangements from three- to two-dimensions. Third, there are several selective neurochemical markers available to reliably stain individual cell types. (However, most of these markers only work reliably in adulthood, rather than early in development.) To see whether the principles of cellular organisation generalise from the retina to other parts of the central nervous system, several experimental challenges must be overcome. For example, we need reliable techniques for identifying and labelling individual cell types. Moving from two- to three-dimensional space will require accurate reconstruction within a volume (Oberlaender et al. 2009). By contrast, most of the theoretical techniques should generalise from the retina to other parts of the CNS (e.g. Prodanov and Feirabend 2007; Prodanov et al. 2007) and into three dimensions (Baddeley et al. 1993). Most of the computational tools are also freely available in either Matlab or R (Rowlingson and Diggle 1993; Baddeley and Turner 2005; Eglen et al. 2008). Finally, aside from investigating developmental mechanisms, the analysis of spatial patterning of neurons in adulthood is also important in several clinical contexts (Diggle et al. 1991; Cotter et al. 2002; Lei et al. 2009). There has been relatively little modelling of spatial patterning in these clinical contexts, but as the technical limitations described above are overcome, I hope that computational modelling will be a useful tool in understanding the generation and perturbation of these patterns.

12.7 Further Reading

- Statistical analysis of spatial point patterns (Diggle 2002). This is a short but comprehensive description of most of the key techniques described in this chapter.
- Principles of computational modelling in neuroscience (Sterratt et al. 2011). Comprehensive textbook on modelling neural systems, including a chapter on neural development.
- Retinal development (Sernagor et al. 2006). Edited collection of articles describing the different stages of vertebrate retinal development.

Acknowledgements This work was supported by funding from the Wellcome Trust and NIH. Experimental data kindly provided by Dr Lucia Galli-Resta and Prof Heinz Wässle. Thanks to Andreas Lønborg for comments on a draft version of this chapter.

References

Ammermüller J, Möckel W, Rugan P (1993) A geometrical description of horizontal cell networks in the turtle retina. Brain Res 616:351–356

Baddeley A, Turner R (2005) Spatstat: an R package for analyzing spatial point patterns. J Stat Softw 12:1–42

Baddeley AJ, Moyeed RA, Howard CV, Boyde A (1993) Analysis of a three-dimensional point pattern with replication. Appl Stat 42:641–668. doi:10.2307/2986181

Cellerino A, Novelli E, Galli-Resta L (2000) Retinal ganglion cells with NADPH-diaphorase activity in the chick form a regular mosaic with a strong dorsoventral asymmetry that can be modelled by a minimal spacing rule. Eur J Neurosci 12:613–620

Cook JE (1996) Spatial properties of retinal mosaics: an empirical evaluation of some existing measures. Vis Neurosci 13:15–30

Cook JE (1998) Getting to grips with neuronal diversity. In: Chaulpa LM, Finlay BL (eds) Development and organization of the retina. Plenum Press, New York, pp 91–120

Cook JE, Chalupa LM (2000) Retinal mosaics: new insights into an old concept. Trends Neurosci 23:26–34

Cotter D, Mackay D, Chana G, Beasley C, Landau S, Everall I (2002) Reduced neuronal size and glial cell density in area 9 of the dorsolateral prefrontal cortex in subjects with major depressive disorder. Cereb Cortex 12:386–394

Diggle PJ (1986) Displaced amacrine cells in the retina of a rabbit: analysis of a bivariate spatial point pattern. J Neurosci Methods 18:115–125

Diggle PJ (2002) Statistical analysis of spatial point patterns, 2nd edn. Edward Arnold, London

Diggle P, Lange N, Benes F (1991) Analysis of variance for replicated spatial point patterns in clinical neuroanatomy. J Am Stat Assoc 86:618–625

Eglen SJ (2006) Development of regular cellular spacing in the retina: theoretical models. Math Med Biol 23:79–99

Eglen SJ, Willshaw DJ (2002) Influence of cell fate mechanisms upon retinal mosaic formation: a modelling study. Development 129:5399–5408

Eglen SJ, Wong JCT (2008) Spatial constraints underlying the retinal mosaics of two types of horizontal cells in cat and macaque. Vis Neurosci 25:209–214

Eglen SJ, van Ooyen A, Willshaw DJ (2000) Lateral cell movement driven by dendritic interactions is sufficient to form retinal mosaics. Netw Comput Neural Syst 11:103–118

Eglen SJ, Diggle PJ, Troy JB (2005) Homotypic constraints dominate positioning of on- and off-centre beta retinal ganglion cells. Vis Neurosci 22:859–871

Eglen SJ, Lofgreen DD, Raven MA, Reese BE (2008) Analysis of spatial relationships in three dimensions: tools for the study of nerve cell patterning. BMC Neurosci 9:68

Finlay BL, Pallas SL (1989) Control of cell number in the developing mammalian visual system. Prog Neurobiol 32:207–234

Frankfort BJ, Mardon G (2002) R8 development in the drosophila eye: a paradigm for neural selection and differentiation. Development 129:1295–1306

Fuerst PG, Koizumi A, Masland RH, Burgess RW (2008) Neurite arborization and mosaic spacing in the mouse retina require DSCAM. Nature 451:470–474

Galli-Resta L (2002) Putting neurons in the right places: local interactions in the genesis of retinal architecture. Trends Neurosci 25:638–643

Galli-Resta L, Resta G, Tan SS, Reese BE (1997) Mosaics of Islet-1-expressing amacrine cells assembled by short-range cellular interactions. J Neurosci 17:7831–7838

Galli-Resta L, Novelli E, Kryger Z, Jacobs GH, Reese BE (1999) Modelling the mosaic organization of rod and cone photoreceptors with a minimal-spacing rule. Eur J Neurosci 11:1461–1469

Galli-Resta L, Novelli E, Viegi A (2002) Dynamic microtubule-dependent interactions position homotypic neurones in regular monolayered arrays during retinal development. Development 129:3803–3814

Goodyear R, Richardson G (1997) Pattern formation in the basilar papilla: evidence for cell rearrangement. J Neurosci 17:6289–6301
Hofer H, Carroll J, Neitz J, Neitz M, Williams DR (2005) Organization of the human trichromatic cone mosaic. J Neurosci 25:9669–9679
Honda H, Tanemura M, Yoshida A (1990) Estimation of neuroblast numbers in insect neurogenesis using the lateral inhibition hypothesis of cell differentiation. Development 110:1349–1352
Jeyarasasingam G, Snider CJ, Ratto GM, Chalupa LM (1998) Activity-regulated cell death contributes to the formation of on and off alpha ganglion cell mosaics. J Comp Neurol 394: 335–343
Lei Y, Garrahan N, Hermann B, Fautsch M, Hernandez MR, Boulton M, Morgan JE (2009) Topography of neuron loss in the retinal ganglion cell layer in human glaucoma. Br J Ophthalmol 93:1676–1679. doi:10.1136/bjo.2009.159210
Mack AF (2007) Evidence for a columnar organization of cones, Müller cells, and neurons in the retina of a cichlid fish. Neuroscience 144:1004–1014
Masland RH (2004) Neuronal cell types. Curr Biol 14:R497–R500. doi:10.1016/j.cub.2004.06.035
Mochizuki A (2002) Pattern formation of cone mosaic in the zebrafish retina: a cell rearrangement model. J Theor Biol 215:345–361
Nathan R, Muller-Landau HC (2000) Spatial patterns of seed dispersal, their determinants and consequences for recruitment. Trends Ecol Evol 15:278–285
Oberlaender M, Dercksen VJ, Egger R, Gensel M, Sakmann B, Hege HC (2009) Automated three-dimensional detection and counting of neuron somata. J Neurosci Methods 180:147–160
O'Leary DDM, Fawcett JW, Cowan WM (1986) Topographic targeting errors in the retinocollicular projection and their elimination by ganglion cell death. J Neurosci 6:3692–3705
Podgorski GJ, Bansal M, Flann NS (2007) Regular mosaic pattern development: a study of the interplay between lateral inhibition, apoptosis and differential adhesion. Theor Biol Med Model 4:43
Prodanov D, Feirabend HKP (2007) Morphometric analysis of the fiber populations of the rat sciatic nerve, its spinal roots, and its major branches. J Comp Neurol 503:85–100. doi:10.1002/cne.21375
Prodanov D, Nagelkerke N, Marani E (2007) Spatial clustering analysis in neuroanatomy: applications of different approaches to motor nerve fiber distribution. J Neurosci Methods 160:93–108. doi:10.1016/j.jneumeth.2006.08.017
Raven MA, Eglen SJ, Ohab JJ, Reese BE (2003) Determinants of the exclusion zone in dopaminergic amacrine cell mosaics. J Comp Neurol 461:123–136
Reese BE, Galli-Resta L (2002) The role of tangential dispersion in retinal mosaic formation. Prog Retin Eye Res 21:153–168
Reese BE, Necessary BD, Tam PPL, Faulkner-Jones B, Tan SS (1999) Clonal expansion and cell dispersion in the developing mouse retina. Eur J Neurosci 11:2965–2978
Ripley BD (1976) The second-order analysis of stationary point processes. J Appl Prob 13: 255–266
Ripley B, Rasson J (1977) Finding the edge of a Poisson forest. J Appl Prob 14:483–491
Rockhill RL, Euler T, Masland RH (2000) Spatial order within but not between types of retinal neurons. Proc Natl Acad Sci USA 97:2303–2307
Rodieck RW (1991) The density recovery profile: a method for the analysis of points in the plane applicable to retinal studies. Vis Neurosci 6:95–111
Roorda A, Metha AB, Lennie P, Williams DR (2001) Packing arrangement of the three cone classes in primate retina. Vis Res 41:1291–1306
Rowlingson BS, Diggle PJ (1993) SPLANCS: spatial point pattern analysis code in S-Plus. Comput Geosci 19:627–655
Ruggiero C, Benvenuti S, Borchi S, Giacomini M (2004) Mathematical model of retinal mosaic formation. Biosystems 76:113–120
Sernagor E, Eglen SJ, Harris WA, Wong ROL (eds) (2006) Retinal development. Cambridge University Press, Cambridge

Shapiro MB, Schein SJ, deMonasterio FM (1985) Regularity and structure of the spatial pattern of blue cones of macaque retina. J Am Stat Assoc 80:803–812

Sterratt D, Graham B, Gillies A, Willshaw D (2011) Principles of computational modelling in neuroscience. Cambridge University Press, Cambridge

Tohya S, Mochizuki A, Iwasa Y (1999) Formation of cone mosaic of zebrafish retina. J Theor Biol 200:231–244

Tohya S, Mochizuki A, Iwasa Y (2003) Difference in the retinal cone mosaic pattern between zebrafish and medaka: cell-rearrangement model. J Theor Biol 221:289–300

van Ooyen A, van Pelt J (1994) Activity-dependent outgrowth of neurons and overshoot phenomena in developing neural networks. J Theor Biol 167:27–43

Wässle H (2004) Parallel processing in the mammalian retina. Nat Rev Neurosci 5:747–757. doi: 10.1038/nrn1497

Wässle H, Riemann HJ (1978) The mosaic of nerve cells in the mammalian retina. Proc R Soc Lond Ser B 200:441–461

Wässle H, Peichl L, Boycott BB (1978) Topography of horizontal cells in the retina of the domestic cat. Proc R Soc Lond Ser B 203:269–291

Wässle H, Boycott BB, Illing RB (1981a) Morphology and mosaic of on-beta and off-beta cells in the cat retina and some functional considerations. Proc R Soc Lond Ser B 212:177–195

Wässle H, Peichl L, Boycott BB (1981b) Morphology and topography of on-alpha and off-alpha cells in the cat retina. Proc R Soc Lond Ser B 212:157–175

Wässle H, Dacey DM, Haun T, Haverkamp S, Grünert U, Boycott BB (2000) The mosaic of horizontal cells in the macaque monkey retina: with a comment on biplexiform ganglion cells. Vis Neurosci 17:591–608

Zhan XJ, Troy JB (2000) Modeling cat retinal beta-cell arrays. Vis Neurosci 17:23–39

Chapter 13
Measuring and Modeling Morphology: How Dendrites Take Shape

Todd A. Gillette and Giorgio A. Ascoli

Abstract Neuronal processes grow under a variety of constraints, both immediate and evolutionary. Their pattern of growth provides insight into their function. This chapter begins by reviewing morphological metrics used in analyses and computational models. Molecular mechanisms underlying growth and plasticity are then discussed, followed by several types of modeling approaches. Computer simulation of morphology can be used to describe and reproduce the statistics of neuronal types or to evaluate growth and functional hypotheses. For instance, models in which branching is probabilistically determined by diameter produce realistic virtual dendrites of most neuronal types, though more complicated statistical models are required for other types. Virtual dendrites grown under environmental and/or functional constraints are also discussed, offering a broad perspective on dendritic morphology.

13.1 Introduction

It is generally understood that for anything that is the result of an evolutionary process, form follows function. A streamlined body in both nature and the world of man-made machines usually indicates an object's function is to move quickly. Even such a seemingly mundane object such as a foot can say a great deal about the foot's owner. Water fowl feet are webbed for paddling, birds of prey have sharp talons for grabbing and killing their prey, many primates have feet with thumbs for traversing trees, and humans have particularly large feet to keep a more upright body well balanced. So, in order to understand function and the developmental process of dendrites, it is highly worthwhile to study their morphological attributes.

T.A. Gillette (✉) • G.A. Ascoli
Center for Neural Informatics, Structures, & Plasticity; and Molecular Neuroscience Department, Krasnow Institute for Advanced Study, MS2A1 George Mason University, Fairfax, VA 22030, USA

N. Le Novère (ed.), *Computational Systems Neurobiology*,
DOI 10.1007/978-94-007-3858-4_13,

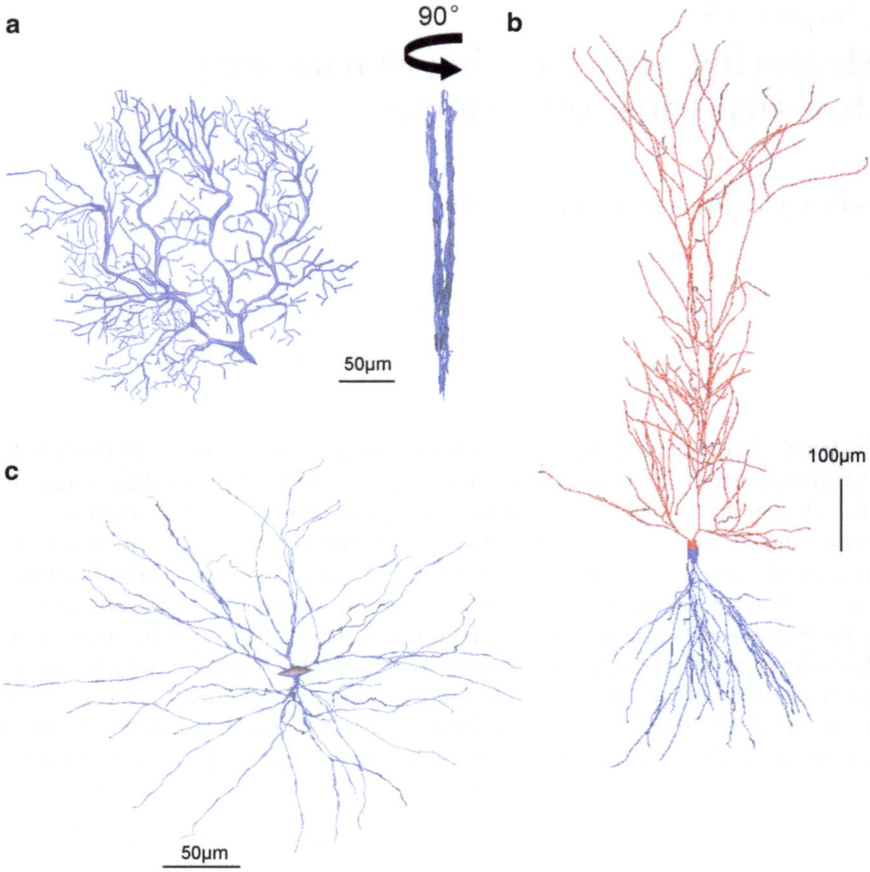

Fig. 13.1 Diversity in neuronal morphology. (**a**) Cerebellar Purkinje cell viewed with its largest area showing (*left*) and edge-on (*right*), exhibiting dense arborization and a planar shape (Rapp et al. 1994). (**b**) Hippocampal CA1 pyramidal cell with polarized basal (*blue*) and apical (*red*) dendrites (Ishizuka et al. 1995). Growth is primarily within a conical region, particularly for the basal dendrites. (**c**) Cortical stellate cell with radial dendritic arborization in all directions (Vetter et al. 2001). Reconstructions retrieved via NeuroMorpho.Org (Ascoli et al. 2007)

A look through the online neuronal reconstruction database NeuroMorpho.Org (Ascoli et al. 2007) will provide one with a clear view of the great variety of dendritic and axonal morphologies. We see differences in size and shape from the planar and densely bifurcating dendrites of the cerebellar Purkinje cell (Fig. 13.1a) to the polarized and conical shape of the pyramidal cell's basal and apical dendrites (Fig. 13.1b). The cortical stellate cell, among many others, exhibits radial growth in all directions (Fig. 13.1c). In the many visually distinctive neuronal classes one can easily see the diversity in neuron size, density, proportion, and complexity. Properties such as taper rates and asymmetry are harder to discern or compare by eye, but are equally important (see Sect. 13.3).

The effort to understand the complexities of dendrites may be largely attributed to the variety of functions they must perform. Different dendritic classes and even individual dendrites are characterized by specific morphological and functional properties. Aside from the particulars of how a dendrite must integrate its synaptic inputs, it must minimize metabolic costs while reaching existing and potential synaptic targets, make appropriate connections to form the network of which it is a part, and be capable of plasticity required for various forms of learning. The morphology of a dendrite has evolved to accomplish these goals, but it may be that in the interest of accomplishing one goal, the morphology is detrimental towards other goals. In such cases, other properties may make up for the counterproductive effects.

For instance, while bifurcations are known to reduce the probability of the continuation of a spike (Spruston et al. 1995; Sieck and Prakash 1995; Williams and Stuart 2000), it is not as clear whether this is useful or detrimental to signal processing. Both channel properties (e.g. density and conductance) and the ratio of parent to daughter segment diameter can impact failure probability (Migliore 1996). If a discovery is made that a greater density of sodium channels or local enlargements of daughter branch diameter are common at bifurcations, then one could argue that neurons have developed mechanisms to reduce the probability of bifurcation spike failure as the increased channel density and daughter branch taper decrease the likelihood of failure. An even more interesting discovery would be of a mechanism providing control over local channel properties or dendritic diameter and therefore propagation failure probability. In Sect. 13.3.1 the relationship between spike propagation and morphology will be further explored. By understanding how particular morphological features affect electrophysiology and how connectivity patterns affect morphology, we can better understand the function of various neuron types as well as how the overall morphology of a neuron and its non-morphological features contribute to function.

A wide variety of techniques are used to explore the relationships between morphology, development, and function. Electrophysiological experiments and models, as well as molecular and other biophysical research, provide functional data. Imaging and reconstruction techniques provide a means to produce rich morphological data. Models fall into and across multiple categories, each with particular strengths for both testing and generating hypotheses of development and function. Morphometrics enable measurement of morphological features which in turn provides the capability to compare cell classes quantitatively and find correlations between morphology and function. Comparisons between cell classes can be extended to morphologies produced by models, providing a mechanism for their validation. This chapter will cover the tools and techniques used to study dendritic morphology, as well as many of the known and proposed underlying mechanisms responsible for growth and the associated functions achieved by dendrites.

13.2 Morphological Data and Computational Capabilities

Before delving into the deeper complexities of morphological models and experiments, it will be useful to understand how the source data is derived and organized. Computing power and algorithmic advances have provided phenomenal opportunities over the last decade for three dimensional digital reconstruction, complex analysis, and simulation of both dendritic morphology and electrophysiology. Applying information technology to the field is also having a great impact, with databases of morphology, physiology, and simulation models such as NeuroMorpho.Org (Ascoli et al. 2007), NeuronDB (Mirsky et al. 1998), and ModelDB (Hines et al. 2004) respectively.

Three dimensional neuronal reconstructions are digitized representations of the morphological structure of neurons. These representations are made up of tapering cylinders (technically cylindrical frustums) that are connected in a tree structure to represent a neuron's soma, dendrites, and/or axon. Some reconstructions may also include dendritic spine data, histological or anatomical boundaries, or other supplementary information. Currently, these reconstructions are created primarily by a human operator working either on a computer equipped microscope or on image stacks produced through a microscope for "offline" tracing (Fig. 13.2). Semi-automated programs are available to assist human users, simplifying the process. These include the freely available Neuromantic (Neuromantic: the Freeware Neuronal Reconstruction Tool), the popular commercial Neurolucida (Glaser and Glaser 1990); (MBF Bioscience: Neurolucida – Neuron Reconstruction), and Bitplane's FilamentTracer (Bitplane: Neuron reconstruction – Automatic neuron tracing, spine detection and analysis in 3D/4D). Fully automated programs exist as well, but are not yet widely employed as they have not been proven reliable on a representative variety of cell types, staining techniques, and microscopy. A competition, the Digital Reconstruction of Axonal and Dendritic Morphology (DIADEM) Challenge, was launched in April 2009 in order to promote development and test the capabilities of such programs (Diadem Competition). Automated reconstruction may yield within a few years an explosion of morphological and connectomic data. Each type of data is useful in its own right, but given the relevance of connectivity to morphology, both types of data will complement each other in both areas of research.

Open access databases, in combination with analytical tools, provide the opportunity for both broad canvases of morphologies and physiological data as well as more targeted analyses using data already acquired for other research. A variety of neuroscience related databases can be found at the Neuroscience Information Framework (Gardner et al. 2008; http://neurogateway.org/). By sharing and making the appropriate data easy to find, the impact of the original research can be multiplied, both scientifically and in terms of citations (Ascoli 2006). Prior to such databases, sharing data was possible but required more time and effort. Now data is available from multiple labs, and as the databases become better known and used, it will become more likely for a researcher to volunteer their data. In the case of NeuroMorpho.Org, this will create the prospect of more up-to-date and complete

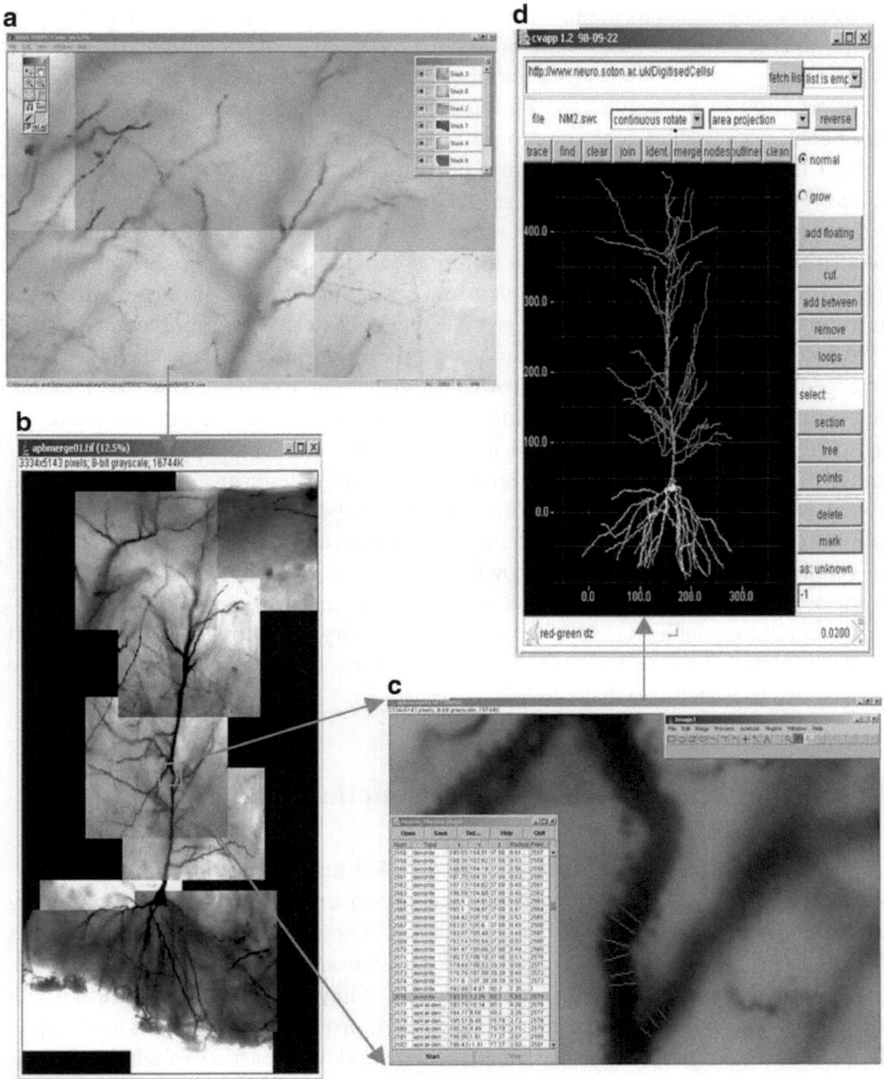

Fig. 13.2 A series of steps using various programs to trace a pyramidal cell and represent it as a digital reconstruction. (**a**) Multiple images are acquired into a stack via the Volume Integration and Alignment System (VIAS). (**b**) The image stack is presented in ImageJ for reconstruction. (**c**) A zoomed in view of the neuron allows for plotting points and determining diameters. (**d**) Once completed, the reconstructed neuron is viewed in Cvapp (Reprinted from Brown et al. 2005 with permission from Springer)

databases with morphologies across the spectrum of species, brain region, cell class, and even experimental conditions such as knock out, age, and behavioral experience. Analytical tools that can be coupled to such databases can multiply their impact.

L-Measure is one such example; a freely available cross-platform tool with the ability to calculate and statistically analyze a broad array of morphometrics for any selected set of morphologies (Scorcioni et al. 2008; http://krasnow.gmu.edu/cn3/).

In addition to molecular and biological experimentation, computer simulation has been another vital tool for morphology research. Aside from electrophysiological models that make use of real dendritic morphologies, growth models in concert with morphological analysis techniques have allowed scientists to test developmental and functional hypotheses. There are many methods of modeling dendritic morphology, each offering different benefits and each capable of addressing different questions. Some models use a particular mechanism of growth, such as bifurcation at terminal tips or growth due to external environmental factors. Other models use statistics to characterize the relationships between initial and emergent properties during the growth process. For example, one such model determines bifurcation and termination probability based on a segment's diameter. In such models "growth" does not necessarily imply developmental growth, but rather the process the model undergoes to generate a dendrite. These models are useful in describing how local morphological properties of a branch predict the downstream dendritic structure. Where relationships are found, new hypotheses can be generated with regard to both growth mechanisms as well as function. Yet other models disregard growth and focus instead of emergent functional properties by using techniques such as graph theory or evolutionary algorithms. Section 13.5 will focus on some of the techniques used in these various types of models and on the discoveries to which they have led.

13.3 Quantitative Measures and Functional Implications

Quantitative morphological measures, also called morphometrics, are the cornerstone of any morphological analysis. Such measures are numerous and many partially overlap each other in their meaning. They can be used to explore the differences between neuron classes, seek correlations with measures of function, and test hypotheses concerning development. They are also useful for growth models, specifically in comparing real neurons to growth models meant to simulate them, or as a basis for statistically based growth models. Measures of whole neurons or the separate domains of axons, dendrites, and sometimes apical dendrites, can be broken down into various categories. Here we discuss them in terms of geometric, size-related, and topological features. Local geometric features, such as taper rate (Fig. 13.3a), tortuosity (Fig. 13.3b), and angle of bifurcation (Fig. 13.3c), are particularly useful for understanding influences on growth. Size measures include total length, surface area, and internal volume, as well as height, width, and depth (determined by a minimal bounding box [Fig. 13.3d] or by principal component analysis). There are also local size measures such as segment length and diameter. Topology deals with bifurcations, their numbers and the distribution of their

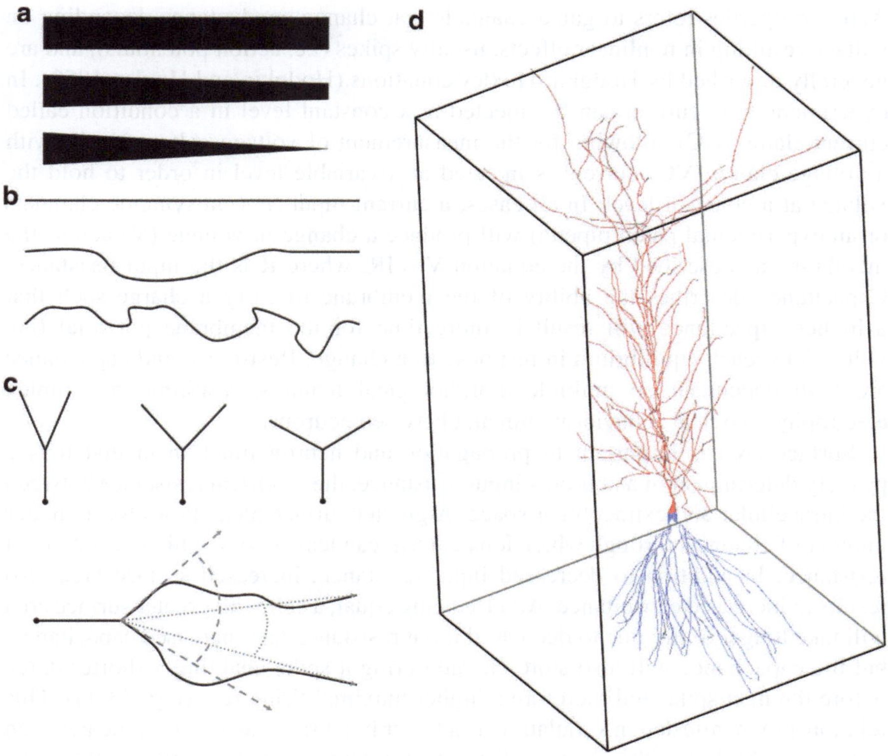

Fig. 13.3 Morphological features. (**a**) Taper rates from *low* to *high* (*top* to *bottom*). (**b**) Increasing tortuosity (*top* to *bottom*). Tortuosity is the ratio between the path length of a segment and the Euclidean distance between its two ends. (**c**) *Top*: Increasing angles of bifurcation. *Bottom*: Such angles can be measured locally as the angle of departure (*dashed lines*) or remotely as the angle between nodes (*dotted lines*). (**d**) The *bounding box* of a pyramidal neuron's dendrites (Cell from Ishizuka et al. 1995, via NeuroMorpho.Org)

resulting segments (i.e. connectivity). Topological features, as well as features that are combinations of topology and size, also fall into the category of morphological complexity. Descriptions of these measures along with some common uses and interpretations follow.

13.3.1 *Size and Geometry*

A number of size and geometric features affect electrophysiological properties at local and global scales. Electrophysiological properties fall into the two categories of passive and active. Passive properties refer to basic resistance and capacitance effects on signal propagation, generally described by Cable Theory (Rall 1969).

Active properties refers to gated channels that change conductance depending on voltage resulting in nonlinear effects, usually spikes (i.e. action potentials), and are generally described by Hodgkin-Huxley equations (Hodgkin and Huxley 1952). In experimentation, current can be injected at a constant level in a condition called current clamp (CC), allowing for the measurement of voltage. Alternatively, with a voltage clamp (VC) current is injected at a variable level in order to hold the voltage at a constant level. In all cases, a current input (*I*) (via synaptic channels or an experimental patch pipette) will produce a change in voltage (V) across the membrane as described by the equation $V = IR$, where R is the input resistance. Capacitance describes the ability of the membrane to carry a charge such that a higher capacitance will result in more time for the membrane potential (i.e. voltage) to reach equilibrium in response to a change. Resistance and capacitance are both dependant on multiple morphological features, resulting in complex electrophysiological behavior within and between neurons.

Surface area is important to propagation and neuron function in that it is a primary determinant of a neuron's input resistance, the electrical resistance between the intracellular and extracellular space. A greater surface area allows for a greater number of channels through which ions/current can leak across, which lowers input resistance. In addition to decreased input resistance, increased surface area also results in increased capacitance. All else being equal, a cell with greater surface area will take longer to fire due to decreased input resistance and increased capacitance, but the capacitance will also store charge during a spike enabling a shorter delay before the next spike and therefore a higher maximal firing rate (Fig. 13.4a). This relationship manifested in simulation as a lower burst rate (i.e. greater time between bursts) but higher within burst spike rate for larger compared to smaller CA3 pyramidal cell dendrites (Krichmar et al. 2002). Greater surface area also allows for a larger number of synapses and therefore more possible connections to other presynaptic neurons, which has implications for structural plasticity and information storage (Chklovskii et al. 2004; Stepanyants and Chklovskii 2005). Surface area is a product of a neuron's path length and diameter, each of which has additional functional implications.

Dendritic length plays several roles in neuronal function. Greater length tends to correspond to greater invaded space. This makes the dendrite available to more incoming connections. The tortuosity of a dendritic branch, or the ratio of the path distance between two consecutive branch points (or a branch point and a termination) and the Euclidean distance, provides information about the dendrite's growth mechanisms and its method of reaching synaptic targets. This and bifurcation angles will be discussed further in Sect. 13.5.

Electrotonically, greater lengths result in greater attenuation of a voltage signal given only passive propagation. This means that, all else being equal, synapses farther from the soma will have a smaller impact on somatic potential than those closer to the soma. In reality, there are a number of mechanisms by which a synapse can have a relatively increased effect, including optimized diameter properties, active dendritic propagation, and distance dependent synaptic strength.

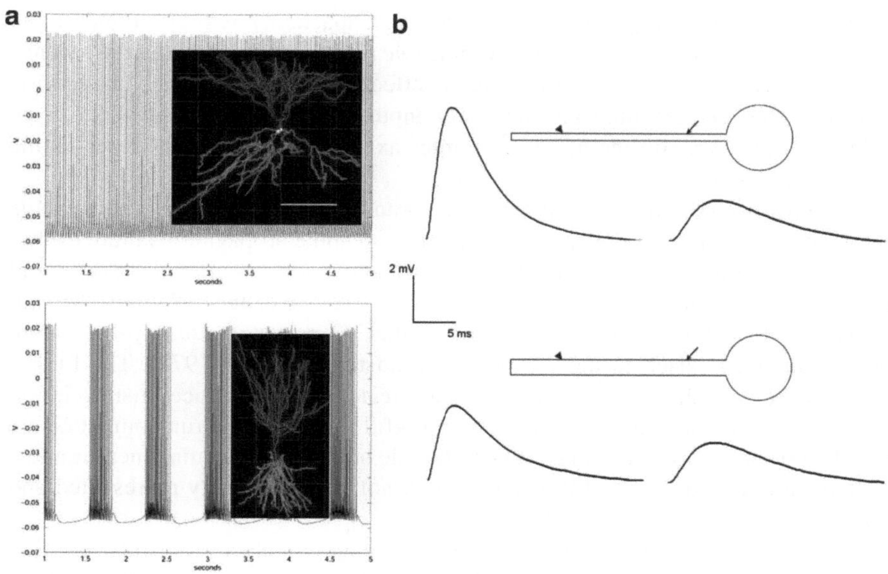

Fig. 13.4 Morphological effects on signal propagation and firing rates. (**a**) A cell with less surface area (*top*) has a lower firing rate than seen within bursts of the larger cell (*bottom*). The faster within burst firing rate and gaps between bursts can both be explained by capacitive effects in concert with channel dynamics (Adapted from Krichmar et al. 2002 with permission from Elsevier). (**b**) Voltage signals at location of synaptic input (*arrowheads* at *left*) and at a point closer to the soma (*arrows* at *right*) in two dendrites with differing diameters. The smaller diameter dendrite (*top*) has a greater input resistance and thus a larger peak voltage and faster rise time at the input relative to that seen in the larger diameter dendrite (*bottom*). The smaller dendrite also displays greater attenuation along the length due to the higher axial resistance. Flattening and delay of signal rise time due to capacitance can be seen between the signals in both dendrites

The attenuation based on increased length is due to input impedance which is the combined effects of the resistance and capacitance along the length of the dendrite. Capacitance is also responsible for signal flattening and delay, or the longer time to peak of a signal (Fig. 13.4b). Rather than representing lengths in microns, particularly path lengths from synapse to soma, lengths are often represented in terms of the space constant λ, or the distance a signal travels before it is attenuated to 1/e of the initial voltage (Rall 1969).

Along with membrane and cytoplasmic resistivity, diameter is the other major morphological factor in determining λ. While a larger diameter necessarily produces greater surface area and therefore locally decreases input resistance, it decreases axial resistance (i.e. resistance along the dendrite) by a greater amount. Axial resistance (per unit length of dendrite) is inversely proportional to the cross-sectional area of the dendrite, which is in turn proportional to the square of the diameter. Membrane resistance is inversely proportional to the circumference (and therefore to the diameter) of the dendrite. Capacitance is proportional to the circumference of the dendrite and is responsible for the dendrite's input impedance and low-pass filter properties, though length effects are generally larger. In dendritic branches with greater diameter, the lower input resistance decreases the voltage achieved by a synaptic input, but the larger axial resistance better preserves the signal (Fig. 13.4b).

The mismatch of impedance between the branches at a bifurcation is responsible for greater attenuation and increased probability of spike propagation failure (Vetter et al. 2001). In order to avoid impedance mismatch at a bifurcation, the parent and daughter branch diameters must conform to the 3/2 power law, whereby the sum of each daughter branch diameter raised to the 3/2 power is equal to the parent branch diameter raised to the 3/2 power (Goldstein and Rall 1974). The further away the relationship is from the rule, the greater the impedance mismatch and attenuation. This rule has been particularly useful for those who run computational simulations of electrophysiology. Where the rule holds (and assuming measurement of axo-somatic properties), bifurcations need not be individually represented and fewer compartments can be used.

The direct impact of dendritic diameter and length on electrotonic properties is seen in synaptic efficacy. One method for accurately measuring and visualizing these effects on synaptic efficacy given passive propagation is the morphoelectrotonic transform (MET) (Zador et al. 1995). The MET visualizes signal attenuation and delay with an "attenogram" and a "delayogram". Each neuron's visualizations are generated given an input site and an input waveform (e.g. a current pulse or a sinusoidal input). The calculations involved in the MET are similar to those for calculating electrotonic length. However, the MET produces an effective length constant (λ_{eff}) that does not assume infinite cylinders as does the classical electrotonic length calculation. With the MET, one can view the electrotonic compactness not just from the soma, but from any individual potential synapse.

When considering active properties, length and diameter can have the exact opposite effect on synaptic efficacy as in the passive case (Komendantov and Ascoli 2009). Simulations showed that higher input resistance at synapses farther from the soma lead to a greater probability of generating dendritic spikes therefore producing greater synaptic efficacy. Synapses closer to the soma with lower input resistance are relatively less likely to contribute to somatic spikes, i.e. they have lower synaptic efficacy. These two analyses on basic morphological properties and synaptic efficacy demonstrate the complex relationship between morphology, physiology, and function, yet there are still many more factors to consider.

13.3.2 Topology and Complexity

In addition to the various and important relationships between size and function, the distribution of those size attributes (i.e. where topologically and spatially length or volume are distributed) also has a vital impact on neuron function. The term

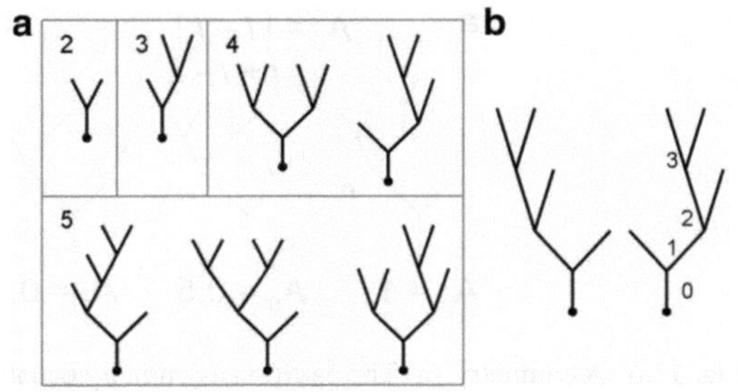

Fig. 13.5 Unlabeled binary tree shapes. (**a**) All tree shapes of degree 2 through 5. (**b**) Two trees that appear different but are the same. They would be different if the nodes were labeled. The numbers refer to branch order. The root node and initial branch are of order 0

"complexity" is often used to encompass measures of the distribution of a tree's size topologically or in space. Complexity measures do have functional implications, but many are more commonly used for classifying neuronal populations and validating models. A particularly common measure for complexity is the number of dendritic bifurcations, which can affect how signals are integrated in terms of impedance mismatch (previously discussed) as well as how a neuron is integrated into its network in terms of space filling and synaptic connections (Wen and Chklovskii 2008).

There are several terms used when discussing topology that are important to know. Many of the terms come from or have synonyms with graph theory, which is a system of describing the ways in which nodes can be connected and how one might find optimal paths given a set of nodes and edges (i.e. connections). A tree is a type of graph in which connections between nodes are unidirectional (i.e. directed edges) from parent to child and in which there are no cycles. In binary trees, nodes are either bifurcations, terminations, or the tree root. Degree refers to the number of terminal nodes in a tree, which is a useful indicator of tree size as it is directly proportional to the number of branches and the number of bifurcations (branches = degree * 2 – 1; bifurcations = degree – 1). A branch is the segment or edge between nodes and a branch's order (i.e. depth) refers to the number of branches between itself and the root node (Fig. 13.5b). A branch's order is equivalent to the order of the bifurcating node that gives rise to it. Child and daughter branch are used interchangeably. These definitions will make it easier to understand topology and complexity.

Ignoring size and geometric features, and given a number of bifurcations, there are a specific number of unique binary trees called unlabeled (also referred to as unordered) binary tree shapes (Harding 1971) as seen in Fig. 13.5. While multifurcations do exist it is common to treat trees as binary. Any multifurcation can be broken into multiple bifurcations, though there are multiple ways to do so which can differentially impact the resulting topology. Methods are available to determine the most likely binary configuration for a multifurcation, assuming probabilities of the binary subtrees are known (Verwer and Van Pelt 1990).

Vertex analysis was an early metric developed to analyze topology in neurons (Berry and Flinn 1984). For binary trees the analysis focuses on the ratio of the number of "primary nodal vertices" (i.e. nodes with two terminating children) to

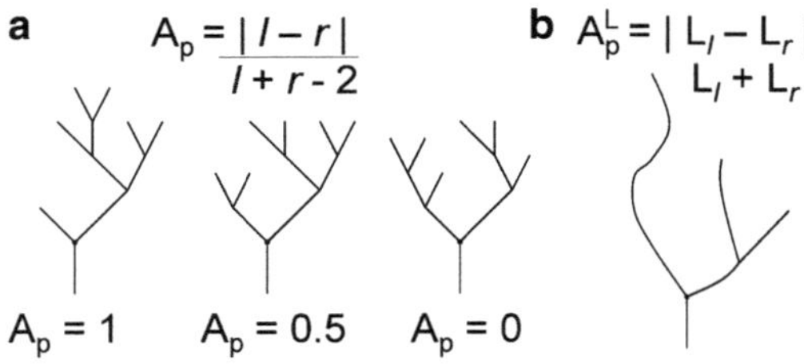

Fig. 13.6 Asymmetry. (**a**) The partition asymmetry equation accompanied by trees (degree 6) with partition asymmetry (A_p) values of the first bifurcation node. The terms *l* and *r* refer to the degree of the *left* and *right* subtrees. (**b**) The equation for partition asymmetry by length accompanied by a tree that has a topological partition asymmetry of 1 but a length partition asymmetry of close to 0 due to the similar total length of *left* and *right* daughter subtrees

the number of "secondary nodal vertices" (i.e. nodes with one bifurcating child and one terminating child). This method, along with asymmetry which is discussed next, was commonly used to determine the probability that a given tree or class of trees was the result of a particular growth process (discussed further in Sect. 13.5.1).

Asymmetry is another property that is highly dependent on the mechanisms of growth. One particular measure of asymmetry, the tree asymmetry index, is measured at the tree level and is based on the partition asymmetry, where 'partition' refers to the balance of size between daughter subtrees measured at a given bifurcation (Van Pelt et al. 1992). The values of partition asymmetry are between 0 and 1 (inclusive) based on the relative distribution (i.e. balance) of terminal nodes between a bifurcation's subtrees. A partition asymmetry value of 0 represents a symmetric distribution and a value of 1 represents a single termination in one subtree with the remaining terminations in the other subtree (Fig. 13.6a). This version of asymmetry can also be referred to as topological asymmetry because it deals strictly with the distributions of terminal degree. Asymmetry can alternatively be calculated using distributions of length or other size attributes rather than degree. These other versions can be defined slightly differently because even in the case of a topologically asymmetric bifurcation the lengths of the branches are unconstrained. In the case of a node of degree 3 it would be possible for the side with the terminal node to have an equal or larger branch length than the combined length of the subtree of degree 2 (Fig. 13.6b). The tree asymmetry index is defined as the average partition asymmetry across all bifurcations in a tree.

Alternative versions of asymmetry can be produced by weighting the partition asymmetry values based on criteria depending on the purpose of the measure. Whether one is interested in electrophysiological effects or differentiation of cell classes, if a particular property is suspected to be more greatly impacted by subtrees with greater size, one might choose to weight nodes by degree. Several alternative methods are raised in (Van Pelt et al. 1992). The first is a scheme in which nodes of degree less than four are not considered. Such partitions are trivial with only one shape each. Another scheme weights nodes by the number of bifurcations in their subtrees (i.e. weight = degree − 2). The third alternative weights nodes by the number of possible distributions of bifurcations among the node's two subtrees (i.e. weight = degree − 3). Van Pelt and colleagues found that all schemes except for the second alternative displayed fairly large variances in tree shape samples and thus best served to differentiate different sets of trees from each other.

The impacts of asymmetry and topology were investigated in an experiment specifically focused on topology and its electrotonic effect. Van Ooyen et al. (2002) found that virtual dendrites with greater mean path length (of dendritic tips) had higher firing rates in response to somatic current injection. The experiment used all possible tree shapes with eight terminal nodes (i.e. of degree 8). Since branch lengths were all set equal to each other, the greater mean path lengths corresponded to trees with greater asymmetry. The tree asymmetry index did correlate with firing frequency, but not as well as mean path length did. However, branches closer to the soma have a greater influence on input resistance, so a weighted asymmetry would be a more representative metric and would be expected to better correlate with firing rate. The relationship between topology and firing rate ultimately is based on input resistance which rises (or input conductance which decreases) with the increased mean path length of the more asymmetric trees. Another way to look at it is that a more symmetric tree has its branches connected in parallel, driving down input resistance, while a more asymmetric tree has its branches connected in series, driving up input resistance. A lower input resistance leads to more current leaking into the dendrites. However, increasing mean path length by increasing branch length has no impact on how the branches interact electrically. Surface area increases with the path length, which in turn decreases input resistance and therefore somatic firing rate.

Another variation on asymmetry, termed "excess asymmetry", was used to determine whether the size of a dendritic branch or subtree is affected by the size of the rest of the dendrite. Excess asymmetry measures the extent to which a dendritic tree's branches grow independently from each other (Samsonovich and Ascoli 2006). One type of dependent growth would result in morphological homeostasis, in which a larger subtree would necessarily be offset by a smaller sibling. The excess partition asymmetry is defined as the difference between the partition asymmetry of a bifurcation and the average partition asymmetry of that bifurcation with its grandchild subtrees shuffled. The rationale is that, assuming morphological homeostasis, a larger subtree will be offset by a smaller sibling subtree and so shuffling the children of those subtrees will result in greater symmetry between the subtrees. Just as with the standard partition asymmetry, excess partition asymmetry can be calculated with regards to various size metrics. It was found that all sets of hippocampal CA1 and CA3 pyramidal cells tested had significantly positive excess asymmetry in terms of degree, length, and surface area. This result complemented a result in the same study that showed morphological homeostasis between entire basal dendritic trees on the same cell, as well as between the basal dendritic arbor (i.e. all basal dendrites) and the apical dendritic arbor.

A topological property that is often visually evident is caulescence, or the prominence of a main path in a tree structure (Brown et al. 2008). Like asymmetry,

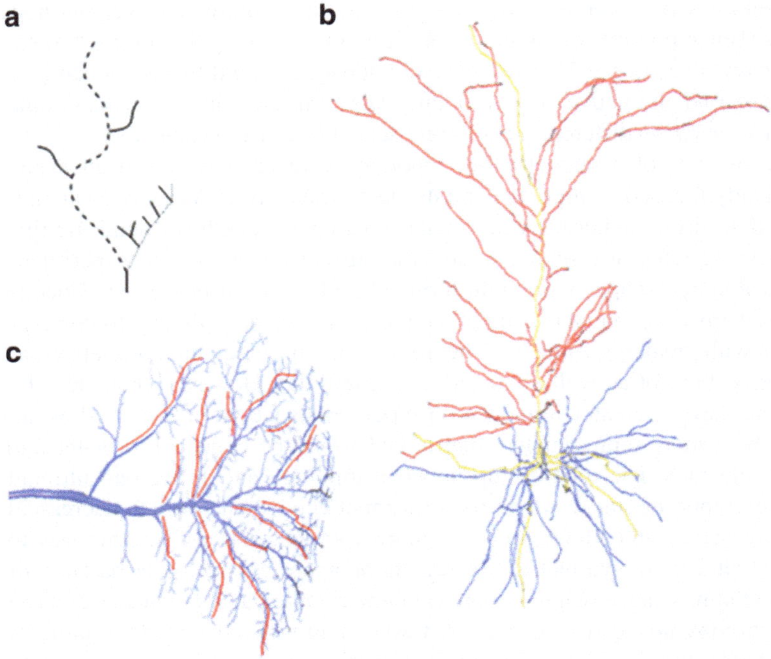

Fig. 13.7 Caulescence. (**a**) Determination of the primary path in a tree by degree (*dotted*) and by total length (*dashed*). (**b**) Pyramidal cell dendrites (from Shepherd and Svoboda 2005, via NeuroMorpho.Org) with the main paths by length (*yellow*) on apical (*red*) and basal (*blue*) dendrites. (**c**) Blowfly tangential cell (from Borst and Haag 1996, via NeuroMorpho.Org) with visually approximated regions of high caulescence highlighted in *red*

caulescence can be calculated using only topology or with any size attribute. Determining the caulescence of a tree is a two step process, the first being the determination of the primary branches (i.e. main path), the second being the determination of the value. In the first step, the primary path starts from a node (usually the soma or more generally a tree's root) and traverses away from the root choosing the larger daughter branch at each bifurcation (Fig. 13.7a). The caulescence is the weighted average of the partition asymmetry along the primary path. As with the partition asymmetry calculation, any size metric can be used to determine which child is larger. It is customary to use the same size metric for both determining the path and calculating the partition asymmetry.

Axons are generally more caulescent than dendrites, but caulescence is also prominent in some classes of pyramidal cell apical dendrites (Fig. 13.7b). Other cell classes may contain dendrites with multiple main paths or with main paths that appear in certain regions but do not start at the soma. Main paths are often found in dendrites in which synaptic targets are clustered at a distance from the soma, such as in the case of apical dendrites and the cluster of synapses in the apical tuft. A blowfly

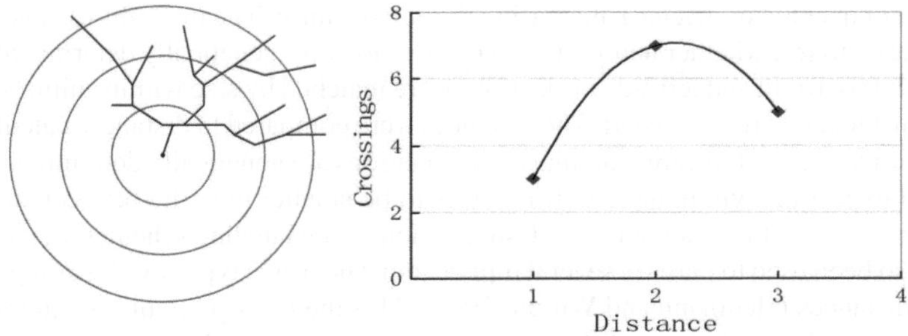

Fig. 13.8 Sholl analysis design and plot. *Concentric circles* (*gray*) denote the distances from the center at which crossings of the sample dendritic tree (*black*) are counted. The graph on the *right* shows a sample Sholl plot based on the example tree. While Sholl analyses usually contain more bins to plot crossings, they do often have an *upward* then *downward* trend as shown

tangential cell dendrite appears fairly symmetric when considered as a whole, but between major bifurcations are segments that appear caulescent (Fig. 13.7c).

Other complexity metrics that integrate topological and size attributes include fractal dimensions and Sholl analysis. Fractal dimension (Df) can be measured in several ways, each method producing an approximation of a form's true Df by performing some measurement at multiple scales (Jelinek and Fernandez 1998). Fractals are common in nature, and while dendrites do not tend to strongly display fractal properties, fractal dimension can serve to differentiate different types of dendrites, including hippocampal neurons (Cannon et al. 1999) and retinal ganglion cells (Caserta et al. 1995). The latter experiment additionally showed correlations between fractal analysis and Sholl analysis. Sholl analysis (Fig. 13.8) is a popular measure that charts the number of branches at set distances from the soma (Sholl 1953). Originally these measures were done only in two dimensions as images of neurons were two-dimensional. With the advent to three-dimensional neuronal reconstructions, many metrics, including Sholl analysis, can now be carried out in three dimensions. Sholl analysis is often used to compare cell types or the effects of some experimental effect on a cell. It provides additional detail when size differences are found between two sets of dendrites. For instance, an experiment focusing on the effects of stress on neurons in the prefrontal cortex exhibited decreases in total length and number of pyramidal cell apical dendritic branches. Sholl analysis showed that within the 240 μm that the dendrites reached, the reduced complexity occurred between 90 and 120 μm of the soma, the region with the highest baseline complexity (most branch crossings) (Radley et al. 2004).

Another metric that has come into more recent use is the tree-edit-distance. Conceptually it is similar in function to the edit distance of strings; however trees contain a topology that strings lack. The distance is the number of insertions and deletions of nodes or branches required to turn one tree into a second tree. Nodes can additionally be labeled with values, perhaps features such as branch length or diameter. If such labels are used in an analysis, matched nodes must be modified (i.e.

those not inserted or deleted), increasing the edit distance. The tree-edit-distance has been used to test whether neuromuscular projections are genetically determined (Lu et al. 2009). Right and left side projections were matched by size within animals, and then distances were calculated. The distances were compared to distances calculated between neurons of different animals. If the neurons are genetically determined, one would expect the within animal differences to be smaller than the between animal differences, but this was not found. Using various node labeling schemes, the metric has also been used to classify several hippocampal neuronal types by clustering pairwise distances (Heumann and Wittum 2009). This metric is particularly interesting in that it only produces a difference between two morphologies as opposed to providing a value for a single tree. Its benefit is the explicit comparison of topology merged with any particular combination of geometric or size attributes at the branch level.

13.3.3 Real Morphologies and Their Properties

All of the discussed metrics can be used to compare dendritic trees in their entirety in order to classify them and understand the diversity within and between classes. A study by Cannon et al. (1999) provides a good example of a wide variety of size, complexity, and electrophysiological characteristics applied to the differentiation of several hippocampal cell classes. Multiple size characteristics sufficiently separated the interneurons, dentate gyrus granule cells, CA1, and CA3 pyramidal cells. Complexity metrics typically had more subtle differences with substantial overlaps in the distributions between classes.

One particular factor that should be kept in mind when using digital reconstructions for modeling or comparing neuronal types is the specific reconstruction methods used to develop the data. While standards and common practices are emerging, some aspects of the process, such as how the diameter of a dendritic segment should be determined, vary from lab to lab and even from person to person. These differences in procedure are in some cases due to resolution limits as well as the different scientific questions being probed. For instance, diameter could be traced at the inside of an irregular edge for the purposes of determining internal dendritic space for organelles to inhabit. Alternatively, tracing might follow the outside edge when considering potential synaptic connectivity based on distances from the dendrite. Assuming surface area to be a key metric in a study, diameter could be traced in the middle to account for the surface area of untraced dendritic spines. As such, one cannot simply assume that all pieces of relevant information needed for a different question will be available. A study examined systematic differences between labs by comparing hippocampal CA3 and CA1 pyramidal cells with a battery of metrics (Scorcioni et al. 2004). The two cell types were differentiated by several metrics, including remote bifurcation angle and maximum branch order, which did not vary consistently between labs. Total volume and diameter were the only two metrics that differentiated both cell types and labs. Local bifurcation angle and contraction (a metric related to tortuosity) most differentiated labs, suggesting that they may be the most variable properties between labs. These results are useful for avoiding misinterpretation of certain metrics while designing or interpreting studies.

Morphometric analysis can also serve to describe and potentially differentiate interneurons, the roles of which are generally less clear than those of principal cells. The results of such an analysis on both dendrites and axons of a set of hippocampal CA3 interneurons used morphological analysis to segregate them into two novel interneuronal classes (Ascoli et al. 2009). Initially distinguished by the number of dendritic bifurcations, further analysis found differences in a variety of local branch properties including branch length, taper, bifurcation angle, and asymmetry. The morphological findings combined with physiological data further suggested feed-forward and lateral inhibitory roles. Specifically, the interneurons target lateral pyramidal cells in CA3. Signals from upstream regions including the dentate gyrus and entorhinal cortex therefore produce feed-forward inhibition on the pyramidal cells. Recurrent collateral inputs from local pyramidal cells onto the interneurons produce lateral inhibition on other nearby pyramidal cells targets, potentially enabling pattern separation.

Complete dendritic reconstructions exhibiting the full complexity of morphological influences are also used frequently in electrophysiological simulation. Simulation studies focusing on channel properties are quite common. With many different channel types exhibiting nonlinear responses to physiological state, there is a wide range of research on the topic that is beyond the scope of this chapter. However, some studies using reconstructions have focused specifically on the effect of morphology by keeping channel properties constant. By controlling for ionic channel type and distributions within dendrites, Mainen and Sejnowski (1996) found that various real morphologies in simulation were able to generate a spectrum of firing patterns which included weak adaptation, strong adaptation, and bursting behavior (Fig. 13.9). These effects were further simulated in two compartment models by varying the coupling strength (i.e. the inverse of the resistance between the compartments) and area ratio between the dendritic and axo-somatic compartments. Some coupling was necessary to achieve bursting, but small changes to either variable had large impacts on cell behavior. Decreasing the area ratio or increasing the coupling strength reduced bursting. Changing channel properties can also alter firing properties, but these results showed how morphological properties alone can have a dramatic impact.

These studies exemplify the importance of morphology in terms of neuronal function as well as morphometrics for producing hypotheses and providing evidence for such function. Size, geometry, and complexity each are relevant pieces of information. In addition to their roles in correlation with functional attributes and comparison between neuronal classes, they can be used as parameters in morphological models (discussed in Sect. 13.5). The above section is far from an exhaustive list of metrics, but it contains the most fundamental and most widely used metrics along with some relatively new and intriguing entries in the field.

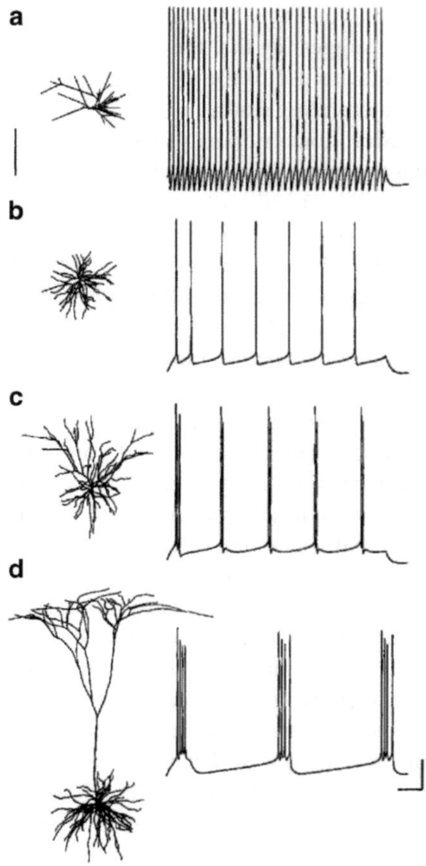

Fig. 13.9 Simulated firing patterns produced by somatic current injection of several types of reconstructed dendrites (**a** shows only connectivity and branch lengths). Channel types and distributions are constant, only morphology is different between simulations. (**a**) Rat somatosensory cortex layer 3 aspiny stellate. (**b–d**) Cat visual cortex. (**b**) Layer 4 spiny stellate. (**c**) Layer 3 pyramidal. (**d**) Layer 5 pyramidal. Scale bars: 250 μm (anatomy), 100 ms, 25 mV (Adapted by permission from Macmillan Publishers Ltd: Mainen and Sejnowski 1996)

13.4 Molecular Underpinnings of Dendritic Development

With dendrites, as with all biological systems, genetic makeup and expression controls protein and molecular interactions. In conjunction with the external environment, these interactions generate a structure whose functional qualities and fitness shape the genetic makeup of successive generations through evolutionary pressures. Here we will discuss the genes and molecular interactions that underlie dendritic development and therefore dendritic morphology. There are of course far too many interactions to cover in this chapter, so we will focus on a few examples of each type of interaction. More complete reviews can be found of transcriptional factors, molecular mechanisms, and signaling mechanisms (Miller and Kaplan 2003; Gao 2007; Parrish et al. 2007). As many molecules and their

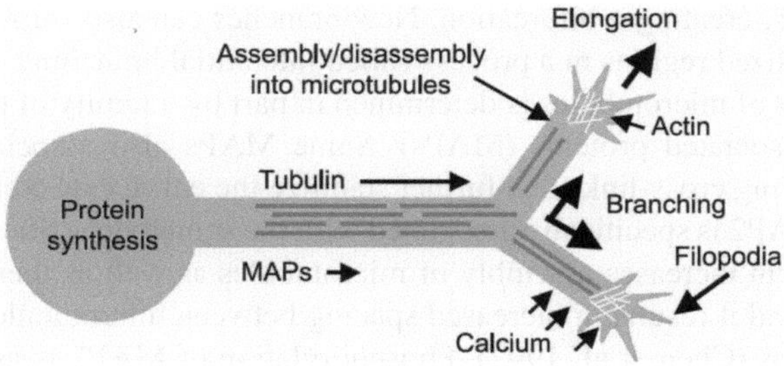

Fig. 13.10 Developmental molecules and mechanisms of dendritic growth and branching. Tubulin and Microtubule associated proteins (MAPs) diffuse and are transported from the soma to the growth cones. They interact to assemble microtubules and stabilize the dendrites as they elongate. The filopodia sample the extracellular environment, elongating the dendrite and causing bifurcations (i.e. branching) in some cases. Calcium influx influences microtubule stability and filopodia behavior (Reprinted from Kiddie et al. 2005 with permission from Elsevier)

interactions are modeled computationally in order to better understand dendritic growth, understanding a sample of the underlying interactions will in turn make it easier to understand the models and their results.

13.4.1 Cytoskeleton

Microtubules form the backbone of both dendrites and axons (Georges et al. 2008; Conde and Caceres 2009). In addition to providing structure, microtubules also act as tracks for the transport of resources, organelles, and signaling molecules by motor proteins in the dynein and kinesin families (Hirokawa 1998). Initially, dendrites and axons both have microtubules with only plus ends distal. During early development, dendrites differentiate, carrying microtubules that face in both directions (Lewis et al. 2009). The microtubules grow at their tail ends in the dendrites, with new tubulin proteins (the building blocks of microtubules) assembling distally into dendritic growth cones to stabilize them (Fig. 13.10). The polarity of microtubules is due to the structure of their alpha and beta tubulin subunits. As molecular motors selectively traverse microtubules in a particular direction, the mixed non-uniform polarity of dendritic microtubules allows for transport of a different set of proteins than seen in axons (Hirokawa and Takemura 2005).

Both microtubules and actin filaments provide stability to dendrites, but actin filaments also play important roles in dendrite growth and in the structure of dendritic spines. Like axons, dendrites have growth cones and filopodia composed of actin filaments that respond to the extracellular environment. Depending on the stability and spacing of the microtubules, strong growth signals in different directions can drive the filopodia to strain the microtubule bundles and split them into two groups, creating a bifurcation. New branches can also form from existing ones in destabilized regions in a process called interstitial branching.

The stability of microtubules is determined in part by a family of proteins called microtubule associated proteins (MAPs). Some MAPs also associate with actin filaments, creating cross-links that further stabilize the entire cytoskeleton (Pedrotti et al. 1994). MAP2 is specifically associated with the somatodendritic region in neurons. The protein increases assembly of microtubules as well as their stabilization via bundling, and it results in increased spacing between microtubules in dendrites relative to axons (Chen et al. 1992). Phosphorylation of MAP2 is associated with reduced microtubule assembly (Yamamoto et al. 1988) and with greater arborization of dendrites due to decreased microtubule stability (Díez-Guerra and Avila 1993). MAP2 has many phosphorylation sites that have varying impact on microtubule interaction (Brugg and Matus 1991), and there are many kinases and phosphatases capable of modulating its phosphorylation state. So controls over MAP2 expression as well as phosphorylation state provide mechanisms for control over dendritic growth and arborization. Calcium-dependent kinases and phosphatases specifically provide mechanisms for activity-dependent modulation of arborization, growth, and retraction (Quinlan and Halpain 1996). While there are many other molecules and signaling pathways that influence actin and microtubule stability (Nikolić 2008), the MAP family and their interactions provide a sufficient example of how dendrites can be effected to grow, bifurcate, and retract.

13.4.2 Regulatory Influences and Signaling Molecules

There are a great many intracellular features and signaling pathways that act both directly and indirectly on MAPs and the dendritic cytoskeleton. For example, in hippocampal neurons the Golgi apparatus (a specialized organelle in charge of post-translational modification and packaging of proteins) has been shown to orient towards the apical dendrite *in vivo* (Horton et al. 2005). In a culture in which no neurite differentiated into an apical dendrite, the Golgi apparatus was oriented toward the largest dendrite. Moreover, Golgi elements were often found in proximal segments of larger dendrites. Horton and colleagues determined that the Golgi apparatus was responsible for the polarization by disrupting the Golgi apparatus organization without functional disruption. This resulted in a decrease in dendritic polarity without significant changes in total dendritic outgrowth. Another influence is the availability, or unavailability, of bifurcation inducing resources, as a function of dendritic volume and distance of travel via diffusion or active transport. That influence is a candidate for explaining how in some dendrites distal dendritic branches and particularly terminating branches tend to be longer than their more proximal counterparts (Graham and van Ooyen 2004).

Dendritic growth is also guided and modulated by a number of different extracellular molecules and their interactions with receptors. Brain derived neurotrophic factor (BDNF), neurotrophin-3 (NT3), neurotrophin-4 (NT4), and nerve growth factor (NGF) are four major proteins of the neurotrophin family that impact dendritic growth by binding to specific members of the tropomyosin-related kinase (Trk) family of receptors and the p75 receptor. These growth factors have different influences depending on cell type, brain region, cortical layer, and even activity patterns (McAllister et al. 1995), and can sometimes have effects that counteract each other (McAllister et al. 1997). These behaviors are made possible by the highly diverse signaling pathways triggered by the various Trk receptors, which include downstream effects on actin and microtubule dynamics (including MAP2) as well as transcriptional modulation of genes involved in dendritic growth. The term neurotrophin is sometimes used to refer to any protein that regulates neuronal growth. For instance, semaphorin3A, traditionally considered an axonal guidance molecule, also acts as an attractant for cortical apical dendrites (Polleux et al. 2000). More neurotrophins and neurotrophic pathways are described in greater detail elsewhere (Donohue and Ascoli 2005b).

Other transcriptional factors and molecules are important in stemming dendritic growth. The gene *sequoia* is at least in part responsible for tiling insect dendritic arborization neurons of the same class, preventing overlap of sensory receptive fields (Grueber et al. 2002). Dendritic arborization neurons are also guided in part by the gene *Dscam*. Via alternative splicing in different neurons (Hattori et al. 2008), the gene produces thousands of different cell surface proteins which enables dendritic self-avoidance (Hughes et al. 2007). These and other related genes are vital for ensuring proper network connectivity.

Activity, or lack thereof, can also trigger growth and retraction of both spines and branches through numerous pathways, greatly impacting neuronal networks (Wong and Ghosh 2002). In *Xenopus laevis* tadpoles, *in vivo* time-lapse imaging showed that optic tectal cells exhibited increased dendritic arborization (i.e. greater elongation and number of branches) when exposed to a lit environment relative to a dark environment (Fig. 13.11) (Sin et al. 2002). The effect was dependent upon a decrease in activation of the signaling protein RhoA and the activation of Rac and Cdc42. These Rho GTPases are involved in a number of activity- and neurotrophin-mediated pathways that influence dendritic and axonal growth (Luo 2000; Miller and Kaplan 2003). Activity during development of hippocampal pyramidal cells also increases dendritic growth, specifically close to the stimulated area (Maletic-Savatic et al. 1999). In some systems, competition between inputs can drive extension and/or retraction of dendrites and axons. For example, activity during development drives multiple olfactory bulb mitral cell dendrites to retract leaving just one to interact with the olfactory glomeruli (Malun and Brunjes 1996). While activity-dependent growth is often mediated by neurotransmitter activity, electric and magnetic fields can also influence the magnitude and direction of growth (Rajnicek et al. 1992; Macias et al. 2000).

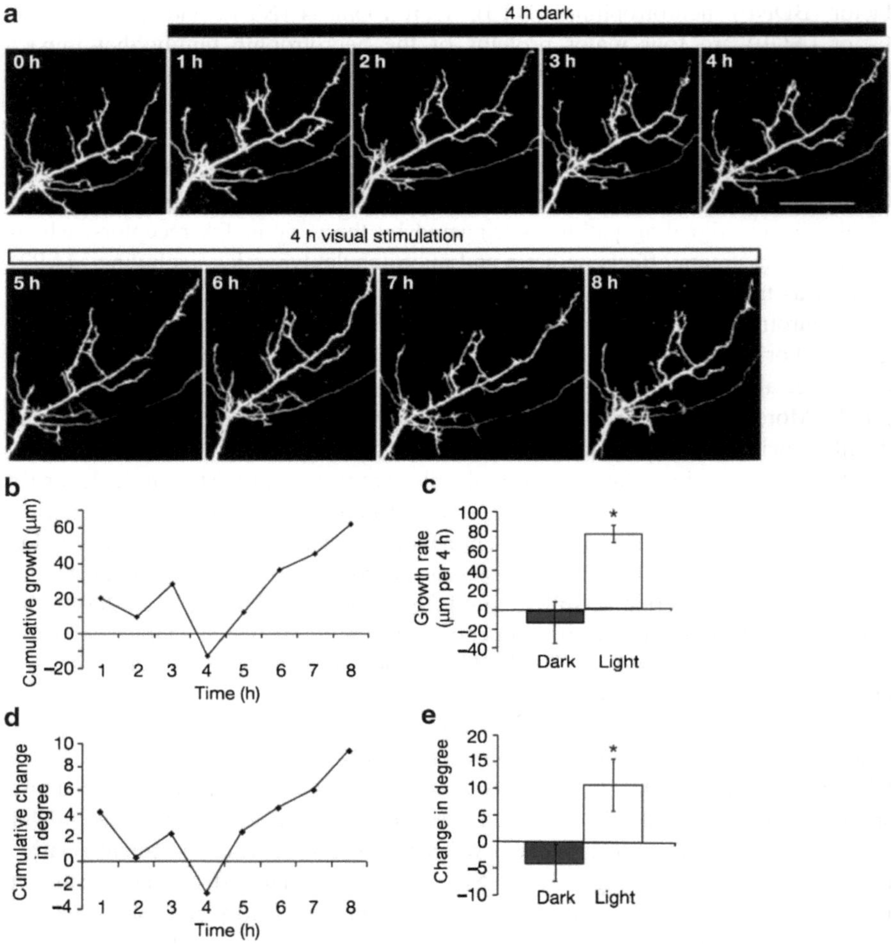

Fig. 13.11 Example of stimulus induced growth. (**a**) Images of an optic tectal neuron taken with two-photon imaging, one per hour over 8 h. The first image (0 h) is taken before light deprivation. Environment was *dark* during the first 4 h, and a light stimulus was present for the latter 4 h. Scale bars, 50 μm. (**b**) Average cumulative branch growth in terms of length. (**c**) Average growth rates during *dark* versus *light* periods. (**d**) Average cumulative change in terms of degree. (**e**) Average change in degree during *dark* versus *light* periods (Adapted by permission from Macmillan Publishers Ltd: Sin et al. 2002)

13.4.3 Biologically Based Models

Clearly there are an enormous number of factors involved in dendritic growth and development. While it is the particular set of interactions that determine cell types, many factors can often be abstracted or simplified for the purposes of modeling.

Aside from the difficulty in making sense of a complicated model where not all factors are fully understood, modeling all interacting factors would be impractical given current computational power. Simplified models can provide insight into the dominant phenomena underlying morphology and help generate hypotheses for the function of proteins, pathways, and more complex mechanisms. Additionally, simplification allows for the isolation of the phenomenon of interest and can in principle ensure that an effect is a direct result of a particular manipulation.

Various biologically based models capture events such as neurite initiation, path finding, and elongation and bifurcation (Kiddie et al. 2005; Graham and van Ooyen 2006). One such example is a model of dendritic growth based on MAP2 concentration and phosphorylation and calcium concentrations (Hely et al. 2001). The model used a constant rate of unbound MAP2 production in the soma, diffusion of MAP2 and calcium, and assumed tubulin availability. The MAP2 could become bound (i.e. associated with microtubules) based on rate equations, and could be phosphorylated and therefore dissociated from microtubules (Fig. 13.12a). The phosphorylation and dephosphorylation rates were partially determined by calcium concentration. In addition to diffusing, the calcium had an influx component determined by each compartment's surface-to-volume ratio, resulting in greater influx in compartments of smaller diameter. At each time step, a terminal compartment would elongate at a rate proportional to the concentrations of bound and phosphorylated MAP2. Additionally, the terminal compartment had a probability of bifurcation based on the bound and unbound MAP2 proportions. In order to generate different morphologies, parameters affecting the reaction rates and sigmoid slopes of the phosphorylation and dephosphorylation functions were varied.

Hely and colleagues produced three different types of morphologies with different parameters. The first type showed increased branching probability and decreased segment length towards the distal ends (Fig. 13.12b). This was a result of a greater dephosphorylation to phosphorylation ratio due to the parameters strongly favoring dephosphorylation at low calcium concentrations. Calcium concentrations rose farther from the soma resulting in increased phosphorylation and branching. The second model had equivalent phosphorylation and dephosphorylation parameters, resulting in a more or less constant ratio of bound to phosphorylated MAP2 from soma to the distal ends (Fig. 13.12c). This model also showed shorter terminal segments, which in both cases could be explained by a decrease in available MAP2 farther away from the soma and diffused throughout a greater total dendritic volume. The last parameter set was intended to produce dendrites similar to pyramidal cell basal dendrites, with few bifurcations and much longer terminal branches than internal branches (Fig. 13.12d). In this case the parameters were nearly the opposite of the first case, with phosphorylation being more dominant at low calcium concentrations. This led to more branching near the soma and less in the more distal region. The smaller volume as a result of fewer total branches allowed sufficient MAP2 for long distal branches. The resulting dendrites compared well to real

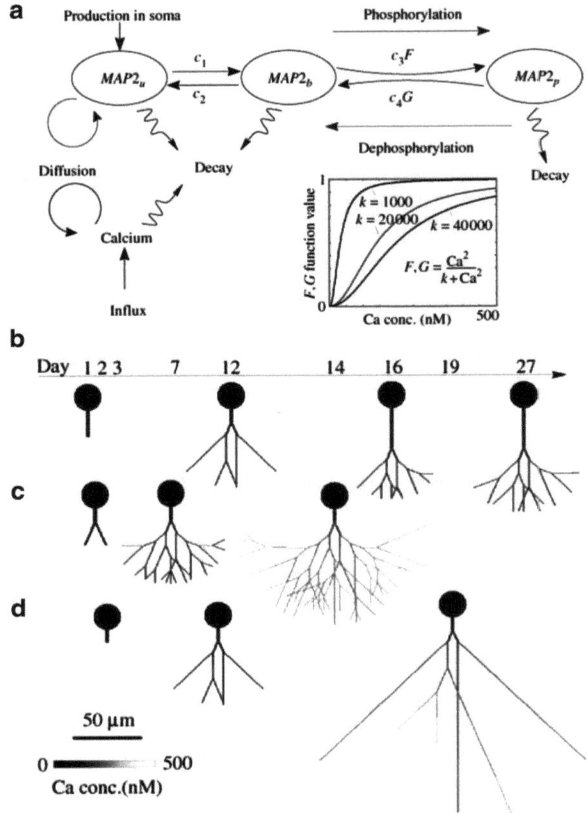

Fig. 13.12 (**a**) Molecular interactions modeled by (Hely et al. 2001). (**b–d**) Dendrites generated with the MAP2 phosphorylation model with different parameters at various times during development. (**b**) *Low* initial branching probability, increasing over time/length. (**c**) Nearly constant branching probability and elongation rate. (**d**) *High* initial branching probability, decreasing with time/length (Adapted from Hely et al. 2001, with permission from Elsevier)

basal dendrites in terms of distributions of degree, asymmetry, and several length properties.

As mentioned, most of the individual molecules and explicit mechanisms are not modeled when generating "virtual" (i.e. computationally derived) dendrites or understanding why dendrites take the shapes they do. In the example above, a small set of properties were modeled and showed how changes in calcium concentration in the context of possibly genetically encoded (de)phosphorylation set points could influence morphology. More generally, the emergent patterns of growth the many molecules and their interactions produce can be used in models to test the influence and relationships of various constraints, functions, and statistical distributions on morphologies of the many neuronal classes.

13.5 Models of Growth and Morphology

The relationships between various morphological and functional properties along with underlying mechanisms of growth and development are just a few pieces of the story of dendritic morphology. A wide variety of models are vital to elucidating how the growth mechanisms interact and how morphological and functional properties are balanced to form different morphologies. Some models mathematically describe and simulate mechanisms of growth based on biophysical theories, either with molecular models as discussed in the previous section or by more abstract relationships between microtubules, dendritic diameter, elongation, and branching (Hillman 1979; Tamori 1993). Other models focus on the results of imposed constraints on growth, such as spatial boundaries (Luczak 2006) or minimization of wiring and synaptic path length (Chklovskii 2000; Cuntz et al. 2007). Still others are statistical in nature, using empirical data to describe the relationships within and between fundamental and emergent features of various types of dendrites (Burke et al. 1992; Ascoli and Krichmar 2000; Samsonovich and Ascoli 2003; Donohue and Ascoli 2008). Not surprisingly, the boundaries between the different types of models are not always clear and many can be considered to fall into multiple categories. After all, it is ultimately the hypothesis to be tested that determines the design requirements for a model.

13.5.1 Determining the Rules of Growth

In an effort to better determine how dendrites grow in terms of where and when bifurcations occur, various models have been developed in which segments are stochastically added to a tree based on certain topological rules. It was proposed that such models could help determine how real dendrites and axons branch by comparing emergent properties of real trees with the emergent properties of trees grown from the models (Berry and Bradley 1976). Since these models work by iteratively adding segments, the total size of the resulting tree could be controlled such that virtual trees of the same size of real dendrites could be generated. Verwer and van Pelt explored the statistical properties of trees generated by terminal growth (i.e. branches added only to terminating branches) or by segmental growth (i.e. branches added to terminating or bifurcating branches), seen in Fig. 13.13. It was found that trees of much greater asymmetry and with more fully asymmetric nodes (i.e. nodes with one terminal child branch) were produced by segmental growth than by terminal growth (Verwer and van Pelt 1983).

Subsequently, the Q-model was developed which provided a single parameter (Q) for controlling the relative probabilities of segmental and terminal growth. In order to integrate the observation that branching probabilities appear to often change with branch order, the S-model was developed. This model included order-dependent branching probabilities defined by the equation $2^{-S\gamma}$, where γ is branch

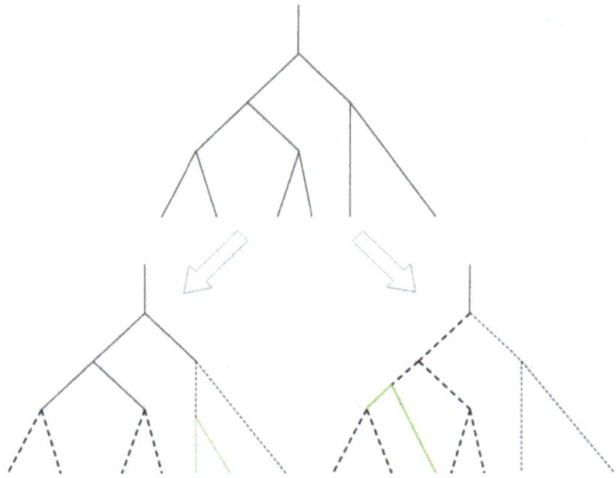

Fig. 13.13 Terminal (*left*) and segmental (*right*) growth. *Green lines* represent the two new branches generated by the growth event (i.e. bifurcation). *Dotted lines* show all of the branches from which the new branch could emerge. Segmental growth gives a greater opportunity for the new segment to grow on the *larger left side* of the tree (*bold*) relative to the *smaller side*, with a 7:3 ratio, while the terminal growth ratio is 2:1. This results in greater asymmetry on average for segmental growth

order. Thus, S values less than zero produce increasing branching probability with increasing branch order, while for values greater than zero branching probability decreases with increasing branch order (Van Pelt and Verwer 1986). The range of parameters allows for possible influences such as the greater space available for growth further from the soma, the greater availability of resources closer to the soma, and microtubule associated interactions such as MAP2 phosphorylation and others described in Sect. 13.4.3. The two parameter model combining the previous two models is referred to as the Q-S model.

The various models were used in conjunction with the tree asymmetry index to test growth hypotheses for several types of dendrites, including basal and apical dendritic trees and multipolar non-pyramidal dendrites (van Pelt et al. 1986). When exploring terminal versus segmental growth alone, random terminal growth was insufficient to explain the asymmetry measures of half of the sets of dendrites, while purely segmental growth could explain none of the sets. Q model optimization resulted in a combination of segmental and terminal growth that fit 4 of the 12 sets of dendrites. However, all sets could be fit by optimized S and Q-S models, highlighting the importance of branch order dependence on branching probabilities. For instance, the distribution of basal dendrites from cortical pyramidal cells was best matched by Q and S values representing primarily terminal growth and a decreasing probability of bifurcation at higher branch order.

Another important factor in dendritic development, plasticity, and pathology is pruning (i.e. the loss/retraction of terminal branches). Applied to the previous models, various forms of pruning had different results (van Pelt 1997). While uniform random pruning on trees grown under random terminal growth had no impact on tree asymmetry index distribution, either uniform random pruning on trees grown with order-dependent branching or non-uniform (i.e. order-dependent) random pruning paired with the terminal growth model resulted in changes to asymmetry index distributions. A set of rat Purkinje cells taken at various ages during periods of growth and retraction was analyzed. The cells showed no change in tree asymmetry index at the different time periods and sizes. The Q-S model parameters that best matched the cells suggested primarily terminal growth with no order dependence. Combined with the results of the pruning model, this suggests that the retraction process was a random uniform pruning process.

Other models have been developed to encompass a wide range of properties in addition to topology and branching probabilities. Hillman put forth an early version of the concept of generating dendrites entirely from a set of fundamental parameters specific to a given class of dendrite. In this case those parameters were determined based on observations of microtubules, diameter, and other morphological properties (Hillman 1979). The theory proposed that a dendrite with an initial diameter would elongate to some length with a particular taper rate, and branch if the diameter was above a particular threshold. The daughter branch diameters would then be determined by Rall's 3/2 power law and a daughter branch ratio parameter. From a biophysics perspective, the relationships between bifurcation angle, diameter, and branch length can be calculated based on the principle of least effective volume, providing fairly accurate predictions of measured relationships for cortical pyramidal cell basal dendrites (Tamori 1993). Both formulations provided a framework for further development and application (discussed in Sect. 13.5.3).

13.5.2 Constraints and Functions

Like all biological systems, dendrites must balance a number of competing priorities under environmental constraints in order to contribute successfully to the neuron and its network. Their success depends on the interaction of their various features. Some priorities are known in principle, such as minimizing metabolic cost and integrating signals from synapses to the soma, but how those priorities are met and balanced is still an open question. A more diverse set of models have been used to answer questions relating to the influence of developmental constraints and emergent function. These questions can be answered by generating virtual dendrites in various ways, constraining either the growth or the final product based on some hypothesis, and comparing the result to real dendrites.

Environmental boundaries act as one of the major constraints on neurons. One particular growth model focused on the effect of generic growth factors within a defined space (Luczak 2006). Luczak hypothesized that the shape of the environment along with the density and origin of neurotrophic factors could determine the morphology of various types of neuron. At initialization, the model contained an initial seed segment (soma) and a set amount of particles (neurotrophic factors) either uniformly distributed or clustered in some location (i.e. layer).

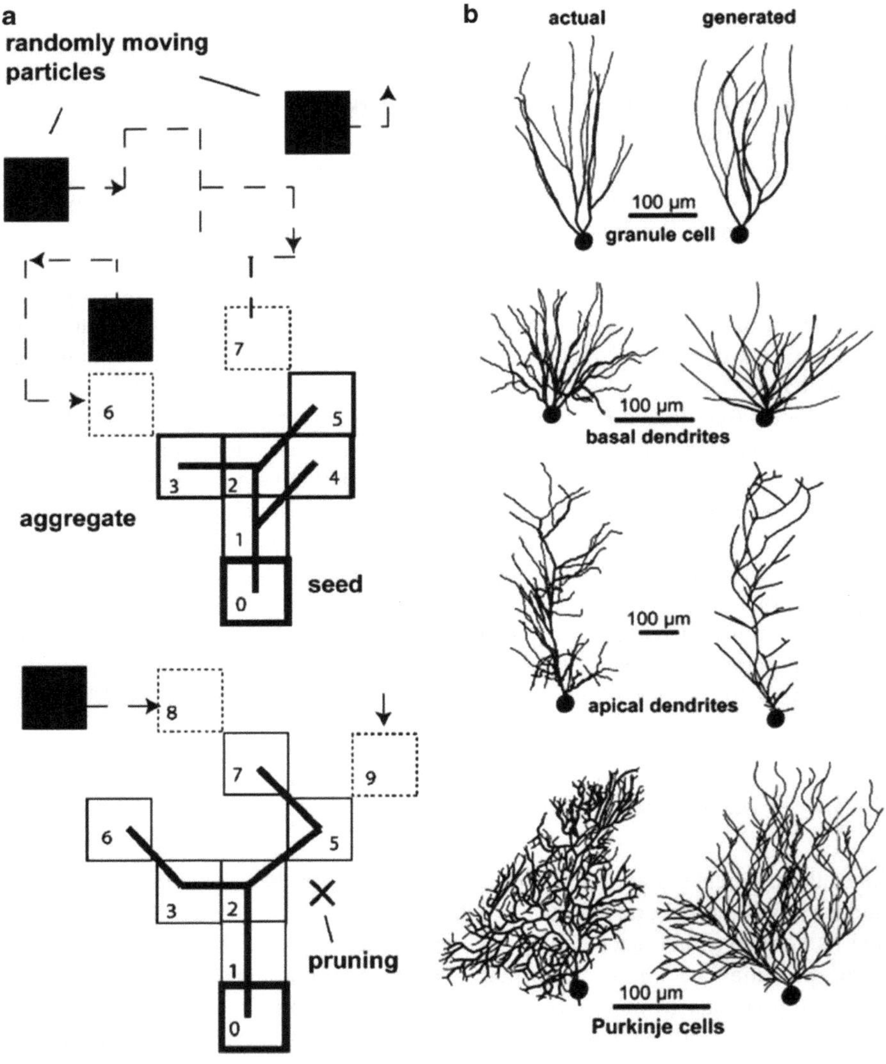

Fig. 13.14 (**a**) Model schematic showing diffusion of particles, growth of the "aggregate", and pruning. The numbers represent the order in which the particles were or will be converted into branches. (**b**) Real and generated dendrites of granule cells, and pyramidal cell basal and apical dendrites (Adapted from Luczak 2006, with permission from Elsevier)

At each iteration the particles diffused randomly, eventually contacting the soma or dendrite and becoming a new segment (Fig. 13.14a). Physiologically this simulates a neurotrophic factor coming in contact with a dendrite, being absorbed, and triggering new growth at the point of contact. The model included a parameter controlling the span of iterations during which new terminal segments had a 40% chance of being pruned. This turned out to be a necessary component for producing realistic dendrites. Initial conditions were created to generate a wide variety of dendrites including granule cells, pyramidal cell apical and basal dendrites, interneurons and Purkinje cells (Fig. 13.14b). The bounding box dimensions generally determined global dendritic shape, while particle concentrations determined branching density.

Multiple trees were grown in one space to observe the results of competition for particles and growth. Given sufficiently small spacing between seeds, the trees that happen to grow more quickly can then block other trees, resulting in larger trees getting larger and smaller trees staying small. Putting limits on growth prevented the problem. One downside to this model was the large size of the particles relative to the space. While computationally necessary, the large size of the particles can have an impact on where they are able to diffuse and prevents realistic generation of cells grown in close proximity. Nevertheless, this model does show that a set of fairly simple and primarily environmental constraints can produce a variety of morphologically distinct dendrites.

In an approach focused on intrinsic constraints, Cuntz et al. (2007) explored the extent to which dendritic morphology can be reproduced solely by minimizing wiring and optimizing synaptic efficacy given passive properties by minimizing path distance. In order to generate a dendrite, their model took a series of nodes, derived from a blowfly lobula plate tangential cell (Fig. 13.15a), and connected them based on graph theory concepts. An initial version of the model minimized only wiring with the minimum spanning tree algorithm, starting with the root node and connecting one node to the dendritic tree one at a time, adding the least wiring possible. The nodes were randomly sampled from the entire dendritic spanning field of a real tangential cell. The results were fairly unlike real dendrites, having some particularly long paths (Fig. 13.15b). In order to minimize path distance in addition to wiring, the minimum spanning tree algorithm was made to consider the path distance of any potential new connection. A balancing factor was used to balance minimization of wiring and path distance. The factor was optimized to produce qualitatively realistic tangential cells (Fig. 13.15c).

In addition to manipulating topology to optimize synaptic efficacy, Cuntz and colleagues altered dendritic diameter. The input resistance of a synapse increases with distance from the soma. Therefore a voltage signal at a synapse will attenuate more at the soma with increased distance given passive dendritic properties. Dendritic diameter profiles were therefore found which optimized the current transfer from any given location to the soma, minimizing attenuation. In order to focus on that particular attribute, the wiring and path length minimization model was run on a set of nodes composed of all bifurcations and terminations of the real dendrite. Figure 13.15d–g show that the dendrite generated by the model minimizing wiring and path length and with the optimized diameters was most successful at optimizing synaptic efficacy and compared well to the real dendrite.

A similar conclusion was reached by Wen and Chklovskii (2008) for Purkinje cells by calculating mesh sizes (i.e. the ratio between a sample area and the total dendritic length within that area) for various types of branching schemes, minimizing wiring and path lengths, and avoiding overlap of potential synapses.

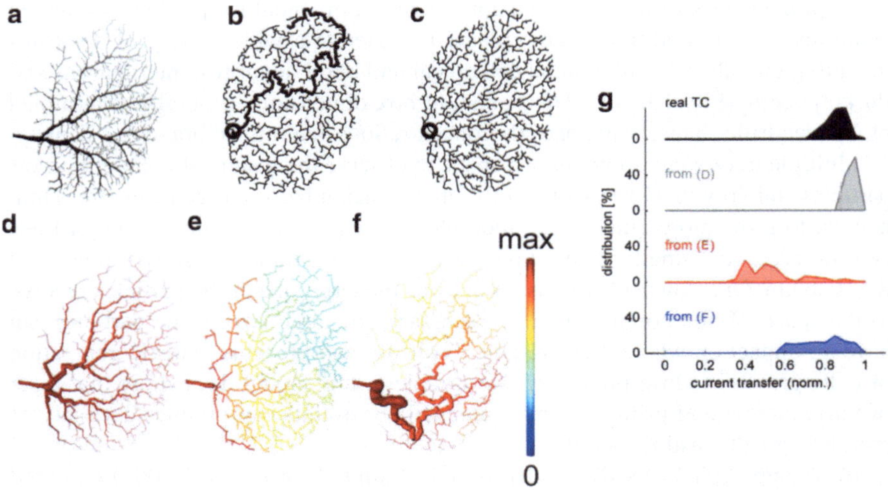

Fig. 13.15 (**a**) Original tangential cell dendrite. (**b**) Virtual dendrite generated by minimizing wiring and connecting randomly sampled points within the spanning field of the dendrite in (**a**). (**c**) Virtual dendrite generated by minimizing wiring and path distance using the same points used in (**b**). (**d–f**) Trees colored by current transfer to the root. (**d**) Virtual dendrite generated by minimizing wiring and path distance and connecting the bifurcation and termination points of the dendrite in (**a**). Diameter has been set to optimize synaptic efficacy. (**e**) The same tree as in (**d**), but all branches have equal diameter. (**f**) Virtual dendrite generated by minimizing wiring, using the same points as in (**d**) and (**e**). (**g**) Distributions of normalized current transfer from the real tangential cell and from trees in (**d**) through (**f**) show that minimizing wiring and path distance and optimizing diameter best reproduces the current transfer profile of the real cell (Merged selection of two figures from Cuntz et al. 2007, with permission from BioMed Central)

Since the role of dendrites is to make synaptic connections, a dendrite's branches should be spaced sufficiently such that one branch does not invade the space in which another branch might form a synapse. The equations developed showed that Purkinje cells' mesh size fit within the theoretical calculations. Purkinje cells were used because they are approximately two-dimensional in shape and exhibited non-overlapping dendrites. These works suggest that efficiency through minimized wiring is a major pressure in determining dendritic morphology.

Another method that has been developed for relating morphology and function produces dendrites through an evolutionary process starting with a particular functional constraint (Stiefel and Sejnowski 2007). With this method generated dendrites can be analyzed to determine the ranges and types of properties necessary to achieve the electrophysiological function, and then compared to real dendrites. The process developed involves parameter sets for creating dendrites, electrophysiological simulation, fitness scoring of the simulation, and selection, mutation, and recombination to produce successive generations with increasing fitness.

The two functions proposed and tested in a proof of concept were linear summation of synaptic inputs and spike-order detection (i.e. A before B, but not B before A). The linear summation selection resulted in neurons with long, thin, and highly polarized dendrites that were very separate electrotonically. Thus, the input current of one synapse would have a minimal impact on the driving force of the other. As predicted, the dendrites looked very much like those of the crocodile and bird nucleus laminaris neurons as well as mammalian medial superior olive neurons, each of which receives auditory input from both ears separately onto polarized dendrites. The polarized and separated morphology of the dendrites enables reaction to simultaneous activation of two different sets of synapses but not to activation of just one set. The spike-order detection selection resulted in two sets of dendrites with different diameters. Thus one set of dendrites acted as strong low-pass filters while the other dendrites were much weaker low-pass filters. This relationship resulted in a greater somatic voltage when the strong low-pass filter dendritic input came first. The authors pointed out that this relationship exists in pyramidal cells between the apical and basal dendrites. Whether spike-order detection is actually used by pyramidal cells is an open question, though the presence of active properties makes answering that question more complicated. With parameters for active properties, synaptic distribution, and simulations of spiking properties, this type of evolutionary model could be useful for generating experimental hypotheses for a wide range of functions.

Electrophysiological effects and efficiency demands appear to be largely responsible for shaping dendritic structure given a spanning field for synaptic locations. Be that the case, the question still remains why synapses are located where they are. A possible component to the answer comes again from considering wiring efficiency, both in terms of total wiring in the brain (Chklovskii et al. 2002) and in the trade-off between axonal and dendritic size (Chklovskii 2000). Consider two sets of neurons with one set being an input (i.e. sending) layer and the other set being an output (i.e. receiving) layer. Now consider that there are twice as many input neurons as output neurons, two input (i.e. presynaptic) and four output (i.e. postsynaptic) as an example, and that each input neuron sends out two axons. This is referred to as the "divergence" (D). Assuming each layer is more or less homogeneous, each output neuron will receive one connection. This is referred to as the "convergence" (C). In this case, in which $D > C$, wiring is minimized by the axons being larger and growing to meet the dendrites of the input neurons (Fig. 13.16). The mathematics used to prove this is based on the spatial density of pre and postsynaptic neurons in a one-dimensional scheme. In two dimensions and with larger numbers of neurons, neurite (i.e. axon or dendrite) "meshes" are used. Chklovskii showed that the optimal ratio of dendritic to axonal length is given by the square-root of the ratio of convergence to divergence. Factoring in differences in axonal and dendritic diameters, the optimal ratio is then adjusted by multiplying the previous ratio by the square root of the axonal to dendritic cross-sectional area ratio. Thus a larger axonal cross-sectional area would result in greater dendritic length and less axonal length. Anatomical data from a variety of experiments on retinal, cerebellar, olfactory bulb, and neocortical neurons support the theory (Chklovskii 2000).

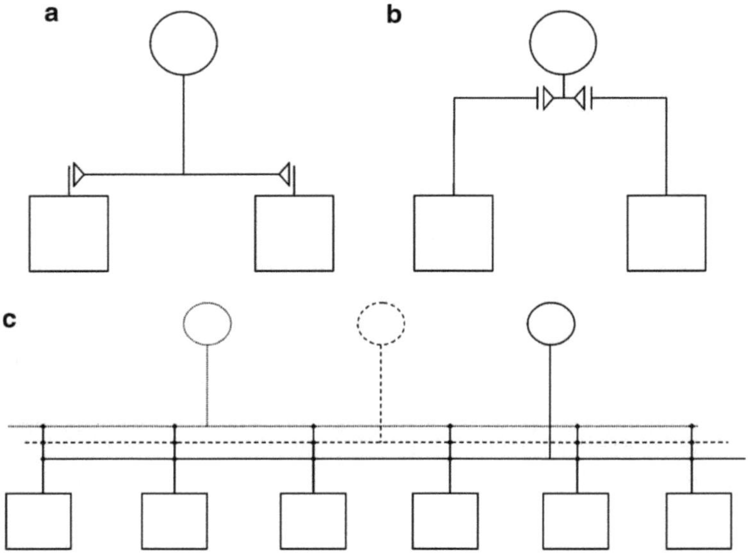

Fig. 13.16 With a convergence of 1 and a divergence of 2, the layout in (**a**) with a longer axon is more efficient than the layout in (**b**) with larger dendrites. (**c**) Example of complete and optimized connectivity with convergence of 3 and divergence of 6

It was further shown that within gray matter optimization theory requires that axonal and dendritic wiring take up 60% (bounded from 50% to 75% depending on certain assumptions) of the volume (Chklovskii et al. 2002). The calculations used independently minimized axonal conduction delay and dendritic attenuation, maximized the number of potential synapses, or minimized wiring length while holding other variables constant. For instance, conduction delays can be reduced by larger axonal diameter, but with sufficiently large diameters the total volume of the gray matter must increase and thus axons must travel farther. Data from neocortex, piriform cortex, and hippocampus show wiring percentages not significantly different from 60%. Thus the evidence suggests that in terms of gross size attributes, axonal and dendritic morphology are results of evolutionary processes optimizing signal propagation (in terms of speed and fidelity), potential synaptic connectivity, and total wiring.

13.5.3 Statistical Models

While statistical models do not provide information on how or why dendrites grow into their various shapes, they are very useful for providing compact descriptions of various morphological classes. These models generally attempt to capture the statistical relationships between known morphological attributes of cell classes, including the variance in those attributes. This is often achieved by using a subset of "fundamental" properties obtained from neuronal reconstructions, such as diameter or branch order, to probabilistically determine other properties, such as length or likelihood of bifurcation or termination. While many models have "growth rules", they do not necessarily simulate developmental growth but rather describe the relationship between parameters in (usually) adult cells in a branch-by-branch manner. One major goal of such models is to expand the store of available morphologies for more physiologically realistic network models. Given the intense amount labor that is currently required to reconstruct individual cells, an accurate statistical model can generate a larger set of realistic morphologies. Thus the need to store large amounts of data is reduced as the morphologies can be generated "on the fly" (Ascoli 2002).

Burke et al. (1992) developed such a model in which branches iteratively bifurcated, terminated, or elongated with probabilities based on observed diameter relationships. Elongation length per iteration and taper rate were set as global parameters while bifurcation and termination probabilities were determined by the current diameter of a segment. Daughter branch diameters satisfied the 3/2 power ratio by an observed margin and were determined by a process that produced the same distribution as seen in the real dendrites. The statistical data determining the probabilities came from reconstructed cat spinal cord α-motoneuron dendrites. By adjusting the global taper rate, the dendritic trees generated from the model had realistic branch order distributions but unrealistic path distance distributions. The addition of path distance (from the soma to the growing segment) as a parameter (in conjunction with diameter) in determining bifurcation and termination probabilities was sufficient to produce realistic path distance distributions, though some discrepancies such as total surface area remained.

In order to enable the further study of various statistical and anatomical models, the program L-Neuron was developed (Ascoli and Krichmar 2000). L-Neuron flexibly implements the concepts of Hillman, Tamori (discussed previously in Sect. 13.5.1), and Burke, allowing an experimenter the option of using statistical distributions or biophysically derived calculations to generate morphologies from fundamental properties. The program additionally includes parameters that allow for path lengths larger than the straight distance between nodes (i.e. tortuosity) and global tropisms simulating external guidance or somatorepulsive effects. Moreover, the program can be fed statistical data of a given neuronal class or digital morphology files in order to generate that data. The models in L-Neuron were evaluated using motoneurons and Purkinje cells (Ascoli et al. 2001). Models based on primarily local constraints, such as diameter, successfully produced virtual dendrites with several emergent parameters, including degree, total length, and asymmetry, which matched both cell types. The global tropic influences improved results in terms of spatial distribution of the dendrites. In general however, variability in virtual dendrites was greater than real dendrites, and certain emergent parameters were better reproduced by certain models. The results suggested that further constraints were necessary and that some combination of the models might produce improvements.

Variations of the Hillman model were explored using L-Neuron for hippocampal CA1 pyramidal cells. It was initially found that constraining branching by diameter alone resulted in excessively large neurons or neurons that failed to stop growing, which was due to selection of taper rates (Donohue et al. 2002). Rather than sampling a taper rate for the entire tree from the entire distribution of trees, a further variation on the model was developed in which taper rates for each branch were sampled based on the branch's diameter (Donohue and Ascoli 2005a). The updated model successfully reproduced the size (i.e. degree) of basal dendrites. Apical dendrites were also less likely to grow out of control, but their sizes were still more varied relative to the distribution seen in real CA1 apical dendrites. In terms of branching patterns, the model actually captured apical dendrite asymmetry fairly closely, and better than it captured basal dendrite asymmetry.

Further work in this direction tested the abilities of branch order and path distance as determinants of local branch properties (i.e. taper rate, bifurcation probability, length, parent-daughter ratio, and daughter ratio) to generate realistic apical, basal, and non-pyramidal dendrites (Donohue and Ascoli 2008). Each determinant produced the best results for some combination of dendrite type and emergent parameter (e.g. degree, asymmetry, surface area, and surface area asymmetry). Given the results, two models were created that merged the influence of branch order, path distance, and diameter in the hopes that a combination of influences would reflect real statistical relationships and produce a more accurate model. The percent mix model merged the sampled values of the local properties with various weights, while "243 Mix" model assigned each local property with one of the determinants. Many weights were run for the percent mix model while for the "243 Mix" all determinant/local property assignments were run to determine the best combination of influence. After normalizing for the large numbers of weights and combinations sampled, the "243 Mix" model performed better than the percent mix or the individual determinants in terms of bifurcation asymmetry and surface area. The particular associations between local basic properties, their determinants, and the emergent properties may provide some idea of how growth mechanisms are determined in the different types of dendrites. For instance, diameter may be a better determinant of bifurcation asymmetry due to the fact that segmental branching tends to result in smaller side branches in terms of diameter and degree. Path distance may serve a role in determining the limits of both elongation and branching via resource availability and transport.

Several of the functional and constraints based models guide growth by primarily extrinsic factors; however there are several potential intrinsic factors that may also determine a dendrite's tropism (i.e. direction of growth). Hippocampal principal cell dendrites have a tendency to grow away from their somas and appear mostly contained within a conical region. Cytoskeletal rigidity may be expected to constrain changes in direction, however when a dendritic branch does turn it seems to usually return to its original direction (Fig. 13.17a). This provided the motivation to determine whether the dendrites are primarily directed by extrinsic factors promoting growth in a particular direction, by intrinsic factors promoting growth away from their soma, or by neither type of influence (Samsonovich

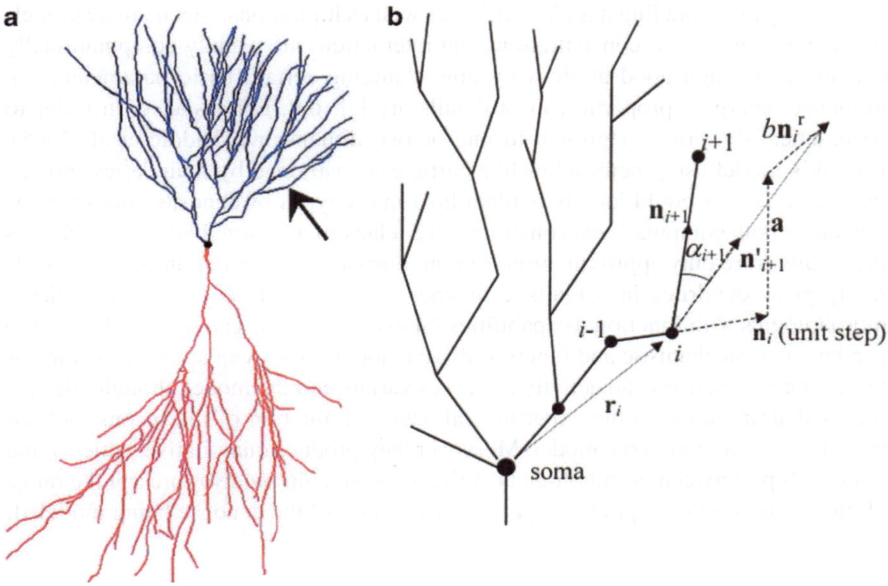

Fig. 13.17 (**a**) Pyramidal cell featuring roughly conical boundaries. *Arrow* points out a clear instance of a turn towards a somatofugal direction. (**b**) Model schematic. Current node is *i*. The three vectors determining the direction of the next elongation are $\mathbf{n}_i$ in the direction of the previous step, **a** in the direction of the constant extrinsic influence, and $b\mathbf{n}_i^{\,r}$ in the direction away from the soma ($\mathbf{r}_i$). The angle α_{i+1} randomly changes the next step, which determined by the three vectors is $\mathbf{n'}_{i+1}$, to the actual next step $\mathbf{n}_{i+1}$ (Adapted from selected figures in Samsonovich and Ascoli 2003 with permission from John Wiley and Sons)

and Ascoli 2003). In order to make the determination a statistical model using Bayesian analysis was developed which represented the growth of non-branching dendritic segments by three vectors of influence with additional random deflection from the combined vector's direction (Fig. 13.17b). The three vectors included a continuation of the direction of the previous segment representing cytoskeletal rigidity, a unidirectional influence representing some extrinsic factor(s), and a radial vector representing a somatorepulsive influence. Using reconstruction data, the magnitudes of the unidirectional and somatorepulsive influences were optimized to minimize the amount of random deflection. The Bayesian analysis showed that the radial tropism was significantly stronger than the fixed direction influence, and that its strength decreased with greater distance from the soma. Moreover, the radial growth was greater in the basal relative to apical dendrites. Similar results were seen in α-motoneuron dendrites but not γ-motoneuron dendrites when analyzing daughter branch orientations relative to parent branch orientation and bifurcation position (Marks and Burke 2007). These results suggest a possible functional benefit of growth away from the soma for certain cell types.

Each type of modeling has clear utility as well as limitations. Mechanistic models that include molecular concentrations and interactions are usually computationally expensive. Though good at showing how changing certain basic parameters can influence emergent properties, most details are left out. For instance, in order to be practical they restrict growth to one or two dimensions (Kiddie et al. 2005). Luczak's model using neutrophin-like particles constrained by boundaries showed that those factors could largely explain how many types of dendrites might grow, but due to both computational constraints and a lack of additional interacting factors the results were only approximations of real morphologies. Function-driven models rarely grow dendrites in a realistic manner as their goal is to relate completed morphologies with functional capabilities. Statistical models can serve to bridge the gap between mechanistic and function-driven models. They can account for various molecular interactions and activity effects as variance in the model, though this also means that any error in the experimental data or limitation of the reconstructions will also be captured in the model. Moreover they produce quantitative values of the relationships between morphological attributes which ultimately represent the range of morphologies that support the particular function of the dendrite being modeled.

13.6 Dendritic Morphology into the Future

There is still room for exploration of the relationship between morphology and electrophysiology, but many aspects have been investigated in great depth. Several of the models discussed focus on the relationship between network connectivity and morphology, of which much more is yet to be discovered. As time passes, molecular neuroscientists will continue to discover pathways and relationships between various growth factors, signaling molecules, and dendritic growth and plasticity. Some may be applicable to all dendrites, but a number will be specific to certain cell classes based on genetic expression. More importantly, many of those will be specifically involved in setting up connections between various anatomical regions and functional networks.

Researchers at the Allen Brain Institute (2009), producers of the Allen Brain Atlas (a map of gene expression patterns in the mouse brain – http://www.brain-map.org/), found gene expression differences in sub-regions of the hippocampus (Thompson et al. 2008). Cell adhesion genes made up a large proportion of the genes that produced the distinct boundaries, suggesting that the genes are at least in part responsible for hippocampal circuitry. Previous tracing studies have provided evidence for some of these regional divisions, but these new results provide an opportunity to further explore the relationships between genetic expression, molecular growth mechanisms, morphology, and network connectivity in very specific subregions.

It is clear that the region of synaptic targets of a neuron play a role in determining the neuron's dendritic morphology. At least one cell class's morphology can be fairly accurately reproduced with knowledge of only the spanning field while minimizing both wiring and synaptic distance to the soma (Cuntz et al. 2008). In fact the general cell class of lobula plate tangential cells is composed of four subclasses that the referenced method suggested were solely differentiated by their spanning fields. Whether this model extends to all classes of dendrite is unclear. Might the balance between minimizing wiring and synaptic difference be shifted in one direction or another in other dendrites? Even if that question is answered, the question of what determines the shape and location of the dendrites' spanning fields remains. As discussed earlier (Sect. 13.5.2), Chklovskii and colleagues' models and equations focused on minimizing wiring in the brain suggest vital components to the answer.

Dendritic morphology is shaped by many forces, both in terms of development as well as function. Varieties of intrinsic genetic, molecular, and electrophysiological factors interact with extrinsic spatial constraints, neurotrophic factors, and network activity to generate the neurons we observe. All of those interactions are influenced by evolutionary pressures to achieve the electrophysiological functionality and network connectivity that enables a neuron to take part in the local and global networks that make up the brain. With advances in imaging such as the individualized cell labeling of Brainbow (Livet et al. 2007) and potential automated reconstruction algorithms, we can expect an explosion of morphological and connectomic data. It will then take the confluence of database management, morphometric analysis, and the broad array of modeling methods to bring us closer to a complete understanding of the many relationships dendritic morphology has with brain function.

References

Allen Institute for Brain Science (2009) Allen Brain Atlas: home. Available at: http://www.brain-map.org/

Ascoli GA (2002) Neuroanatomical algorithms for dendritic modelling. Network 13:247–260

Ascoli GA (2006) Mobilizing the base of neuroscience data: the case of neuronal morphologies. Nat Rev Neurosci 7:318–324

Ascoli GA, Krichmar JL (2000) L-Neuron: a modeling tool for the efficient generation and parsimonious description of dendritic morphology. Neurocomputing 32–33:1003–1011

Ascoli GA, Krichmar JL, Scorcioni R, Nasuto SJ, Senft SL, Krichmar GL (2001) Computer generation and quantitative morphometric analysis of virtual neurons. Anat Embryol 204: 283–301

Ascoli GA, Donohue DE, Halavi M (2007) NeuroMorpho.Org: a central resource for neuronal morphologies. J Neurosci 27:9247–9251

Ascoli GA, Brown KM, Calixto E, Card JP, Galván E, Perez-Rosello T, Barrionuevo G (2009) Quantitative morphometry of electrophysiologically identified CA3b interneurons reveals robust local geometry and distinct cell classes. J Comp Neurol 515:677–695

Berry M, Bradley P (1976) The application of network analysis to the study of branching patterns of large dendritic fields. Brain Res 109:111–132

Berry M, Flinn R (1984) Vertex analysis of Purkinje cell dendritic trees in the cerebellum of the rat. Proc R Soc Lond B Biol Sci 221:321–348

Bitplane: neuron reconstruction – automatic neuron tracing, spine detection and analysis in 3D/4D. Available at: http://www.bitplane.com/go/products/filamenttracer

Borst A, Haag J (1996) The intrinsic electrophysiological characteristics of fly lobula plate tangential cells: I. Passive membrane properties. J Comput Neurosci 3:313–336
Brown K, Donohue D, D'Alessandro G, Ascoli G (2005) A cross-platform freeware tool for digital reconstruction of neuronal arborizations from image stacks. Neuroinformatics 3: 343–359
Brown KM, Gillette TA, Ascoli GA (2008) Quantifying neuronal size: summing up trees and splitting the branch difference. Semin Cell Dev Biol 19:485–493
Brugg B, Matus A (1991) Phosphorylation determines the binding of microtubule-associated protein 2 (MAP2) to microtubules in living cells. J Cell Biol 114:735–743
Burke R, Marks W, Ulfhake B (1992) A parsimonious description of motoneuron dendritic morphology using computer simulation. J Neurosci 12:2403–2416
Cannon RC, Wheal HV, Turner DA (1999) Dendrites of classes of hippocampal neurons differ in structural complexity and branching patterns. J Comp Neurol 413:619–633
Caserta F, Eldred WD, Fernandez E, Hausman RE, Stanford LR, Bulderev SV, Schwarzer S, Stanley HE (1995) Determination of fractal dimension of physiologically characterized neurons in two and three dimensions. J Neurosci Methods 56:133–144
Chen J, Kanai Y, Cowan NJ, Hirokawa N (1992) Projection domains of MAP2 and tau determine spacings between microtubules in dendrites and axons. Nature 360:674–677
Chklovskii DB (2000) Optimal sizes of dendritic and axonal arbors in a topographic projection. J Neurophysiol 83:2113–2119
Chklovskii DB, Schikorski T, Stevens CF (2002) Wiring optimization in cortical circuits. Neuron 34:341–347
Chklovskii DB, Mel BW, Svoboda K (2004) Cortical rewiring and information storage. Nature 431:782–788
Conde C, Caceres A (2009) Microtubule assembly, organization and dynamics in axons and dendrites. Nat Rev Neurosci 10:319–332
Cuntz H, Borst I, Segev I (2007) Optimization principles of dendritic structure. Theor Biol Med Model 4:21
Cuntz H, Forstner F, Haag J, Borst A (2008) The morphological identity of insect dendrites. PLoS Comput Biol 4:e1000251
Diadem competition. Available at: http://www.diademchallenge.org/
Díez-Guerra FJ, Avila J (1993) MAP2 phosphorylation parallels dendrite arborization in hippocampal neurones in culture. Neuroreport 4:419–422
Donohue DE, Ascoli GA (2005a) Local diameter fully constrains dendritic size in basal but not apical trees of CA1 pyramidal neurons. J Comput Neurosci 19:223–238
Donohue DE, Ascoli GA (2005b) Models of neuronal outgrowth. In: Koslow SH, Subramaniam S (eds) Databasing the brain: from data to knowledge. Wiley, Hoboken, pp 304–323
Donohue DE, Ascoli GA (2008) A comparative computer simulation of dendritic morphology. PLoS Comput Biol 4:e1000089
Donohue DE, Scorcioni R, Ascoli GA (2002) Generation and description of neuronal morphology using L-Neuron: a case study. In: Ascoli GA (ed) Computational neuroanatomy: principles and methods. Humana Press, Totowa, pp 49–70
Gao F (2007) Molecular and cellular mechanisms of dendritic morphogenesis. Curr Opin Neurobiol 17:525–532
Gardner D, Akil H, Ascoli G, Bowden D, Bug W, Donohue D, Goldberg D, Grafstein B, Grethe J, Gupta A et al (2008) The neuroscience information framework: a data and knowledge environment for neuroscience. Neuroinformatics 6:149–160
Georges P, Hadzimichalis N, Sweet E, Firestein B (2008) The yin–yang of dendrite morphology: unity of actin and microtubules. Mol Neurobiol 38:270–284
Glaser JR, Glaser EM (1990) Neuron imaging with Neurolucida–a PC-based system for image combining microscopy. Comput Med Imaging Graph 14:307–317
Goldstein SS, Rall W (1974) Changes of action potential shape and velocity for changing core conductor geometry. Biophys J 14:731–757
Graham BP, van Ooyen A (2004) Transport limited effects in a model of dendritic branching. J Theor Biol 230:421–432

Graham B, van Ooyen A (2006) Mathematical modelling and numerical simulation of the morphological development of neurons. BMC Neurosci 7(Suppl 1):S9

Grueber WB, Jan LY, Jan YN (2002) Tiling of the Drosophila epidermis by multidendritic sensory neurons. Development 129:2867–2878

Harding EF (1971) The probabilities of rooted tree-shapes generated by random bifurcation. Adv Appl Probab 3:44–77

Hattori D, Millard SS, Wojtowicz WM, Zipursky SL (2008) Dscam-mediated cell recognition regulates neural circuit formation. Cell Dev Biol 24:597–620

Hely TA, Graham B, Ooyen AV (2001) A computational model of dendrite elongation and branching based on MAP2 phosphorylation. J Theor Biol 210:375–384

Heumann H, Wittum G (2009) The tree-edit-distance, a measure for quantifying neuronal morphology. Neuroinformatics 7:179–190

Hillman D (1979) Neuronal shape parameters and substructures as a basis of neuronal form. In: The neurosciences, 4th study program, MIT Press, Cambridge, pp 477–498

Hines ML, Morse T, Migliore M, Carnevale NT, Shepherd GM (2004) ModelDB: a database to support computational neuroscience. J Comput Neurosci 17:7–11

Hirokawa N (1998) Kinesin and dynein superfamily proteins and the mechanism of organelle transport. Science 279:519–526

Hirokawa N, Takemura R (2005) Molecular motors and mechanisms of directional transport in neurons. Nat Rev Neurosci 6:201–214

Hodgkin AL, Huxley AF (1952) A quantitative description of membrane current and its application to conduction and excitation in nerve. J Physiol (Lond) 117:500–544

Horton AC, Rácz B, Monson EE, Lin AL, Weinberg RJ, Ehlers MD (2005) Polarized secretory trafficking directs cargo for asymmetric dendrite growth and morphogenesis. Neuron 48: 757–771

Hughes ME, Bortnick R, Tsubouchi A, Bäumer P, Kondo M, Uemura T, Schmucker D (2007) Homophilic Dscam interactions control complex dendrite morphogenesis. Neuron 54: 417–427

Ishizuka N, Cowan WM, Amaral DG (1995) A quantitative analysis of the dendritic organization of pyramidal cells in the rat hippocampus. J Comp Neurol 362:17–45

Jelinek HF, Fernandez E (1998) Neurons and fractals: how reliable and useful are calculations of fractal dimensions? J Neurosci Methods 81:9–18

Kiddie G, McLean D, Van Ooyen A, Graham B (2005) Biologically plausible models of neurite outgrowth. Prog Brain Res 147:67–80

Komendantov AO, Ascoli GA (2009) Dendritic excitability and neuronal morphology as determinants of synaptic efficacy. J Neurophysiol 101:1847–66

Krichmar JL, Nasuto SJ, Scorcioni R, Washington SD, Ascoli GA (2002) Effects of dendritic morphology on CA3 pyramidal cell electrophysiology: a simulation study. Brain Res 941: 11–28

Lewis TL, Mao T, Svoboda K, Arnold DB (2009) Myosin-dependent targeting of transmembrane proteins to neuronal dendrites. Nat Neurosci 12:568–576

Livet J, Weissman TA, Kang H, Draft RW, Lu J, Bennis RA, Sanes JR, Lichtman JW (2007) Transgenic strategies for combinatorial expression of fluorescent proteins in the nervous system. Nature 450:56–62

Lu J, Tapia J, White O, Lichtman J (2009) The interscutularis muscle connectome. PLoS Biol 7:265–277

Luczak A (2006) Spatial embedding of neuronal trees modeled by diffusive growth. J Neurosci Methods 157:132–141

Luo L (2000) RHO GTPASES in neuronal morphogenesis. Nat Rev Neurosci 1:173–180

Macias MY, Battocletti JH, Sutton CH, Pintar FA, Maiman DJ (2000) Directed and enhanced neurite growth with pulsed magnetic field stimulation. Bioelectromagnetics 21:272–286

Mainen ZF, Sejnowski TJ (1996) Influence of dendritic structure on firing pattern in model neocortical neurons. Nature 382:363–366

Maletic-Savatic M, Malinow R, Svoboda K (1999) Rapid dendritic morphogenesis in CA1 hippocampal dendrites induced by synaptic activity. Science 283:1923–1927

Malun D, Brunjes PC (1996) Development of olfactory glomeruli: temporal and spatial interactions between olfactory receptor axons and mitral cells in opossums and rats. J Comp Neurol 368:1–16
Marks WB, Burke RE (2007) Simulation of motoneuron morphology in three dimensions. I. Building individual dendritic trees. J Comp Neurol 503:685–700
MBF bioscience: neurolucida – neuron reconstruction. Available at: http://www.mbfbioscience.com/neurolucida
McAllister AK, Lo DC, Katz LC (1995) Neurotrophins regulate dendritic growth in developing visual cortex. Neuron 15:791–803
McAllister A, Katz LC, Lo DC (1997) Opposing roles for endogenous BDNF and NT-3 in regulating cortical dendritic growth. Neuron 18:767–778
Migliore M (1996) Modeling the attenuation and failure of action potentials in the dendrites of hippocampal neurons. Biophys J 71:2394–2403
Miller FD, Kaplan DR (2003) Signaling mechanisms underlying dendrite formation. Curr Opin Neurobiol 13:391–398
Mirsky JS, Nadkarni PM, Healy MD, Miller PL, Shepherd GM (1998) Database tools for integrating and searching membrane property data correlated with neuronal morphology. J Neurosci Methods 82:105–121
Neuromantic: the freeware neuronal reconstruction tool. Available at: http://www.rdg.ac.uk/neuromantic
Nikolić M (2008) The Pak1 kinase: an important regulator of neuronal morphology and function in the developing forebrain. Mol Neurobiol 37:187–202
Parrish JZ, Emoto K, Kim MD, Jan YN (2007) Mechanisms that regulate establishment, maintenance, and remodeling of dendritic fields. Annu Rev Neurosci 30:399–423
Pedrotti B, Colombo R, Islam K (1994) Microtubule associated protein MAP1A is an actin-binding and crosslinking protein. Cell Motil Cytoskeleton 29:110–116
Polleux F, Morrow T, Ghosh A (2000) Semaphorin 3A is a chemoattractant for cortical apical dendrites. Nature 404:567–573
Quinlan EM, Halpain S (1996) Emergence of activity-dependent, bidirectional control of microtubule-associated protein MAP2 phosphorylation during postnatal development. J Neurosci 16:7627–7637
Radley JJ, Sisti HM, Hao J, Rocher AB, McCall T, Hof PR, McEwen BS, Morrison JH (2004) Chronic behavioral stress induces apical dendritic reorganization in pyramidal neurons of the medial prefrontal cortex. Neuroscience 125:1–6
Rajnicek A, Gow N, McCaig C (1992) Electric field-induced orientation of rat hippocampal neurones in vitro. Exp Physiol 77:229–232
Rall W (1969) Time constants and electrotonic length of membrane cylinders and neurons. Biophys J 9:1483–1508
Rapp M, Segev I, Yarom Y (1994) Physiology, morphology and detailed passive models of guinea-pig cerebellar Purkinje cells. J Physiol 474:101–118
Samsonovich AV, Ascoli GA (2003) Statistical morphological analysis of hippocampal principal neurons indicates cell-specific repulsion of dendrites from their own cell. J Neurosci Res 71:173–187
Samsonovich AV, Ascoli GA (2006) Morphological homeostasis in cortical dendrites. Proc Natl Acad Sci USA 103:1569–1574
Scorcioni R, Lazarewicz MT, Ascoli GA (2004) Quantitative morphometry of hippocampal pyramidal cells: differences between anatomical classes and reconstructing laboratories. J Comp Neurol 473:177–193
Scorcioni R, Polavaram S, Ascoli GA (2008) L-Measure: a web-accessible tool for the analysis, comparison and search of digital reconstructions of neuronal morphologies. Nat Protoc 3: 866–876
Shepherd GMG, Svoboda K (2005) Laminar and columnar organization of ascending excitatory projections to layer 2/3 pyramidal neurons in rat barrel cortex. J Neurosci 25:5670–5679

Sholl DA (1953) Dendritic organization in the neurons of the visual and motor cortices of the cat. J Anat 87:387–4061
Sieck GC, Prakash YS (1995) Fatigue at the neuromuscular junction. Branch point vs. presynaptic vs. postsynaptic mechanisms. Adv Exp Med Biol 384:83–100
Sin WC, Haas K, Ruthazer ES, Cline HT (2002) Dendrite growth increased by visual activity requires NMDA receptor and Rho GTPases. Nature 419:475–480
Spruston N, Schiller Y, Stuart G, Sakmann B (1995) Activity-dependent action potential invasion and calcium influx into hippocampal CA1 dendrites. Science 268:297–300
Stepanyants A, Chklovskii D (2005) Neurogeometry and potential synaptic connectivity. Trends Neurosci 28:387–394
Stiefel K, Sejnowski T (2007) Mapping function onto neuronal morphology. J Neurophysiol 98:513–526
Tamori Y (1993) Theory of dendritic morphology. Phys Rev E 48:3124
Thompson C, Pathak S, Jeromin A, Ng L, Macpherson C, Mortrud M, Cusick A, Riley Z, Sunkin S, Bernard A (2008) Genomic anatomy of the hippocampus. Neuron 60:1010–1021
Van Ooyen A, Duijnhouwer J, Remme M, van Pelt J (2002) The effect of dendritic topology on firing patterns in model neurons. Netw Comput Neural Syst 13:311–325
van Pelt J (1997) Effect of pruning on dendritic tree topology. J Theor Biol 186:17–32
Van Pelt J, Verwer R (1986) Topological properties of binary trees grown with order-dependent branching probabilities. Bull Math Biol 48:197–211
van Pelt J, Verwer RW, Uylings HB (1986) Application of growth models to the topology of neuronal branching patterns. J Neurosci Methods 18:153–165
Van Pelt J, Uylings H, Verwer R, Pentney R, Woldenberg M (1992) Tree asymmetry—a sensitive and practical measure for binary topological trees. Bull Math Biol 54:759–784
Verwer RW, van Pelt J (1983) A new method for the topological analysis of neuronal tree structures. J Neurosci Methods 8:335–351
Verwer R, Van Pelt J (1990) Analysis of binary trees when occasional multifurcations can be considered as aggregates of bifurcations. Bull Math Biol 52:629–641
Vetter P, Roth A, Hausser M (2001) Propagation of action potentials in dendrites depends on dendritic morphology. J Neurophysiol 85:926–937
Wen Q, Chklovskii DB (2008) A cost-benefit analysis of neuronal morphology. J Neurophysiol 99:2320–2328
Williams SR, Stuart GJ (2000) Action potential backpropagation and somato-dendritic distribution of ion channels in thalamocortical neurons. J Neurosci 20:1307–1317
Wong ROL, Ghosh A (2002) Activity-dependent regulation of dendritic growth and patterning. Nat Rev Neurosci 3:803–812
Yamamoto H, Saitoh Y, Fukunaga K, Nishimura H, Miyamoto E (1988) Dephosphorylation of microtubule proteins by brain protein phosphatases 1 and 2A, and its effect on microtubule assembly. J Neurochem 50:1614–1623
Zador A, Agmon-Snir H, Segev I (1995) The morphoelectrotonic transform: a graphical approach to dendritic function. J Neurosci 15:1669–1682

Chapter 14
Axonal Growth and Targeting

Duncan Mortimer, Hugh D. Simpson, and Geoffrey J. Goodhill

Abstract The growth and guidance of axons is an undertaking of both great complexity and great precision, involving processes at a range of length and time scales. Correct axonal guidance involves directing the tips of individual axons and their branches, interactions between branches of a single axon, and interactions between axons of different neurons. In this chapter, we describe examples of models operating at and between each of these scales.

14.1 Introduction

The modeling of information processing by neural networks has had a long and fruitful history (see, for example, Chap. 10). In contrast, relatively little is understood about the computational principles involved in initially wiring such networks. In the developing human brain, hundreds of billions of neurons form hundreds of trillions of connections by extending their axons over sometimes vast distances (on the cellular scale). How do these axons "know" where to grow? This is the axon guidance problem. In this chapter, we describe various mathematical modeling approaches that have been taken, and how they have informed our understanding of this crucial process.

D. Mortimer
Electronics and Computer Science, Faculty of Physical and Applied Sciences, University of Southampton, Southampton, SO17 1BJ United Kingdom
e-mail: dm3@ecs.soton.ac.uk

H.D. Simpson
The Queensland Brain Institute, The University of Queensland, QLD 4072, Brisbane, Australia
e-mail: hughdsimpson@gmail.com

G.J. Goodhill (✉)
The School of Mathematics and Physics, The Queensland Brain Institute, The University of Queensland, QLD 4072, Brisbane, Australia
e-mail: g.goodhill@uq.edu.au

N. Le Novère (ed.), *Computational Systems Neurobiology*,
DOI 10.1007/978-94-007-3858-4_14,

Understanding axon guidance requires studying processes that occur on a variety of time and length scales. The tips of axons, growth cones, are semi-autonomous structures, capable of responding to multiple cues in their environment, and have hence been the target of focused research. However, on a larger scale, guidance involves the entire axon; for example it may involve the selective pruning or promotion of axonal branches to achieve a specific aim. On a larger scale again, wiring up the nervous system is a collective problem, as axons interact with each other and with other cells. At this level, growth cones, axons and their substrates cooperate to shape guidance. There may also be a role for electrical activity in guiding this stage of nervous system wiring, as well as its more established role in refining and patterning established neuronal architecture.

A wide variety of modeling approaches have been used to tackle axon guidance, ranging from models taking a detailed simulation perspective, attempting to reflect the biophysical basis of axon guidance in as much detail as possible, through to highly abstract models focussing on how particular strategies for specifying guidance routes might allow for the formation of a complex nervous system. (For other reviews, see van Ooyen 2003; Maskery and Shinbrot 2005; Graham and van Ooyen 2006; Simpson et al. 2009.)

In this chapter, we examine models that operate at each of the scales previously identified: ranging from the behavior of individual growth cones, through the dynamics of entire axons and their branches, to the dynamics of populations of axons. We first review models of the growth cone—both how it moves, and how it is guided by external cues. We then consider the axon as a functional guidance unit, looking at models that study how resources (e.g. cytoskeletal proteins) are allocated between different branches of the same axon, or between the cell body and growth cones. Finally, we look at the issues encountered when considering the guidance of populations of axons, such as how interactions between axons can act to improve the robustness and specificity of projections.

14.2 Guidance for the Tip: Models of the Axonal Growth Cone

The embryonic environment is awash with chemical signals that direct the intricacies of nervous system development. In order to guide axons, these chemical signals are detected by special structures at the tips of growing axons, known as growth cones (Gordon-Weeks 2000) (Fig. 14.1). First identified and described by Ramon ý Cajal in the late nineteenth century, these complex, motile "battering rams" read the information provided by the environment, and transduce it into decisions about the direction in which the axon should grow. They are micromachines with sensory and motor capabilities, tasked with wiring the nervous system. Thus, to understand axon guidance, it is of key importance to understand the growth cone, and how it senses and responds to the chemical signals that provide the map for axonal pathfinding.

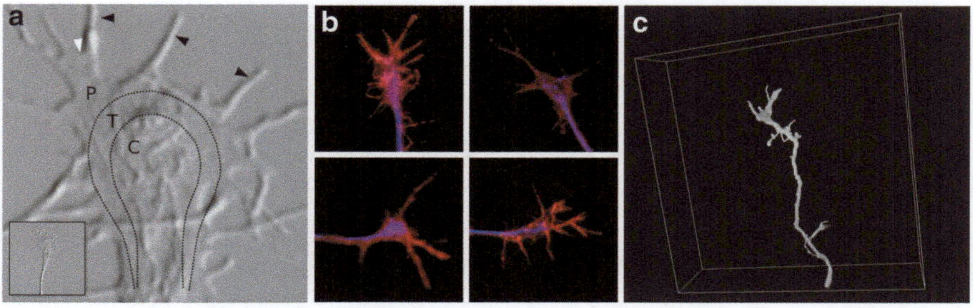

Fig. 14.1 Growth cones are complex structures on the tips of developing neurites: (**a**) Anatomically, the growth cone can be roughly partitioned into three sections: the central zone (*C*), transitional zone (*T*) and peripheral zone (*P*). Magnified region marked on inset in *red*. The *arrowheads* indicate filopodia (*black*) and a lamellipodium (*white*) (**b**) Growth cones display widely varying morphology, even when grown on the same substrate, from the same tissue source (in this case, rat dorsal root ganglion neurons, grown on laminin and stained for actin (*red*) and tubulin (*blue*)). (**c**) In three dimensions, growth cones tend to take on a more "streamlined", filopodially dominated morphology (All images courtesy of Z. Pujic, Goodhill lab)

Although growth cones can show great variability in their morphology, three distinct regions can be defined, as illustrated in Fig. 14.1a: a central region containing organelles and rich in microtubules, a thinner peripheral domain predominantly consisting of a network of actin filaments, and a narrow transitional domain between the previous two regions.

As with many other motile biological structures, actin dynamics are crucial for growth cone motility and morphology. Actin filaments tend to be oriented with their "barbed" ends—the ends at which unpolymerized actin monomers (G-actin) are most easily incorporated into the filament—towards the outside of the growth cone. Actin polymerization thus tends to push against and expand the outer membrane. On two dimensional substrates (where all modeling work so far has occurred), these dynamics lead to the formation of two distinct classes of structure at the growth cone leading edge: filopodia and lamellipodia (Gordon-Weeks 2000) (Fig. 14.1b,c).

14.2.1 Models of Growth Cone Motility

14.2.1.1 Descriptive Models

A large part of modeling work has been devoted to simply describing how growth cones behave, and to extracting rules about their behavior which can be incorporated into more explanatory models. A number of attempts to capture the dynamics of growth cones were made throughout the 1990s by Buettner and colleagues (e.g. Buettner et al. 1994; Buettner 1996; Odde et al. 1996). Growth cones were filmed while growing on different substrates, and the results analyzed to extract statistical regularities. Buettner et al. (1994), described the growth cone using a

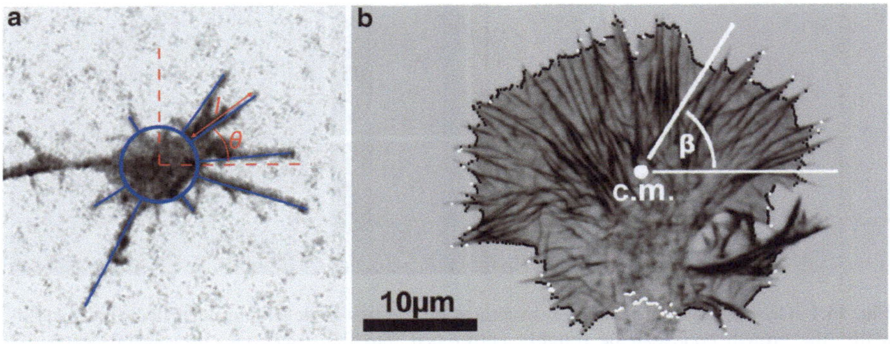

Fig. 14.2 Example methods for describing growth cones: (**a**) Buettner (1996) modeled filopodia searching for target tissue. In this case, modeling the growth cone as a circular central region, with radially extending filopodia was found to be sufficient. (**b**) Betz et al. (2006, 2009) studied the dynamics of protrusion and retraction of elements of the growth cone boundary, and how these related to actin polymerization and retraction within the growth cone itself. They used radial profiles of actin staining intensity, and radial displacement of the growth cone boundary to quantify these phenomena. *c.m.* = center of mass (**Panel b** from Betz et al. 2006)

hybrid system: a contour detailing the shape of the lamellipodia, and a series of "sticks" extending radially from the centroid of the growth cone representing the filopodia (Fig. 14.2a). These models were fit by hand to movies, and the statistics of the resulting parameters examined. For example, it was observed that filopodia tended to extend and retract with a constant rate, switching between the extension and retraction phases according to a gamma-distribution. The statistical models of growth cone behavior obtained in this manner informed estimates of how easily a growth cone could cross a gap between two permissive substrates, and also the likelihood of it contacting a locally expressed guidance cue.

Advances in experimental technique allowed the motion of the growth cone as a whole to be compared to the underlying dynamics of the cytoskeleton. For example, Odde et al. (1996) analyzed the joint statistics of growth cone motion and microtubule dynamics, finding that the two were coupled with a slight delay, giving credence to the idea that growth cone advance involves the active coordination of actin and microtubule dynamics.

Betz et al. (2006, 2009) measured the extension of the membrane from the central region as a function of time (Fig. 14.2b), and then fit the motion of a small section of membrane with a one-dimensional random walk in a potential field (Fig. 14.3). They found that the membrane dynamics both of growth cones, and of cell-lines made to express growth-cone-like characteristics, could be well described if the field consisted of two shallow wells (Fig. 14.3b). Intriguingly, however, real growth cones showed a form of "stochastic resonance", in which the noise properties of the effective random walk were tuned to the potential field, such that small modulations in the steepness of the underlying potential surface had a strong effect on the resulting dynamics.

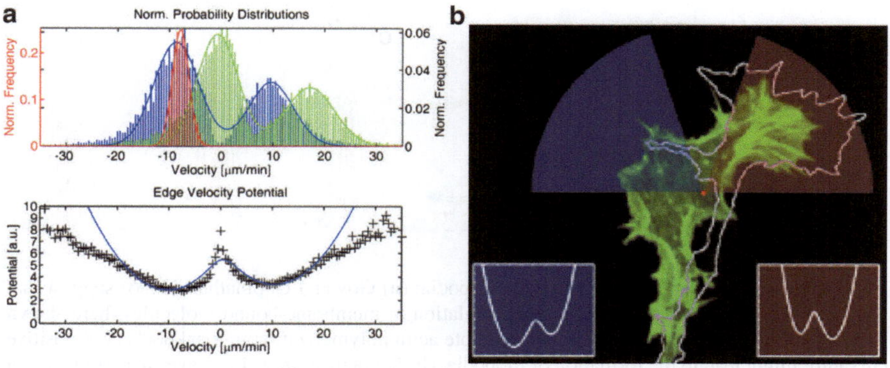

Fig. 14.3 Growth cone membrane dynamics tend to hop stochastically between extension and retraction: In Betz et al. (2006, 2009), the extension or retraction rates of small patches around the growth cone periphery were measured by calculating the change in distance from the *center* of the growth cone to the periphery in subsequent video frames. (**a**) In the *top panel*, the distribution of rates of rearward actin flow is shown in *red*, the distribution of edge velocity in *blue*, and the inferred distribution of the rate of actin polymerization in *green*. The distribution of, and temporal correlation in, the edge velocities can be captured as a random walk in a bimodal potential field, shown in the *lower panel*. The "hump" between the two dips in the *lower panel* controls the rate at which protrusion switches to retraction, and vice-versa. (**b**) In a turning growth cone, the potential field is biased towards protrusion on one side, and retraction on the other (Images from Betz et al. 2009)

14.2.1.2 Regulation of the Cytoskeleton, and Growth Cone Morphology

Though not specific to growth cones, a series of models have studied how the interaction between the plasma membrane and actin cytoskeleton (mediated or tuned by membrane-associated regulatory proteins) lead to the formation, and control the dynamics, of filopodia and lamellipodia. Gov and Gopinathan (2006) studied the linkage between actin and membrane dynamics caused by the preferential localization of membrane-associated actin regulatory proteins with regions of specific membrane curvature. The authors developed a partial-differential equation model describing the coupling between the curvature-dependent diffusion of membrane-bound molecules, and feedback onto membrane curvature through the modulation of actin dynamics by those molecules. Two distinct behaviors were observed, depending on whether the regulatory proteins favoured regions of positive (i.e. outward-bulging) or negative (i.e. inward) curvature. For positive-curvature-preferring regulatory proteins, small positive curvature fluctuations in the cell membrane tended to be amplified by the attraction of regulatory proteins to those regions (Fig. 14.4a), and the subsequent local promotion of actin polymerization. Depending on the relationships between the rate of diffusion of the regulatory proteins, the rate at which they promote actin polymerization, and the membrane tension, this amplification effect could lead to dynamic instabilities, in which the bump would continue to grow, suggesting a possible mechanism for the formation

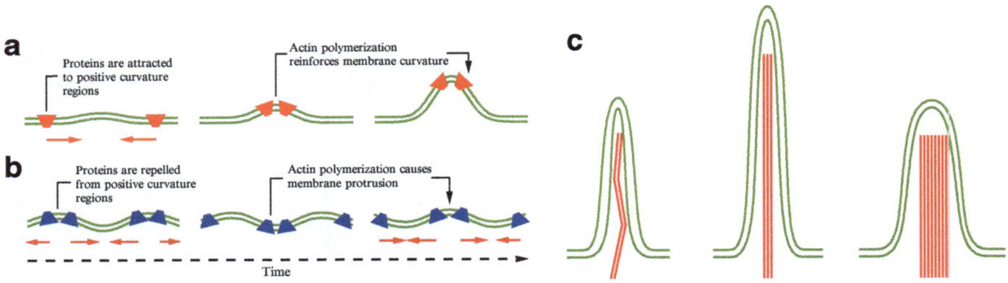

Fig. 14.4 Formation and growth limits to filopodia: (**a**) Gov and Gopinathan (2006) suggest that a positive feedback loop involving the accumulation of membrane-bound molecules (here shown in *red*) that prefer positive curvature and promote actin polymerization in local pockets of positive curvature might lead to the formation of filopodia. (**b**) In contrast, traveling waves may occur when these membrane-bound proteins (now in *blue*) prefer regions of negative curvature. (**c**) When there are too few actin filaments, tension causes a filopodium to buckle and collapse. In constrast, when there are too many, g-actin cannot be delivered to the filopodium tip fast enough to overcome depolymerization (Figures **a** and **b** adapted from Gov and Gopinathan 2006)

of filopodia. In contrast, when the regulatory proteins prefer regions of negative curvature, traveling waves of actin polymerization resulted, reminiscent of dynamics observed in lamellipodial structures (Fig. 14.4b).

Given that a filopodium has begun to form, it is of interest to know how far and how fast it can grow (thus potentially limiting the sensory range of a growth cone), and how it might interact with other nearby filopodia. Atilgan et al. (2006) used energy-minimisation arguments to demonstrate that at least two bundled actin filaments are required to overcome membrane elasticity and initiate filopodium growth, and that membrane deformation induces an effective attractive force between nearby filopodia. Mogilner and Rubinstein (2005) showed that filopodial length is limited on the one hand by the number of filaments in the cross-section of the actin fibre bundle (i.e. its strength), and on the other by the rate at which unpolymerized actin can be delivered to the filopodium tip by diffusion. If the bundle contains too few filaments, it buckles under strain from the membrane (Fig 14.4c, left). However, if it contains too many, then the rate at which new actin must be delivered to the tip is higher than can be achieved by diffusion, limiting the ultimate length of the filopodium (Fig. 14.4c, right).

14.2.1.3 The Generation of Traction Force/Growth Cone Advance

Concurrently with actin polymerization at the growth cone periphery, the actin cytoskeleton is withdrawn towards the central region of the growth cone, probably through the combination of myosin action and pressure from the membrane, where it depolymerizes (Medeiros et al. 2006). Free G-actin moves to the leading edge of the growth cone either through diffusion or active transport, where it is again incorporated into the F-actin cytoskeleton. This cycle of actin polymerization and cell membrane extension at the leading edge, retrograde actin flow, and actin depolymerization in the central region acts as an engine that can be harnessed to

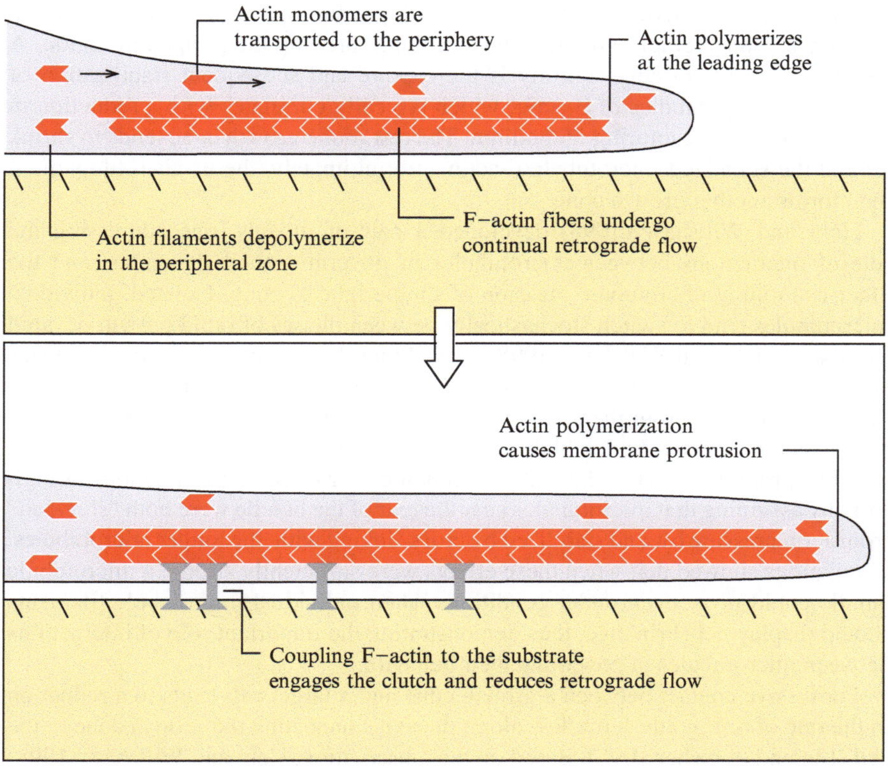

Fig. 14.5 The actin treadmill and molecular clutch: In the *top panel*, actin fibers within the growth cone are uncoupled from the substrate. They undergo continual retrograde flow, depolymerization in the peripheral zone (*left* of the figure), and polymerization at the leading edge. Coupling the F-actin to the substrate (*lower panel*) reduces the rate of retrograde flow, allowing polymerization at the leading edge to push the membrane forward

provide motility (Suter and Forscher 2000). However, in order to provide forward movement, and not to simply cycle on the spot, the actin cytoskeleton must be coupled to a permissive substrate. The idea that growth cones can regulate their motion by modulating this coupling has been called a "clutch mechanism", and provides a useful conceptual framework for analysing growth cone motility (Fig. 14.5). For an excellent review of models of actin-treadmill-driven motility (not specifically focussed on growth cones), see Mogilner (2009).

Microtubules also play a crucial role in growth cone motility (Gordon-Weeks 2004). They form a thick bundle in the neurite shaft that extends from the cell body to the growth cone. This bundle provides stability for the growing neurite, and acts as a scaffold along which materials from the cell body may be transported to the growth cone. Microtubules penetrate into the central region of the growth cone, sometimes remaining bundled, at other times splaying out to explore the boundary between the central and peripheral domains. Within the growth cone, the microtubules demonstrate "dynamic instability" (Mitchison and Kirschner 1984), rapidly growing and shrinking, probing the peripheral region. A key event in growth cone motility is the capture and subsequent stabilization of microtubules by bundles of F-actin, which typically correlates with a reduction in F-actin flow in the capturing filopodium. The rest of the growth cone tends to shrink around the stabilized microtubule/F-actin core; ultimately, the invaded filopodium then forms another growth cone.

Hely and Willshaw (1998) developed a pair of models, one addressing the role of interactions between microtubules in governing their dynamics, and the other examining microtubule invasion of the peripheral zone. Isolated, individual microtubules tend to switch stochastically between phases of rapid growth or rapid shrinkage. Hely and Willshaw (1998) argued that the rate at which these switches occur in bundled microtubules in the growth cone is modulated by two effects: a backward pressure from the actin cytoskeleton which tends to promote shrinkage, and crosslinking between neighboring microtubules which tends to enhance stability and thus promote growth. The authors accounted for these effects phenomenologically by assuming that microtubules near the rear of the bundle were both "shielded" from actin pressure by, and stabilized by cross-linking with, the longer microtubules. Simulations showed that when these effects were sufficiently strong, a microtubule bundle could grow even under conditions when individual microtubule filaments should display net shrinkage, thus demonstrating the important role of interactions between microtubules in governing their behavior.

Permissive contact between a growth cone and a target cell leads to a reduction in the rate of retrograde actin flow along the axis connecting the growth cone to the cell (Lin and Forscher 1995). Based on this observation, Hely and Willshaw (1998) analyzed the degree to which such a reduction was sufficient to enhance microtubule invasion into the region of the peripheral zone closest to the target cell. In this model, the authors assumed that microtubules within the growth cone extended in random directions, undergoing dynamic instability with the probability of switching depending on the local rate of actin flow and the proximity to other microtubules. However, they found that even under unrealistically favourable conditions, very few microtubules invaded the target region of the peripheral zone, suggesting that microtubules are actively directed within the growth cone, rather than relying solely on random search.

14.2.2 *Models of Guidance*

The growth cone has a sensory task in addition to, and closely associated with, its motor task of extending the axon: detecting and responding to chemical and mechanical cues in its environment in order to guide axon growth. Our understanding of axon guidance has undergone an explosion in the last few decades, thanks to the discovery and cloning of many of the proteins involved; in particular, the "guidance cue" molecules (Tessier-Lavigne and Goodman 1996). Growth cones detect these cues through specialized chemical receptors expressed on their surface. Interactions between these receptors and the guidance cues lead to changes in conformation of the receptors, and subsequent downstream signalling (Lauffenburger and Linderman 1993). If the cue is expressed in a graded manner, then this downstream signalling is in turn asymmetric across the growth cone's spatial extent, with the potential to regulate asymmetric remodeling of the cytoskeleton.

Often, guidance cues are highly expressed on the surface of individual cells. In this case, the contact of a single filopodium with such a cell can completely reorient the growth cone (O'Connor et al. 1990). The problem the growth cone faces in this situation is thus that of searching out such targets. A few models have addressed how this search might be undertaken efficiently. Buettner (1996) studied how the parameters governing filopodial initiation, growth and eventual collapse influenced the probability of contacting a single such cell. Taking a more abstract approach, Maskery et al. (2004) investigated the interplay between deterministic and random growth cone behavior in searching for, and responding to, a guidance cue expressing cell. They found that there was an optimal balance between the two types of behavior, in which the growth cone could both efficiently search (requiring a degree of random wandering), and effectively respond (requiring a deterministic component) to such localized cues.

Guidance cues can also be present in long-range gradients, potentially produced by the diffusion of secreted molecules from a localized source or graded expression in the substrate (Dickson 2002). One challenge we face in understanding how growth cones respond to such gradients is that the growth cone's sensory system is inherently noisy: receptor-ligand interactions are stochastic events (Bialek and Setayeshgar 2005; Mortimer et al. 2009; 2010a), receptor signalling involves the addition of further noise (Ueda and Shibata 2007; Mortimer et al. 2010a), and the gradient itself will be subject to thermal fluctuations further degrading the signal (Bialek and Setayeshgar 2005). Furthermore, if the gradient is to provide guidance over an appreciable range, it cannot be too steep as the growth cone can only effectively respond to the graded signal within a reasonably narrow range of concentrations (Mortimer et al. 2009).

How can growth cones detect and respond to shallow gradients reliably given their noisy sensory apparatus? In addition to models applied directly to growth cones, models of other gradient sensing cells are also likely to be of use in understanding this phenomenon. In the interests of focus and brevity, we will constrain ourselves to models dealing specifically with growth cones, but provide further references to related modeling work in other systems (see e.g. Bialek and Setayeshgar 2005; Ueda and Shibata 2007; Herzmark et al. 2007; Endres and Wingreen 2008). Two methods by which this problem has been attacked are: directly modeling the molecules or mechanisms thought to be involved in the growth cone response; and, seeking to understand the limits to gradient sensing imposed by noise in the growth cone's sensory systems. We give examples of each of these model classes in the following sections.

14.2.2.1 Mechanistic Models

Many molecular mechanisms are involved in growth cone guidance. A variety of guidance cues and their cognate receptors have been identified and these in turn act through a range of internal signalling pathways. These pathways include calcium signalling (both from intracellular and extracellular sources), active redistribution of receptors, cyclic AMP and cyclic GMP pathways, the phosphatidylinositol pathway, MAP Kinases, the rho-GTPase pathways, the directed transport of vesicles for subsequent exocytosis (and potential autocrine signalling) and even asymmetrically localized protein synthesis within the growth cone (reviewed in Zheng and Poo 2007; Mortimer et al. 2008; Lowery and van Vactor 2009). Although no models currently exist incorporating all of these mechanisms, models have been developed which attempt to capture some subset of them. More recently, advances in experimental techniques allowing for visualization and quantification of protein levels in different regions of the growth cone have provided inspiration for progressively more complete models (see Chap. 3 for a discussion of modelling signalling pathways in general).

An early model in this vein was developed by Aeschlimann and Tettoni (2001). Their aim was to obtain a biophysically plausible model of growth cone movement and neurite extension that could at least roughly mimic experimentally observed behavior. In their model, filopodia took on a central role, as both the sensory organs and primary motor units of the growth cone. Each filopodium was assumed to produce a small pulling force in the direction in which it extended. Depending on the size of the net force generated in this manner, the distal axon segment was assumed to stretch or, if the force was above some threshold, lengthen through inelastic extension. The growth cone was able to respond to external guidance cues via calcium signals produced at the bases of filopodia: through an unspecified mechanism, contact of a filopodium with an external cue would lead to the opening of calcium channels at base of the filopodium, thus producing an influx of calcium. Calcium dynamics were modeled through standard diffusion equations. Modeling calcium concentration provided a link between the sensory and motor systems, as the probability of initiating a filopodium at a given angular location was determined by the calcium concentration at that location. With this model, the authors were able to qualitatively reproduce the kinds of trajectories traced out by growth cones in the presence of a gradient formed by diffusion from a point source.

In Goodhill et al. (2004), filopodia again played a central role in both the sensory and motor behavior of the modeled growth cone. As with Aeschlimann and Tettoni (2001), filopodia were assumed to be the driving force of growth cone motility, though the relationship between filopodial force and growth cone movement was modeled implicitly. At each timestep, the growth direction was determined by taking a weighted average of the direction of net filopodial force and the current direction of growth. Again, in this model, the direction of filopodial initiation was assumed to be biased by external cues. A key issue that this model tackled was that direct proportionality between external concentration and the probability of filopodial initiation is not sufficient to produce a turning response to shallow external gradients. Rather, some degree of internal amplification of the external gradient must be performed, sharpening the distribution of filopodial initiation. The performance of the growth cone in responding to the gradient was found to depend on the strength of this amplification. The model also predicted that the dependence of gradient sensing performance on concentration would be different for attractive, as opposed to repulsive, gradients—a prediction that has not yet been tested.

Xu et al. (2005) presented a related model, in which at each timestep, the growth cone's movement was calculated by averaging its current heading with an estimate of the gradient direction determined by input from surface receptors. In contrast to Goodhill et al. (2004), these authors focussed on the role of temporal and spatial averaging of receptor generated signals, rather than on force transduction through filopodia. They directly simulated a random-walk based model, in which growth cone response to a gradient is mediated through the production of a second messenger by bound receptors. Temporal and spatial averaging occurred through the dynamics of decay and diffusion of the second messenger. The authors fitted their model to a corpus of experimental data (Rosoff et al. 2004), finding that a good fit could be obtained when the diffusion rate and decay time of the second messenger were such that signals diffused over roughly a third of the growth cone, over a lifetime of about 3 min.

The observation that, preceding growth cone turning, bound GABA receptors are actively trafficked to the up-gradient side of a growth cone exposed to a gradient of GABA (Bouzigues et al. 2007) has inspired two recent models. Causin and Facchetti's (2009) assumed that a similar mechanism might play a role in growth cone response to netrin, through the DCC receptor. Netrin-DCC binding was assumed to drive the activation of L-type calcium channels (LCCs); these activated LCCs then indirectly recruited bound DCC receptors through an unspecified mechanism, modelled as a convective force proportional to the vector gradient of calcium channel activation. The authors incorporated these mechanisms into a partial differential equation model, with the aim of showing that they were sufficient to produce an asymmetric distribution of receptors in the presence of an external gradient. A crucial parameter of their model turned out to be the strength of the coupling between the gradient of activated LCCs and the transport of DCC receptors: a stable asymmetric distribution of DCC was only achieved when the coupling strength exceeded a threshold dependent on the diffusion rate of the receptors, and the degree of amplification in the signalling cascade leading from netrin-DCC binding to channel activation. Bouzigues et al. (2010) developed an explicit stochastic partial differential equation model linking the transport of bound GABA receptors to microtubule growth. The tip of each microtubule was assumed to exert a pull, modelled as a localised reduction in a potential energy function, on GABA receptors diffusing on the surface of the growth cone. The microtubules themselves were assumed to be biased in their growth by the distribution of activated GABA receptors: as with Causin and Facchetti's (2009) model, this leads to a positive feedback loop by which localised receptor activation leads to receptor recruitment. The receptor diffusion coefficient again emerged as an important parameter in this model: large values tended to reduce the time taken for an asymmetry to form, but tended also to reduce the strength of the asymmetry for a given gradient strength.

14.2.2.2 Abstract Models

Goodhill and Urbach (1999) and Mortimer et al. (2009) sought to understand the limitations imposed by noise to a growth cone's gradient sensing abilities. Following the seminal work of Berg and Purcell (1977), these studies argued that the perception of a guidance cue gradient by a growth cone is essentially a problem of signal estimation: on the basis of noisy measurements from its receptors, the growth cone must estimate the direction of the external gradient (for a more general discussion of the role of noise in nervous system function see Chap. 8). The reliability with which such estimates can be made is ultimately limited by the steepness of the gradient, the time the growth cone has to make its decision and the amount of noise associated with measuring concentration via the binding of receptors.

Goodhill and Urbach (1999) directly applied the results of Berg and Purcell (1977) to analyze two methods by which a growth cone might detect and respond to a gradient: "temporal" and "spatial" gradient sensing. Under a temporal gradient sensing strategy, the growth cone is assumed to measure the concentration at a given point in space, then move to a nearby point and again measure the concentration. Comparing these two measurements then gives the growth cone some idea of whether it is tending to move up, or down, the gradient. In contrast, under a spatial strategy, concentration measurements at two different spatially-separated points on the growth cone itself are compared. In each of these cases, the ability of the growth cone to detect the gradient is limited by the difference in concentration ΔC between the two measurement points (determined by the distance between the points, and the steepness of the gradient), and the noise associated with concentration measurement σ. This latter value is determined by quantities such as the diffusivity of the guidance cue and the number of receptors involved in making the measurement. In the case of a temporal sensing mechanism, all receptors on the growth cone can be used to measure the concentration at both of the measurement points. However, for the spatial sensing strategy, at most only half of the receptors can be used at each of the measurement locations.

Mortimer et al. (2009, 2010b, 2011) extended these ideas, and directly compared the results with experimental data on growth cone chemotaxis. In this study, it was assumed that growth cones implement some form of spatial strategy, and techniques from statistical decision theory were used to determine the optimal form for this strategy. Mortimer et al. (2009) constructed a statistical model of the probability of observing a particular pattern of receptor binding under any given gradient conditions. They then applied Bayes' theorem to invert this model in order to obtain the gradient conditions with the highest probability of causing an observed pattern of receptor binding. Mathematical analysis of this strategy revealed that a growth cone's performance in a gradient sensing task should be proportional to the gradient steepness, and depend in a defined way on the background concentration. The

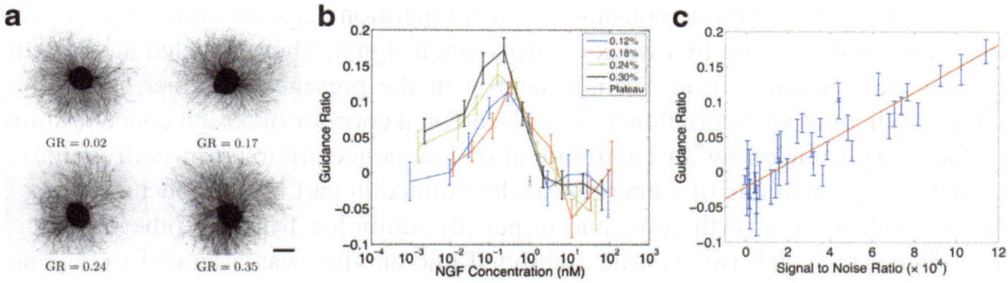

Fig. 14.6 A Bayesian model of axon guidance is consistent with guided growth of rat dorsal root ganglion neurites: (**a**) Example images of rat dorsal root ganglion explants, illustrating the "guidance ratio" (GR), which quantifies the degree of asymmetric growth. In these examples, explants were exposed to an upwards-pointing gradient of nerve growth factor. (**b**) The magnitude of the guidance response depends on the steepness of the gradient (expressed as the percentage change in concentration across 10 μm) and average background concentration. Guided growth tends to increase with increasing steepness, and to be biphasic in concentration. The model in Mortimer et al. (2009) predicts a linear relationship between the guidance ratio, and a "signal-to-noise ratio" which depends on the background concentration and gradient steepness. This relationship is demonstrated in (**c**), based on the data in (**b**). Error bars are SEMs

response of rat dorsal root ganglion neurons to gradients of nerve growth factor (NGF) displayed performance consistent with this prediction over a large range of gradient conditions (Fig. 14.6). Surprisingly, however, the authors were unable to detect any tendency for neurites to correct their direction of growth by turning in the direction of the gradient (Mortimer et al. 2010b). Rather, the results were consistent with a model in which neurites modulate their growth rate in response to the gradient, speeding up when heading toward higher concentrations of NGF, and slowing down when heading toward lower concentrations. A possible benefit of this kind of strategy when responding to shallow gradients is that concentration comparisons might be made over (for example) the length of the axon, rather than being confined to either side of the growth cone, thus increasing any observed differences in concentration.

14.3 Models of the Entire Axon

Axons often display not one, but multiple growing branches each with its own growth cone. This raises the possibility that axon guidance should rightly be considered a behavior of the entire axon, rather than of the growth cone. For example, it might be that axon branches that are detected to be growing in a good direction are favoured in their growth at the expense of other, less optimal branches. This idea along with early observations that implicated a form of competition between branches of the same neurite (reviewed by Smalheiser and Crain 1984) has influenced several modeling studies, which investigated the potential effects of such competition, along with the mechanism by which it might be mediated.

Li et al. (1992, 1995) demonstrated that competition between branches can lead to a sharpened response to a weak environmental signal. They modeled the growth of branching neurites from a single neuron in the presence of either a smooth, linear gradient of an external factor, or placed on a corridor of a high concentration of factor, surrounded by an environment of low concentrations. In both of these models, the growth rate of a neurite was determined in part by the concentration of factor local to the growth cone, and in part by inhibition from the other neurites. The degree to which two neurites inhibited one another was assumed to depend on the number of branch points separating them, so that neurites which are more distant had less influence than neurites that are closer. As a result, neurites at lower concentrations of growth factor tended to face more inhibition (as well as slower intrinsic growth) than neurites at high concentrations. Thus, the relative effect of any asymmetry in the environment on neurite growth rate was amplified.

How might competition between neurite branches be mediated? Neurite elongation tends to occur at the growth cone. However, the raw materials to support this growth are largely manufactured in the cell body, and must be transported to the growth cone to be incorporated into the growing axon as it extends. This suggests one way in which such competition might occur: through competition of individual axon branches for resources to support their continued growth.

Tubulin is a strong candidate for such a resource. Microtubules (polymerized tubulin) form the backbone of neurites. Tubulin polymerizes and is incorporated into the cytoskeleton predominantly at the growth cone, but it is manufactured in the cell body. Several models have thus focused on the dynamics of tubulin transport and assembly, and the role of competition between different neurites for tubulin. Van Veen and Van Pelt (1994) developed an ODE model of tubulin transport and polymerization in neurite growth, examining the results for an unbranched neurite, a neurite with one branch, and for neurons with complex arbors. They assumed that tubulin was synthesized in the soma at a constant rate, was transported to the neurite tips through diffusion (active transport was judged to be insignificant), and that the rate of neurite growth was directly proportional to the rate of tubulin polymerization at the growth cone. For a single neurite, they found that tubulin concentration in the growth cone reached a constant value, while the neurite length and the tubulin concentration in the soma increased linearly with time. In contrast when multiple neurites were involved, they found that competition occurred depending on the local rates of tubulin polymerization in the growth cones. Despite the simplicity of this model, it was able to reproduce a number of experimental results: that growing neurites tend to extend at a constant rate, that individual neurites in an arbor occasionally retracted in favour of other growing neurites, and that dormant growth cones would occasionally activate and begin growing some time after their formation.

Van Veen and Van Pelt (1994) made the simplification of directly modeling tubulin concentrations only at branch points, the soma and in the growth cones, treating diffusion along neurite segments at steady-state. Clearing the way for further studies, Graham and van Ooyen (2001) extended the compartmental models commonly used to study the propagation of electrical activity in mature neurons to the case of axonal development, in which neuronal morphology can change over time. They highlighted the challenges associated with developing such models: in particular, that large artificial transients can arise depending on choices made in how the system is "recompartmentalized" as it grows (see also Kiddie et al. 2005).

McLean et al. (2004) and Graham et al. (2006) examined the dynamics of tubulin transport and polymerization in more detail for the growth of a single neurite. They employed a partial differential-equation approach, allowing them to model how tubulin concentration varies along the length of the axon. This model included both active and diffusive tubulin transport, along with tubulin degradation, and simulations demonstrated non-trivial interactions between the three mechanisms. When the rate of tubulin degradation was set to zero, the results of Van Veen and Van Pelt (1994) were reproduced: namely, the neurite grew at a constant rate due to a constant tubulin concentration in the growth cone, with tubulin concentration in the soma increasing linearly with time. These results were independent of whether active transport, diffusion or a combination of both transport mechanisms were included (though the shape of the tubulin concentration gradient along the length of the neurite was affected by the choice of transport mechanism). However, when tubulin degradation was included, the neurite no longer grew indefinitely: rather, it eventually reached an equilibrium length which depended on the rates of transport, degradation and tubulin production. Graham et al. (2006) studied the sensitivity of this final equilibrium length to other parameters of the model, with the particular aim of understanding how easily neurite length could be regulated by a cell. They found that for short neurites, the length was essentially insensitive to variations in the rate of active transport, while long neurites showed insensitivity to the diffusion rate.

Despite the theoretical advantages of competition between neurites from the same neuron, there is experimental evidence suggesting that such competition does not, in fact, occur (Lamoureux et al. 1998). This suggests that the cell must actively regulate the production of key cytoskeletal molecules (such as tubulin) in order to support uniform growth independent of the number of neurite tips. First steps towards understanding self-regulation of tubulin production were taken by Graham et al. (2006), though only in the context of a single growing neurite; future extensions of this model to include multiple branches may shed light on how competition between branches is avoided.

14.4 Guidance at the Systems Level

Modeling the guidance of neuronal projections involving sometimes large numbers of axons adds another layer of complexity to considerations for individual growth cones and/or axons. Within projections, axons may provide scaffolds for each other, compete with each other, and interact in other more complex ways. A key step in understanding the behavior and targeting of neural projections is to focus on the interactions between growing axons and/or growth cones. Insights gained from this can then be combined with knowledge of individual axon and growth cone guidance.

Modeling the targeting of large numbers of axons has been undertaken within a number of biological contexts; such as within specific parts of the spinal cord and brain, or more abstractly, such as the development of sheets of neurons or neural networks. One of the most useful paradigms for studying this problem is the development of retinotopic maps; the primary examples being retinotectal/retinocollicular maps (from retina to midbrain) or retinogeniculocortical maps (maps from retina to thalamus to visual cortex). Of these two, the influences on neuronal guidance have been better established for retinotectal maps, an area which has also received a large amount of theoretical attention. As such, focusing on the retinotectal map gives us a broad spectrum of modeling to discuss with a minimum of background.

14.4.1 Background: The Retinotectal Map

Neural connections between the retina in the eye and the visual centers of the brain are referred to as visual maps. They are described as topographic, or retinotopic, if the spatial relationships between cells in the retina are preserved in their pattern of termination elsewhere in the brain. The retinotectal (also retinocollicular) map is the neural representation of visual space formed by projections arising in the retina (from the retinal ganglion cells or RGCs) and terminating in the midbrain (more specifically, the target is the optic tectum in fish, frogs and other lower vertebrates; and the superior colliculus/SC in mammals) (Fig. 14.7).

A number of different mechanisms have been postulated to play roles in retinotectal map development. The most important of these are as follows:

- Chemoaffinity: The idea, proposed by Sperry (1963), that gradients of molecular markers can be used to specify axial positions, and thus one or more in each of two areas is sufficient to specify a topographic map between these areas.
- Competition and other axon–axon interactions: Competition between axons is usually for a limiting resource, such as target space, neurotrophins, or synaptic input. Axons can also exert a range of other influences on each other in addition to competition.
- Branching: Multiple interacting agents are generated by branching processes, and the interactions of branches with each other and with molecular cues can effect different forms of guidance. Types of branching include:
  - Growth cone splitting: bifurcation of the leading edge of a growing axon.
  - Backbranching: Branching, usually associated with retraction of the primary axon, that occurs just proximal to the growth cone.
  - Interstitial branching: branching that may occur anywhere along the entire length of the axon, usually at right angles (or nearly so) to the main axon shaft.

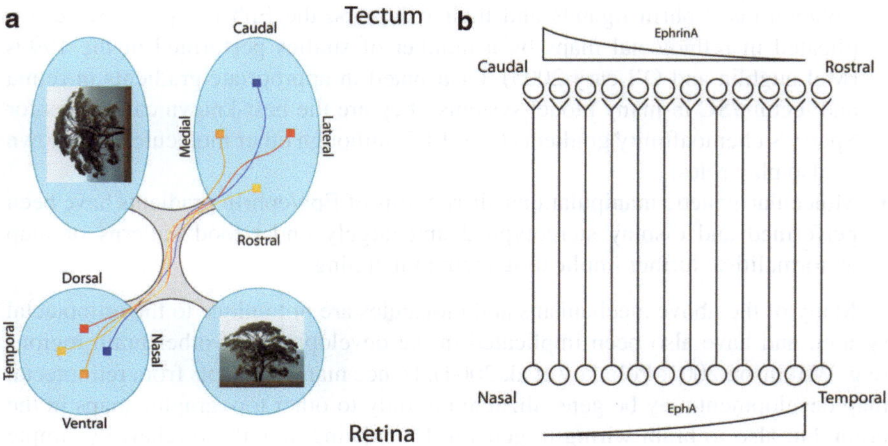

Fig. 14.7 The retinotectal map: a paradigm example of topographic map development. During development, the retinal ganglion cells (RGCs)—the output layer of the retina—send out axonal projections to the midbrain tectum. If two cells are close to each other in the retina, their terminations are similarly close in the tectum. In this way, an image of visual space is transmitted faithfully to the brain. (**a**) Nasal retina maps to caudal tectum, and temporal retina maps to rostral tectum. Similarly, dorsal retina maps to lateral tectum, and ventral retina maps to medial tectum. (**b**) The patterns of terminations are in part achieved by the use of chemoaffinity gradients; i.e. gradients of molecules that can specify unique positions along the gradient axis by unique levels of that molecule. In this example we show one axis of the retinotectal map: that of nasotemporal retina to caudorostral tectum. A row of cells is depicted in the retina, with positions identified with EphA receptor level. Similarly a row of tectal postsynaptic cells is shown marked by ephrinA ligands. EphA-bearing cells are typically repelled by ephrinA ligands, so that here high EphA maps to low ephrinA and vice versa. (Note this is a simplified example, in that gradients of EphA receptor are also present in the tectum, and similarly gradients of ephrinA ligand are present in the retina.) The dorsoventral to mediolateral axis is controlled in an analogous way by EphB-ephrinB interactions, although this interaction is attractive rather than repulsive (not pictured)

- Marker induction/regulation: It has been hypothesized that ingrowing axons can upregulate various molecular cues in the target which can in turn influence the guidance of the ingrowing axons (Willshaw and von der Malsburg 1979).
- Neural activity/synaptic modulation: Neural activity can change the strength of synaptic connections between neurons, and hence can have a range of effects on reforming neuronal projections; but plays more of a refining rather than defining role in retinotectal mapping by increasing the precision of maps generated by the above mechanisms. We do not consider models of neural activity in this section.

A large amount of data has been gathered over more then 50 years on retinotectal map development, which can be summarized as follows:

- Systems manipulations: In the 1940s–1980s, before roles for specific molecules were identified, gross anatomical manipulations of the retina and/or tectum were performed, including ablations and surgical grafting (Udin and Fawcett 1988).

- Eph/ephrins: Ephrin ligands and their receptors, the Eph receptors, were implicated in retinotectal maps by a number of studies performed in the 1990s (McLaughlin and O'Leary 2005). Distributed in appropriate gradients in retina and tectum/SC in many model systems, they are the best-known candidates for Sperry's chemoaffinity gradients (Fig. 14.7) although other molecules are known to also play roles.
- Molecular-genetic manipulations: disruptions of Eph/ephrin gradients have been performed and display stereotypical and largely understood patterns of map abnormalities, further implicating them in mapping.

Many of the above mechanisms and molecules are not unique to the retinotectal system, and have also been implicated in the development of other brain regions (e.g. Wilkinson 2001; Poliakov et al. 2004). Hence many concepts from retinotectal map development may be generalizable not only to other topographic maps in the brain, but also to brain wiring in general. Elucidating how these relatively simple patterns form can thus substantially contribute to our understanding of strategies that the CNS uses to wire itself up.

14.4.2 *Approaches to Modeling the Guidance of Multiple Axons*

The retinotectal system has been a target for computational models since the 1970s, and generations of models of map development have generally followed generations of experiments. Instead of analyzing one of the many models in great detail, or all models only briefly, we will focus on specific illuminating examples of problems that arise in modeling the retinotectal system, and the methods used to approach these problems.

Terminology used to describe topographic map models and development varies from model to model, and also over time. We refer to 'projections' of neurons and their axons as arising from an origin and projecting to a target. Origin and target can be either discrete or continuous, but usually the origin is a discrete array and the target is a discrete lattice or continuous sheet. We refer to individual members of an array as growth cones, axons, or branches, or generally as 'agents'.

14.4.2.1 Competition in the Development of Targeted Neural Populations

Competition is a central concept in developmental neuroscience, as it is in biology in general (see van Ooyen 2001 for a review). Competition describes an interaction whereby multiple agents seek to exclusively control a limiting resource. In axonal growth and targeting the most common applications of this idea are competition for physical space, for neurotrophins/growth factors, or for synaptic input/activity.

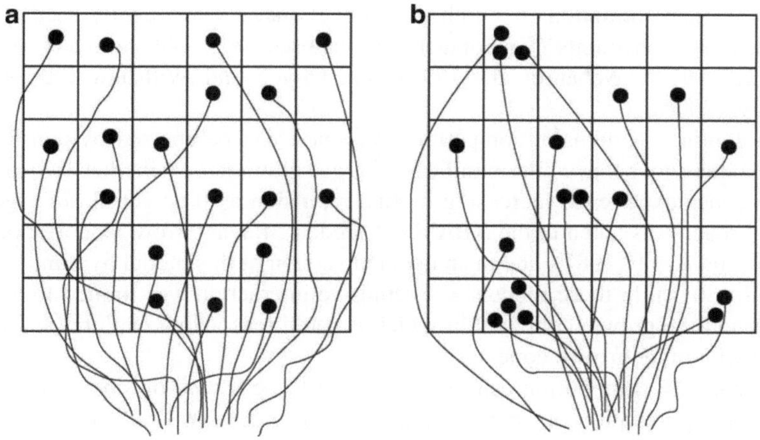

Fig. 14.8 Competition through synaptic normalization. Two 6×6 arrays representing schematized postsynaptic sites on the target are shown. Eighteen RGCs, half the number of available lattice points, are depicted projecting to each array. Axons are shown as projecting from bottom of picture and their terminations are represented by *black discs* on the target. (**a**) A one-to-one mapping is enforced, so that axons compete for space, and as a result, spread out over more of the target. (**b**) The number of contacts that can be made with each lattice point is now not fixed, and there is no competition, and in this case, less coverage of the target

A straightforward way to include competition in models is to use a discrete mapping model where contacts between the projecting array and the targeted array are one-to-one (Fig. 14.8). This models a competition for space or synaptic contacts, as only one axon can occupy each lattice point at a given time. The term synaptic normalization is often applied to models that limit pre- and post-synaptic cells to one synaptic contact each, although it can also be used to refer to cases where each can make a small fixed number of contacts, greater than one. Examples of this type of model are Hope et al. (1976) and Koulakov and Tsigankov (2004). Although this is simple to implement in models, it may also oversimplify (it neglects the potential of multiple inputs/outputs to single agents), and that the discrete nature of the target array may lead to unrealistic axon/growth cone motion, as axons hop from lattice point to lattice point (though this latter issue can be mitigated by increasing the density of lattice points).

A more realistic approach is to allow a small fixed number of contacts (greater than one) to be made with each lattice point by axons. An early exploration of this idea was carried out by Prestige and Willshaw (1975). These authors considered a mapping between discrete arrays and showed that when the number of contacts between arrays was unlimited, a more rigid type of chemoaffinity assumption needed to be made, whereas if the number of contacts was fixed, simpler chemoaffinity assumptions could be made. This model demonstrated how different forms of chemoaffinity could lead to topography depending on competitive constraints, and has been influential in our understanding of how these mechanisms interact in map formation. A number of models have subsequently used similar normalization constraints (for contacts or synapses), e.g. Willshaw and von der Malsburg (1979), Weber et al. (1997), van Ooyen and Willshaw (2000), and Willshaw (2006).

In a continuous domain, competition for space can be enacted by considering growth cones to behave like similarly charged particles, such that two nearby growth cones experience increasing mutual repulsion as they get closer together. This approach was taken in the XBAM (eXtended Branch Arrow Model) model of Overton and Arbib (1982), and in an updated version of this model by Simpson and Goodhill (2011). In these models, the repulsive interaction was limited to a small area around the growth cone within which it could reasonably explore (and hence interact with other growth cones).

Alternatively competition can be considered to be a type of smoothing mechanism for a population of interacting neurons. Honda (2003) added this type of constraint to his model, whereby initially a map was set up using chemoaffinity rules, and then a smoothing process was undertaken in which axons were moved from areas of higher density to areas of lower density. This kind of algorithm promotes uniformity or smoothness of the map.

14.4.2.2 Axon–Axon Interactions as a Sorting Process

An appealing strategy is to consider the ordering of topographic maps as a sorting process. An example of this is the work of Hope et al. (1976), who considered the movements of discrete axons on a lattice. It was assumed that a pair of interacting axons could compare a scalar attribute common to all axons (e.g.: some molecular label). Given the axons' relative positions and relative molecular label, the axons could opt to exchange places according to a particular rule. Although this is a relatively strong assumption from a biological point of view, it is an effective algorithm in that it can reproduce several phenomena important in the field of retinotectal mapping. This was demonstrated in simulations of normal and surgically altered development.

Overton and Arbib (1982) developed this approach further by moving to a continuous domain. In this model the positions of axons are continuously-valued, and hence more specific assumptions about the algorithm need to be made. Instead of switching places, axons experience a vector push towards or away from each other (again depending in some way on relative position and molecular label). This model produced more realistic axon growth and targeting, while maintaining the sorting process as a key feature. More recently, Simpson and Goodhill (2011) updated this model to 2D, and modified the sorting process to reflect a more realistic type of axon–axon interaction based on repulsive guidance receptor signaling.

Koulakov and Tsigankov (2004) used a lattice sorting approach similar to that of Hope et al. (1976). In this model, pairs of axons were considered iteratively, and an exchange probability calculated which depended proportionally on differences in molecular labels between the axons. This model was updated by the same

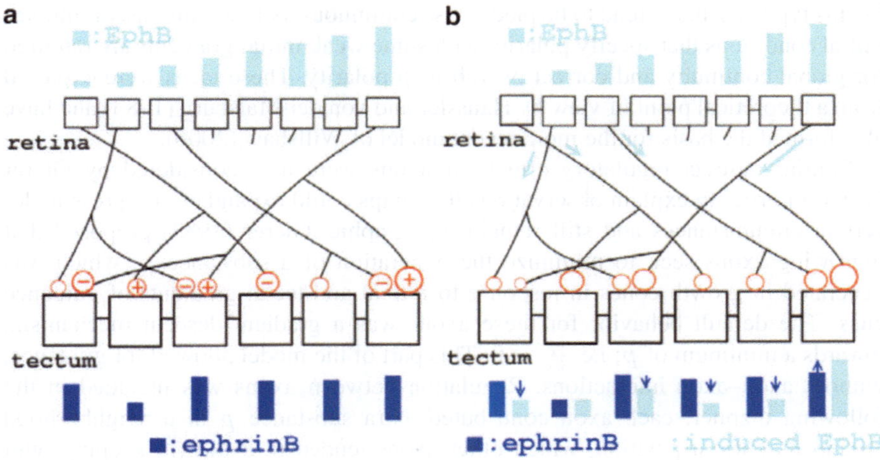

Fig. 14.9 The retinal induction mechanism. A one-dimensional mapping is depicted from retina to tectum. The dorsoventral to mediolateral axis is pictured, which is governed by EphB-ephrinB interactions. A row of cells in the retina bearing EphB receptors projects to a row of cells in the tectum bearing ephrinB ligands. One or more synapses are made by each RGC (shown in *red*). (**a**) Updating the synaptic strengths. Synapses are strengthened or weakened according to how similar the amount of EphB in the presynaptic cell is to the amount of ephrinB is in the postsynaptic cell. If the amounts are more similar, the connection is strengthened, whereas if the amounts are quite dissimilar, the connections are weakened. (**b**) Induction/regulation. Levels of EphB are up- or down-regulated in postsynaptic cells in proportion to the amount of EphB in presynaptic contacts and the strength of their respective synapses. Single postsynaptic cells may receive contributions from more than one presynaptic cell. Levels of ephrinB in the postsynaptic cells are constantly adjusted to be closer to the level of induced EphB in that cell (Figure from Willshaw (2006))

authors by adding an additional constraint involving correlated activity. This was combined with differences in molecular labels to calculate an energy for the entire map (Tsigankov and Koulakov 2006). Depending on whether this energy would be increased or decreased by the swap, a probability of the swap occurring was again calculated.

14.4.2.3 Axonal Regulatory Effects

Willshaw and von der Malsburg (1979) proposed that ingrowing axons can up- or down-regulate certain molecular markers (e.g. axon guidance molecules) in cells of the target array (Fig. 14.9). These molecules in turn influence axonal behavior, and hence axons exert indirect influence on each other through this regulatory process. Allowing axons to influence their guidance environment in this manner gave this model a rich dynamics and some unique characteristics; most notably, a 'neighbor matching' property that may be compared with models of correlated neural activity (Willshaw and Price 2003). A consequence of this property is that maps generated by this type of scheme tend to be piece-wise continuous, but globally discontinuous. Initial conditions that specify polarity with some weak initial gradients are required for global continuity and correct overall map polarity. These ideas were explored from a theoretical point of view by Häussler and von der Malsburg (1983) and have also formed the basis for the more recent model of Willshaw (2006).

Similar indirect regulatory effects of axons were also considered by Gierer (1983) in order to explain observations that maps could expand or compress under certain circumstances and still remain topographic. Gierer (1983) proposed that ingrowing axons seek to minimize the generation of a substance p, which was generated in growth cones in response to retinal and tectal gradients of guidance cues. The default behavior for these axons was a gradient descent mechanism, towards a minimum of p; i.e. $\frac{\partial p}{\partial x} = 0$. This part of the model allowed for guidance, without axon–axon interactions. Regulation between axons was included in the following manner: each axon contributed extra substance p in a neighborhood around its current position, which other axons tended to avoid (in keeping with the assumption that axons seek to minimize this p). As functions of the tectal x-axis position and retinal u-axis position, and time t, Gierer modeled this extra contribution as $r(x,t)$, and the contribution from gradients as $g(x,u)$ (which could be chosen largely arbitrarily), so that the total p was:

$$p = g(x,u) + r(x,t) \tag{14.1}$$

$$\frac{\partial r}{\partial t} = \varepsilon \rho(x,t) \tag{14.2}$$

Here $\rho(x,t)$ is the local density of fiber terminals, and ε is a constant, so that increased densities of axons causes increased contributions to the local level of p, and hence a greater tendency to move away from this local area. This mechanism tended to smooth the mapping out and give it a target filling property.

14.4.2.4 Branching

So far our discussion has been limited to problems involving only one growth cone per axon, however mature biological neural networks are complex interconnected structures made up of highly branched nerve fibers. Branches can interact with each other intra-axonally and inter-axonally, so that branching can be considered a generator of multiple interacting agents. Systems of branching neurons can display their own unique behaviors, including pathfinding and dynamic branch regulation. The former idea has been explored by representing the addition of new branches as new contacts or synapses (Willshaw and von der Malsburg 1979; Willshaw 2006). By making the generation of new contacts more probable in areas where there are existing (successful) synapses, a kind of pathfinding or guidance is achieved. This mechanism alone might simply lead to clustering of contacts, but combined with other mechanisms (retinal induction in this case), branching of this type can help order the map.

More detailed mechanisms for patterns of branching in retinotectal axon pathfinding were considered by Gierer (1987). Growth cone splitting, back-branching, and random interstitial branching were employed, and their effect on pathfinding was examined. The main differences noted between these three cases were variations in trajectories and axonal arbor morphology. The pathfinding potential of interstitial branching was also investigated in the model of Yates et al. (2004), based on previous work showing that interstitial branching was biased by gradients of chemoaffinity-type markers/molecules (Fig. 14.10). It was shown by these authors that differential branching according to chemoaffinity-type gradients can generate topography similarly to axonal guidance models. Each axonal arbor has its own internal dynamics, in which branching depends in part on the rest of the arbor, and as such represents intra-axonal interactions between branches. Inter-axon interactions were also included by assuming that as branching occurs and arbors grow, the markers that they carry contribute to the gradients of markers already on the tectum. This added marker has additional more complex effects on arbor refinement and targeting. A similar biased branching approach was taken by Godfrey et al. (2009), but this time a different form of gradient-based chemoaffinity was employed, which was also influenced by growth factors. An additional 'resource' component was added to the model to constrain growth of axonal arbors (also influenced by growth factors), introducing a more explicit form of intra-axonal interaction to the model (See Sect. 14.3).

14.4.2.5 Molecular Mechanism Models

Some models of how growth cones use their receptors to navigate in ligand gradients also give rise to ways in which axons can interact based on similar processes. For example, the servomechanism model of Honda (1998, 2003) (based on the observations of Nakamoto et al. 1996) proposed that ingrowing axons seek a specific level of 'signal'. This signal, S, was produced in an axon i as a function of its receptor level, R_i, and local ligand level $L(\mathbf{x})$ (where $\mathbf{x}$ is a position vector). Using first order mass action kinetics this becomes $S_i(\mathbf{x}) = R_i L(\mathbf{x})$. A 'servomechanism' was proposed along with this idea such that an axon seeks to minimize the difference $\Delta S = S_{0,i} - S_i(\mathbf{x})$, where $S_{o,i}$ is a reference signal. In the first stage of the model this was simply used to move individual growth cones, but later during a smoothing stage of the model ΔS was used to compare growth cones in locally dense areas, and move ones which had higher values of ΔS. In this way growth cones interacted indirectly through this relative affinity (cf. Prestige and Willshaw 1975, which goes into more detail on types of affinity; also discussed in Goodhill and Xu 2005).

Models utilizing gradient-based chemoaffinity rules have tended to focus on interactions between guidance receptors on ingrowing RGCs and ligands in the target. However, both these sets of receptors and ligands are also present on RGC growth cones, and efforts have been made to quantify how growth cones may interact with each other based on these molecules. Such a model was suggested by Reber et al. (2004) following on from the work of Brown et al. (2000). Here it was

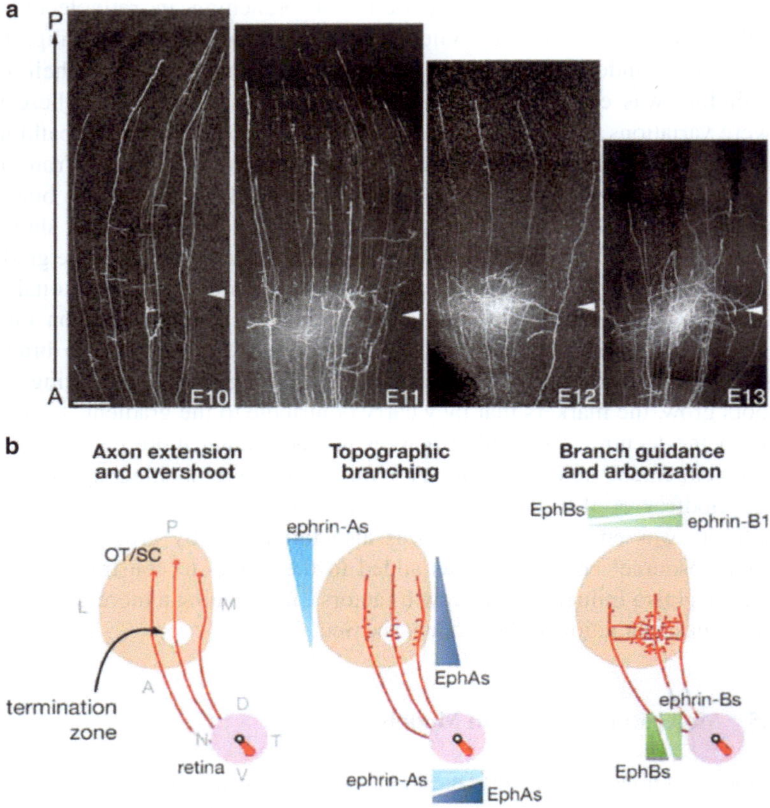

Fig. 14.10 Mechanisms of interstitial branching in retinotopic map development. (**a**) The retino-tectal projection in chick optic tectum from embryonic day 10 to day 13. Vertical axis of the picture corresponds to anterior (A, or rostral) to posterior (P, or caudal) tectal axis. *Arrowhead* indicates appropriate termination zone of these RGC axons which have been labeled with a single DiI injection into temporal retina. RGC axons initially overshoot their target, then send out interstitial branches roughly at *right* angles to the primary axon centred on their target. Topographic specificity increases with time until a topographically correct termination zone is present at E13. Figure from Yates et al. (2001); scale bar 250 μm. (**b**) This process is modeled conceptually in the following manner. RGC axons initially overshoot their desired termination zone in a non-guided manner. Then interstitial branching is controlled by opposing inhibitory gradients of EphA and ephrinA in the optic tectum/superior colliculus (OT/SC), which interact with the same molecules present in the retina to determine an optimal region for branching. Then opposing gradients of EphB/ephrinB molecules control medial/lateral guidance of branches towards the termination zone. Note that in contrast to some previous examples, this model relies on there being dual gradients (of Eph receptors and ephrin ligands) present in both retina and tectum/superior colliculus (Figure from McLaughlin and O'Leary (2005))

noted that specific relationships appeared to exist between RGCs based solely on comparisons of EphA receptor level between axons. When, under the conditions of certain genetic manipulations, the receptor ratios between nearby axons fell below a certain threshold, receptor-based topographic organizing behavior was significantly altered. It remains unclear what the exact mechanisms are behind these receptor ratio comparisons but two recent studies present suggestions (Tsigankov and Koulakov 2010; Simpson and Goodhill 2011).

14.4.2.6 Integrating Multiple Constraints

A problem common to most models of retinotectal mapping that involve axon–axon interactions is how to combine information from interactions with other axons with information from environmental cues (e.g. physical cues, guidance cues, neural activity, etc.). The integration of this information may occur within the growth cone, the axon, or the whole cell including the cell body. However, the individual signal transduction pathways are often not well described, nor how each individual pathway is integrated with the others. (There are exceptions to this however, with some pathways/systems representing good candidates for modeling signal integration, such as the Rho GTPases; for example see Giniger 2002; Sakumura et al. 2005.) As a result, most of the approaches have been more abstract or phenomenological in nature, attempting to model the effects of signal integration rather than specific biophysical events in how it occurs.

A straightforward way to integrate multiple signals is the approach taken by Overton and Arbib (1982) and Simpson and Goodhill (2011). In these models, independent influences on growth cone motion are modeled as vector contributions, which are combined in a linear sum. This is obviously a simplification of growth cone signal transduction machinery, nonetheless both studies demonstrated the ability of this model to display complex behavior, including the reproduction of specific experimental observations.

Fraser and Perkel (1990) took another approach, motivated by the idea that adhesion between projecting axons and their targets may be central in regulating targeting of projections. These authors defined an 'adhesive free energy' which was an additive combination of contributions from other axons, from target-based adhesive markers, competition for physical space, and effects of correlated neural activity. The model considered the movement of terminal arbors subject to the above energetic constraints. At each step, arbors randomly moved to new nearby positions. The new positions were either accepted or rejected with probability dependent on the adhesive energy of the new configuration. A similar approach was employed by Tsigankov and Koulakov (2006). Independent contributions to an energy function from chemoaffinity and neural activity were combined linearly for each axon, and then these were summed to calculate a total map energy. Similar to the switching/sorting algorithm of Hope et al. (1976), pairs of axons were chosen at random, and their positions exchanged with a probability that was a function of the map energy.

Some models use the more familiar concept of synaptic modulation as a common pathway for fusing various mechanisms of axonal guidance. These models use the creation and modulation of synapses to achieve targeting and mostly do not specifically consider the movement or targeting of growth cones. A prominent example of this type of model is the series of models based on Cowan's model. The first major model in this series (Whitelaw and Cowan 1981) used an ODE to update synaptic strengths in proportion to the product of two axons' chemospecific affinity and their coincident activity. The most recent version of the model (Weber et al. 1997) takes a more typical synaptic modulation scheme; where the additive influences of gradient-based chemoaffinity cues, chemical affinity for other axons, and neural activity combine to alter synaptic strengths. In contrast, models such as Willshaw (2006) also use a form of synaptic modulation, but instead assume that it is dependent on the similarity of molecular label on the presynaptic and postsynaptic cells.

14.5 Summary

Throughout this chapter, we have discussed many of the aspects involved in the correct wiring of the axons of the nervous system. In particular, we have illustrated that axon guidance is a field of study with facets that give rise to problems on a variety of spatial and temporal scales.

- At a sub-cellular scale, the growth cone integrates sensory information, and directs axon growth. Models at this level have provided insight into the biochemical events involved in axon extension, and the processes by which cues in the embryonic environment are interpreted.
- At the cellular scale, the processes of axon extension and branching determine the neuron morphology. Models at this level have suggested a functional role for interaction between multiple branches of the same axon, and have begun to explore the machinery regulating the manufacture of the raw materials required for axon extension.
- Finally, at the systems scale, interactions between axons play an important role. Through the example of retinotectal map formation, we have discussed models incorporating such interactions. These models have demonstrated that axon–axon interactions help to make the topographic structure of the retinotectal projection robust to a range of experimental manipulations. Future work in other systems may yield similar insights.

An ultimate understanding of axon guidance would incorporate elements at each scale, and would enable us to link defects in (for example) growth cone cytoskeletal remodeling to gross defects in wiring between brain regions. Developing such an understanding is a grand challenge for the future.

References

Aeschlimann M, Tettoni L (2001) Biophysical model of axonal pathfinding. Neurocomp 38–40:87–92

Atilgan E, Wirtz D, Sun SX (2006) Mechanics and dynamics of actin-driven thin membrane protrusions. Biophys J 90:65–76

Berg HC, Purcell EM (1977) Physics of chemoreception. Biophys J 20:193–219

Betz T, Lim D, Käs JA (2006) Neuronal growth: a bistable stochastic process. Phys Rev Lett 96:098103

Betz T, Koch D, Lim D, Käs JA (2009) Stochastic actin polymerization and steady retrograde flow determine growth cone advancement. Biophys J 96:5130–5138

Bialek W, Setayeshgar S (2005) Physical limits to biochemical signaling. Proc Natl Acad Sci USA 102:10040–10045

Bouzigues C, Morel M, Triller A, Dahan M (2007) Asymmetric redistribution of GABA receptors during GABA gradient sensing by nerve growth cones analyzed by single quantum dot imaging. Proc Natl Acad Sci USA 104:11251–11256

Bouzigues C, Holcman D, Dahan M (2010) A mechanism for the polarity formation of chemoreceptors at the growth cone membrane for gradient amplification during directional sensing. PLoS One 5(2):e9243

Brown A, Yates PA, Burrola P, no DO, Vaidya A, Jessell TM, Pfaff SL, O'Leary DD, Lemke G (2000) Topographic mapping from the retina to the midbrain is controlled by relative but not absolute levels of EphA receptor signaling. Cell 102(1):77–88

Buettner H (1996) Analysis of cell-target encounter by random filopodial projections. AICHE J 42:1127

Buettner HM, Pittman RN, Ivins JK (1994) A model of neurite extension across regions of nonpermissive substrate: Simulations based on experimental measurements of growth cone motility and filopodial dynamics. Dev Biol 163:407–422

Causin P, Facchetti G (2009) Autocatalytic loop, amplification and diffusion: a mathematical and computational model of cell polarization in neural chemotaxis. PLoS Comp Biol 5:e1000479

Dickson BJ (2002) Molecular mechanisms of axon guidance. Science 298:1959–1964

Endres RG, Wingreen NS (2008) Accuracy of direct gradient sensing by single cells. Proc Natl Acad Sci USA 105:15749–15754

Fraser SE, Perkel DH (1990) Competitive and positional cues in the patterning of nerve connections. J Neurobiol 21(1):51–72

Gierer A (1983) Model for the retino-tectal projection. Proc R Soc Lond B Biol Sci 218(1210):77–93

Gierer A (1987) Directional cues for growing axons forming the retinotectal projection. Development 101(3):479–489

Giniger E (2002) How do rho family gtpases direct axon growth and guidance? a proposal relating signaling pathways to growth cone mechanics. Differentiation 70(8):385–396

Godfrey KB, Eglen SJ, Swindale NV (2009) A multi-component model of the developing retinocollicular pathway incorporating axonal and synaptic growth. PLoS Comput Biol 5(12):e1000600

Goodhill GJ, Urbach JS (1999) Theoretical analysis of gradient detection by growth cones. J Neurobiol 41:230–241

Goodhill GJ, Xu J (2005) The development of retinotectal maps: a review of models based on molecular gradients. Network 16(1):5–34

Goodhill GJ, Gu M, Urbach JS (2004) Predicting axonal response to molecular gradients with a computational model of filopodial dynamics. Neural Comput 16:2221–2243

Gordon-Weeks PR (2000) Neuronal growth cones. Cambridge University Press, Cambridge

Gordon-Weeks PR (2004) Microtubules and growth cone function. J Neurobiol 58:70–83

Gov NS, Gopinathan A (2006) Dynamics of membranes driven by actin polymerization. Biophys J 90:454–469

Graham BP, van Ooyen A (2001) Compartmental models of growing neurites. Neurocomputing 38–40:31–36

Graham BP, van Ooyen A (2006) Mathematical modelling and numerical simulation of the morphological development of neurons. BMC Neurosci 7(1):S9

Graham BP, Lauchlan K, McLean DR (2006) Dynamics of outgrowth in a continuum model of neurite elongation. J Comput Neurosci 20:43–60

Häussler A, von der Malsburg C (1983) Development of retinotopic projections: an analytical treatment. J Theoret Neurobiol 2:47–73

Hely TA, Willshaw DJ (1998) Short-term interactions between microtubules and actin filaments underlie long-term behaviour in neuronal growth cones. Proc R Soc Lond B 265:1801–1807

Herzmark P, Campbell K, Wang F, Wong K, El-Samad H, Groisman A, Bourne HR (2007) Bound attractant at the leading vs. the trailing edge determines chemotactic prowess. Proc Natl Acad Sci USA 104:13349–13354

Honda H (1998) Topographic mapping in the retinotectal projection by means of complementary ligand and receptor gradients: a computer simulation study. J Theor Biol 192(2):235–246

Honda H (2003) Competition between retinal ganglion axons for targets under the servomechanism model explains abnormal retinocollicular projection of Eph receptor-overexpressing or ephrin-lacking mice. J Neurosci 23(32):10368–10377

Hope RA, Hammond BJ, Gaze RM (1976) The arrow model: retinotectal specificity and map formation in the goldfish visual system. Proc R Soc Lond B Biol Sci 194(1117):447–466

Kiddie G, McLean D, Van Ooyen A, Graham B (2005) Biologically plausible models of neurite outgrowth. Progr Brain Res 147:67–80

Koulakov AA, Tsigankov DN (2004) A stochastic model for retinocollicular map development. BMC Neurosci 5:30

Lamoureux P, Buxbaum RE, Heidemann SR (1998) Axonal outgrowth of cultured neurons is not limited by growth cone competition. J Cell Sci 111:3245–3252

Lauffenburger DA, Linderman JL (1993) Receptors: models for binding, trafficking and signaling. Oxford university press, Oxford

Li GH, Qin CD, Wang ZS (1992) Neurite branching pattern formation: modeling and computer simulation. J Theor Biol 157:463–486

Li GH, Qin CD, Wang LW (1995) Computer model of growth cone behavior and neuronal morphogenesis. J Theor Biol 174:381–389

Lin CH, Forscher P (1995) Growth cone advance is inversely proportional to retrograde F-actin flow. Neuron 14:763–771

Lowery LA, van Vactor D (2009) The trip of the tip: understanding the growth cone machinery. Nat Rev Mol Cell Biol 10:332–343

Maskery S, Shinbrot T (2005) Deterministic and stochastic elements of axonal guidance. Annu Rev Biomed Eng 7:187–221

Maskery S, Buettner H, Shinbrot T (2004) Growth cone pathfinding: a competition between deterministic and stochastic events. BMC Neurosci 5:22

McLaughlin T, O'Leary DDM (2005) Molecular gradients and development of retinotopic maps. Annu Rev Neurosci 28:327–355

McLean DR, van Ooyen A, Graham BP (2004) Continuum model for tubulin-driven neurite elongation. Neurocomp 58–60:511–516

Medeiros NA, Burnette DT, Forscher P (2006) Myosin II functions in actin-bundle turnover in neuronal growth cones. Nature Cell Biol 8:215–226

Mitchison T, Kirschner M (1984) Dynamic instability of microtubule growth. Nature 312:237–242

Mogilner A (2009) Mathematics of cell motility: have we got its number? J Math Biol 58:105–134

Mogilner A, Rubinstein B (2005) The physics of filopodial protrusion. Biophys J 89:1–14

Mortimer D, Fothergill T, Pujic Z, Richards LJ, Goodhill GJ (2008) Growth cone chemotaxis. Trends Neurosci 31:90–98

Mortimer D, Feldner J, Vaughan T, Vetter I, Pujic Z, Rosoff WJ, Burrage K, Dayan P, Richards LJ, Goodhill GJ (2009) A bayesian model predicts the response of axons to molecular gradients. Proc Natl Acad Sci USA 106(25):10296–10301

Mortimer D, Dayan P, Burrage K, Goodhill G (2010a) Optimizing chemotaxis by measuring unbound-bound transitions. Physica D 239:477–484

Mortimer D, Pujic Z, Vaughan T, Thompson AW, Feldner J, Vetter I, Goodhill GJ (2010b) Axon guidance by growth-rate modulation. Proc Natl Acad Sci USA 107:5202–5207

Mortimer D, Dayan P, Burrage K, Goodhill GJ (2011) Bayes-optimal chemotaxis. Neural Comput 23:336–373

Nakamoto M, Cheng HJ, Friedman GC, McLaughlin T, Hansen MJ, Yoon CH, O'Leary DD, Flanagan JG (1996) Topographically specific effects of ELF-1 on retinal axon guidance in vitro and retinal axon mapping in vivo. Cell 86(5):755–766

O'Connor TP, Duerr JS, Bentley D (1990) Pioneer growth cone steering decisions mediated by single filopodial contacts in situ. J Neurosci 10:3935

Odde D, Tanaka E, Hawkins S, Buettner H (1996) Stochastic dynamics of the nerve growth cone and its microtubules during neurite outgrowth. Biotechnol Bioeng 50:452–461

Overton KJ, Arbib MA (1982) The extended branch-arrow model of the formation of retino-tectal connections. Biol Cybern 45(3):157–175

Poliakov A, Cotrina M, Wilkinson DG (2004) Diverse roles of eph receptors and ephrins in the regulation of cell migration and tissue assembly. Dev Cell 7(4):465–480

Prestige MC, Willshaw DJ (1975) On a role for competition in the formation of patterned neural connexions. Proc R Soc Lond B Biol Sci 190(1098):77–98

Reber M, Burrola P, Lemke G (2004) A relative signalling model for the formation of a topographic neural map. Nature 431(7010):847–853

Rosoff WJ, Urbach JS, Esrick MA, McAllister RG, Richards LJ, Goodhill GJ (2004) A new chemotaxis assay shows the extreme sensitivity of axons to molecular gradients. Nat Neurosci 7(6):678–682

Sakumura Y, Tsukada Y, Yamamoto N, Ishii S (2005) A molecular model for axon guidance based on cross talk between rho gtpases. Biophys J 89(2):812–822

Simpson HD, Goodhill GJ (2011) A simple model can unify a broad range of phenomena in retinotectal map development. Biol Cybern 104(1):9–29

Simpson HD, Mortimer D, Goodhill GJ (2009) Theoretical models of neural circuit development. In: Hobert O (ed)The development of neural circuitry. Current topics in developmental biology, vol 87. Elsevier, Amsterdam, pp 1–51

Smalheiser NR, Crain SM (1984) The possible role of "sibling neurite bias" in the coordination of neurite extension, branching, and survival. J Neurobiol 15:517–529

Sperry R (1963) Chemoaffinity in the orderly growth of nerve fiber patterns and connections. Proc Natl Acad Sci USA 50:703–710

Suter DM, Forscher P (2000) Substrate-cytoskeletal coupling as a mechanism for the regulation of growth cone motility and guidance. J Neurobiol 44:97–113

Tessier-Lavigne M, Goodman CS (1996) The molecular biology of axon guidance. Science 274:1123

Tsigankov D, Koulakov AA (2010) Sperry versus hebb: topographic mapping in isl2/epha3 mutant mice. BMC Neurosci 11:155

Tsigankov DN, Koulakov AA (2006) A unifying model for activity-dependent and activity-independent mechanisms predicts complete structure of topographic maps in ephrin-A deficient mice. J Comput Neurosci 21(1):101–114

Udin SB, Fawcett JW (1988) Formation of topographic maps. Annu Rev Neurosci 11:289–327

Ueda M, Shibata T (2007) Stochastic signal processing and transduction in chemotactic response of eukaryotic cells. Biophys J 93:11–20

van Ooyen A (2001) Competition in the development of nerve connections: a review of models. Network 12(1):R1–47

van Ooyen A (ed) (2003) Modeling Neural Development. MIT Press, Cambridge

van Ooyen A, Willshaw DJ (2000) Development of nerve connections under the control of neurotrophic factors: parallels with consumer-resource systems in population biology. J Theor Biol 206(2):195–210

Van Veen MP, Van Pelt J (1994) Neuritic growth rate described by modeling microtubule dynamics. Bull Math Biol 56:249–273

Weber C, Ritter H, Cowan J, Klaus Obermayer K (1997) Development and regeneration of the retinotectal map in goldfish: a computational study. Philos Trans 352(1361):1603–1623

Whitelaw VA, Cowan JD (1981) Specificity and plasticity of retinotectal connections: a computational model. J Neurosci 1(12):1369–1387

Wilkinson DG (2001) Multiple roles of Eph receptors and ephrins in neural development. Nat Rev Neurosci 2(3):155–164

Willshaw D (2006) Analysis of mouse EphA knockins and knockouts suggests that retinal axons programme target cells to form ordered retinotopic maps. Development 133(14):2705–2717

Willshaw DJ, von der Malsburg C (1979) A marker induction mechanism for the establishment of ordered neural mappings: its application to the retinotectal problem. Philos Trans R Soc Lond B Biol Sci 287(1021):203–243

Willshaw DJ, Price DJ (2003) Models for topographic map formation. In: van Ooyen A (ed) Modeling neural development. MIT Press, Cambridge, pp 213–244

Xu J, Rosoff WJl, Urbach JS, Goodhill GJ (2005) Adaptation is not required to explain the long-term response of axons to molecular gradients. Development 132(20):4545–4552

Yates PA, Roskies AL, McLaughlin T, O'Leary DD (2001) Topographic-specific axon branching controlled by ephrin-As is the critical event in retinotectal map development. J Neurosci 21(21):8548–8563

Yates PA, Holub AD, McLaughlin T, Sejnowski TJ, O'Leary DDM (2004) Computational modeling of retinotopic map development to define contributions of EphA-ephrinA gradients, axon–axon interactions, and patterned activity. J Neurobiol 59(1):95–113

Zheng JQ, Poo MM (2007) Calcium signaling in neuronal motility. Ann Rev Cell Dev Biol 23:375–404

Chapter 15
Encoding Neuronal Models in SBML

Sarah M. Keating and Nicolas Le Novère

Abstract Encoding computational models in a standard format permit to share and re-use them in a variety of contexts. The Systems Biology Markup Language (SBML) is the de facto standard open format for exchanging models between software tools in systems biology. Neuronal models can often be encoded using this format, thus providing the modeler with access to a large variety of software packages that support SBML. We give a brief overview of the main constructs of SBML Level 3 Version 1 Core (the latest version of SBML). We provide practical examples of encoding particular neuronal models using SBML, illustrate the results of using the SBML encoding to simulate the models, and demonstrate the correspondance of results produced by the original modelers and the exchangeable encoding of the model in SBML.

15.1 Systems Biology and the Need to Exchange Models

Systems Biology of neurobiological systems needs to take into account the interactions between a very large number of physical entities, and the analysis of many parameters. As seen in the previous chapters, the quantitative relationships between entities, their interactions, are often described using mathematical models. It is therefore crucial to be able to encode those models in a standard way to foster their exchange and re-use. Additionally using a standard format permits further processing, for instance merging of models, and relating them to other types of information. Early modelers and software developers in systems biology quickly realized that if their efforts were to be of benefit to the wider community it must be possible to share and re-use the models. The best way to facilitate this, and to

S.M. Keating (✉) • N. Le Novère
EMBL-EBI, Cambridge, UK
e-mail: skeating@ebi.ac.uk; lenov@ebi.ac.uk

N. Le Novère (ed.), *Computational Systems Neurobiology*,
DOI 10.1007/978-94-007-3858-4_15,

enable concurrent use of multiple software packages with different capabilities, was to agree a common format for describing the models. There are many ways to describe models in a standardized manner. One can use natural languages, graphical languages, sets of equations, logical relationships between elements etc. In this chapter, as in the chapter (Chap. 16), we focus on the representation of the variables representing physical entities, their relationships and the necessary parameters, encoded in a text file.

15.1.1 A Bit of History

The need for a language to exchange models became manifest at the end of the last century, with efforts starting in the field of metabolic networks (see history in Kell and Mendes 2008) and physiology modeling (Hedley et al. 2001). A similar need was expressed during the first Workshop on Software Platforms for Systems Biology, held at the California Institute of Technology in early 2000. The Systems Biology Markup Language (SBML) (Hucka et al. 2003) is a machine-readable model definition language based on XML, the eXtensible Markup Language (Bray et al. 2000). An SBML document contains all the information pertaining to the structure of a model, including the list of symbols, both variables and constants, the list of mathematical relationships linking them, and all the numbers needed to instantiate simulations. SBML was originally viewed as being aimed towards models of molecular pathways (Strömbäck and Lambrix 2005). However, its versatility means that SBML can be, and today is being, used in a variety of modeling contexts. For instance, BioModels Database (Le Novère et al. 2006) contains SBML representations of models including cell signaling (Goldbeter 1991), metabolism (Curto et al. 1998), gene regulation (Elowitz and Leibler 2000) and neuronal models, some of which are described in detail later in this chapter. In general, SBML enables the encoding of any mathematical model based on pools of entities and processes that modify them. This versatility is currently expanding, towards rule-based modeling, reaction-diffusion etc. Since its creation in 2000, SBML has continued to evolve as an international community effort, and has grown in terms of the levels of acceptance; to the point where, at the time of this writing, it is used by over 200 software packages worldwide and required as a format for model encoding by many journals.

15.1.2 Levels and Versions of SBML

SBML is being developed in stages, with specifications released at the end of each development stage. This approach, which effectively freezes SBML development at incremental points, allows users to work with stable standards and gain experience with the standard before further development. Future development can then benefit from the practical experiences of users and developers.

Major editions of SBML are termed *levels* and represent substantial changes to the composition and structure of the language. The latest level being developed is Level 3 (Hucka et al. 2009); representing a major evolution of the language through Level 2 (Hucka et al. 2008) from the introduction of Level 1 in the year 2001 (Hucka et al. 2001, 2003). SBML Level 3 is being developed as a modular language, with a central core comprising a self-sufficient model definition language, and extension packages layered on top of this core to provide additional, optional sets of features. Only the core will be described in this chapter.

The separate levels of SBML are intended to coexist. All of the constructs of Level 1, i.e. the elements and attributes of the SBML representation, can be mapped to Level 2; likewise, the majority of the constructs from Level 2 can be mapped to Level 3 Core.[1] In addition, a subset of Level 3 constructs can be mapped to Level 2, and a subset of Level 2 constructs can be mapped to Level 1. However, the levels remain distinct; a valid SBML Level 1 document is not a valid SBML Level 2 document, and so on.

Minor revisions of SBML are termed *versions*, and constitute changes within a Level to correct, adjust and refine language features. All examples used here will be from the latest stable version of SBML; that is, SBML Level 3 Version 1 Core. It should be noted that SBML Level 3 is a recent development and as yet not many software tools support it.

15.2 Structure of SBML

SBML is a structured language, with a strict syntax and very precise semantics. In this section we will present the most common constructs of SBML. However, a serious understanding of the language can only be achieved through the SBML specification document (Hucka et al. 2009).

An SBML document is essentially an XML document containing an **sbml** element which declares the *namespace*, *level* and *version* of SBML. The **sbml** element MUST contain a **model** element which itself consists of lists of one or more components. The SBML snippet shown illustrates an **sbml** element containing a **model** element.

```
<?xml version="1.0" encoding="UTF-8"?>
<sbml xmlns="http://www.sbml.org/sbml/level3/version1/core"
      level="3" version="1">
  ...
  <model ...>  ... </model>
</sbml>
```

[1]The SpeciesType and CompartmentType contructs which appear in Level 2 Versions 2–4 were removed in Level 3 Core as it was considered they were better suited to an extension package.

Table 15.1 Components of an SBML **model** element

Component	Description
compartment	A container of finite size for well-stirred substances
species	A pool of undistinguishable entities
parameter	A quantity of whatever type is appropriate
reaction	A statement describing some transformation, transport or binding process that can change one or more species
rule	A mathematical expression that is added to the model equations
event	A set of mathematical formulas evaluated at specified moments in the time evolution of the system
initialAssignment	A mathematical formula to assign the initial value of a component
functionDefinition	A named mathematical function that can be used in place of repeated expressions
constraint	A mathematical formula for stating the assumptions under which the model is designed to operate
unitDefinition	A name for a unit used in the expression of quantities in a model

Table 15.1 lists all the components defined by SBML; with a brief description of the semantics of each.

Some components in SBML represent items that have a numerical value, that may be constant or may vary throughout a simulation. The constructs that represent possible variables are compartment, species and parameter. In all these cases, the *id* attribute of the component is used throughout the model to represent the numerical value of that component at the point in time specified by any simulation/analysis that is being undertaken. It is also possible to introduce variable stoichiometry into reactions but this is beyond the scope of the current text.

Other components represent mathematical constructs that define some level of interaction between the components that can be varied. These constructs include the reaction, rules and event components.

The remaining constructs: initialAssignment, functionDefinition, constraint and unitDefinition; provide methods of adding further information or mathematical detail to a model. It is however possible to construct a complete model without using these components, which will not be explored further here.

15.2.1 Compartment

The **compartment** component in SBML represents a container of finite size for well-stirred substances where the species defined in the model are located. Biologically speaking this may represent for instance a body fluid, a cell or a subcellular compartment, but SBML does not require the **compartment** to represent an actual structure, either inside or outside a biological system.

The **compartment** has attributes that specify its *spatialDimensions*, its *size* and the corresponding *units*, plus a *constant* attribute that determines whether the size can change or not during a simulation. The following SBML snippet represents a constant, 3D compartment with volume 2.3 l.

```
<model>
  ...
  <listOfCompartments>
    <compartment id="cell" spatialDimensions="3"
                 size="2.3" units="litre" constant="true"/>
  </listOfCompartments>
  ...
</model>
```

The *id* attribute can be used elsewhere in the model to represent the numerical value of the size of the **compartment**.

15.2.2 Species

The **species** component in SBML does not represent a single molecule but rather a pool, that is an ensemble of indistinguishable entities, represented by its concentration or amount in a **compartment**. The environment is well-stirred and thus no concentration gradients need to be considered.[2] Regardless of whether species within models are specified using amount or concentration, the value of the attribute *id* of a **species** refers to concentration when it is used in a mathematical context, UNLESS the **species** has been declared as being in units of amount by either using the *hasOnlySubstanceUnits* attribute or locating the **species** in a **compartment** with *spatialDimensions* of zero.

It is also common for a species to exist on the boundary of the system being modeled; in which situation the quantity is unchanged by reactions it may be involved in. The *boundaryCondition* attribute implies that whilst the species may be a product or reactant within reactions, its quantity is not determined by those reactions.

The SBML snippet illustrates a variable species located in the compartment with id ‘cell’. It is not on the boundary, it is to be used as a concentration and possesses an initial amount of 4.6 mol.

```
<model>
  ...
  <listOfSpecies>
    <species id="s"
             compartment="cell"
             initialAmount="4.6"
             substanceUnits="mole"
```

[2]Discussions are under way to propose one or more Level 3 package that will address this issue.

```
                hasOnlySubstanceUnits="false"
                boundaryCondition="false"
                constant="false"/>
    </listOfSpecies>
    ...
</model>
```

The species specified above has an initial amount of 4.6 mol. However, the *hasOnlySubstanceUnits* attribute has a value of false, indicating that whenever the *id* of the species appears in the model it refers to concentration. Thus for any analysis, it may be necessary to convert between amount and concentration using

$$concentration = \frac{amount}{size}$$

where size refers to the *size* of the **compartment** in which the **species** is located.

It is possible to create models in SBML without the need to consider units and thus units have largely been ignored within this text. However, in the situation where a model uses species that have been located within a compartment whose size is not unity the issue of concentration and amount **must** be considered.

15.2.3 Parameter

The **parameter** component in SBML can represent anything with a numerical value that the modeler wishes to include. This may be the rate constant of a rate equation, the potential of a membrane or the current used to induce spiking in a neuronal model. Therefore, it is a component of particular importance for neurobiological models. The **parameter** has attributes that specify the *value*, *units* and whether the value is fixed.

The SBML snippet illustrates a variable parameter with value 3,000 and a constant parameter with value 8,000.

```
<model>
    ...
    <listOfParameters>
      <parameter id="p1"
                 value="3000"
                 constant="false"/>
      <parameter id="p2"
                 value="8000"
                 constant="true"/>
    </listOfParameters>
    ...
</model>
```

15.2.4 Reaction

A **reaction** in SBML represents any kind of process that can change the quantity of one of more **species**. It may be a mass action reaction, or involve transport, catalysis, or any process that changes the species involved (note that transport changes species because they are located in compartments). It is necessary to define the participating reactants and/or products. This is done using a **speciesReference** component that identifies the species from the model's **listOfSpecies** and assigns a *stoichiometry* value to that species role within the reaction. Species that merely influence a reaction, such as a catalyst, are listed as objects of type **modifierSpeciesReference**. This construct is similar to **speciesReference** without the *stoichiometry* attribute. Attributes for a **reaction** object allow the modeler to specify whether the reaction is *reversible* or *fast*. The mathematics describing the velocity of the reaction can be encoded in the **kineticLaw** component. SBML uses a subset of the MathML 2.0 standard (Ausbrooks et al. 2003) to encode mathematical formula directly within SBML components. **LocalParameter** objects can be included within a **kineticLaw**. These localParameters have constant values and are local to the enclosing kineticLaw; they cannot be used elsewhere in the model. Note that an SBML **kineticLaw** represents the extent of the reaction per unit of time, and not the rate of the reaction. In other words, the result is not a concentration per time, but a quantity per time. This is why the rate is multiplied by the volume in the **kineticLaw**.

The SBML snippet shows the description of the reaction

$$S0 \xrightarrow{k} S1$$

with a rate of

$$k \times S0 \times S2$$

where $S0$ and $S1$ are two species residing in a compartment V, and $S2$ is a catalyst.

```
<model>
  ...
  <listOfReactions>
    <reaction reversible="false"
              fast="false">
      <listOfReactants>
        <speciesReference species="S0"
                          stoichiometry="1"/>
      </listOfReactants>
      <listOfProducts>
        <speciesReference species="S1"
                          stoichiometry="1"/>
      </listOfProducts>
      <listOfModifiers>
        <modifierSpeciesReference species="S2"/>
      </listOfModifiers>
      <kineticLaw>
```

```
          <math xmlns="http://www.w3.org/1998/Math/MathML">
            <apply>
              <times/>
              <ci> k </ci>
              <ci> S0 </ci>
              <ci> S2 </ci>
              <ci> V </ci>
            </apply>
          </math>
          <listOfLocalParameters>
            <localParameter id="k" value="0.1"/>

          </listOfLocalParameters>
        </kineticLaw>
      </reaction>
    </listOfReactions>
    ...
  </model>
```

15.2.5 Rule

In SBML, **rules** provide additional ways of defining values of variables in a model, their relationships, and the dynamical behaviors of those variables. For example, a model may wish to calculate the total concentration of a chemical that appears as a part of several compounds within the model or vary a parameter representing voltage. There are three subclasses of **rules**: **algebraicRules**, **assignmentRules** and **rateRules**. In the current discussion, we will only consider the latter two types, which have the following form:

$$Assignment: \ x = f(V)$$

$$Rate: \ \frac{dx}{dt} = f(W)$$

where

x variable
f some arbitrary function
V vector of variables not including x
W vector of variables that may include x

Rules included in an SBML model are considered to hold true at all times. Therefore they must be included in the set of equations that define the model for simulation or other purposes.

The SBML snippet here shows two rules that describe the following equations.

$$y = 2x + 1$$

$$\frac{dg}{dt} = g - 1$$

```
<model>
  ...
  <listOfRules>
    <assignmentRule variable="y">
      <math xmlns="http://www.w3.org/1998/Math/MathML">
        <apply>
          <plus/>
          <apply>
            <times/>
            <cn> 2 </cn>
            <ci> x </ci>
          </apply>
          <cn> 1 </cn>
        </apply>
      </math>
    </assignmentRule>
    <rateRule variable="g">
      <math xmlns="http://www.w3.org/1998/Math/MathML">
        <apply>
          <minus/>
          <ci> g </ci>
          <cn> 1 </cn>
        </apply>
      </math>
    </rateRule>
  </listOfRules>
  ...
</model>
```

15.2.6 Event

An **event** in SBML describes the time and form of an explicit discontinuous state change in the model. For example this could be a situation where the voltage is reset when it reaches a given threshold.

The description of an **event** involves a **trigger**; a mathematical statement that determines when the event is fired; and a **listOfEventAssignments** that determine the action to be executed.

SBML does provide a **delay** for describing delayed events, when there is a period of time between when an event is 'fired' and when the event is 'executed'. There is also a **priority** component for assigning a priority to the **event**. Neither **delay** or **priority** are discussed here.

The SBML snippet[3] describes an **event** that resets the value of **parameter** Vthres to −60 when the value of a second **parameter** V exceeds a value of 30.

```
<model>
  ...
  <listOfParameters>
    <parameter id="Vthres"
               value="30"
               constant="false"/>
  </listOfParameters>
  ...
  <listOfEvents>
    <event useValuesFromTriggerTime="true">
      <trigger initialValue="true" persistent="true">
        <math xmlns="http://www.w3.org/1998/Math/MathML">
          <apply>
            <gt/>
            <ci> V </ci>
            <cn> 30 </cn>
          </apply>
        </math>
      </trigger>
      <listOfEventAssignments>
        <eventAssignment variable="Vthres">
          <math xmlns="http://www.w3.org/1998/Math/MathML">
            <cn> -60 </cn>
          </math>
        </eventAssignment>
      </listOfEventAssignments>
    </event>
  </listOfEvents>
  ...
</model>
```

15.2.7 Metadata

In addition to the model semantics, that is the variables and their mathematical relationships, SBML provides two mechanisms to add a layer of biological semantics on top of each component of the model.

Firstly, an attribute *sboTerm* allows any element to be linked to a single term of the *Systems Biology Ontology* (http://www.ebi.ac.uk/sbo, Le Novère et al. 2007). SBO is a controlled vocabulary (an "ontology") tailored specifically for the kinds of problems being faced in Systems Biology, especially in the context of computational modeling. SBO is made up of different vocabularies covering

[3]In order to show valid SBML a number of attributes are shown within the snippet. These are not referred to in the text as they represent a level of complexity beyond the scope of this text.

quantitative parameters, modeling frameworks, type of mathematical expressions, biological interactions, types of entities etc. The *sboTerm* enables unambiguous identification of the type of concept being dealt with by the model.

Secondly, one can add biological information by linking an SBML component to external resources, either terms from controlled vocabularies or entries in biological databases. In order to precisely relate SBML components and annotations, the links are encoded using existing semantic web technologies such as the *Resource Description Framework* (Manola and Miller 2004).

The following SBML snippet described a species that corresponds to a "protein complex" (SBO:0000297). It is made up of two parts, the protein calmodulin (described by the UniProt entry P62158) and the divalent calcium cation (ChEBI term CHEBI:29108).

```
<species id="Ca_calmodulin" metaid="cacam" sboTerm="SBO:0000297"
         compartment="C" hasOnlySubstanceUnits="false"
         boundaryCondition="false" constant="false">
  <annotation>
    <rdf:RDF xmlns:rdf="http://www.w3.org/1999/02/22-rdf-syntax-ns#"
             xmlns:bqbiol="http://biomodels.net/biology-qualifiers/">
      <rdf:Description rdf:about="#cacam">
        <bqbiol:hasPart>
          <rdf:Bag>
            <rdf:li rdf:resource="urn:miriam:uniprot:P62158"/>
            <rdf:li rdf:resource="urn:miriam:obo.chebi:CHEBI%3A29108"/>
          </rdf:Bag>
        </bqbiol:hasPart>
      </rdf:Description>
    </rdf:RDF>
  </annotation>
</species>
```

15.3 Creating Neuronal Models in SBML

The SBML structures discussed above provide a syntax sufficiently rich to encode many basic neuronal models. In particular, models involving metabolic networks, signaling pathways or gene regulatory networks are covered. In addition, "single compartment" electrical models can also be encoded in SBML. In this section, we give a couple of concrete examples. For encoding multi-compartment electrical models ("Rall models"), other representations such as NeuroML (Gleeson et al. 2010; see also Chap. 16) are more suitable.

15.3.1 Integration of Dopamine and Glutamate Signals

The integration of different neurochemical signals (Cohen 1992) is one of the fundamental basis of neuronal function and plasticity (Bhalla and Iyengar 1999).

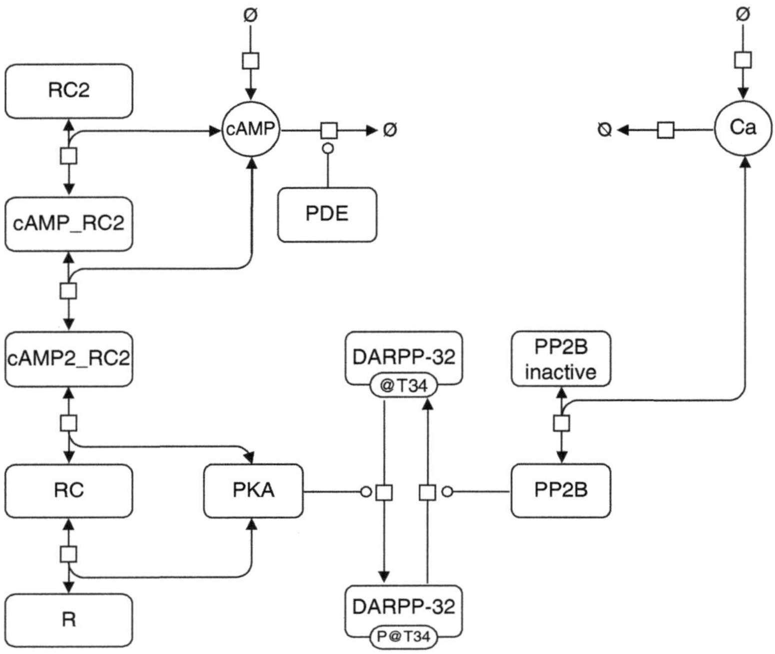

Fig. 15.1 Graphical representation of the model in the Process Description language of the Systems Biology Graphical Notation (Le Novère et al. 2009)

Figure 15.1 depicts a simple integration of glutamate and dopamine signals by the phosphatase inhibitor DARPP-32. The model is derived from a more comprehensive one described by Fernandez et al. (2006). It is possible to create a quantitative model of the reactions involved in the phosphorylation and dephosphorylation of DARPP-32, and study the dynamic behavior using calcium and cAMP as inputs to represent the response to glutamate and dopamine respectively. The complete model can be found in BioModels Database with the accession BIOMD0000000152.

15.3.1.1 Mathematical Model of the Biochemistry

Here we construct a reduced version of the model concentrating on the threonine 34 phosphorylation site of DARPP-32, the phosphorylation due to protein kinase A (PKA) and the dephosphorylation by calcineurin (PP2B). The resulting reactions are listed below in Table 15.2. The parameter values are modified slightly from the original model in order to retain the main behavior despite a drastic simplification.

Table 15.2 Reactions in the model

Type	Reactions
Phosphorylations	$D + PKA \rightleftharpoons D\_PKA$ $D\_PKA \rightarrow D34 + PKA$ $D34 + PP2B \rightleftharpoons D34\_PP2B$ $D34\_PP2B \rightarrow D + PP2B$
PP2B activation	$PP2B_{inactive} + 2Ca \rightleftharpoons PP2B$
PKA activation	$R2C2 + cAMP \rightleftharpoons cAMP\_R2C2$ $cAMP\_R2C2 + cAMP \rightleftharpoons cAMP2\_R2C2$ $cAMP2\_R2C2 + cAMP \rightleftharpoons cAMP3\_R2C2$ $cAMP3\_R2C2 + cAMP \rightleftharpoons cAMP4\_R2C2$ $cAMP4\_R2 + PKA \rightleftharpoons cAMP4\_R2C$ $cAMP4\_R2C + PKA \rightleftharpoons cAMP4\_R2C2$
cAMP degradation	$cAMP + PDE \rightleftharpoons cAMP\_PDE$ $cAMP\_PDE \rightarrow PDE + \emptyset$
Calcium input/destroy	$\emptyset \rightarrow Ca$ $Ca \rightarrow \emptyset$

15.3.1.2 Encoding the Model in SBML

The model consists of **species** that represent the chemicals being altered, **parameters** that define the rate constants, and **reactions** that determine how the species are being altered. All reactions are modeled using Mass-Action Law, and the enzymatic processes are decomposed in elementary steps. Consider the first equation listed:

$$D + PKA \rightleftharpoons D\_PKAs \tag{15.1}$$

with forward rate equation:

$$v_f = k1_f \times D \times PKA \tag{15.2}$$

and backward rate equation:

$$v_b = k1_b \times D\_PKA \tag{15.3}$$

The SBML model to encode this reaction must contain a **compartment**, in which the species are located and definitions for each of the **species**. Although biochemically speaking the reaction is reversible, since the rate equations for the forward and reverse reactions differ it is sometimes more convenient to define this as two irreversible SBML **reactions**. In the following, we use reversible reactions for reasons of compactness.

```
<listOfCompartments>
  <compartment id="Spine" size="1e-15" spatialDimension="3"
     constant="true"/>
</listOfCompartments>

<listOfSpecies>
 <species id="D" compartment="Spine"
          initialConcentration="4.98e-06" boundaryCondition="false"
          hasOnlySubstanceUnits="false" constant="false"/>
 <species id="PKA" compartment="Spine"
          initialConcentration="0" boundaryCondition="false"
          hasOnlySubstanceUnits="false" constant="false"/>
 <species id="D_PKA" compartment="Spine"
          initialConcentration="0" boundaryCondition="false"
          hasOnlySubstanceUnits="false" constant="false"/>
  ...
</listOfSpecies>

<listOfReactions>
  <reaction name="D_PKA_binding" reversible="true" fast="false">
    <listOfReactants>
      <speciesReference species="D" stoichiometry="1"
          constant="true"/>
      <speciesReference species="PKA" stoichiometry="1"
          constant="true"/>
    </listOfReactants>
    <listOfProducts>
      <speciesReference species="D_PKA" stoichiometry="1"
          constant="true"/>
    </listOfProducts>
    <kineticLaw>
      <math xmlns="http://www.w3.org/1998/Math/MathML">
        <apply>
          <minus/>
          <apply>
            <times/>
            <ci> Spine </ci>
```

```
          <ci> kon </ci>
          <ci> D </ci>
          <ci> PKA </ci>
        </apply>
        <apply>
          <times/>
          <ci> Spine </ci>
          <ci> koff </ci>
          <ci> D_PKA </ci>
        </apply>
      </apply>
    </math>
    <listOfLocalParameters>
      <localParameter id="kon" value="5600000"/>
      <localParameter id="koff" value="10.8"/>
    </listOfLocalParameters>
  </kineticLaw>
</reaction>
...
</listOfReactions>
```

This model includes two reactions intended to regulate the level of calcium, that is, the input and output of calcium from the system, represented as:

$$\emptyset \rightarrow Ca$$

$$Ca \rightarrow \emptyset$$

Obviously calcium does not just magically appear and disappear; however the input and output are not relevant to the system being modeled. If one wants to explicitly represent the fact that there are source and sink, one can make use of the *boundaryCondition* attribute on a **species** ‘Empty’, and define the reactions as follows. Labeling a species as being a *boundaryCondition* implies that it remains unaffected by any reactions it takes part in.

```
<listOfSpecies>
  <species id="Empty" compartment="Spine" boundaryCondition="true"
           hasOnlySubstanceUnits="false" constant="true"/>
  ...
</listOfSpecies>

<listOfReactions>
  <reaction name="Ca_in" reversible="false" fast="false">
    <listOfReactants>
      <speciesReference species="Empty" stoichiometry="1"
          constant="true"/>
    </listOfReactants>
    <listOfProducts>
      <speciesReference species="Ca" stoichiometry="1"
          constant="true"/>
    </listOfProducts>
```

```
      <kineticLaw>
        <math xmlns="http://www.w3.org/1998/Math/MathML">
          <apply>
            <times/>
            <ci> Spine </ci>
            <ci> kin </ci>
          </apply>
        </math>
      </kineticLaw>
    </reaction>

    <reaction name="Ca_out" reversible="false" fast="false">
      <listOfReactants>
        <speciesReference species="Ca" stoichiometry="1"
           constant="true"/>
      </listOfReactants>
      <listOfProducts>
        <speciesReference species="Empty" stoichiometry="1"
           constant="true"/>
      </listOfProducts>
      <kineticLaw>
        <math xmlns="http://www.w3.org/1998/Math/MathML">
          <apply>
            <times/>
            <ci> Spine </ci>
            <ci> kout </ci>
            <ci> Ca </ci>
          </apply>
        </math>
        <listOfLocalParameters>
          <localParameter id="kout" value="1.7"/>
        </listOfLocalParameters>
      </kineticLaw>
    </reaction>
    ...
  </listOfReactions>
```

Since the value of the species "Empty" is neither used nor displayed, it could equally be ignored altogether. Indeed a reaction in SBML is only required to have at least one reactant or one product. The following example represents the regulation of calcium with one reversible reaction.

```
  <listOfReactions>
    <reaction name="Ca_reg" reversible="true" fast="false">
      <listOfProducts>
        <speciesReference species="Ca" stoichiometry="1"
           constant="true"/>
      </listOfProducts>
      <kineticLaw>
        <math xmlns="http://www.w3.org/1998/Math/MathML">
          <apply>
            <minus/>
            <apply>
```

```
            <times/>
            <ci> Spine </ci>
            <ci> kin </ci>
          </apply>
          <apply>
            <times/>
            <ci> Spine </ci>
            <ci> kout </ci>
            <ci> Ca </ci>
          </apply>
        </apply>
      </math>
    </kineticLaw>
  </reaction>
  ...
</listOfReactions>
```

Note that in the reactions above, the parameter 'kin' is a global parameter, the value of which will affect the amount of calcium being added to the system. Since the purpose of this model is to study the effect of different amounts of calcium, the parameter 'kin' can be varied to simulate different situations. The model uses **events** to add a pulse of cAMP and calcium in the course of the simulation. For the cAMP, this is done by setting a high concentration at a given time, while for the calcium, we increase the input (conductance of calcium channels) at a particular point and return it to a lower value after a delay. The **events** to produce a spike of calcium are shown below. Note there is an assumption that 'kin' is initially low.

```
<listOfEvents>
 <event id="event_2" name="ca_on1" useValuesFromTriggerTime="true">
   <trigger initialValue="true" persistent="false">
     <math xmlns="http://www.w3.org/1998/Math/MathML">
        <apply>
          <geq/>
          <csymbol encoding="text"
            definitionURL="http://www.sbml.org/sbml/symbols/time">
            time </csymbol>
          <apply>
            <plus/>
            <ci> cAMP_delay </ci>
            <ci> cAMP_Ca_delay </ci>
          </apply>
        </apply>
      </math>
    </trigger>
    <listOfEventAssignments>
      <eventAssignment variable="kin">
        <math xmlns="http://www.w3.org/1998/Math/MathML">
          <ci> kon_high </ci>
        </math>
      </eventAssignment>
    </listOfEventAssignments>
  </event>
```

```
<event id="event_4" name="ca_off" useValuesFromTriggerTime="true">
  <trigger initialValue="true" persistent="false">
    <math xmlns="http://www.w3.org/1998/Math/MathML">
      <apply>
        <geq/>
        <csymbol encoding="text"
          definitionURL="http://www.sbml.org/sbml/symbols/time">
          time </csymbol>
        <apply>
          <plus/>
          <ci> cAMP_delay </ci>
          <ci> cAMP_Ca_delay </ci>
          <ci> spike_duration </ci>
        </apply>
      </apply>
    </math>
  </trigger>
  <listOfEventAssignments>
    <eventAssignment variable="kin">
      <math xmlns="http://www.w3.org/1998/Math/MathML">
       <ci> kon_low </ci>
      </math>
    </eventAssignment>
  </listOfEventAssignments>
</event>
...
</listOfEvents>
```

Since the chemical of interest in this model is DARPP-32 phosphorylated on Thr34, and that some of these molecules may be bound to other species, such as PP2B, the SBML adds an **assignmentRule** which calculates the total number of D34 molecules present.

```
<assignmentRule variable="parameter_1">
  <math xmlns="http://www.w3.org/1998/Math/MathML">
    <apply>
      <plus/>
      <apply>
        <times/>
        <ci> D34 </ci>
        <csymbol encoding="text"
          definitionURL="http://www.sbml.org/sbml/symbols/
             avogadro">
          Na </csymbol>
        <ci> Spine </ci>
      </apply>
      <apply>
        <times/>
        <ci> D34_PP2B </ci>
        <csymbol encoding="text"
          definitionURL="http://www.sbml.org/sbml/symbols/
             avogadro">
          Na </csymbol>
```

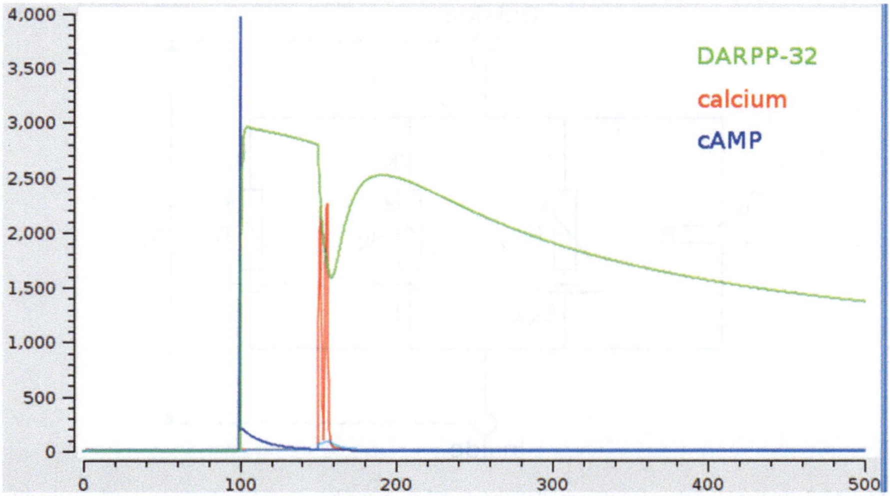

Fig. 15.2 Simulation of a model of glutamate and dopamine signal integration. The plot represent the temporal evolution of DARPP-32 phosphorylated on Thr34. After 100 s, cAMP molecules are added, and after a further delay, two square increases of conductance cause calcium spikes

```
        <ci> Spine </ci>
      </apply>
    </apply>
  </math>
</assignmentRule>
```

15.3.1.3 Simulation of a Time Course

We can use any simulator supporting SBML's reactions, assignmentRules and events to simulate the model and obtain the time course of the different model's variables. The result obtained with COPASI (Hoops et al. 2006) is shown in Fig. 15.2. The description of the simulation experiment, that is what to do with the model and how, is not encoded in SBML. Another language is under development to cover this part of the model life-cycle, the Simulation Experiment Description Markup Language (SED-ML) (Köhn and Le Novère 2008).

15.3.2 Hodgkin-Huxley Axon Model

In the 1930s, Alan Hodgkin and Andrew Huxley started a series of experiments and modeling to elucidate the flow of electric current through an axonal membrane. This led to the formulation of the Hodgkin-Huxley model in 1952 (Hodgkin and Huxley 1952), a model that had major influence on our understanding of neuronal function

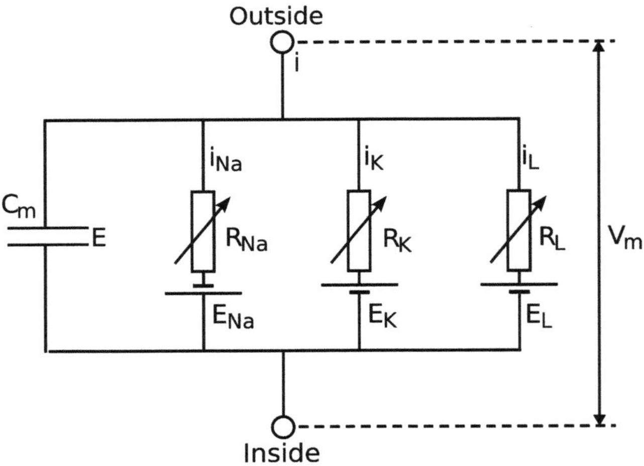

Fig. 15.3 In the Hodgkin-Huxley model, the membrane can be represented as an electrical circuit. Ionic current through the membrane can be divided into three components: potassium current (i_K), sodium current (i_{Na}), and a small leakage current (i_L) caused by other ions. The total current is calculated assuming that these components plus the capacity current are in parallel

and for which they received the Nobel Prize in 1963. The development of this model can also be seen as the birth of Systems Biology, since the authors described the emergence of a system's dynamic behavior using computational simulations of its components. The complete model can be found in BioModels Database with the accession BIOMD0000000020.

15.3.2.1 Mathematical Model of the Axon

Using the equivalent electrical circuit to represent a patch of membrane (Fig. 15.3) Hodgkin and Huxley derived four main equations that describe how the system, that is the probabilities of channel openings and the membrane voltage, varies with time.

$$\frac{dV}{dt} = \frac{I - (i_{Na} + i_K + i_L)}{C_m}$$
$$\frac{dm}{dt} = \alpha_m(1 - m) - \beta_m \cdot m$$
$$\frac{dh}{dt} = \alpha_h(1 - h) - \beta_h \cdot h$$
$$\frac{dn}{dt} = \alpha_n(1 - n) - \beta_n \cdot n \qquad (15.4)$$

where

V membrane depolarization voltage
I **constant** applied current
i_{Na} sodium current
i_K potassium current
i_L leakage current
C_m **constant** membrane capacitance
m sodium channel activation coefficient
h sodium channel inactivation coefficient
n potassium channel activation coefficient

and the α and β parameters are rate coefficients for the opening and closure of the gates, dependent on the instantaneous value of the membrane potential.

$$\begin{aligned}
\alpha_m &= \frac{0.1(V+25)}{\exp\left(\frac{V+25}{10}\right)-1}\\
\alpha_n &= \frac{0.01(V+10)}{\exp\left(\frac{V+10}{10}\right)-1}\\
\alpha_h &= 0.07\exp\left(\frac{V}{20}\right)\\
\beta_m &= 4\exp\left(\frac{V}{18}\right)\\
\beta_n &= 0.125\exp\left(\frac{V}{80}\right)\\
\beta_h &= \frac{1}{\exp\frac{V+30}{10}+1}
\end{aligned} \tag{15.5}$$

The individual current values also depend on the difference between the equilibrium potential of each ion and the membrane potential, and the conductance of each channel (e.g. g_{Na} for the sodium channel).

$$\begin{aligned}
i_{Na} &= g_{Na}\cdot m^3\cdot h\cdot (V-V_{Na})\\
i_K &= g_K\cdot n^4\cdot (V-V_K)\\
i_L &= g_L\cdot (V-V_L)
\end{aligned} \tag{15.6}$$

15.3.2.2 Encoding the Model in SBML

Since none of the variables under consideration in the model represent **species**, the SBML encoding of the model defines all these, and a number of other constants as parameters. Similarly the model does not contain any reactions, and all the equations are encoded using **rules**. For instance, the temporal evolution of the system is described with **rateRules**:

```
<rateRule variable="V">
  <math xmlns="http://www.w3.org/1998/Math/MathML">
    <apply>
      <divide/>
      <apply>
        <minus/>
        <ci> I </ci>
        <apply>
          <plus/>
          <ci> i_Na </ci>
          <ci> i_K </ci>
          <ci> i_L </ci>
        </apply>
      </apply>
      <ci> Cm </ci>
    </apply>
  </math>
</rateRule>

<rateRule variable="m">
  <math xmlns="http://www.w3.org/1998/Math/MathML">
    <apply>
      <minus/>
      <apply>
        <times/>
        <ci> alpha_m </ci>
        <apply>
          <minus/>
          <cn> 1 </cn>
          <ci> m </ci>
        </apply>
      </apply>
      <apply>
        <times/>
        <ci> beta_m </ci>
        <ci> m </ci>
      </apply>
    </apply>
  </math>
...
</rateRule>
```

The rate coefficients and the current, depending on the instantaneous voltage are encoded in SBML using **assignmentRules**. For instance:

```
<assignmentRule variable="alpha_n">
  <math xmlns="http://www.w3.org/1998/Math/MathML">
    <apply>
      <divide/>
      <apply>
```

```
            <times/>
            <cn> 0.01 </cn>
            <apply>
              <plus/><ci> V </ci><cn> 10 </cn>
            </apply>
          </apply>
          <apply>
            <minus/>
            <apply>
              <exp/>
              <apply>
                <times/>
                <cn> 0.1 </cn>
                <apply>
                 <divide/>
                 <apply>
                  <plus/><ci> V </ci><cn> 10 </cn>
                 </apply>
                 <cn> 10 </cn>
              </apply>
              </apply>
            </apply>
            <cn> 1 </cn>
          </apply>
        </apply>
      </math>
    </assignmentRule>
    ...
    <assignmentRule variable="i_Na">
      <math xmlns="http://www.w3.org/1998/Math/MathML">
        <apply>
          <times/>
          <ci> g_Na </ci>
          <apply>
            <power/><ci> m </ci><cn> 3 </cn>
          </apply>
          <ci> h </ci>
          <apply>
            <minus/><ci> V </ci><ci> V_Na </ci>
          </apply>
        </apply>
      </math>
    </assignmentRule>
```

15.3.2.3 Simulation of an Action Potential

Any simulator supporting SBML's assignment and rate rules can be used to simulate the behavior of the model. The results obtained with SBMLodeSolver (Machné et al. 2006) (see Fig. 15.4 reproducing Fig. 12 from the original paper).

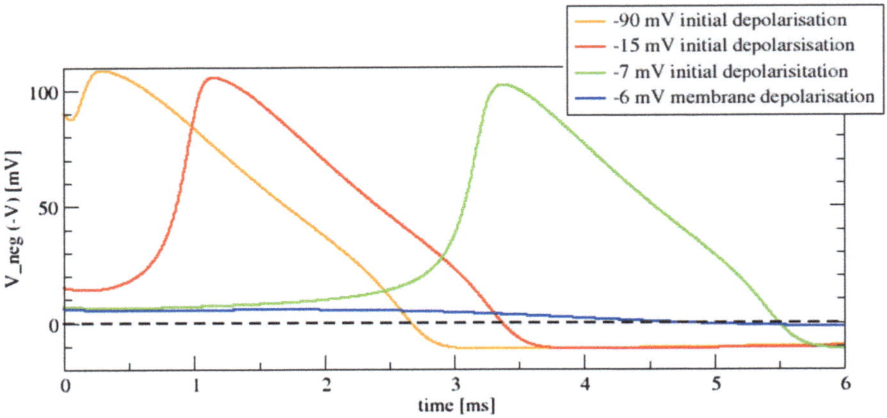

Fig. 15.4 Simulation output of Hodgkin-Huxley model using SBML ODESolver

15.3.3 *Cortical Spiking Neurons*

While Sects. 15.3.1 and 15.3.2 describe mechanistic models based on experimental measurements of components' properties, whether biochemical or electrophysiological, one can also describe the behavior of neurons by developing phenomenological models. This is particularly useful when one does not possess enough molecular or morphological details about the neuron of interest to reconstruct large-scale neuronal networks. Examples of such models are the spiking neurons, discussed in more details in chapter (Chap. 6). Because one can encode any mathematical description in SBML rules, those models can be easily encoded.

In his paper of 2003 (Izhikevich 2004a), Eugene Izhikevich proposed a simple model of spiking neuron able to reproduce the behavior of many spiking and bursting cortical neurons. In a further paper, he explored the response of this model to various inputs (Izhikevich 2004b). Here we will construct an SBML version of the model with a choice of parameters that makes it a class 1 excitability neuron, that is a neuron firing at low frequency for lower inputs. The complete model can be found in BioModels Database with the accession BIOMD0000000141.

15.3.3.1 Model of a Spiking Neuron

Izhikevitch's model is a 2-dimensional FitzHughNagumo class model (FitzHugh 1961) described with the following equations:

$$\frac{dv}{dt} = 0.04v^2 + 5v + 140 - u + I$$

$$\frac{du}{dt} = a(b - u) \tag{15.7}$$

the system being reset when $v \geq 30$ as:

$$\begin{aligned} v &= c \\ u &= u + d \end{aligned} \tag{15.8}$$

where

v membrane potential
u membrane recovery variable
I input current
a time scale of u
b sensitivity of u to fluctuations of v
c after-spike reset value of v
d after-spike reset value of u

15.3.3.2 Encoding a Model with Conditional Assignment

As with the previous example, none of the variables under consideration in the model represent **species** or **compartments**. The SBML encoding of the model defines all these, and a number of other constants as **parameters**. Similarly the model does not contain any reactions, and all the equations are encoded using **rules**. Note in some cases the parameters will be constant. Others must have the *constant* attribute set to 'false' as they will be controlled by other elements of the model, in this case through **rules** and **events**.

```
<listOfParameters>
  <parameter id="a"       value="0.02" constant="true"/>
  <parameter id="b"       value="-0.1" constant="true"/>
  <parameter id="c"       value="-55"  constant="true"/>
  <parameter id="d"       value="6"    constant="true"/>
  <parameter id="Vthresh" value="30"   constant="true"/>
  <parameter id="I"       value="0"    constant="false"/>
  <parameter id="flag"    value="0"    constant="false"/>
  <parameter id="v"       value="-60"  constant="false"/>
  <parameter id="u"       value="6"    constant="false"/>
</listOfParameters>
```

Equations 15.7 can be encoded using **rateRules** as follows.

```
<rateRule variable="v">
  <math xmlns="http://www.w3.org/1998/Math/MathML">
    <apply>
      <plus/>
      <apply>
        <minus/>
```

```
          <apply>
            <plus/>
            <apply>
              <times/>
              <cn> 0.04 </cn>
              <apply>
                <power/> <ci> v </ci> <cn> 2 </cn>
              </apply>
            </apply>
            <apply>
              <times/> <cn> 4.1 </cn> <ci> v </ci>
            </apply>
            <cn type="integer"> 108 </cn>
          </apply>
          <ci> u </ci>
        </apply>
        <ci> i </ci>
      </apply>
    </math>
  </rateRule>

  <rateRule variable="u">
    <math xmlns="http://www.w3.org/1998/Math/MathML">
      <apply>
        <times/>
        <ci> a </ci>
        <apply>
          <minus/>
          <apply>
            <times/>
            <ci> b </ci>
            <ci> v </ci>
          </apply>
          <ci> u </ci>
        </apply>
      </apply>
    </math>
  </rateRule>
```

Equation 15.8 can be represented by using an **event**.

```
  <event useValuesFromTriggerTime="true">
    <trigger initialValue="true" persistent="false">
      <math xmlns="http://www.w3.org/1998/Math/MathML">
        <apply>
          <gt/>
          <ci> v </ci>
          <ci> Vthresh </ci>
        </apply>
      </math>
    </trigger>
    <listOfEventAssignments>
      <eventAssignment variable="v">
        <math xmlns="http://www.w3.org/1998/Math/MathML">
          <ci> c </ci>
```

```
      </math>
    </eventAssignment>
    <eventAssignment variable="u">
      <math xmlns="http://www.w3.org/1998/Math/MathML">
        <apply>
          <plus/>
          <ci> u </ci>
          <ci> d </ci>
        </apply>
      </math>
    </eventAssignment>
  </listOfEventAssignments>
</event>
```

The Class 1 Excitable neurons can encode the strength of the input into their firing rate. In order to illustrate this, we want to encode an input current that is zero until a certain point and then increases steadily with time. Mathematically, this can be described as a discontinuous function:

$$I(t) = \begin{cases} 0 & t < 30 \\ 0.075(t - 30) & t \geq 30 \end{cases} \tag{15.9}$$

To encode Eq. 15.9, we can use of an **assignmentRule** containing a piecewise construct:

```
<assignmentRule variable="I">
  <math xmlns="http://www.w3.org/1998/Math/MathML">
    <piecewise>
      <piece>
        <cn type="integer"> 0 </cn>
        <apply>
          <lt />
          <csymbol encoding="text"
                   definitionURL="http://www.sbml.org/sbml/
                      symbols/time" />
          <cn type="integer"> 30 </cn>
        </apply>
      </piece>
      <otherwise>
        <apply>
          <times/>
          <cn> 0.075 </cn>
          <apply>
            <minus/>
            <csymbol encoding="text"
                     definitionURL="http://www.sbml.org/sbml/
                        symbols/time" />
            <cn type="integer"> 30 </cn>
          </apply>
        </apply>
      </otherwise>
    </piecewise>
  </math>
</assignmentRule>
```

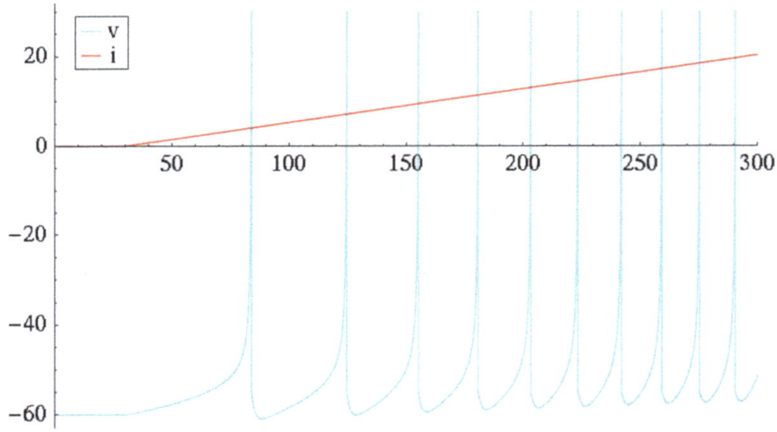

Fig. 15.5 Simulation of 300 ms of the class I excitable neuron of Izhikevich's model using MathSBML. The monotonic input is shown as well as the resulting increasing frequency of discharge

15.3.3.3 Simulation of a Firing Pattern

We can use any simulator supporting SBML's piecewise assignmentRules to simulate the model and obtain the time course of the different model's variables. The result obtained with MathSBML (Shapiro et al. 2004) as shown in Fig. 15.5 reproduces figure 1G from the original paper.

15.3.4 Conclusion and Perspectives

For the modeling of biological systems, SBML represented a breakthrough, by enabling for the first time interoperability between modeling and simulation tools. Since the number, size and complexity of computational models in biology has increased in line with the rise of Systems Biology, it is illusory to imagine that people could re-implement models they required, as was the case during the last century. SBML provided a means for researchers to consistently encode, exchange and re-use models. Its creation has revolutionized the modeling process.

A frequent misconception about SBML is that only models of biochemical reactions using chemical kinetic approaches can be encoded. The versatility of the language enables the encoding of a wide diversity of models, either based on processes affecting pools of entities (where the entities are not necessarily biomolecules), or variables described by differential or algebraic equations. The existence of discrete conditional events allows the introduction of perturbations, discontinuous behaviors and event-driven model elements.

Some types of models cannot currently be encoded in SBML, such as compartmental approximations of the cable theory and state-transition representations of ion channels. Other languages, like NeuroML, are more suitable for such a purpose. As this chapter is written, the NeuroML and SBML communities are working conjointly to make those languages interoperable, so that hybrid models containing biochemical and electrophysiological components may be exchanged. Such a cooperation may show the way for other collaborations, for instance with the nascent NineML language developed under the guidance of the International Neuroinformatics Coordination Facility (INCF) for encoding large neuronal networks.

References

Ausbrooks R, Buswell S, Carlisle D, Dalmas S, Devitt S, Diaz A, Froumentin M, Hunter R, Ion P, Kohlhase M, Miner R, Poppelier N, Smith B, Soiffer N, Sutor R, Watt S (2003) Mathematical Markup language (MathML) version 2.0, 2nd edn. Available via the World Wide Web at http://www.w3.org/TR/MathML2/

Bhalla US, Iyengar R (1999) Emergent properties of networks of biological signaling pathways. Science 283(5400):381–387

Bray T, Paoli J, Sperberg-McQueen CM, Maler E (2000) Extensible Markup Language (XML) 1.0, W3C recommendation 6-October-2000, 2nd edn. Available via the World Wide Web at http://www.w3.org/TR/1998/REC-xml-19980210

Cohen P (1992) Signal integration at the level of protein kinases, protein phosphatases and their substrates. Trends Biochem Sci 17(10):408–413

Curto R, Voit EO, Sorribas A, Cascante M (1998) Mathematical models of purine metabolism in man. Math Biosci 151(1):1–49

Elowitz MB, Leibler S (2000) A synthetic oscillatory network of transcriptional regulators. Nature 403(6767):335–338

Fernandez E, Schippa R, Girault J, Le Novère N (2006) DARPP-32 is a robust integrator of dopamine and glutamate signals. PLoS Comput Biol 2(12):e176

FitzHugh R (1961) Impulses and physiological states in theoretical models of nerve membrane. Biophys J 1:445–466

Gleeson P, Crook S, Cannon RC, Hines ML, Billings GO, Farinella M, Morse TM, Davison AR, Ray S, Bhalla UP, Barnes SR, Dimitrova YD, Silver RA (2010) NeuroML: A language for describing data driven models of neurons and networks with a high degree of biological detail. PLoS Comput Biol 6:e1000815 (in press)

Goldbeter A (1991) A minimal cascase model for the mitotic oscillator involving cyclin and cdc2 kinase. Proc Natl Acad Sci USA 88(20):9107–9111

Hedley W, Nelson MR, Bullivant DP, Nielsen PF (2001) A short introduction to CellML. Philos Trans R Soc Lond A 359(5):1073–1079

Hodgkin AL, Huxley AF (1952) A quantitative description of membrane current and its application to conduction and excitation in nerve. J Physiol 117(5):500–544

Hoops S, Sahle S, Gauges R, Lee C, Pahle J, Simus N, Singhal M, Xu L, Mendes P, Kummer U (2006) COPASI–a COmplex PAthway Simulator. Bioinformatics 22(24):3067–3074

Hucka M, Finney A, Sauro HM, Bolouri H (2001) Systems Biology Markup Language (SBML) Level 1: structures and facilities for basic model definitions. Available via the World Wide Web at http://www.sbml.org/Documents/Specifications

Hucka M, Finney A, Sauro HM, Bolouri H, Doyle JC, Kitano H, Arkin AP, Bornstein BJ, Bray D, Cornish-Bowden A, Cuellar AA, Dronov S, Gilles ED, Ginkel M, Gor V, Goryanin II, Hedley WJ, Hodgman TC, Hofmeyr JH, Hunter PJ, Juty NS, Kasberger JL, Kremling A,

Kummer U, Le Novère N, Loew LM, Lucio D, Mendes P, Minch E, Mjolsness ED, Nakayama Y, Nelson MR, Nielsen PF, Sakurada T, Schaff JC, Shapiro BE, Shimizu TS, Spence HD, Stelling J, Takahashi K, Tomita M, Wagner J, Wang J (2003) The Systems Biology Markup Language (SBML): a medium for representation and exchange of biochemical network models. Bioinformatics 19(4):524–531

Hucka M, Hoops S, Keating SM, Le Novère N, Sahle S, Wilkinson DJ (2008) Systems Biology Markup Language (SBML) Level 2: structures and facilities for model definitions. Available via the World Wide Web at http://www.sbml.org/Documents/Specifications

Hucka M, Bergmann F, Hoops S, Keating SM, Sahle S, Wilkinson DJ (2009) The Systems Biology Markup Language (SBML) Language specification for Level 3 Version 1 Core. Available via the World Wide Web at http://www.sbml.org/Documents/Specifications

Izhikevich EM (2004a) Simple model of spiking neurons. IEEE Trans Neural Netw 14(6):1569–1573

Izhikevich EM (2004b) Which model to use for cortical spiking neurons? IEEE Trans Neural Netw 15(5):1063–1070

Kell D, Mendes P (2008) The markup is the model: Reasoning about systems biology models in the semantic web era. J Theor Biol 252:538–543

Köhn D, Le Novère N (2008) SED-ML – an XML format for the implementation of the MIASE guidelines. In: Heiner M, Uhrmacher A (eds) Proceedings of the 6th conference on Computational Methods in Systems Biology, Rostock. Lecture Notes in Bioinformatics, vol 5307. Springer, Berlin/New York, pp 176–190

Le Novère N, Bornstein B, Broicher A, Courtot M, Donizelli M, Dharuri H, Li L, Sauro H, Schilstra M, Shapiro B, Snoep JL, Hucka M (2006) BioModels Database: A Free Centralized Database of Curated, Published, Quantitative Kinetic Models of Biochemical and Cellular Systems. Nucleic Acids Res 34:D689–D691

Le Novère N, Courtot M, Laibe C (2007) Adding semantics in kinetics models of biochemical pathways. In: Proceedings of the 2nd international symposium on experimental standard conditions of enzyme characterizations, Rüdesheim. Logos Verlag, Berlin, pp 137–153

Le Novère N, Hucka M, Mi H, Moodie S, Schreiber F, Sorokin A, Demir E, Wegner K, Aladjem MI, Wimalaratne SM, Bergman FT, Gauges R, Ghazal P, Kawaji H, Li L, Matsuoka Y, Villéger A, Boyd SE, Calzone L, Courtot M, Dogrusoz U, Freeman TC, Funahashi A, Ghosh S, Jouraku A, Kim S, Kolpakov F, Luna A, Sahle S, Schmidt E, Watterson S, Wu G, Goryanin I, Kell DB, Sander C, Sauro H, Snoep JL, Kohn K, Kitano H (2009) The Systems Biology Graphical Notation. Nature biotechnol 27(8):735–741

Machné R, Finney A, Müller S, Lu J, Widder S, Flamm C (2006) The SBML ODE Solver Library: a native API for symbolic and fast numerical analysis of reaction networks. Bioinformatics 22(11):1406–1407

Manola F, Miller E (2004) RDF primer. Tech. rep., W3C, URL http://www.w3.org/TR/2004/REC-rdf-primer-20040210/

Shapiro BE, Hucka M, Finney A, Doyle J (2004) MathSBML: a package for manipulating SBML-based biological models. Bioinformatics 20(16):2829–2831

Strömbäck L, Lambrix P (2005) Representations of molecular pathways: an evaluation of SBML, PSI ML and BioPAX. Bioinformatics 21(24):4401–4407

Chapter 16
NeuroML

Padraig Gleeson, Volker Steuber, R. Angus Silver, and Sharon Crook

Abstract NeuroML is a language based on XML for describing detailed neuronal models, which can contain multiple active conductances and complex morphologies. Networks of such cells positioned and synaptically connected in 3D can also be described. In this chapter we present an overview of the history of NeuroML, a brief description of the current version of the language, plans for future developments and the relationship to other standardisation initiatives in the wider computational neuroscience field. We also present a list of NeuroML resources which are currently available, such as language specifications, services on the NeuroML website, examples of models in this format, simulation platform support, and other applications for generating and visualising highly detailed neuronal networks. These resources illustrate how NeuroML can be a key part of the toolchain for researchers addressing complex questions of neuronal system function.

P. Gleeson (✉) • R. Angus Silver
Department of Neuroscience, Physiology and Pharmacology, University College London, London, UK
e-mail: p.gleeson@ucl.ac.uk; a.silver@ucl.ac.uk

V. Steuber
School of Computer Science, Science and Technology Research Institute, University of Hertfordshire, Hatfield, UK
e-mail: v.steuber@herts.ac.uk

S. Crook
School of Mathematical and Statistical Sciences, School of Life Sciences, and Center for Adaptive Neural Systems, Arizona State University, Tempe, AZ, USA
e-mail: sharon.crook@asu.edu

N. Le Novère (ed.), *Computational Systems Neurobiology*,
DOI 10.1007/978-94-007-3858-4_16,

16.1 Introduction

The complexity of the nervous system means that a variety of experimental and theoretical approaches are required to investigate the way that information is processed across the many biological scales involved. The use of computational models of neuronal systems has been steadily growing in recent years and they are increasingly seen as essential tools towards a greater understanding of brain function. Such models have been developed using a variety of programming approaches, from "home grown" simulators in single or closely collaborating labs (for example Rhodes and Llinas 2001; Traub et al. 2005), to the use of general purpose simulation environments for analysing dynamical systems (Ermentrout 2002), and dedicated neuronal simulation packages (Carnevale and Hines 2006; Bower and Beeman 1997).

While a publication describing a modelling investigation should ideally provide enough detail to reproduce the results, the complexity of the systems involved often means in practice that the original scripts have to be consulted for the full details of the model. A number of detailed cell and network models have been made available to the community (for example at ModelDB Hines et al. 2004). However, the wide variety of scripting languages used to express them and the different programming styles used can form a barrier to researchers who want to reuse all or part of the models. A model created for the NEURON simulation environment (Carnevale and Hines 2006) will not execute in the GENESIS simulator (Bower and Beeman 1997), and the specification of a cell model created in any simulator may be spread across scripts which also include simulation control, data analysis and visualisation of results. Ease of access to the essential neurophysiological elements of a model is needed to allow individual components to be validated and reused, and will be crucial for gaining a deeper understanding of the properties of the system being modelled.

A number of recent reviews of the neuroinformatics field (Brette et al. 2007; Cannon et al. 2007; De Schutter 2008) have highlighted these issues, including a detailed analysis of the state of tools for large scale neuronal modelling (Djurfeldt and Lansner 2007) by the International Neuroscience Coordinating Facility (INCF). In this chapter we describe NeuroML,[1] a language for describing biophysically detailed cell and network models, which has been developed to address a number of these issues. A recent publication (Gleeson et al. 2010) has provided a detailed introduction to the latest version of the language. In this chapter we present an overview of the history of NeuroML, a brief description of the current version of the language and plans for future developments and the relationship to other standardisation initiatives in the wider computational neuroscience field. We present a list of NeuroML resources which are currently available, such as the language specifications, services on the NeuroML website, examples of models in this format,

[1]http://www.neuroml.org

simulation platform support, and other applications for generating and visualising highly detailed neuronal network models. These resources illustrate how NeuroML can be a key part of the toolchain for researchers addressing complex questions of neuronal system function.

16.2 The Development of the NeuroML Model Description Language

The current version of NeuroML (Gleeson et al. 2010) has reached a stage of maturity where a number of freely available software tools support all or part of the language and a range of detailed cell and network models from different brain regions are available in this format. Here we outline a brief history of the NeuroML project, discuss the current version of the language, relate developments to other modelling language and standardization initiatives, and present ongoing development work towards the next version of the language.

16.2.1 Early History

A series of meetings attended by parties interested in developing a common language for specifying computational neuroscience models in XML (Extensible Markup Language, Bray et al. 1998) led to a publication describing the initial aims of the NeuroML initiative (Goddard et al. 2001). A number of these goals were in line with earlier work in this area (Gardner et al. 2001). This early effort provided a set of templates for describing neuronal models at the channel, cellular and network levels, which subsequently led to software implementations by some of the original NeuroML contributors, including NeoSim (Howell et al. 2003) and the NeuroML Development Kit (NDK). While several other software projects adopted these templates and used the NDK (for example KInNeSS,[2] Virtual RatBrain,[3]) use of this version of the NeuroML language was lower than expected in the growing computational neuroscience software development community. Some of the potential reasons for this were lack of communication between the various simulator communities, the relatively low number of detailed models which were publicly available at the time, and other, more pressing priorities among simulator developers, for example improving simulator performance and adding new features including graphical interfaces.

---

[2]http://symphony.bu.edu

[3]http://www.virtualratbrain.org

16.2.2 NeuroML Version 1.x

A new approach for the NeuroML initiative was adopted following meetings in 2004 and 2005 at the GENESIS and NEURON user group workshops. At this time a new language for describing neuronal morphologies in XML (MorphML) was under development (Qi and Crook 2004; Crook et al. 2007). Independently, neuroConstruct (Gleeson et al. 2007), an application for generating neuronal simulations for the NEURON and GENESIS simulators, was being developed. neuroConstruct had its own internal simulator independent representation for morphologies, channel and networks. It was agreed that these efforts should be merged under the banner of NeuroML, and the current structure of the NeuroML language was created. This new structure was split into three Levels comprising MorphML, ChannelML and NetworkML (see Sect. 16.2.2.2), providing greater modularity of the language and giving application developers the freedom to choose to support only part of the language as needed.

This modular approach focused on the elements of biophysically detailed neuronal models which can be analysed in isolation and are targets for reuse between models. The description of a typical biophysically detailed model includes: the structure of a neuron's dendritic/axonal arborisation in 3D (its morphology); the distribution of ion channels across this morphology; the kinetics of these ion channels; the properties of the synaptic mechanisms associated with the neuron; and the positions and connectivity patterns of multiple cells in 3D networks.

The priority for this new version of NeuroML (Gleeson et al. 2010) was to create a format for expressing these core model elements in a language which could be mapped to the most widely used neuronal simulators at that time, NEURON (Carnevale and Hines 2006) and GENESIS (Bower and Beeman 1997), although a number of other simulators and visualisation tools have since added NeuroML support (see Sect. 16.3). This resulted in a language focused on conductance based (multi-) compartmental models of neurons, although it includes support for some basic types of abstract neuronal models. The specification of simulation parameters (run time, integration method, etc.) was not included, as the language was designed to specify the models themselves as opposed to the details of how the simulations were run.

As mentioned in Sect. 16.2.4, SBML and CellML have a lower level representation of model behaviour, where all the dynamical behaviour of the system is described in XML. At the time of development of NeuroML version 1.x it was felt that there was much to be gained by having a concise interchange format specific for neuroscience applications which shared high level concepts such as cell morphologies, active conductances and 3D network structure. The next version of NeuroML will allow greater access to the underlying behaviour of NeuroML model components, allow greater interoperability with, and mappings to and from, models specified in these languages (see Sect. 16.2.5).

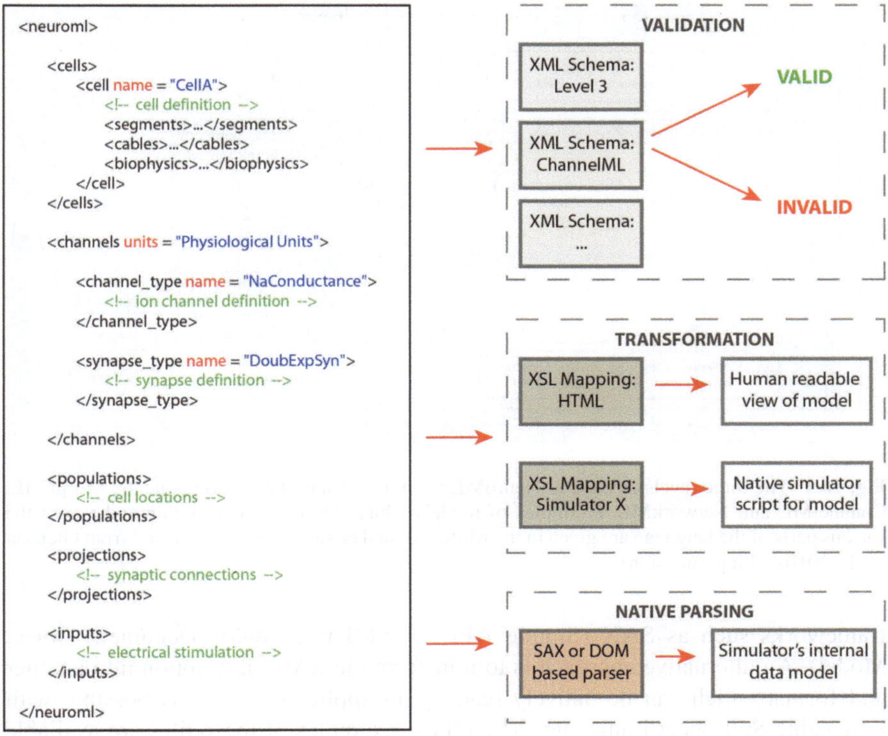

Fig. 16.1 An example of a section of a NeuroML file. The validity of the file can automatically be checked against the NeuroML specification contained in XML Schema Description (XSD) files. Extensible Stylesheet Language (XSL) files can be used to transform the contents of the file into formats which can be read more easily by humans or by other applications (for example a simulator's native file format). Alternatively an application can parse the XML using one of the standard XML parsing techniques and transform the contents into its own data representation. Note that a complex NeuroML model such as this could also be split between individual files for each cell, channel, synapse and one for the network structure

16.2.2.1 Technical Approach to Language Specification

A NeuroML document consists of the XML elements describing the physiological components of the neuronal model. Figure 16.1 shows an example of a NeuroML file containing various elements such as ***cells***, ***channels*** and ***populations***. The structure of a valid NeuroML document is defined using XML Schema Definition (XSD) files. Using these, standard XML handling libraries can be used to check the validity of an XML file against the various modules of the language. The XML Schema files used for the language are discussed further in Sect. 16.2.2.2.

Once an XML file is known to be in valid NeuroML format, the contents of the file can be transformed into other formats in a number of different ways. An application can read the XML natively using one of the commonly used parsing

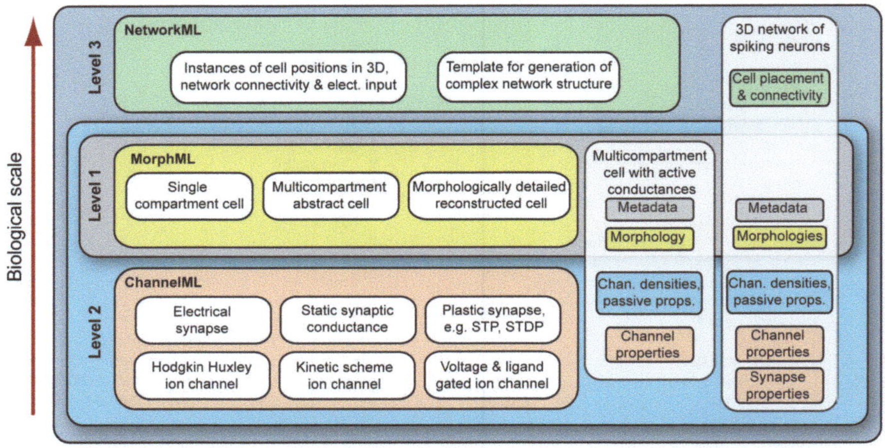

Fig. 16.2 The three levels as used in NeuroML version 1.0, and the subcomponents, MorphML, ChannelML and NetworkML. Examples of models which can be created within and across the various parts of the language are given in the white rounded boxes (Image reproduced from Gleeson et al. (2010) with permission)

frameworks such as SAX (Simple API for XML) or DOM (Document Object Model). An alternative approach is to transform the XML description into another text format which can be natively read by an application. This is possible with Extensible Stylesheet Language (XSL) files. Examples of these files are available for mapping NeuroML files onto HTML format (for making a more human readable web page description of the model), and a number of simulators' own script formats, including NEURON, GENESIS and PSICS (Cannon et al. 2010). This approach has the advantage that applications need not be reimplemented to natively support NeuroML, but can still have access to models in the format.

16.2.2.2 Levels in NeuroML Version 1.x

Three Levels are defined in NeuroML v1.x to facilitate modular use of the language, see Fig. 16.2. These are related to the different biological scales present in the systems being modelled. As models of single neurons are at the core of most of the systems to be described, specifications for the structure of individual cells form the core of Level 1. Level 2 builds on this by including definitions of the electrical properties of these cells, allowing for specifications of spiking cell models. Level 3 is used for networks of these cell models in 3D.

Level 1: Morphological Descriptions of Cells and Metadata

NeuroML Level 1 is used to define neuronal morphologies (MorphML) and to describe metadata, where the metadata provide additional information about model components at any subsequent Level.

MorphML: Neuronal morphologies are described (inside the ***cell*** element, see Fig. 16.1) using a series of connected ***segment*** elements, with each element containing the information needed to describe the 3D location and shape of that segment, as well as information about how segments should be connected. Optional ***cable*** elements can be used to group these and to assist in defining regions of interest on the cell such as apical/basal dendrites. The details of the mappings between the elements that comprise MorphML and the data structures of other applications that handle neuronal morphologies such as Neurolucida, NEURON and GENESIS have been described previously (Crook et al. 2007). Another advantage of MorphML over other formats is that it also allows for the description of additional histological details from cell reconstructions such as points of interest or perceived boundaries in the tissue (Crook et al. 2007).

Metadata: NeuroML files at any Level can be annotated with structured metadata. This can be used to provide information on the authors of the original models, who translated the model to NeuroML, relevant publications, source of the data used to construct the model from databases such as ModelDB (Hines et al. 2004) and NeuroMorpho.org (Ascoli et al. 2007). A ***status*** element can also be defined to provide useful information on the stability and limitations of various model components, for example, by indicating if additional experimental data are required to constrain free parameters of model components.

Figure 4 of Gleeson et al. (2010) shows the various elements allowed for describing cells in NeuroML at Level 1 and above. Supporting Text 1 of that paper provides a detailed description of each element shown.

Level 2: Channel and Synapse Descriptions; Biophysical Cell Properties

NeuroML Level 2 describes the electrical properties of the excitable membrane that underlie neural signalling and is composed of two main parts: ChannelML and a description of the biophysical properties of cells defined in Level 1.

ChannelML: Two main types of conductances can be described by ChannelML: conductances distributed over the plasma membrane (contained in the ***channel_type*** element), such as voltage-gated conductances or conductances gated by intracellular ions (for example [Ca^{2+}] dependent K^+ conductances); and conductances at synaptic contacts (***synapse_type*** element). Distributed ion channels can be described using both the traditional Hodgkin-Huxley formalism and by more biophysically detailed state-based kinetic (Markov) models. A wide range of voltage-gated conductances are supported by ChannelML, including those underlying fast and persistent Na^+ currents, delayed rectifier, A- and M-type K^+ currents, H-currents and high voltage activated (HVA, such as L-type and P-type) and low voltage activated (LVA, T-type) Ca^{2+} currents. [Ca^{2+}] dependent BK and SK type K^+ channels can also be expressed. There is some limited support for simplified neuron models, for example leaky integrate and fire cells, but the primary focus of v1.x is on conductance based neuron models.

ChannelML also supports both phenomenological models of neurotransmitter gated chemical synapses and gap junction conductances at electrical synapses. Supported models of conductances at chemical synapses include simple linear ohmic types used to model most α-amino-3-hydroxy-5-methyl-4-isoxazolepropionic acid (AMPA) and γ-aminobutyric acid-A ($GABA_A$) receptor mediated synapses and nonlinear voltage dependent components such as those used to model Mg^{2+} block of N-methyl-D-aspartate (NMDA) receptor mediated synaptic components.

Moreover, a short-term plasticity mechanism (Tsodyks et al. 2000; Tsodyks and Markram 1997) and a spike timing dependent plasticity (STDP) mechanism (Song et al. 2000) are supported allowing some commonly used models of activity dependent synaptic plasticity to be incorporated in networks.

Biophysical properties of membranes: Level 2 also allows extension of the morphological description of Level 1 to describe the passive electrical properties of the cells (specific capacitance, axial resistance, passive membrane conductances), and how active conductances are distributed across the membrane. These properties are contained in the ***biophysics*** element. While NeuroML Level 2 is required for defining a full spiking neuron model, the components of the model can be split across individual morphology and channel files facilitating the exchange and reuse of individual model components.

Figure 5 of Gleeson et al. (2010) shows the various elements allowed for specifying channels and synapses in ChannelML. Supporting Text 1 of that paper provides a detailed description of these elements, and the appendix to that document provides explanation of the equations behind channel and synapse models in ChannelML, and an example of a Level 2 cell model with active conductances.

Level 3: Networks

Level 3 of NeuroML allows for the specification of the 3D anatomical structure and synaptic connectivity of a network of neurons, together with the properties of the external input used to drive the network. NetworkML, the main component of Level 3, contains the elements needed to describe networks and external input. Level 3 also allows extension of Level 2 cells to specify the sub-regions of the cell membrane where specific synaptic connections may be located (***connectivity*** element).

NetworkML: Three core elements are used for describing networks in NetworkML: ***population*** represents a group of cells of a specific type; ***projection*** defines the set of synaptic connections between populations and specifies the types of synapses present; and ***input*** describes an external electrical input into the network, for example a current pulse delivered by model electrodes or random synaptic stimulation.

There are two possible ways to describe networks in NetworkML: an explicit list of cell positions and synaptic connections (the instance based representation), and templates for describing how instances of the network should be generated, for example by placing 100 cells randomly in a particular 3D region (template based representation). Currently, NetworkML supports a limited range of network templates; however, these templates are being updated and expanded for the next version of NeuroML (see Sect. 16.2.5).

Figure 6 of Gleeson et al. (2010) shows the elements in NetworkML. Supporting Text 1 of that paper provides a detailed description of these elements, and the appendix to that document provides an example of a simple network constructed in NetworkML.

Schemas for Levels

The NeuroML specification is split into a number of XML Schema documents. For Level 1, the morphological elements are defined in **MorphML_vXXX.xsd**, where the XXX is replaced by the current version number, for example **Metadata_v1.8.1.xsd**. Metadata are defined in **Metadata_vXXX.xsd**. The Schema for MorphML imports the Schema for Metadata, so elements of Metadata will be in the namespace associated with that Schema, for example ***meta:notes***. A file valid against **Metadata_vXXX.xsd** will have root element ***morphml***. Another schema file has been defined, **NeuroML_Level1_vXXX.xsd**, which imports both of these Schemas, and files valid against this will have root element ***neuroml***.

Level 2 adds a Schema for ChannelML (**ChannelML_vXXX.xsd**), which also imports Metadata and defines the structure of files having root ***channelml***. **Biophysics_vXXX.xsd** cannot be used for standalone files, but contains details of the elements for passive electrical properties and channel distributions of cells and is imported into **NeuroML_Level2_vXXX.xsd** (along with the Schemas for Metadata, MorphML and ChannelML) to define Level 2 cells (or a file containing cells and channel definitions) with root element ***neuroml***.

Level 3 deals with network descriptions and contains a Schema for standalone network descriptions (**NetworkML_vXXX.xsd**, with root element ***networkml***), and one for Level 3 files (**NeuroML_Level3_vXXX.xsd**, root element ***neuroml***), one of which could contain all the elements for the cells, channels, synapses, positions and synaptic connections in a complex 3D network (see Fig. 16.1).

Figure 16.3 shows a screenshot of the online validator application (see Sect. 16.3.2) showing the results of the validation of a ChannelML file against these Schemas.

16.2.3 Organisational Structure of the NeuroML Initiative

The NeuroML language has been developed as a project on the SourceForge website[4] since 2005. The mailing lists available there are the main source of

---

[4]http://sourceforge.net/projects/neuroml

XML Schema	Links	Result
Level1		
Metadata_v1.8.1.xsd (HTML, XML)		FAILED More...
MorphML_v1.8.1.xsd (HTML, XML)		FAILED More...
NeuroML_Level1_v1.8.1.xsd (HTML, XML)		FAILED More...
Level2		
Biophysics_v1.8.1.xsd (HTML, XML)		FAILED More...
ChannelML_v1.8.1.xsd (HTML, XML)	Convert NeuroML to Updated format Convert NeuroML to NEURONmod format Convert NeuroML to NEURONChanBuild format Convert NeuroML to PSICS format Convert NeuroML to HTML format Convert NeuroML to GENESIStab format	PASSED
NeuroML_Level2_v1.8.1.xsd (HTML, XML)	Convert NeuroML to HTML format	PASSED
Level3		
NetworkML_v1.8.1.xsd (HTML, XML)		FAILED More...
NeuroML_Level3_v1.8.1.xsd (HTML, XML)	Convert NeuroML to HTML format	PASSED

Fig. 16.3 The validation of a ChannelML file using the online validator application. As can be seen, the file is not valid against any of the Level 1 Schemas, is valid against the ChannelML and Level 2 Schemas, as well as the main Level 3 Schema (which includes ChannelML) but not NetworkML. Various options are provided for transforming the ChannelML into other formats (via XSL mapping files)

information on activities in the project. NeuroML is an open, community based project, and participation is welcome, also from the wider computational biology and experimental neuroscience communities.

The organizational structure of the NeuroML initiative was given a more formal structure at the first NeuroML Development Workshop in March 2009 in London. The NeuroML Team was formed, which consisted of Robert Cannon, Sharon Crook, Padraig Gleeson and Angus Silver. At the NeuroML Development Workshop in March 2011 in London, this core team was expanded into a 10 member Scientific Committee to drive forward the development of NeuroML. Initial membership of this consisted of the four NeuroML Team members along with Upi Bhalla, Avrama Blackwell, Hugo Cornelis, Andrew Davison, Lyle Graham and Michael Hines.

16.2.4 Relationship to Other Standardization Initiatives

The following are a number of initiatives also involved in standardization and interoperability in computational neuroscience and the wider systems biology field.

PyNN

PyNN (Davison et al. 2008) is a Python package for simulator independent neural network development. This language allows simulators which have a Python based scripting interface (for example NEURON Carnevale and Hines 2006; NEST

Diesmann and Gewaltig 2002; Brian Goodman and Brette 2008) to use the same set of scripting commands to create large scale neural networks. The PyNN approach to simulator independent model specification differs from that of NeuroML, as it is a language for the procedural description of model creation whereas a model specified in NeuroML is a declarative specification of the model structure. PyNN has also to date concentrated on descriptions of large scale networks of simplified neurons, whereas NeuroML has mainly been used for smaller networks of more biologically detailed cells. Thus, the two approaches are complementary. See Sect. 16.3.8 for more on interoperability between PyNN and NeuroML.

BrainML

BrainML[5] is an initiative to develop standards in XML for exchanging neuroscience data. This initiative initially concentrated on standards for the annotation of experimental data as opposed to computational models, and so there has not to date been a significant overlap with the scope of NeuroML. Much of the work on BrainML and the related Neurodatabase.org project (Gardner 2004) have fed into other National Institute of Health (NIH) funded initiatives, such as the Neuroscience Information Framework (see section below on NeuroLex).

INCF Multi-scale Modelling Program

The International Neuroscience Coordinating Facility (INCF) was formed in 2004 through the Global Science Forum of the OECD with the aim of promoting international collaboration in the area of neuroinformatics. Its activities include a number of themed Programs which concentrate on areas of interest to the community, arrange meetings and develop standards and guidelines to facilitate collaborative research. One of the Programs which has been set up is on Multi-scale Modelling of Neuronal Systems. Current work in this area involves development of NineML,[6] a layered language for describing large scale models of spiking neurons. NeuroML Team members are present on the Oversight Committee and Task Force of this Program and future developments in NeuroML will be closely aligned with progress in this initiative.

NeuroLex

The Neuroscience Information Framework[7] (NIF) is an initiative of the NIH Blueprint for Neuroscience Research and aims to be a single point of access to multiple data resources related to neuroscience. NeuroLex is neuroscience specific lexicon used by the NIF which provides a set of terms for uniquely identifying

[5]http://www.brainml.org

[6]http://www.nineml.org

[7]http://www.neuinfo.org

concepts in neuroscience (such as cell types, brain regions, organisms), to facilitate annotating and searching for information across multiple resources. This type of information can currently be added to NeuroML files as custom metadata (for example in the ***properties*** or ***annotation*** elements), and there are plans for greater native support for NeuroLex in version 2.0 of NeuroML.

Model Description Languages in Systems Biology

Standardized model descriptions in the field of Systems Biology are at a more mature stage compared to those in computational neuroscience. The Systems Biology Markup Language (SBML, Hucka et al. 2003) and CellML (Lloyd et al. 2004) are two popular languages for describing systems of interacting entities in Systems Biology, although both systems can be used for describing more generic dynamical models. NeuroML differs from these languages in that it is a domain specific model description language, and neuroscience concepts such as cells, ion channels and synaptic connections are built into the language. While there is overlap in the types of models that NeuroML and SBML/CellML can describe (for example a single compartment cell with Hodgkin Huxley like conductances), NeuroML focusses on providing a concise format for the neuronal model elements which can be readily understood by software applications using these as core concepts.

As described in Sect. 16.2.5, NeuroML version 2 will have greater interaction with these languages, with SBML being an initial focus of the work. This will allow, for example, complex (intra- or extracellular) signalling pathways to be expressed in one of these formats (with the added bonus of tool support and model repositories) with the rest of the cell and network model specified in NeuroML.

16.2.5 Work Towards NeuroML Version 2.0

The current version of NeuroML can be used to specify a wide range of neuronal models ranging from single cells to complex networks in 3D containing multiple cell types. Models have been developed in this format of cells and microcircuits from multiple brain regions (see Sect. 16.3.3). There are however a number of ways in which the language can be further developed to increase the scope of models which can be specified. This has led to plans for a version 2.0 of the language which will extend the scope of NeuroML through greater flexibility in the types of models which can be created and through links to model components in external markup languages.

Changes over version 1.x include:

- A mechanism for defining reusable component types, which will serve as the basis for model components (LEMS, the Low Entropy Model Specification language, see http://www.neuroml.org/lems for more details)
- Import and export functions for models in SBML and CellML, including detailed cellular signalling pathways

- Greater support for templates of network structure, in line with work in the INCF Multi-scale Modelling Program
- A library for reading/writing NeuroML files in multiple languages (libNeuroML)
- Links to structured annotation formats (for example Systems Biology Ontology,[8] Gene Ontology,[9] NeuroLex, etc.)

More details on the future directions of the NeuroML initiative are in Sect. 16.4.

16.3 NeuroML Resources

The following is a list of resources which can be used by researchers who wish to get more details about the NeuroML specifications, find example models and tools supporting this format, convert their own models to NeuroML or who want to adapt their applications to import/export the format. An updated list of software applications with NeuroML support is available at http://www.neuroml.org/tool_support.

16.3.1 NeuroML Language Specifications

The latest version of the NeuroML language specifications can be obtained online.[10] As outlined in Sect. 16.2.2.1, the language is defined in a set of XML Stylesheet Document (XSD) files. These files can be downloaded individually, or can be viewed in a web browser converted to a more readable format.[11] A detailed explanation of the elements allowed in NeuroML files at each Level is contained in Supporting Text 1 of Gleeson et al. (2010).

The very latest version of the specification files, as well as previous versions, can be obtained from the NeuroML version control repository on SourceForge (which uses Subversion[12]). The XSD files for version 1.x of the language as described in Sect. 16.2.2 can be found in the directory *trunk/web/NeuroMLFiles/Schemata*.

This resource can be useful for:

- Finding detailed definitions of NeuroML elements
- Ensuring your application generates valid NeuroML code

[8]http://www.ebi.ac.uk/sbo

[9]http://www.geneontology.org

[10]http://www.neuroml.org/specifications

[11]forexampleseehttp://neuroml.org/NeuroMLValidator/Latest.jsp?spec=MorphML

[12]http://neuroml.svn.sourceforge.net/viewvc/neuroml

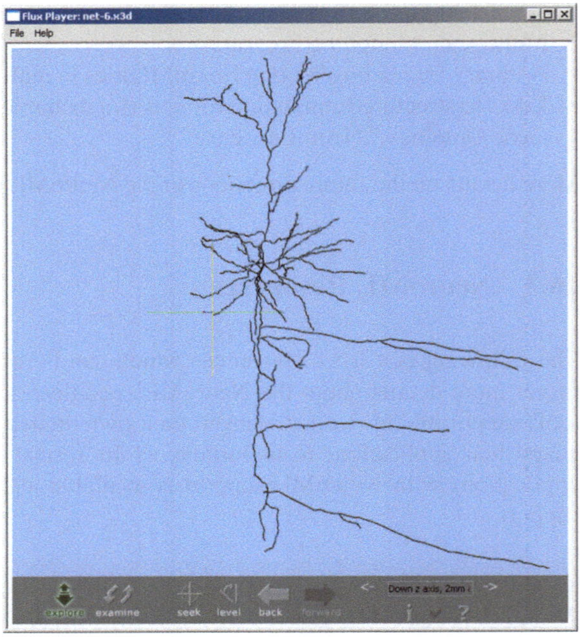

Fig. 16.4 A pyramidal cell morphology originally in MorphML was validated using the NeuroML validator and then mapped to X3D format. It is visualised here using a web browser plugin (Flux Player)

16.3.2 NeuroML Validator

A useful tool for validating XML files against the NeuroML specifications is available at http://www.neuroml.org/NeuroMLValidator/Validation.jsp. The contents of a NeuroML file can be pasted into the text box provided and the application will validate the file against each of the NeuroML Schemas. An example of the results of the validation of a ChannelML file is given in Fig. 16.3.

Once a file has been successfully validated in this way, a number of options are given for converting the file to other formats using XSL mappings. For MorphML files, mappings to a HTML description of the cell or NEURON and GENESIS morphology files are provided. ChannelML files can be mapped to HTML, NEURON (either NMODL or ChannelBuilder format as appropriate), GENESIS or PSICS. NetworkML files can be mapped to a description in HTML of the structure of the network.

Cell morphologies or instance based network descriptions can also be converted to X3D[13] format to visualize the structure of the cell or network in an X3D compatible browser plug-in. While this functionality is more limited than applications which read the NeuroML files natively and have inbuilt visualisation capability, it is useful for providing a quick 3D representation of the model (see Fig. 16.4).

[13]http://www.web3d.org/x3d

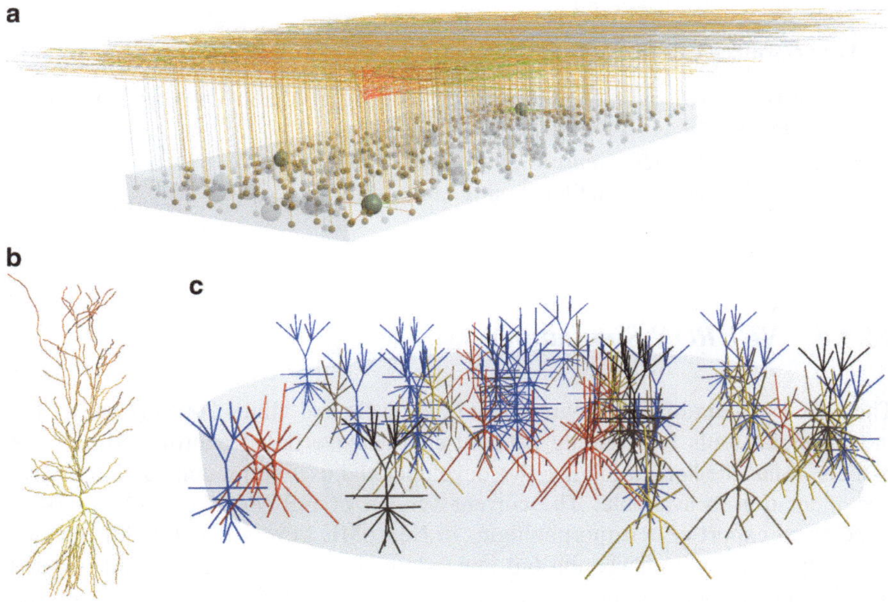

Fig. 16.5 A number of models which are available in NeuroML. (**a**) Network model of the granule cell layer of cerebellum. (**b**) CA1 pyramidal cell model. (**c**) Layer 2/3 cortical network model (Images in **b** and **c** modified from Gleeson et al. (2010) with permission)

This resource can be useful for:

- Quick validation of NeuroML files without the need to install any extra software
- Visualisation of the contents of NeuroML files

16.3.3 NeuroML Example Models

A number of published cell and network models have been converted to NeuroML format over the past number of years and many of these are available from http://www.neuroml.org/models. Figure 16.5 illustrates some of these examples. These are available as a zip file containing all the NeuroML files, or as a neuroConstruct (Sect. 16.3.6) project which can be used to view the model and generate code for multiple simulators. Greater interaction with ModelDB (Hines et al. 2004) is in development to make it easier to search for models of specific cell types and brain regions. A number of NeuroML models are also available at http://www.neuroConstruct.org/samples to illustrate the functionality of that application.

This resource can be useful as:

- A repository of published models in valid NeuroML to serve as example implementations
- A test suite for new NeuroML compliant tools
- A set of models to benchmark simulator performance

16.3.4 NEURON Simulator

The NEURON simulation environment (Carnevale and Hines 2006) is one of the most popular tools for creating detailed conductance based neuron and network models. It has the benefit of an active developer and user community and extensive documentation is available. The current version of NEURON natively supports import and export of cell morphologies in NeuroML Levels 1 and 2. All releases of the application from version 6.0 onwards include these features, and updated files for this support can be retrieved from the NeuroML SourceForge repository. Details of the locations of the relevant files can be found at http://www.neuroml.org/neuron_tools.

The NeuroML export function in NEURON can be accessed when ModelView is open (Main NEURON Menu → Tools → ModelView). This functionality works best when just one morphologically detailed cell has been created from a cell template, as is the case with many single cell models on ModelDB. When exporting as a Level 2 file, the densities of channels and passive properties of the cell are included. To this end, the groups of sections with common electrical properties as generated by ModelView (ModelViewParamSubsets) are used as section groups, and a ***biophysics*** element is added to the exported NeuroML file.

The option for importing NeuroML morphologies is available via Main NEURON Menu → Build → Cell Builder → Management → Import → Import Button → NeuroML. Figure 16.6 illustrates a cell in MorphML which has been imported into NEURON.

ChannelML files can be converted to NMODL files using the latest XSL file for this mapping, for example **ChannelML_vXXX_NEURONmod.xsl**. This converts the XML into a mod file which can be compiled for use in NEURON. This conversion can be done with any XML tool for handling XSL file transformations, and a short script in Python to facilitate this is available.[14] This conversion is also possible via the NeuroML Validator web application (Sect. 16.3.2, Fig. 16.3). The flexibility of the NMODL language has meant that all channel and synaptic mechanisms covered by NeuroML to date are supported by NEURON. Level 1 cell

[14]http://www.neuroml.org/neuron_tools

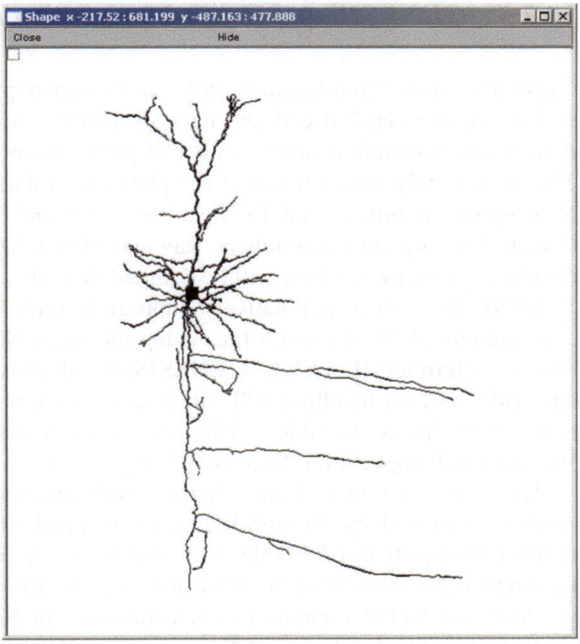

Fig. 16.6 A pyramidal cell originally described in MorphML cell which has been loaded into the NEURON simulator and visualized with a shape plot

morphologies (MorphML files) can also be converted to NEURON hoc files via that site with an XSL mapping (for example **MorphML_vXXX_NEURON.xsl**), but use of an interactive tool like neuroConstruct (Sect. 16.3.6), which can import Level 1–3 cells, allows visualization and editing of the cells and can export to NEURON and other formats, is recommended.

The latest version of the NEURON simulation environment can be obtained from http://www.neuron.yale.edu/neuron.

This resource can be useful as:

- One of the main target simulation platforms for executing NeuroML based cell and network models
- Many published models for NEURON (and other simulators) are available from ModelDB at: http://senselab.med.yale.edu/ModelDB, a large number of which could potentially be converted to NeuroML
- A version of NEURON which can run across multiple processors is available allowing simulations of networks with numbers of neurons approaching biological scales

16.3.5 GENESIS and Related Simulation Environments

GENESIS (Bower and Beeman 1997) is another popular platform for developing and simulating detailed cell and network models. It too has an active user community, documentation and a number of publications using models in this format. The most widely used version of the platform to date has been GENESIS 2. The Neurospaces (Cornelis and De Schutter 2003) and MOOSE (Multiscale Object Oriented Simulation Environment, Ray and Bhalla 2008) communities are actively developing new tools which will be compatible with scripts for this simulator.

GENESIS 2 does not natively support NeuroML. ChannelML files can be converted to GENESIS script files using the latest XSL file for this mapping, for example **ChannelML_vXXX_ GENESIStab.xsl**. This conversion can be done with any XML tool for handling XSL file transformations, but a short script in Python to facilitate this is available.[15] This conversion is also possible via the NeuroML Validator web application (Sect. 16.3.2, Fig. 16.3).

The majority of (non kinetic scheme based) ion channels and non plastic synapse models supported by NeuroML can be mapped to GENESIS 2. There are a number of objects in GENESIS 2 for implementing plastic synapses (for example facsynchan, hebbsynchan), but these are based on different models of plasticity than the STP and STDP synaptic mechanisms used in NeuroML. There are ongoing contacts with the developers of Neurospaces, MOOSE, and other interested parties such as the PyNN developers, to reach community wide agreement regarding a core set of synaptic plasticity models for future support.

Level 1 cell morphologies (MorphML files) can also be converted to GENESIS script files via the validator website with an XSL mapping (for example **MorphML_vXXX_ GENESIS.xsl**), but use of an interactive tool like neuroConstruct (Sect. 16.3.6) is recommended.

The most recent version of the GENESIS 2 simulation environment can be obtained from http://www.genesis-sim.org/GENESIS.

The Neurospaces project is developing major components of the GENESIS 3 platform. This will be a modular reimplementation of the core of GENESIS into a number of components for model loading and editing (including parsers for GENESIS 2 scripts and import functions for NeuroML cell models), a number of compartmental solvers, a scheduler for managing simulations, and command line and graphical interfaces. More details of the current developments towards GENESIS 3 can be found at http://www.genesis-sim.org and http://neurospaces.sourceforge.net.

The MOOSE platform also has extensive support for loading GENESIS 2 scripts. Some native support for loading NeuroML has been developed, and the C++ library for this can potentially be reused by other applications. The latest details on MOOSE development are available at http://moose.sourceforge.net.

---

[15] http://www.neuroml.org/neuron_tools

This resource can be useful as:

- One of the main target simulation platforms for running NeuroML based cell and network models
- MOOSE and Neurospaces/GENESIS 3 are contributing to the ongoing development of NeuroML which will help ensure compatibility of these tools with future features of the language

16.3.6 neuroConstruct

neuroConstruct (Gleeson et al. 2007) is a graphical application to facilitate the development of networks of biophysically detailed neurons in 3D. Once cell and network models are created through the GUI, scripts to execute simulations of them can be generated for a number of applications including NEURON, GENESIS, MOOSE, PSICS and PyNN. neuroConstruct has been developed in parallel with the NeuroML specifications and there is native support for all parts of NeuroML.

The internal representation of cells in neuroConstruct is closely based on MorphML. For historical reasons, cells in neuroConstruct consist of segments grouped in sections, whereas in MorphML the equivalent entities are segments grouped in cables. Most cells can be exported in NeuroML Level 3 and re-imported with no loss of information (including group information and non uniform channel distributions). Figure 16.7 shows the neuroConstruct interface with a cell visualised in 3D which can be exported/imported in NeuroML Level 1–3 formats.

neuroConstruct uses ChannelML files for channel and synapse models. When generating the scripts for a particular simulator, neuroConstruct applies the corresponding XSL mapping to make a native representation of the model component and the generated file is compiled if necessary (as in the case of NEURON mod files). Cell mechanisms can also be included in neuroConstruct in the native simulator scripts (File Based Cell Mechanisms) and this can be useful in the process of converting a channel from one simulator's native format to ChannelML, as cells with two versions of a channel can be run side by side and compared directly.[16] There are a number of inbuilt features in neuroConstruct for generating plots from the contents of ChannelML files, for example of the voltage dependences of the steady-state activation and inactication variables and time constants or the synaptic conductance waveforms (Tab Cell Mechanism → (select a ChannelML based Cell Mechanism) → Edit selected Cell Mechanism → Generate associated plots).

NetworkML is used by neuroConstruct for storing and reloading the generated network structure. When a network is generated (at tab Generate) it can be stored

[16]seehttp://www.neuroConstruct.org/docs/importneuronformoredetails

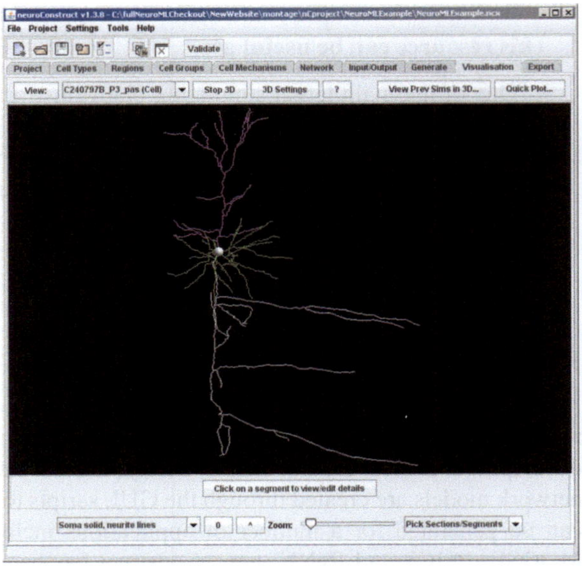

Fig. 16.7 A pyramidal cell visualised in neuroConstruct from a MorphML cell description

for future use in neuroConstruct or other application by pressing Save NetworkML. Options are present for storing in: XML text files (files can be viewed with text browser; produces large files; slower to generate and reload); zipped XML files (produces smaller files; slightly slower to save and reload); or HDF5 format (faster to save and reload; up to 90% smaller files; special software is needed to view and edit these types of files outside of neuroConstruct, for example HDFView[17]).

There are also a number of options in neuroConstruct for exporting and importing NeuroML files combining elements from a number of different Levels. At tab Export → NeuroML, in addition to options for exporting only the cells in Levels 1, 2 or 3, the cells, channels, synapses and generated network structure can be exported in NeuroML, either as a set of separate files, or as a single NeuroML Level 3 file. When exported as one Level 3 file, there is an option to include annotations with neuroConstruct specific information (for example information on regions, cell group colors, plots, simulation configurations, etc.), which can be read when the Level 3 file is imported into a new, empty neuroConstruct project, facilitating model exchange between neuroConstruct users. The exported file is still in valid Level 3 format, and other NeuroML compliant applications can read the file, ignore the neuroConstruct specific annotations and just import the cells, channels, populations, etc.

A Level 3 file (generated by any NeuroML compliant application) containing a mixture of cells, channels and network information can be imported into neuroCon-

[17] http://www.hdfgroup.org/products/hdf5_tools

struct and a new project created, ready for export to supported simulators. This can be done through the GUI, or at command line:

Windows: *nC.bat -neuroml MyNeuroML.xml*
Linux/Mac: *./nC.sh -neuroml MyNeuroML.xml*

The latest version of neuroConstruct and related documentation can be obtained from http://www.neuroConstruct.org.

This resource can be useful for:

- A user friendly graphical application for creating, importing and exporting NeuroML Levels 1–3
- Visualizing various elements of NeuroML based models, including 3D cell structure, network layouts and plotting channel rates and synapse waveforms
- Comparing NeuroML cell and network model behaviour across simulators

16.3.7 PSICS Simulator

The recently developed neuronal simulator PSICS (Parallel Stochastic Ion Channel Simulator) allows simulation of detailed neuronal models which include stochastic ion channel transitions, and so can be used to examine the effect of low numbers of ion channels on neuronal firing behaviour (Cannon et al. 2010). This simulator has had an initial focus on single cell modelling, and while support for networks of cells is in development, NeuroML models incorporating synapses and networks cannot currently be run on this platform.

Level 1 morphologies can be imported natively by PSICS. PSICS does not have an internal representation of cables, so MorphML ***cable*** elements (used for grouping segments) are only used to assign labels to points (taken from the ***segment*** elements), which can then be used for channel allocation. Figure 16.8 shows a cell morphology in PSICS.

PSICS natively reads a large subset of channel specifications in ChannelML 1.8.1 but does not support ligand gated channels (for example [Ca^{2+}] dependent K^{+} channels), synapses or integrate-and-fire mechanisms. Supported ChannelML models can be converted to PSICS format using the latest XSL file for this mapping, for example **ChannelML_vXXX_PSICS.xsl**. This converts the XML into a PSICS compatible XML file and reports if the ChannelML file uses an unsupported construct. The conversion can be done with any XML tool for handling XSL file

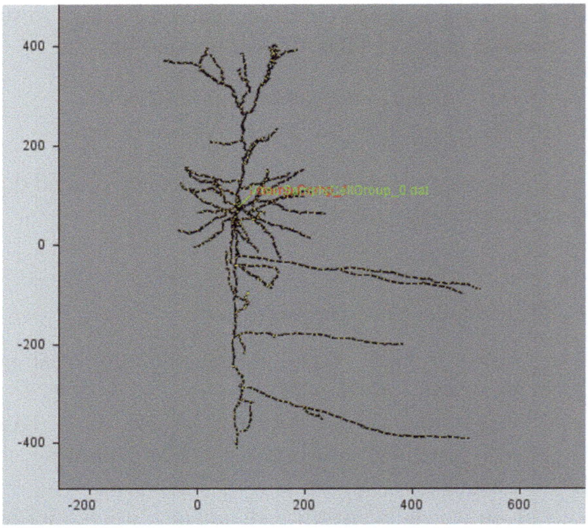

Fig. 16.8 A pyramidal cell visualised in PSICS from a MorphML cell description. The figure is automatically generated by PSICS after running a simulation with the cell

transformations, but a short script in Python to facilitate this is available.[18] This conversion is also possible via the NeuroML Validator web application (Sect. 16.3.2, Fig. 16.3).

PSICS, associated documentation and the ICING visualization application are available from http://www.psics.org.

This resource can be useful because:

- PSICS handles cellular morphologies and channel distributions differently than NEURON and GENESIS, and so a NeuroML cell model can be tested on this platform to see how firing behaviour is influenced by a simulator's representation of the physiology
- Both deterministic and stochastic simulations based on individual channel opening and closing for cell models expressed in NeuroML format are possible with this simulator

16.3.8 PyNN

A number of neuronal network simulators have recently introduced scripting interfaces based on the Python scripting language. The PyNN initiative, which

[18]http://www.neuroml.org/neuron_tools

started as part of the EU FACETS project (and is continuing in the EU BrainScaleS project), seeks to create a specification for a set of common Python commands for setting up neuronal network simulations. Simulators which currently support the language include NEURON, NEST, Brian, MOOSE and PCSIM[19] and there is also work to support the running of such networks on VLSI neuromorphic hardware created by the FACETS project. PyNN is a specification incorporating procedural descriptions of network structure and is complementary to the declarative model specifications being developed in the NeuroML initiative.

PyNN scripts are intended to be used on multiple simulators with little or no modification. The one change that is usually needed is the line at the start of a PyNN script: *from pyNN.neuron import* * should be replaced with *from pyNN.nest import* * to use the simulator NEST instead of NEURON, etc. There is an initial implementation of a NeuroML module in PyNN which can be used to export the structure of the network created in the PyNN script, together with cell and synapse properties, to a NeuroML compliant file, as opposed to executing the network on the specified simulator.

A subset of models specified in NeuroML can be converted to valid PyNN scripts. Currently this is enabled by export of NeuroML based models from neuroConstruct. Due to the scope of models which can currently be expressed in PyNN (the target simulators for PyNN were initially large scale integrate and fire based network simulators) only neuroConstruct projects containing the following can currently be mapped onto PyNN: single compartment cells containing only a passive conductance together with an integrate and fire based conductance; networks connected by alpha or single exponential waveform conductances (which could include STP or STDP based plasticity mechanisms); random spiking inputs.

Members of the PyNN and NeuroML development communities are involved in the INCF Program on Multi-scale modelling (Sect. 16.2.4), and updates of network representations (population layouts, connectivity schemes, etc.), as well as support for more generic representations of abstract cells in both PyNN and future versions of NeuroML will be coordinated through this forum.

The latest version of PyNN is available from http://neuralensemble.org/PyNN.

This resource can be useful because:

- NeuroML to date has focussed on detailed conductance based cell models, and these preliminary interactions with PyNN are a step towards using NeuroML with large scale network simulation tools like NEST
- PyNN development is closely linked with a number of other (mainly Python based) tools for simulation, data analysis and simulation management (see http://www.neuralensemble.org)

[19]http://www.lsm.tugraz.at/pcsim

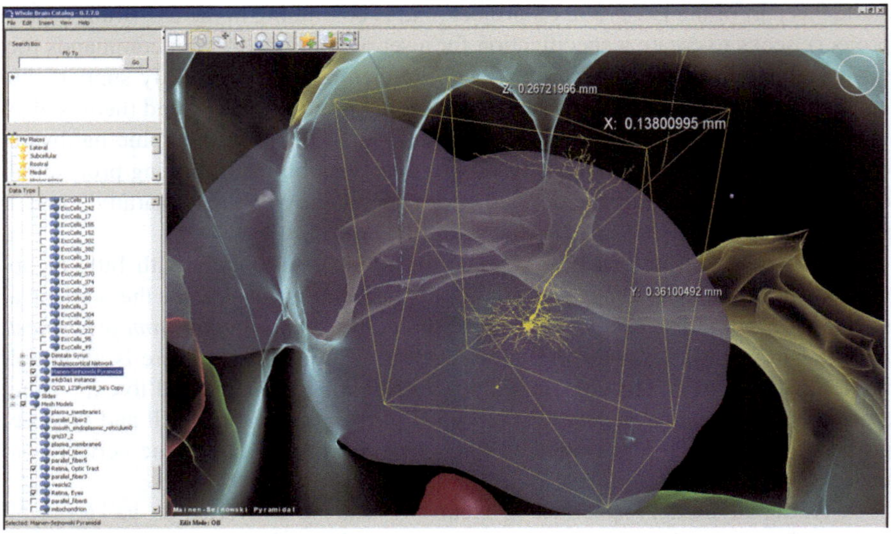

Fig. 16.9 A screenshot of the whole brain catalog interface showing a pyramidal cell. The cell can be selected and downloaded in NeuroML format

16.3.9 Whole Brain Catalog

This project is closely related to the Neuroscience Information Framework (Sect. 16.2.4) and has developed an application with a 3D graphical interface for accessing a range of neuroscience resources, including mouse brain atlases, imaging data and neuronal reconstructions. Data from these resources can be visualised at multiple scales, and links are provided to the original data sources. Objects on display are tagged with terms from the NeuroLex lexicon which facilitates linking data across resources.

Models in NeuroML format can be uploaded to the system for visualisation though the interface (currently a username and password are required), see Fig. 16.9. Both MorphML and NetworkML data can be imported. Cells already present in the system can be selected and downloaded in NeuroML.

The latest version of the Whole Brain Catalog interface can be downloaded from http://www.wholebraincatalog.org.

This resource can be useful because:

- NeuroML cell and network models can be viewed within the framework of a 3D mouse brain
- This graphical application is closely linked with a number of other neuroinformatics initiatives and offers intriguing possibilities for obtaining NeuroML files rich in structured metadata

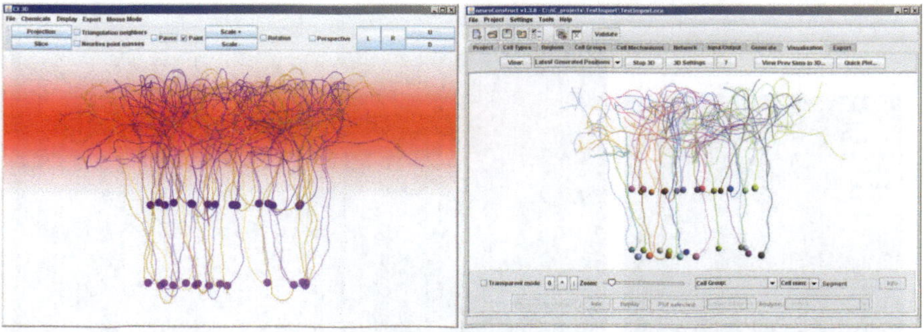

Fig. 16.10 A screenshot of a network generated in CX3D (*left*). The axons of both the *upper* and *lower* cell groups are attracted to the chemical substance whose concentration is shown in *red*. The network was exported in NeuroML format and loaded into neuroConstruct (*right*)

- Ongoing work integrating this application with neuroConstruct aims to allow execution of submitted NeuroML models on high performance computing resources and visualisation of the results through this interface

16.3.10 CX3D

This application can be used for simulating the growth of neurons in 3D, both stochastically and in response to the presence of neurochemical attractors. The application can export the grown cells and networks in a NeuroML Level 3 file (see Fig. 16.10).

The latest version of CX3D is available from http://www.ini.uzh.ch/projects/cx3d.

This resource can be useful as:

- A source of stochastically generated NeuroML cells and networks for use in other visualisation/modelling environments
- Export of the network structure at fixed intervals during its growth, followed by simulations of the cells or networks generated at different stages, can be used to investigate how cell and network behaviour alters during development

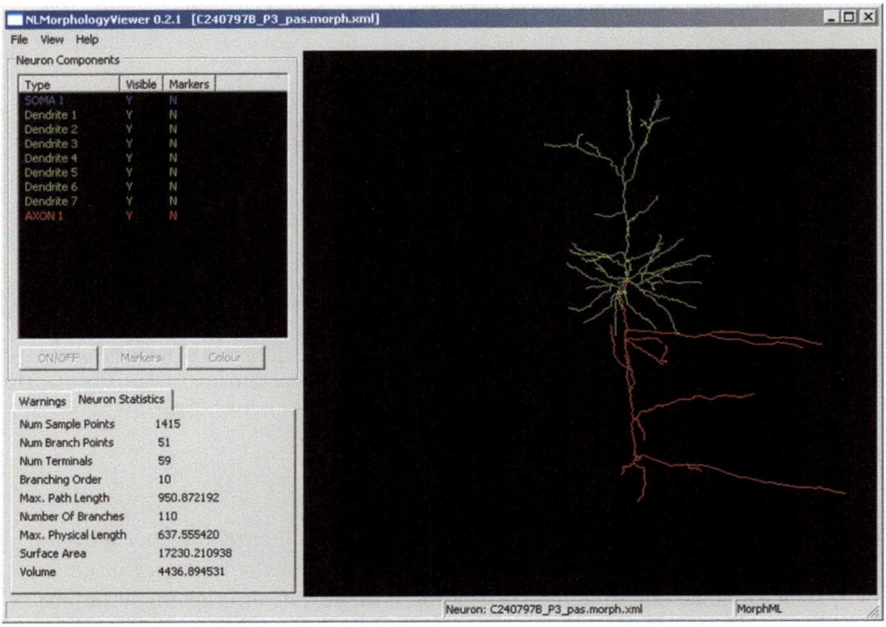

Fig. 16.11 A MorphML description of a pyramidal cell loaded and visualised in the NLMorphologyViewer

16.3.11 NeuronLand

NeuronLand develops free software tools for the experimental and computational neuroscience communities. Currently two tools for handling 3D neuronal morphology files are available: NLMorphologyConverter, which allows conversion between many commonly used morphology file formats, including MorphML, Neurolucida and SWC; and NLMorphologyViewer, which is a graphical interface for viewing the structures of cells in these formats (see Fig. 16.11), and can also import and export in multiple formats.

These applications are available from http://neuronland.org.

This resource can be useful for:

- Converting a wide range of neuronal morphology formats to and from MorphML

16.4 The Future of NeuroML

The NeuroML initiative has evolved over a number of years into its current modular form which is supported by an increasing number of applications from many areas of computational neuroscience. A growing community is committed to the development of the language, adding greater tool support and making their own and other published models available in this format.

In parallel with further development of the language, the range of cell and network models available in the format will be extended. Many well known cortical, cerebellar and hippocampal models are available already in NeuroML and systems will be put in place to allow researchers to collaboratively build on and improve these models, providing a valuable resource for both experimentalists and theoreticians to use in their work.

The next version of the NeuroML language will significantly extend the range of models which can be expressed. Greater support for abstract cell models such as integrate and fire (I&F) models, Izhikevich neurons and adaptive exponential I&F models, etc., will be included. The new system for defining the behaviour of extensible component types will also allow more complex ion channel and synapse models to be created. Links to models expressed in SBML and CellML will allow detailed subcellular signalling pathways to be included in cell models, enabling true multi-scale modelling. There will be greater support for HDF5 versions of morphologies and network structure to cope with the data demands of connectomics initiatives. The support for network templates will also be significantly improved allowing more compact representations of complex network structure to be exchanged between applications for generating, visualising and simulating large scale networks.

All of these advances support our vision of an ecosystem of computational neuroscience tools each with specialised functionality interacting through a common model exchange language, allowing researchers the freedom to ask new and interesting questions about the functioning of the nervous system.

Acknowledgements The NeuroML initiative has involved contributions from a large number of researchers over many years. Please see http://www.neuroml.org/contributors.php for full details on contributors. Support for PG and RAS came from the MRC (Program grant G0400598 to RAS and a Special Research Training Fellowship to PG), the EU (EUSynapse, LSHM-CT-2005-019055) and a Wellcome Trust Biomedical Resources Grant (086699). The BBSRC (005490) funded collaboration in this area between PG, VS and RAS. RAS is the recipient of a Wellcome Trust Principal Research Fellowship (095667) and a European Research Council Advanced Grant. Contributions of SC were supported by NIH R01MH081905.

References

Ascoli GA, Donohue DE, Halavi M (2007) NeuroMorpho.org: a central resource for neuronal morphologies. J Neurosci 27(35):9247–9251

Bower J, Beeman D (1997) The book of GENESIS: exploring realistic neural models with the GEneral NEural SImulation system. Springer, New York

Bray T, Paoli J, Sperberg-McQueen CM (1998) Extensible Markup Language (XML) 1.0. Http://www.w3.org/TR/REC-xml

Brette R, Rudolph M, Carnevale T, Hines M, Beeman D, Bower JM, Diesmann M, Morrison A, Goodman PH, Harris JFC, Zirpe M, Natschlager T, Pecevski D, Ermentrout B, Djurfeldt M, Lansner A, Rochel O, Vieville T, Muller E, Davison AP, El Boustani S, Destexhe A (2007) Simulation of networks of spiking neurons: a review of tools and strategies. J Comput Neurosci 23(3):349–398

Cannon R, Gewaltig MO, Gleeson P, Bhalla U, Cornelis H, Hines M, Howell F, Muller E, Stiles J, Wils S, De Schutter E (2007) Interoperability of neuroscience modeling software: current status and future directions. Neuroinformatics 5(2):127–138

Cannon RC, O'Donnell C, Nolan MF (2010) Stochastic ion channel gating in dendritic neurons: morphology dependence and probabilistic synaptic activation of dendritic spikes. PLoS Comput Biol 6(8):e1000886

Carnevale NT, Hines ML (2006) The NEURON book. Cambridge University Press, Cambridge

Cornelis H, De Schutter E (2003) NeuroSpaces: separating modeling and simulation. Neurocomputing 52(4):227–231

Crook S, Gleeson P, Howell F, Svitak J, Silver RA (2007) MorphML: Level 1 of the NeuroML standards for neuronal morphology data and model specification. Neuroinformatics 5(2):96–104

Davison AP, Bruderle D, Eppler J, Kremkow J, Muller E, Pecevski D, Perrinet L, Yger P (2008) PyNN: a common interface for neuronal network simulators. Front Neuroinf 2:11

De Schutter E (2008) Why are computational neuroscience and systems biology so separate? PLoS Comput Biol 4(5):e1000078

Diesmann M, Gewaltig MO (2002) NEST: An Environment for Neural Systems Simulations, vol Forschung und wisschenschaftliches Rechnen, Beitrage zum Heinz-Billing-Preis 2001. Gottingen: Ges. fur Wiss. Datenverarbeitung

Djurfeldt M, Lansner A (2007) Workshop report: 1st INCF workshop on large-scale modeling of the nervous system. Accessed May 2012

Ermentrout B (2002) Simulating, analyzing, and animating dynamical systems: a guide to XPPAUT for researchers and students. Society for Industrial and Applied Mathematics, Philadelphia

Gardner D (2004) Neurodatabase.org: networking the microelectrode. Nat Neurosci 7(5):486–487

Gardner D, Knuth KH, Abato M, Erde SM, White T, DeBellis R, Gardner EP (2001) Common data model for neuroscience data and data model exchange. J Am Med Inform Assoc 8(1):17–33

Gleeson P, Steuber V, Silver RA (2007) neuroConstruct: a tool for modeling networks of neurons in 3D space. Neuron 54(2):219–235

Gleeson P, Crook S, Cannon RC, Hines ML, Billings GO, Farinella M, Morse TM, Davison AP, Ray S, Bhalla US, Barnes SR, Dimitrova YD, Silver RA (2010) NeuroML: a language for describing data driven models of neurons and networks with a high degree of biological detail. PLoS Comput Biol 6(6):e1000815

Goddard NH, Hucka M, Howell F, Cornelis H, Shankar K, Beeman D (2001) Towards NeuroML: model description methods for collaborative modelling in neuroscience. Philos Trans R Soc Lond B Biol Sci 356(1412):1209–1228

Goodman D, Brette R (2008) Brian: a simulator for spiking neural networks in Python. Front Neuroinformatics 2:5

Hines ML, Morse T, Migliore M, Carnevale NT, Shepherd GM (2004) ModelDB: a database to support computational neuroscience. J Comput Neurosci 17(1):7–11

Howell F, Cannon R, Goddard N, Bringmann H, Rogister P, Cornelis H (2003) Linking computational neuroscience simulation tools–a pragmatic approach to component-based development. Neurocomputing 52–54:289–294

Hucka M, Finney A, Sauro HM, Bolouri H, Doyle JC, Kitano H, Arkin AP, Bornstein BJ, Bray D, Cornish-Bowden A, Cuellar AA, Dronov S, Gilles ED, Ginkel M, Gor V, Goryanin I, Hedley WJ, Hodgman TC, Hofmeyr JH, Hunter PJ, Juty NS, Kasberger JL, Kremling A, Kummer U, Le Novere N, Loew LM, Lucio D, Mendes P, Minch E, Mjolsness ED, Nakayama Y, Nelson MR, Nielsen PF, Sakurada T, Schaff JC, Shapiro BE, Shimizu TS, Spence HD,

Stelling J, Takahashi K, Tomita M, Wagner J, Wang J (2003) The Systems Biology Markup Language (SBML): a medium for representation and exchange of biochemical network models. Bioinformatics 19(4):524–531

Lloyd CM, Halstead MD, Nielsen PF (2004) CellML: its future, present and past. Prog Biophys Mol Biol 85(2–3):433–450

Qi W, Crook S (2004) Tools for neuroinformatic data exchange: an XML application for neuronal morphology data. Neurocomputing 58–60:1091–1095

Ray S, Bhalla US (2008) PyMOOSE: interoperable scripting in Python for MOOSE. Front Neuroinformatics (2:6)

Rhodes PA, Llinas RR (2001) Apical tuft input efficacy in layer 5 pyramidal cells from rat visual cortex. J Physiol 536(1):167–187. DOI 10.1111/j.1469-7793.2001.00167.x

Song S, Miller KD, Abbott LF (2000) Competitive Hebbian learning through spike-timing-dependent synaptic plasticity. Nat Neurosci 3:919–926

Traub RD, Contreras D, Cunningham MO, Murray H, LeBeau FE, Roopun A, Bibbig A, Wilent WB, Higley MJ, Whittington MA (2005) Single-column thalamocortical network model exhibiting gamma oscillations, sleep spindles, and epileptogenic bursts. J Neurophysiol 93(4):2194–2232

Tsodyks MV, Markram H (1997) The neural code between neocortical pyramidal neurons depends on neurotransmitter release probability. Proc Natl Acad Sci USA 94:719–723

Tsodyks M, Uziel A, Markram H (2000) Synchrony generation in recurrent networks with frequency-dependent synapses. J Neurosci 20:RC50

Chapter 17
XPPAUT

Bard Ermentrout

Abstract This chapter describes the XPPAUT software package. XPPAUT is a general purpose tool for numerically solving and analyzing dynamical systems. It combines a graphical interface, many numerical routines, two- and three-dimensional graphics, and an interface to AUTO.

17.1 Introduction

XPPAUT is a general numerical tool for simulating, animating, and analyzing dynamical systems. These can range from discrete finite state models to stochasic Markov models, to discretization of partial differential and integrodifferential equations. *XPPAUT* was not specifically developed for systems biology simulations but because of its ability to provide a complete numerical analysis of the dependence of solutions on parameters ("bifurcation diagrams") it is widely used by the community of computational and theoretical biologists. While it can be used for modest sized networks, it is not specifically designed for this purpose and due to its history, there are limits on the size of problems which can be solved (about 2,000 differential equations is the current limit). Rather than a pure simulator, *XPPAUT* is a tool for understanding the equations and the results of simulating the equations. *XPPAUT* uses a highly optimized parser to produce a pseudocode which is interpreted and runs very fast – at about half the speed of directly compiled code. Since no compiler is required, *XPPAUT* is a stand alone program and runs on all platforms which have an X11 interface available (UNIX, MAC OSX, Windows, etc.). Versions are also available for the iPad and iPhone. The program is open source and available in source and various binary versions. *XPPAUT* includes a version of AUTO Doedel

B. Ermentrout (✉)
University of Pittsburgh, Pittsburgh, PA 15260, USA
e-mail: bard@pitt.edu

N. Le Novère (ed.), *Computational Systems Neurobiology*,
DOI 10.1007/978-94-007-3858-4_17,

(1981) which is a continuation and bifurcation package. *XPPAUT* can be compiled on 64-bit platforms, however, the AUTO component does not run properly on 64-bit machines unless compiled in 32 bit mode.

XPPAUT can be run interactively (the preferred method) but can also be run in batch mode with no GUI with the results dumped to one or more files. Graphic output as postscript, GIF, PBM, and animated GIF is possible. (There are codecs available for AVI format but these are not generally included in the compiled versions.) Recent changes in *XPPAUT* have greatly extended the batch capabilities and it is now possible to easily set it up to run on multiple threads through a friendly Java interface. Numerous packages for controlling *XPPAUT* have been written, some stand-alone such as JigCell (http://jigcell.cs.vt.edu/) and others using Matlab or Python. (All of these frontends can be found on the website http://www.math.pitt.edu/~bard/xpp/xpp.html.)

There are no limits as far as the form of the equations is concerned since the actual equations that you desire to solve are written down like you would write them in a paper. For example the voltage equation for a conductance-based model would be written as:

```
dv/dt = (-gl*(v-el) - gna*m^3*h*(v-ena)-gk*n^4*(v-ek))/cm
```

There is a method for writing indexed networks as well, so that one does not have to write every equation. Special operators exist for speeding up network functions like discrete convolutions and implementation of the stochastic Gillespie algorithm. Furthermore, the user can link the right-hand sides of differential equations to external C libraries to solve complex equations (for example, equation-free firing rate models Laing et al. 2010). Because it is a general purpose solver, the user can mix different types of equations for example stochastic discrete time events with continuous ODEs. Event driven simulations are also possible and can be performed in such as way that output occurs only when an event happens. There are many ways to display the results of simulations including color-coded plots showing space-time behavior and a built-in animation language.

17.2 Types of Problems that Can Be Solved

XPPAUT solves a variety of types of problems. The most common class of equations is *initial value problems* of the form:

$$X' = F(X,t), \quad X(t_0) = X_0$$

where $F : R^m \times R \to R^m$ and $X_0 \in R^m$. (Here, and throughout this chapter, X' means the time derivative of X.) *XPPAUT* solves first order differential equations so that systems like $x'' = -x$ have to be converted into a system of first order equations, e.g., $x' = y, y' = -x$. *XPPAUT* does not handle complex variables, so that a system of equations on C^m must be converted to real equations, e.g., $z' = z - |z|^2 z$ is written as $x' = x - x(x^2 + y^2), y' = y - y(x^2 + y^2)$. *XPPAUT* will also solve discrete dynamical systems of the form

$$X_{n+1} = F(X_n, n), \quad X(n_0) = X_0.$$

Equilibria, stability, and periodic solutions can be found as parameters vary for sufficiently smooth right-hand sides using the continuation package AUTO which is incorporated into *XPPAUT* .

XPPAUT solves a variety of functional differential equations, such as delay equations and some classes of Volterra integro-differential equations. An example of the former is the delayed negative feed-back equation

$$\frac{dx}{dt} = -x(t) + f(I - bx(t - \tau)). \tag{17.1}$$

While not so common in systems biology, Volterra integral equations do arise on occasion. For example, consider the nonlinear equation:

$$x(t) = a \int_0^t e^{-(t-s-\tau)^2} x(s)(1 - x(s))\, ds. \tag{17.2}$$

This cannot be converted into an ODE so the full integral equation must be solved.

Equations need not have smooth or even continuous right-hand sides. The adaptive quadratic integrate and fire model is an example of such an equation:

$$V' = V^2 + I - w, \qquad w' = a(bV - w) \tag{17.3}$$

along with the condition that if $V(t^-) = V_{th}$ then $V(t^+) = V_{reset}, w(t^+) = w(t^-) + d$. Another example is a simple cell growth model due to John Tyson Fall et al. (2002) (Chap. 10):

$$\begin{aligned} u' &= k_4(v - u)(a + u^2) - k_6 u \\ v' &= k_1 m - k_6 u \\ m' &= bm \end{aligned} \tag{17.4}$$

along with the reset condition. Whenever u falls below 0.2, the mass, m is halved.

XPPAUT also solves boundary-value problems (BVPs) in which conditions are given at both ends of an integration interval. For example, Eq. 17.3 has a periodic solution for I large enough. This solution can be found by integration, but it is hard to automatically compute the period as parameters vary (e.g. through continuation) since it is discontinuous. However, the "periodic" solution can be treated as a BVP. Indeed, it must satisfy, $V(0) = V_{reset}$, $V(P) = V_{th}$ and $w(0) = w(P) + d$. There are three equations, but only two ODEs. The period, P is a free parameter, so by adding the equation $P' = 0$, we now have three equations and three unknowns.

Stochastic simulations are also possible in *XPPAUT* as it includes a method for continuos time Markov processes such as seen in channel simulations. For example, many molecular motors are based on exploiting thermal noise along with an asymmetric periodic potential that randomly flashes on and off. The net effect of these two random processes is the performance of work. Let $V(x)$ be a potential function. Assume that it is periodic on the line with period 1 and that it is asymmetric. Let z be a two state Markov process that determines the magnitude of the potential – when $z = 0$ the potential has zero magnitude and when $z = 1$ it is maximal. Finally assume a certain amount of Brownian motion. Then when the potential is on, the probability distribution builds up proportionally to $P(x) = \exp(-V(x)/\sigma)$. When the potential is turned off, there is diffusional drift but it favors one direction more than the other because of the asymmetric potential. The result is that there is a net flux in one direction. Many molecules work in this fashion by changing configuration (due to the hydrolysis of ATP) to produce in effect a molecular ratchet driven by the noise. There are many realistic models of this but they all work on the same principle illustrated by the above example. The equations for this are:

$$dx = -zV'(x)dt + \sigma d\xi, \tag{17.5}$$

where z is a two state Markov process that has switching rates α,β for $0 \to 1$ and $1 \to 0$ respectively.

Partial differential equations and integro-differential equations can also be solved with *XPPAUT* by discretizing space into a set of ODEs. For example, consider the Brusslator reaction-diffusion equation:

$$\begin{aligned} U_t &= A - (B + 1)U + U^2V + D_U U_{xx} \\ V_t &= BU - U^2V + D_V V_{xx} \end{aligned} \tag{17.6}$$

with, say, periodic boundary conditions, $U(L,t) = U(0,t), V(L,t) = V(0,t)$. Here, U_t means the partial derivative with respect to t, etc. An integral equation arising in neuroscience is the Ermentrout-McCleod Ermentrout and McLeod (1993) equation

$$U_t = -U + F(\int_{-\infty}^{\infty} K(x - y)U(y,t)\,dy). \tag{17.7}$$

XPPAUT has special operators for dealing with spatial convolutions.

XPPAUT can solve other types of problems we will not illustrate here but which can be useful for computational biologists. For example, in many cellular systems, there are small numbers of molecules, so that approximating them as continuous chemical concentrations is not really valid. Thus, one typically runs Monte Carlo simulations and the most accurate way to implement continuous time discrete state Markov models is the Gillespie algoithm. *XPPAUT* can automate this process quite easily; the user need only provide the reaction rates. Some types of differential equations involve nonlinear dependendence on the derivatives or additional algebraic conditions. *XPPAUT* has solvers for these so-called differential-algebraic equations.

17.3 Running XPP

Running *XPPAUT* requires that the user first create an *ODE file*, that is, a plain text file that describes the model to be solved. ODE files can also contain parameters for the numerical methods, graphics, and output. Once these are created, either by the user or automatically, then *XPPAUT* can be run with the ODE file as an input. *XPPAUT* is usually run interactively so that the user can change parameters, integration time, initial data, etc. However, since much of the information needed to run the ODE can be included in the file, *XPPAUT* can also be run without a GUI using the "-silent" option. The result is a file or series of files with the output of the simulation as a series of columns of numbers. In the next few subsections, we illustrate how to solve the example problems described in Sect. 17.2.

17.3.1 Ordinary Differential Equations

We start with the familiar Hodgkin-Huxley equation, which is a four-dimensional dynamicl system:

$$
\begin{aligned}
CV' &= I - g_L(V - E_L) - g_K n^4(V - E_K) - g_{Na} m^3 h(V - E_{Na}) \\
m' &= M(m, v) \\
h' &= H(h, v) \\
n' &= N(m, v)
\end{aligned}
$$

where, M, N, H are the gating functions for the channels. The complete ODE file for *XPPAUT* is

```
#  hh.ode
init v=-65  m=.05  h=0.6  n=.317
par i=0
par ena=50  ek=-77  el=-54.4  gna=120  gk=36  gl=0.3
  c=1  phi=1
am(v)=phi*.1*(v+40)/(1-exp(-(v+40)/10))
bm(v)=phi*4*exp(-(v+65)/18)
ah(v)=phi*.07*exp(-(v+65)/20)
bh(v)=phi*1/(1+exp(-(v+35)/10))
an(v)=phi*.01*(v+55)/(1-exp(-(v+55)/10))
bn(v)=phi*.125*exp(-(v+65)/80)
v'=(I - gna*h*(v-ena)*m^3-gk*(v-ek)*n^4-gl*(v-el))/c
m'=am(v)*(1-m)-bm(v)*m
h'=ah(v)*(1-h)-bh(v)*h
n'=an(v)*(1-n)-bn(v)*n
done
```

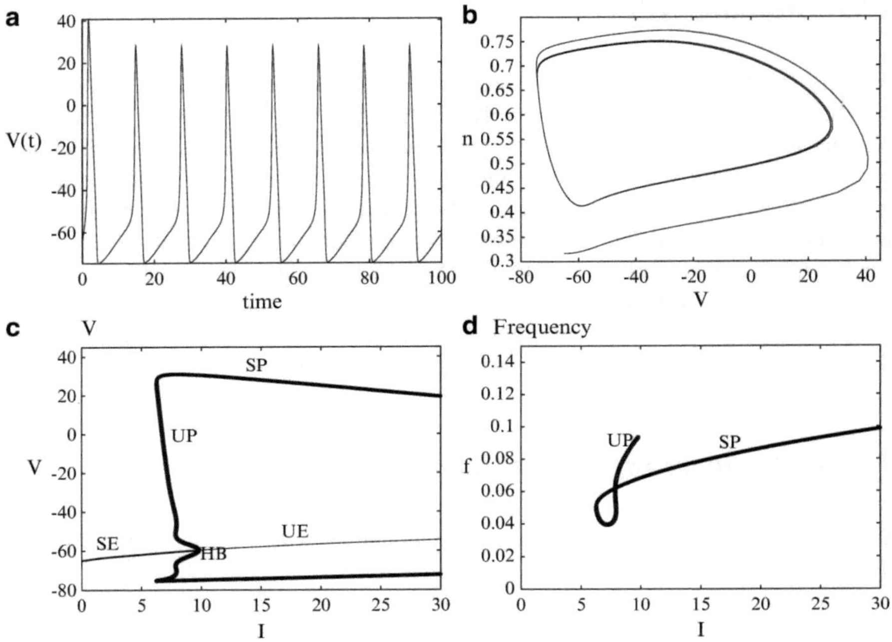

Fig. 17.1 Numerical solution to the Hodgkin-Huxley equations. (**a**) Time series plot of the voltage when $I = 15$; (**b**) Projection into the (V, n)—plane; (**c**) bifurcation diagram with current, I varying (*SE* stable equilibria, *UE* unstable equilibria, *SP* stable periodic, *UP* unstable perriodic); (**d**) frequency vs. current

Running *XPPAUT* on this file, we can look at a variety of aspects of the ODE. Figure 17.1 shows an example run. Panel A shows the voltage as a function of time when $I = 15$. Panel b shows a projection of the same solution in the (V, n)—plane. *XPPAUT* provides a very simple interface to AUTO; panels c and d of Fig. 17.1 show an example run as the injected current I varies. *XPPAUT* first plots the equilibria and shows where they are stable. It also shows bifurcations; here HB denotes a Hopf bifurcation. A branch of unstable periodic orbits (UP) arises from this bifurcation point, but stabilizes (SP) to a large amplitude periodic solution. Both the maximum and minimum can be shown. The frequency of the periodic solutions can also be plotted.

If an ODE has a discontinuity or jump condition, it is necessary to use event checking. *XPPAUT* incorporates such "global" events using a simple crossing criterion. Here is the ODE file for the Tyson model:

```
# tyson.ode
init u=.0075,v=.48,m=1
par k1=.015,k4=200,k6=2,a=.0001,b=.005
u'=  k4*(v-u)*(a+u^2) - k6*u
v'= k1*m - k6*u m'= b*m
```

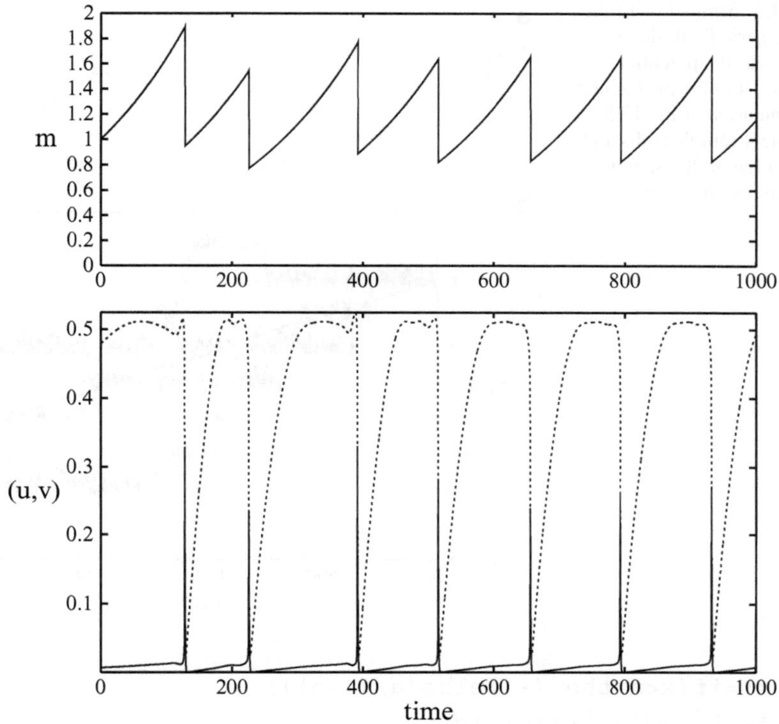

Fig. 17.2 Simulation of the discontinuous Tyson model. *Upper panel* shows the mass; *lower*, (u, v) (u is *dashed* and v *solid*)

```
global -1 {u-.2} {m=.5*m}
@ total=1000,meth=8,dt=.25
@ xhi=1000,ylo=0,yhi=2,yp=m
done
```

This file contains `global` which tells *XPPAUT* to look for an event and then if it occurs to do something; in this case, the mass is halved. In addition, there are lines starting with @ which tell *XPPAUT* various numerical and graphing parameters. These could all be done within the program, but this way, it is ready to run without any user interaction. Figure 17.2 shows a sample simulation that appears to approach a periodic solution after a sufficient amount of time is reached. Note that $m(t)$ jumps discontinuously at certain times due to u decreasing through $u = 0.2$.

We conclude this section with a simulation of the Brownian ratchet, Eq. 17.5. The ODE file for this is

```
# ratchet
wiener w
par a=.8,sig=.05,alpha=.1,beta=.1
```

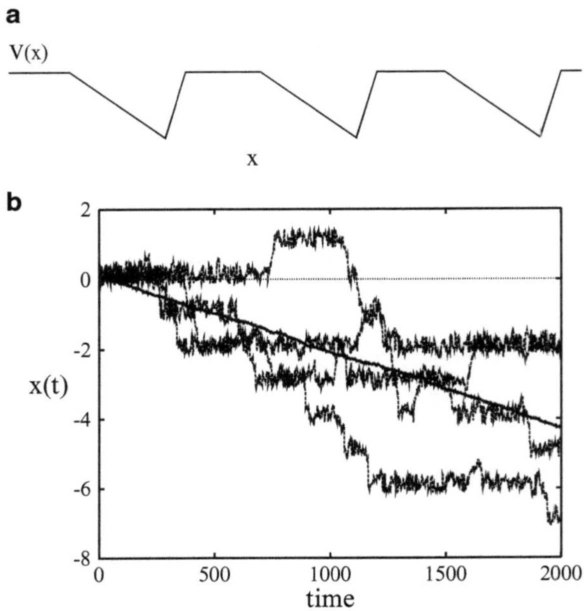

Fig. 17.3 Simulation of the infinite periodic flashing ratchet. (**a**) the potential fuction, $V(x)$ for the ratchet; (**b**) simulation of Eq. 17.5 showing 4 sample paths and the average of 200 sample paths (*thick olid line*)

```
f(x)=if(x<a)then(-1)else(a/(1-a))
x'=z*f(mod(x,1))+sig*w
markov z 2
{0} {alpha}
{beta} {0}
@ meth=euler,dt=.1,total=2000,njmp=10
@ xhi=2000,yhi=8,ylo=-8
done
```

The construction `wiener w` tells *XPPAUT* that this is a white noise process so that it has to be numerically scaled correctly. `markov` defines a multi-state Markov process and the following lines are the transition matrix. Diagonal entries are ignored since the transition probabilities will sum to 1 for each row. Figure 17.3 shows both the potential function (top) and sample paths as well as the mean in the bottom panel. The simulation shows that there is net movement in the direction towards the smaller slope.

17.3.1.1 Boundary Value problems

XPPAUT solves boundary value problems (BVPs) by either using shooting (within *XPPAUT*) or using AUTO. The latter has the advantage of being able to adaptively compute the solution to the BVP as parameters change; however, AUTO requires a good starting solution. The *XPPAUT* BVP solver is more forgiving and thus can be used as a first step. Consider Eq. 17.3. The ODE file for the corresponding BVP is

```
v'=p*(v^2+I-w)
w'=p*a*(b*v-w)
p'=0
b v'-vth
b v-vreset
b w -(w'+d)
par I=1,a=.05,b=1,d=1,vth=20,vreset=-1
@ total=1,dt=.01
init v=-1,w=1.89,p=15
done
```

The first three equations are the ODEs. Note that time is rescaled by the period, P. The next three lines (all starting with "b") define the boundary conditions. Here `v'` denotes the value of `v` at the end of the time interval. In *XPPAUT*, you run the boundary value solver on this and, if you have made a good guess, the solution will converge. Once you have found a solution, you can use AUTO to continue it as some parameter varies. For example, you might want to find the frequency of periodic solutions as a function of the applied current.

17.3.2 Functional Equations

Equation 17.1 can be solved by creating the following ODE file:

```
x'=-x+f(i-b*delay(x,tau))
f(x)=1/(1+exp(-x))
par i=5,b=10,tau=3
@ delay=10,total=50
aux xtau=delay(x,tau)
done
```

Here, I have added an additional plottable quantity, `xtau` which is the delayed version of $x(t)$. Figure 17.4 shos the time series as well as the projection of $x(t)$ and $x(t-\tau)$. Volterra integral equations are solvable. The simplest kind, convolution equations, as is Eq. 17.2, can be solved by creating the file:

```
x(t)=c*exp(-b*t)+a*int{exp(-(t-tau)^2)#x*(1-x)}
par a=2.05,tau=5,c=.1,b=8
@  maxstor=100000,total=150,dt=.005,trans=50
@ xlo=50,xhi=150,ylo=0
done
```

XPPAUT recognizes this is a Volterra integral equation due to the symbols `int` and `#` embedded in the code. These tell *XPPAUT* there is a convolution in time.

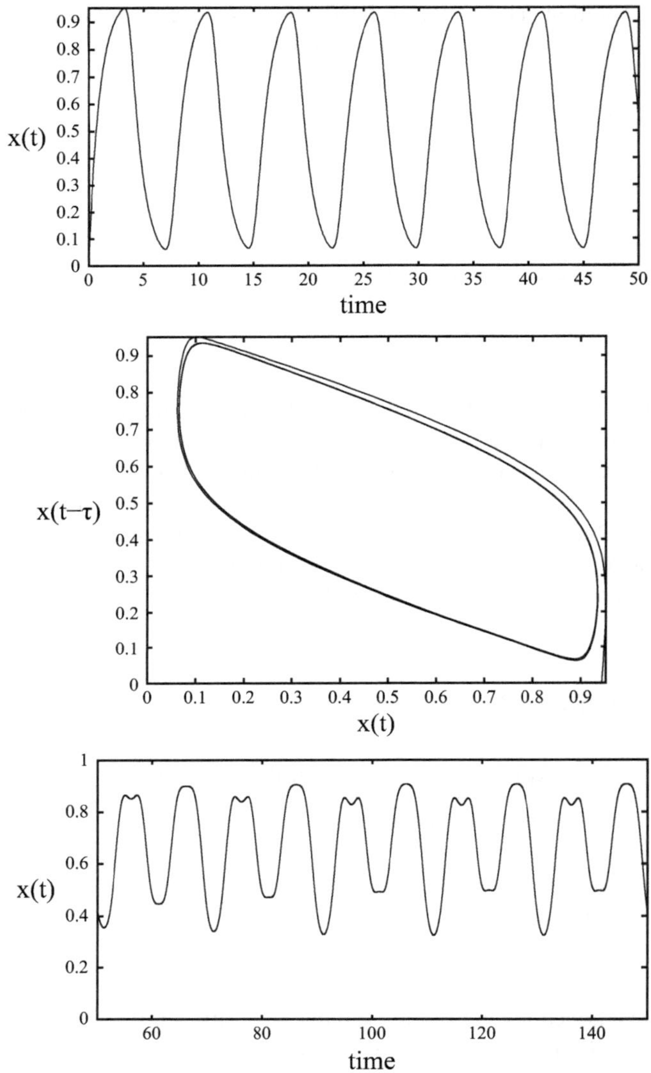

Fig. 17.4 Simulation of functional equations. *Upper panel* shows a solution to Eq. 17.1 as a time series. *Middle panel* is a plot of $x(t), x(t-\tau)$ showing convergence to a periodic solution. The *bottom panel* shows a complex periodic solution to (17.2)

The additional terms `c*exp(-b*t)` are included to push $x(t)$ away from the equilibrium point, $x = 0$ since there are no initial data for Volterra equations. The bottom of Fig. 17.4 shows a simulation of this equation – there is a complex but apparently periodic solution.

17.3.3 Spatial Problems

It is pretty simple to discretize Eqs. 17.6 or 17.7 in order to create a large system of ODEs. For example, the convolution in Eq. 17.7 can be evaluated using an approximate Riemann sum over a lage enough domain and the derivatives in Eq. 17.6 can be approximated by finite differences. The problem then is how to create the set of ODEs in an efficient manner; no one want to write down 100 or more equations. *XPPAUT* runs a pre-parser that will automate creation of repeated sets of ODEs. In addition, *XPPAUT* has some nice graphical methods for plotting the results. The ODE file for the Brusselator has the form

```
# brusselator PDE
u0=u200
u201=u1
v0=v200
v201=v1
par a=1,b=2.5,du=4,dv=80
u[1..200](0)=ran(1)
v[1..200](0)=1
f(u,v)=a-(b+1)*u+u^2*v
g(u,v)=b*u-u^2*v
u[1..200]'=f(u[j],v[j])+du*(u[j-1]+u[j+1]-2*u[j])
  v[1..200]'=g(u[j],v[j])+dv*(v[j-1]+v[j+1]-2*v[j])
@ total=200,maxstor=10000
@ meth=cvode,bandup=2,bandlo=2
done
```

The first four lines set the boundary conditions since this is a PDE. Here they are periodic. The line `u[1..200](0)=ran(1)` sets the initial data of the $u-$variable to be random. The rest of the file is similar to what we have already described except that the ODEs represent an abbreviated way to write 400 differential equations in just two lines of text. These will be expanded out to the 400 corresponding ODEs. Figure 17.5 left shows the evolution in time of solutions to the discretized PDE. Synchronous oscillatory solutions lose stability and become a spatially periodic pattern.

The neural network, (17.7) could be discretized and the convolutions approximated by sums. *XPPAUT* again makes this very easy to do using look-up tables and the `conv` operator:

```
f(u)=1/(1+exp(-u))
table w % 101 -50 50 exp(-abs(t)/5)/10
special k=conv(even,301,50,w,u0)
par a=20,b=5
@ total=50
u[0..10](0)=1
u[0..300]'=-u[j]+f(a*k([j])-b)
done
```

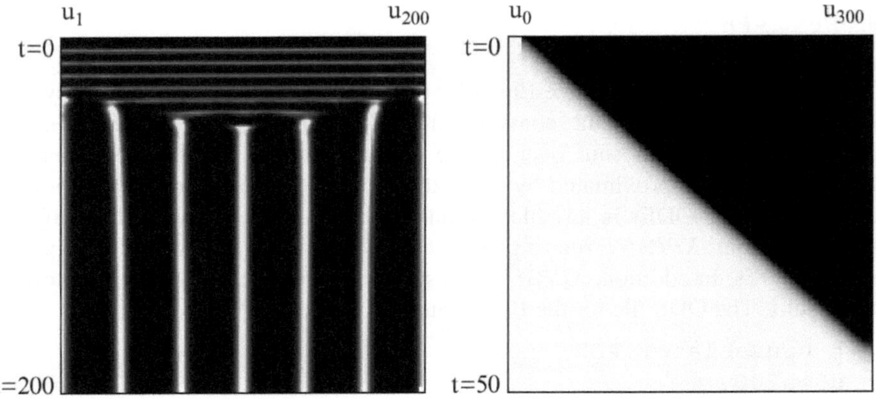

Fig. 17.5 Space-time plots of solutions to Eqs. 17.6 and 17.7 respectively. Time increases downward and space runs horizontally. *Grey* scale corresponds to the value of u in both graphs

In this simulation, the first 11 units are set to 1 and the rest to 0. Theory proves that these conditions result in a traveling wave in which the "1's" take over the "0's". Figure 17.5 right shows the simulation.

17.4 Inside the Box

XPPAUT provides a variety of numerical methods for solving differential equations, stochastic systems, delay equations, Volterra integral equations, and boundary-value problems (BVP). The numerical integrators are very robust and vary from the simple Euler method to the standard method for solving stiff differential equations, CVODE. The latter allows the user to specify whether the system is banded and thus can improve calculation speed by up to two orders of magnitude. Rather than pure simulation, *XPPAUT* should be regarded a tool for understanding what is going on in the system of equations. Tools for analyzing dynamical properties such as equilibria, basins of attraction, Lyapunov exponents, Poincare maps, embeddings, and weakly coupled oscillators are all available via menus. Some statistical analysis of simulations is possible; power spectra, mean and variance, correlation analysis and histograms are also included in the package. There is a very robust parameter fitting algorithm (Marquardt-Levenburg) which allows the user to find parameters and initial conditions which best approximate specified data.

As we saw in Sect. 17.3.1, *XPPAUT* contains a version of the continuation and bifurcation package, AUTO. This package allows the user to track equilibria, limit cycles, and solutions to boundary-value problems as parameters vary. With AUTO users can track the entire qualitative behavior of a differential equation.

For planar differential equations, *XPPAUT* includes algorithms for computing the nullclines and allows users to draw direction fields and flows. There are many others such as the ability to create animations whose elements are dependent on the dynamics. The interested user should download the package and read the 100+ pages of documentation.

17.5 Further Reading

The most complete source for information on *XPPAUT* is in Ermentrout (2002) which gives hundreds of examples and illustrations of the capabilities. Many books such as Fall et al. (2002) use *XPPAUT* extensively and contain tutorials on how to use the program. There are many web tutorials for the program as well; the most complete is found on the main *XPPAUT* web site.

References

Doedel E (1981) AUTO: a program for the automatic bifurcation analysis of autonomous systems. In: Proceedings of the tenth manitoba conference on numerical mathematics and computing, vol. I, Winnipeg, 1980, vol 30, pp 265–284

Ermentrout B (2002) Simulating, analyzing, and animating dynamical systems, software, environments, and tools, vol 14. Society for Industrial and Applied Mathematics (SIAM), Philadelphia, a guide to XPPAUT for researchers and students

Ermentrout GB, McLeod JB (1993) Existence and uniqueness of travelling waves for a neural network. Proc R Soc Edinb Sect A 123(3):461–478

Fall CP, Marland ES, Wagner JM, Tyson JJ (eds) (2002) Computational cell biology. Interdisciplinary applied mathematics, vol 20. Springer, New York

Laing CR, Frewen T, Kevrekidis IG (2010) Reduced models for binocular rivalry. J Comput Neurosci 28(3):459–476. doi:10.1007/s10827-010-0227-6, http://dx.doi.org/10.1007/s10827-010-0227-6

Chapter 18
NEST by Example: An Introduction to the Neural Simulation Tool NEST

Marc-Oliver Gewaltig, Abigail Morrison, and Hans Ekkehard Plesser

Abstract The neural simulation tool NEST can simulate small to very large networks of point-neurons or neurons with a few compartments. In this chapter, we show by example how models are programmed and simulated in NEST.

18.1 Introduction

NEST is a simulator for networks of point neurons, that is, neuron models that collapse the morphology (geometry) of dendrites, axons, and somata into either a single compartment or a small number of compartments (Gewaltig and Diesmann, 2007). This simplification is useful for questions about the dynamics of large neuronal networks with complex connectivity. In this text, we give a practical introduction to neural simulations with NEST. We describe how network models are defined and simulated, how simulations can be run in parallel, using multiple cores or computer clusters, and how parts of a model can be randomized.

The development of NEST started in 1994 under the name SYNOD to investigate the dynamics of a large cortex model, using integrate-and-fire neurons (Diesmann et al., 1995). At that time the only available simulators were NEURON (Hines

M.-O. Gewaltig (✉)
EPFL/Blue Brain Project, CH-1015 Lausanne, Switzerland
e-mail: marc-oliver.gewaltig@epfl.ch

A. Morrison
Functional Neural Circuits Group, Bernstein Center Freiburg, Hansastr. 9A, 79104 Freiburg, Germany
e-mail: morrison@bcf.uni-freiburg.de

H.E. Plesser
Department of Mathematical Sciences and Technology, Norwegian University of Life Sciences, PO Box 5003, 1432 Aas, Norway
e-mail: hans.ekkehard.plesser@umb.no

N. Le Novère (ed.), *Computational Systems Neurobiology*,
DOI 10.1007/978-94-007-3858-4_18,

and Carnevale, 1997) and GENESIS (Bower and Beeman, 1995), both focussing on morphologically detailed neuron models, often using data from microscopic reconstructions.

Since then, the simulator has been under constant development. In 2001, the Neural Simulation Technology Initiative was founded to disseminate our knowledge of neural simulation technology. The continuing research of the member institutions into algorithms for the simulation of large spiking networks has resulted in a number of influential publications. The algorithms and techniques developed are not only implemented in the NEST simulator, but have also found their way into other prominent simulation projects, most notably the NEURON simulator (for the Blue Brain Project: Migliore et al., 2006) and IBM's C2 simulator (Ananthanarayanan et al., 2009).

Today, in 2010, there are several simulators for large spiking networks to choose from (Brette et al. 2007), but NEST remains the best established simulator with the largest developer community.

A NEST simulation consists of three main components:

Nodes Nodes are all neurons, devices, and also sub-networks. Nodes have a dynamic state that changes over time and that can be influenced by incoming *events*.

Events Events are pieces of information of a particular type. The most common event is the spike event. Other event types are voltage events and current events.

Connections Connections are communication channels between nodes. Only if one node is connected to another node, can they exchange events. Connections are weighted, directed, and specific to one event type. Directed means that events can flow only in one direction. The node that sends the event is called *source* and the node that receives the event is called *target*. The weight determines how strongly an event will influence the target node. A second parameter, the *delay*, determines how long an event needs to travel from source to target.

In the next sections, we will illustrate how to use NEST, using examples with increasing complexity. Each of the examples is self-contained. We suggest that you try each example, by typing it into Python, line by line. Additionally, you can find all examples in your NEST distribution.

18.2 First Steps

We begin by starting Python. For interactive sessions, we recommend the IPython shell (Pérez and Granger, 2007). It is convenient, because you can edit the command line and access previously typed commands using the up and down keys. However, all examples in this chapter work equally well without IPython. For data analysis and visualization, we also recommend the Python packages Matplotlib (Hunter, 2007) and NumPy (Oliphant, 2006).

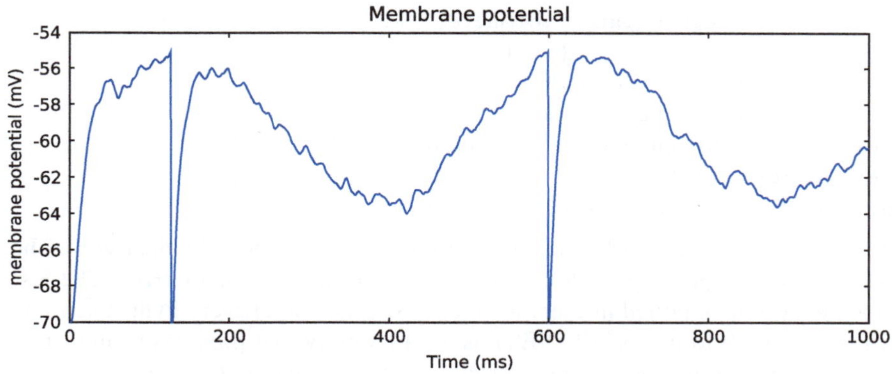

Fig. 18.1 Membrane potential of a neuron in response to an alternating current as well as random excitatory and inhibitory spike events. The membrane potential roughly follows the injected sine current. The small deviations from the sine curve are caused by the excitatory and inhibitory spikes that arrive at random times. Whenever the membrane potential reaches the firing threshold at –55 mV, the neuron spikes and the membrane potential is reset to –70 mV. In this example this happens twice: once at around 110 ms and again at about 600 ms

Our first simulation investigates the response of one integrate-and-fire neuron to an alternating current and Poisson spike trains from an excitatory and an inhibitory source. We record the membrane potential of the neuron to observe how the stimuli influence the neuron (see Fig. 18.1).

In this model, we inject a sine current with a frequency of 2 Hz and an amplitude of 100 pA into a neuron. At the same time, the neuron receives random spiking input from two sources known as Poisson generators. One Poisson generator represents a large population of excitatory neurons and the other a population of inhibitory neurons. The rate for each Poisson generator is set as the product of the assumed number of neurons in a population and their average firing rate.

The small network is simulated for 1,000 ms, after which the time course of the membrane potential during this period is plotted (see Fig. 18.1). For this, we use the pylab plotting routines of Python's Matplotlib package. The Python code for this small model is shown below.

```
import nest
import nest.voltage_trace
import pylab
neuron = nest.Create('iaf_neuron')
sine = nest.Create('ac_generator', 1,
                   {'amplitude': 100.0,
                    'frequency': 2.0})
noise= nest.Create('poisson_generator', 2,
                   [{'rate': 70000.0},
                    {'rate': 20000.0}])
```

```
11 voltmeter = nest.Create('voltmeter',1,
12                         {'withgid': True})
13 nest.Connect(sine, neuron)
14 nest.Connect(voltmeter, neuron)
15 nest.ConvergentConnect(noise, neuron, [1.0, −1.0], 1.0)
16 nest.Simulate(1000.0)
17 nest.voltage_trace.from_device(voltmeter)
```

We will now go through the simulation script and explain the individual steps. The first two lines **import** the modules nest and its sub-module **voltage_trace**. The nest module must be imported in every interactive session and in every Python script in which you wish to use NEST. NEST is a C++ library that provides a simulation kernel, many neuron and synapse models, and the simulation language interpreter SLI. The library which links the NEST simulation language interpreter to the Python interpreter is called PyNEST (Eppler et al., 2009).

Importing nest as shown above puts all NEST commands in the *namespace* nest. Consequently, all commands must be prefixed with the name of this namespace.

In line 4, we use the command **Create** to produce one node of the type iaf_neuron. As you see in lines 5, 8, and 11, **Create** is used for all node types. The first argument, 'iaf_neuron', is a string, denoting the type of node that you want to create. The second parameter of **Create** is an integer representing the number of nodes you want to create. Thus, whether you want one neuron or 100,000, you only need one call to **Create**. nest.Models() provides a list of all available node and connection models.

The third parameter is either a dictionary or a list of dictionaries, specifying the parameter settings for the created nodes. If only one dictionary is given, the same parameters are used for all created nodes. If an array of dictionaries is given, they are used in order and their number must match the number of created nodes. This variant of **Create** is used in lines 5, 8, and 11 to set the parameters for the Poisson noise generator, the sine generator (for the alternating current), and the voltmeter. All parameters of a model that are not set explicitly are initialized with default values. You can display them with nest.GetDefaults(*model_name*). Note that only the first parameter of **Create** is mandatory.

Create returns a list of integers, the global identifiers (or GID for short) of each node created. The GIDs are assigned in the order in which nodes are created. The first node is assigned GID 1, the second node GID 2, and so on.

In lines 13–15, the nodes are connected. First we connect the sine generator and the voltmeter to the neuron. The command **Connect** takes two or more arguments. The first argument is a list of source nodes. The second argument is a list of target nodes. **Connect** iterates these two lists and connects the corresponding pairs.

A node appears in the source position of **Connect** if it sends events to the target node. In our example, the sine generator is in the source position because it injects an alternating current into the neuron. The voltmeter is in the source position, because it polls the membrane potential of the neuron. Other devices may be in the target position, e.g., the spike detector which receives spike events from a neuron. If in doubt about the order, consult the documentation of the respective nodes, using

NEST's help system. For example, to read the documentation of the voltmeter you can type nest.help('voltmeter').

Next, we use the command **ConvergentConnect** to connect the two Poisson generators to the neuron. **ConvergentConnect** is used whenever a node is to be connected to many sources at the same time. The third and fourth arguments are the weights and delays, respectively. For both it is possible to supply either an array with values for each connection, or a single value to be used for all connections. For the weights, we supply an array, because we create one excitatory connection with weight 1.0 and one inhibitory connection with weight -1.0. For the delay, we supply only one value, consequently all the connections have the same delay.

After line 15, the network is ready. The following line calls the NEST function **Simulate** which runs the network for 1,000 ms. The function returns after the simulation is finished. Then, function **voltage_trace** is called to plot the membrane potential of the neuron. If you are running the script for the first time, you may have to tell Python to display the figure by typing pylab.show(). You should then see something similar to Fig. 18.1.

If you want to inspect how your network looks so far, you can print it using the command **PrintNetwork**():

```
>>>nest.PrintNetwork()
+-[0] root dim=[5]
   |
   +-[1] iaf_neuron
   +-[2] ac_generator
   +-[3]...[4] poisson_generator
   +-[5] voltmeter
```

If you run the example a second time, NEST will leave the existing nodes intact and will create a second instance for each node. To start a new NEST session without leaving Python, you can call nest.**ResetKernel**(). This function will erase the existing network so that you can start from scratch.

18.3 Example 1: A Sparsely Connected Recurrent Network

Next we discuss a model of activity dynamics in a local cortical network proposed by Brunel (2000). We only describe those parts of the model which are necessary to understand its NEST implementation. Please refer to the original paper for further details.

The local cortical network consists of two neuron populations: a population of N_E excitatory neurons and a population of N_I inhibitory neurons. To mimic the cortical ratio of 80% excitation and 20% inhibition, we assume that $N_E = 8,000$ and $N_I = 2,000$. Thus, our local network has a total of 10,000 neurons.

For both the excitatory and the inhibitory population, we use the same integrate-and-fire neuron model with current-based synapses. Incoming excitatory and in-

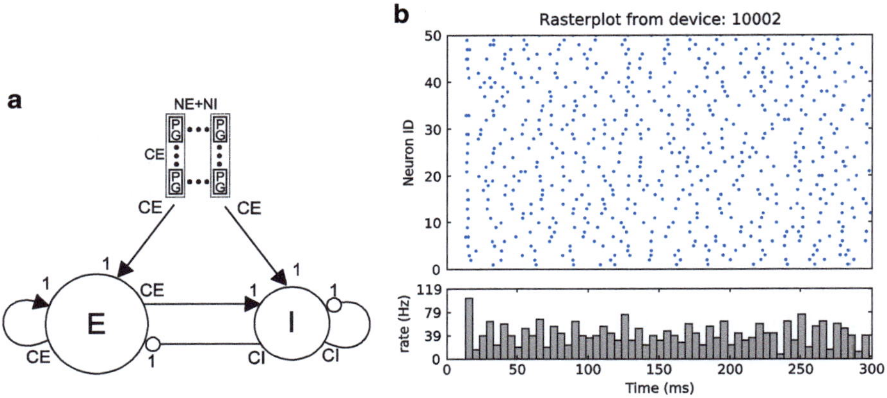

Fig. 18.2 Sketch of the network model proposed by Brunel (2000). (**a**) The network consists of three populations: N_E excitatory neurons (circle labeled *E*), N_I inhibitory neurons (circle labeled *I*), and a population of identical, independent Poisson processes (*PGs*) representing activity from outside the network. Arrows represent connections between the network nodes. Triangular arrow-heads represent excitatory and round arrow-heads represent inhibitory connections. The numbers at the start and end of each arrow indicate the multiplicity of the connection. See also Table 18.1. (**b**) Spiking activity of 50 neurons during the first 300 ms of simulated time as a raster plot. Time is shown on the x-axis, neuron id on the y-axis. Each dot corresponds to a spike of the respective neuron at the given time. The histogram below the raster plot shows the population rate of the network

hibitory spikes displace the membrane potential V_m by J_E and J_I, respectively. If V_m reaches the threshold value V_{th}, the membrane potential is reset to V_{reset}, a spike is sent with delay $D = 1.5$ ms to all post-synaptic neurons, and the neuron remains refractory for $\tau_{\text{rp}} = 2.0$ ms.

The neurons are mutually connected with a probability of 10%. Specifically, each neuron receives input from $C_E = 0.1 \cdot N_E$ excitatory and $C_I = 0.1 \cdot N_I$ inhibitory neurons (see Fig. 18.2a). The inhibitory synaptic weights J_I are chosen with respect to the excitatory synaptic weights J_E such that

$$J_I = -g \cdot J_E \tag{18.1}$$

with $g = 5.0$ in this example.

In addition to the sparse recurrent inputs from within the local network, each neuron receives excitatory input from a population of C_E randomly firing neurons, mimicking the input from the rest of cortex. The randomly firing population is modeled as C_E independent and identically distributed Poisson processes with rate ν_{ext}. Here, we set ν_{ext} to twice the rate ν_{th} that is needed to drive a neuron to threshold asymptotically. The details of the model are summarized in Tables 18.1 and 18.2.

Figure 18.2b shows a raster-plot of 50 excitatory neurons during the first 300 ms of simulated time. Time is shown along the x-axis, neuron id along the y-axis. At $t = 0$, all neurons are in the same state $V_m = 0$ and hence there is no spiking

Table 18.1 Summary of the network model, proposed by Brunel (2000)

A	Model summary
Populations	Three: excitatory, inhibitory, external input
Topology	—
Connectivity	Random convergent connections with probability $P = 0.1$ and fixed in-degree of $C_E = PN_E$ and $C_I = PN_I$.
Neuron model	Leaky integrate-and-fire, fixed voltage threshold, fixed absolute refractory time (voltage clamp)
Channel models	—
Synapse model	δ-current inputs (discontinuous voltage jumps)
Plasticity	—
Input	Independent fixed-rate Poisson spike trains to all neurons
Measurements	Spike activity

B		Populations
Name	Elements	Size
E	Iaf neuron	$N_{\mathrm{E}} = 4N_{\mathrm{I}}$
I	Iaf neuron	N_{I}
$\mathrm{E}_{\mathrm{ext}}$	Poisson generator	$C_E(N_{\mathrm{E}} + N_{\mathrm{I}})$

C			Connectivity
Name	Source	Target	Pattern
EE	E	E	Random convergent $C_{\mathrm{E}} \rightarrow 1$, weight J, delay D
IE	E	I	Random convergent $C_{\mathrm{E}} \rightarrow 1$, weight J, delay D
EI	I	E	Random convergent $C_{\mathrm{I}} \rightarrow 1$, weight $-gJ$, delay D
II	I	I	Random convergent $C_{\mathrm{I}} \rightarrow 1$, weight $-gJ$, delay D
Ext	$\mathrm{E}_{\mathrm{ext}}$	$\mathrm{E} \cup \mathrm{I}$	Non-overlapping $C_{\mathrm{E}} \rightarrow 1$, weight J, delay D

D	Neuron and synapse model
Name	Iaf neuron
Type	Leaky integrate-and-fire, δ-current input
Subthreshold dynamics	$\tau_m \dot{V}_m(t) = -V_m(t) + R_m I(t)$ if not refractory $(t > t^* + \tau_{\mathrm{rp}})$ $V_m(t) = V_{\mathrm{reset}}$ while refractory $(t^* < t \leq t^* + \tau_{\mathrm{rp}})$ $I(t) = \frac{\tau_m}{R_m} \sum_{\tilde{t}} w\delta(t - (\tilde{t} + D))$
Spiking	If $V_m(t-) < V_\theta \wedge V_m(t+) \geq V_\theta$ 1. set $t^* = t$ 2. emit spike with time-stamp t^*

E	Input
Type	Description
Poisson generators	Fixed rate ν_{ext}, C_{E} generators per neuron, each generator projects to one neuron

F Measurements
Spike activity as raster plots, rates and "global frequencies", no details given

Table 18.2 Summary of the network parameters for the model, proposed by Brunel (2000)

G	Network parameters
Parameter	Value
Number of excitatory neurons N_E	8,000
Number of inhibitory neurons N_I	2,000
Excitatory synapses per neuron C_E	800
Inhibitory synapses per neuron C_E	200

H	Neuron parameters
Parameter	Value
Membrane time constant τ_m	20 ms
Refractory period τ_{rp}	2 ms
Firing threshold V_{th}	20 mV
Membrane capatitance C_m	1 pF
Resting potential E_L	0 mV
Reset potential V_{reset}	10 mV
Excitatory PSP amplitude J_E	0.1 mV
Inhibitory PSP amplitude J_I	-0.5 mV
Synaptic delay D	1.5 ms
Background rate η	2.0

activity. The external stimulus rapidly drives the membrane potentials towards the threshold. Due to the random nature of the external stimulus, not all the neurons reach the threshold at the same time. After a few milliseconds, the neurons start to spike irregularly at roughly 40 spikes/s. In the original paper, this network state is called the *asynchronous irregular state* (Brunel, 2000).

18.3.1 NEST Implementation

We now show how this model is implemented in NEST. Along the way, we explain the required steps and NEST commands in more detail so that you can apply them to your own models.

The first three lines import NEST, a NEST module for raster-plots, and the plotting package pylab. We then assign the various model parameters to variables.

```
import nest
import nest.raster_plot
import pylab
g      = 5.0
eta    = 2.0
delay  = 1.5
tau_m  = 20.0
V_th   = 20.0
N_E = 8000
```

```
10 N_I = 2000
11 N_neurons = N_E+N_I
12 C_E      = N_E/10
13 C_I      = N_I/10
14 J_E   = 0.1
15 J_I   = -g*J_E
16 nu_ex    = eta*V_th/(J_E*C_E*tau_m)
17 p_rate = 1000.0*nu_ex*C_E
```

In line 16, we compute the firing rate nu_ex (ν_{ext}) of a neuron in the external population. We define nu_ex as the product of a constant eta times the threshold rate ν_{th}, i.e. the steady state firing rate which is needed to bring a neuron to threshold. The value of the scaling constant eta is defined in line 5.

In line 17, we compute the population rate of the whole external population. With C_E neurons, the population rate is simply the product nu_ex*C_E. The factor 1000.0 in the product changes the units from spikes per ms to spikes per second.

```
18 nest.SetKernelStatus({'print_time': True})
19 nest.SetDefaults('iaf_psc_delta',
20                  {'C_m': 1.0,
21                   'tau_m': tau_m,
22                   't_ref': 2.0,
23                   'E_L': 0.0,
24                   'V_th': V_th,
25                   'V_reset': 10.0})
```

Next, we prepare the simulation kernel of NEST (line 18). The command **SetKernelStatus** modifies parameters of the simulation kernel. The argument is a Python dictionary with *key*:*value* pairs. Here, we set the NEST kernel to print the progress of the simulation time during simulation.

In lines 19–25, we change the parameters of the neuron model we want to use from the built-in values to the defaults for our investigation. **SetDefaults** expects two parameters. The first is a string, naming the model for which the default parameters should be changed. Our neuron model for this simulation is the simplest integrate-and-fire model in NEST's repertoire: ' iaf_psc_delta '. The second parameter is a dictionary with parameters and their new values, entries separated by commas. All parameter values are taken from Brunel's paper (Brunel, 2000) and we insert them directly for brevity. Only the membrane time constant tau_m and the threshold potential V_th are read from variables, because these values are needed in several places.

```
26 nodes   = nest.Create('iaf_psc_delta',N_neurons)
27 nodes_E= nodes[:N_E]
28 nodes_I= nodes[N_E:]
29 nest.CopyModel('static_synapse_hom_wd',
30                'excitatory',
31                {'weight':J_E,
32                 'delay':delay})
```

```
33 nest.RandomConvergentConnect(nodes_E, nodes, C_E,
34                              model='excitatory')
35 nest.CopyModel('static_synapse_hom_wd',
36                'inhibitory',
37                {'weight':J_I,
38                 'delay':delay})
39 nest.RandomConvergentConnect(nodes_I, nodes, C_I,
40                              model='inhibitory')
```

In line 26 we create the neurons. **Create** returns a list of the global IDs which are consecutive numbers from 1 to N_neurons. We split this range into excitatory and inhibitory neurons. In line 27 we select the first N_E elements from the list nodes and assign them to the variable nodes_E. This list now holds the GIDs of the excitatory neurons.

Similarly, in line 28 we assign the range from position N_E to the end of the list to the variable nodes_I. This list now holds the GIDs of all inhibitory neurons. The selection is carried out using standard Python list commands. You may want to consult the Python documentation for more details.

The next two commands generate the connections to the excitatory neurons. In line 29, we create a new connection type 'excitatory' by copying the built-in connection type 'static_synapse_hom_wd' while changing its default values for *weight* and *delay*. The command **CopyModel** expects either two or three arguments: the name of an existing model, the name of the new model, and optionally a dictionary with the new default values of the new model.

The connection type 'static_synapse_hom_wd' uses the same values of weight and delay for all synapses. This saves memory for networks in which these values are identical for all connections. In Sect. 18.5 we use a different connection model to implement randomized weights and delays.

Having created and parameterized an appropriate synapse model, we draw the incoming excitatory connections for each neuron (line 33). The function **RandomConvergentConnect** expects four arguments: a list of source nodes, a list of target nodes, the number of connections to be drawn, and finally the type of connection to be used. **RandomConvergentConnect** iterates over the list of target nodes (nodes). For each target node it draws the required number of sources (C_E) from the list of source nodes (nodes_E) and connects these to the current target with the selected connection type (excitatory).

In lines 35–40 we repeat the same steps for the inhibitory connections: we create a new connection type and draw the incoming inhibitory connections for all neurons.

Next, we create and connect the external population and some devices to measure the spiking activity in the network.

```
41 noise=nest.Create('poisson_generator',1,{'rate': p_rate})
42 nest.DivergentConnect(noise,nodes,model='excitatory')
43 spikes=nest.Create('spike_detector',2,
44                    [{'label': 'brunel-py-ex'},
45                     {'label': 'brunel-py-in'}])
```

```
46 spikes_E=spikes[:1]
47 spikes_I=spikes[1:]
48 N_rec   = 50
49 nest.ConvergentConnect(nodes_E[:N_rec],spikes_E)
50 nest.ConvergentConnect(nodes_I[:N_rec],spikes_I)
```

In line 41, we create a device known as a poisson_generator, which produces a spike train governed by a Poisson process at a given rate. We use the third parameter of **Create** to initialize the rate of the Poisson process to the population rate p_rate which we previously computed in line 17.

If a Poisson generator is connected to n targets, it generates n independent and identically distributed spike trains. Thus, we only need one generator to model an entire population of randomly firing neurons.

In the next line (42), we use **DivergentConnect** to connect the Poisson generator to all nodes of the local network. Since these connections are excitatory, we use the 'excitatory' connection type.

We have now implemented the Brunel model and we could start to simulate it. However, we won't see any results unless we record something from the network. To observe how the neurons in the recurrent network respond to the random spikes from the external population, we create two spike detectors in line 43; one for the excitatory neurons and one for the inhibitory neurons. By default, each detector writes its spikes into a file whose name is automatically generated from the device type and its global id. We use the third argument of **Create** to give each spike detector a 'label', which will be part of the name of the output file written by the detector. Since two devices are created, we supply a list of dictionaries.

In line 46, we store the GID of the first spike detector in a one-element list and assign it to the variable spikes_E. In the next line, we do the same for the second spike detector that is dedicated to the inhibitory population.

Our network consists of 10,000 neurons, all of which have the same activity statistics due to the random connectivity. Thus, it suffices to record from a representative sample of neurons, rather than from the entire network. Here, we choose to record from 50 neurons and assign this number to the variable N_rec. We then connect the first 50 excitatory neurons to their spike detector. Again, we use standard Python list operations to select N_rec neurons from the list of all excitatory nodes. Alternatively, we could select 50 neurons at random, but since the neuron order has no meaning in this model, the two approaches yield qualitatively the same results. Finally, we repeat this step for the inhibitory neurons.

Now everything is set up and we can run the simulation.

```
51 simtime=300
52 nest.Simulate(simtime)
53 events = nest.GetStatus(spikes,'n_events')
54 rate_ex= events[0]/simtime*1000.0/N_rec
55 print "Excitatory rate   : %.2f 1/s" % rate_ex
56 rate_in= events[1]/simtime*1000.0/N_rec
57 print "Inhibitory rate   : %.2f 1/s" % rate_in
58 nest.raster_plot.from_device(spikes_E, hist=True)
```

In line 51, we select a simulation time of 300 ms and assign it to a variable. Next, we call the NEST command **Simulate** to run the simulation for 300 ms. During simulation, the Poisson generators send spikes into the network and cause the neurons to fire. The spike detectors receive spikes from the neurons and write them to a file, or to memory.

When the function returns, the simulation time has progressed by 300 ms. You can call **Simulate** as often as you like and with different arguments. NEST will resume the simulation at the point where it was last stopped. Thus, you can partition your simulation time into small epochs to regularly inspect the progress of your model.

After the simulation is finished, we compute the firing rate of the excitatory neurons (line 54) and the inhibitory neurons (line 56). Finally, we call the NEST function **raster_plot** to produce the raster plot shown in Fig. 18.2b. **raster_plot** has two modes. **raster_plot**.from_device expects the global ID of a spike detector. **raster_plot**. from_file expects the name of a data file. This is useful to plot data without repeating a simulation.

18.4 Parallel Simulation

Large network models often require too much time or computer memory to be conveniently simulated on a single computer. For example, if we increase the number of neurons in the previous model to 100,000, there will be a total of 10^9 connections, which won't fit into the memory of most computers. Similarly, if we use plastic synapses (see Sect. 18.7) and run the model for minutes or hours of simulated time, the execution times become uncomfortably long.

To address this issue, NEST has two modes of parallelization: multi-threading and distribution. Multi-threaded and distributed simulation can be used in isolation or in combination (Plesser et al., 2007), and both modes allow you to connect and run networks more quickly than in the serial case.

Multi-threading means that NEST uses all available processors or cores of the computer. Today, most desktop computers and even laptops have at least two processing cores. Thus, you can use NEST's multi-threaded mode to make your simulations execute more quickly whilst still maintaining the convenience of interactive sessions. Since a given computer has a fixed memory size, multi-threaded simulation can only reduce execution times. It cannot solve the problem that large models exhaust the computer's memory.

Distribution means that NEST uses many computers in a network or computer cluster. Since each computer contributes memory, distributed simulation allows you to simulate models that are too large for a single computer. However, in distributed mode it is not currently possible to use NEST interactively.

In most cases, writing a simulation script to be run in parallel is as easy as writing one to be executed on a single processor. Only minimal changes are required, as described below, and you can ignore the fact that the simulation is actually executed by more than one core or computer. However, in some cases your knowledge about the distributed nature of the simulation can help you improve efficiency even further. For example, in the distributed mode, all computers execute the same simulation script. We can improve performance if the script running on a specific computer only tries to execute commands relating to nodes that are represented on the same computer. An example of this technique is shown below in Sect. 18.6.

To switch NEST into multi-threaded mode, you only have to add one line to your simulation script:

```
nest.SetKernelStatus({'local_num_threads': n})
```

Here, n is the number of threads you want to use. It is important that you set the number of threads *before* you create any nodes. If you try to change the number of threads after nodes were created, NEST will issue an error.

A good choice for the number of threads is the number of cores or processors on your computer. If your processor supports hyperthreading, you can select an even higher number of threads.

The distributed mode of NEST is particularly useful for large simulations for which not only the processing speed, but also the memory of a single computer are insufficient. The distributed mode of NEST uses the Message Passing Interface (MPI, MPI Forum 2009), a library that must be installed on your computer network when you install NEST. For details, please refer to NEST's website at www.nest-initiative.org.

The distributed mode of NEST is also easy to use. All you need to do is start NEST with the MPI command mpirun:

```
mpirun -np m python script.py
```

where m is the number of MPI processes that should be started. One sensible choice for m is the total number of cores available on the cluster. Another reasonable choice is the number of physically distinct machines, utilizing their cores through multithreading as described above. This can be useful on clusters of multi-core computers.

In NEST, processes and threads are both mapped to *virtual processes* (Plesser et al., 2007). If a simulation is started with m MPI processes and n threads on each process, then there are m×n virtual processes. You can obtain the number of virtual processes in a simulation with

```
nest.GetKernelStatus('total_num_virtual_procs')
```

The virtual process concept is reflected in the labeling of output files. For example, the data files for the excitatory spikes produced by the network discussed here follow the form brunel−py−ex−x−y.gdf, where x is the id of the data recording node and y is the id of the virtual process.

18.5 Randomness in NEST

NEST has built-in random number sources that can be used for tasks such as randomizing spike trains or network connectivity. In this section, we discuss some of the issues related to the use of random numbers in parallel simulations. In Sect. 18.6, we illustrate how to randomize parameters in a network.

Let us first consider the case that a simulation script does not explicitly generate random numbers. In this case, NEST produces identical simulation results for a given number of virtual processes, irrespective of how the virtual processes are partitioned into threads and MPI processes. The only difference between the output of two simulations with different configurations of threads and processes resulting in the same number of virtual processes is the result of query commands such as **GetStatus**. These commands gather data over threads on the local machine, but not over remote machines.

In the case that random numbers are explictly generated in the simulation script, more care must be taken to produce results that are independent of the parallel configuration. Consider, for example, a simulation where two threads have to draw a random number from a single random number generator. Since only one thread can access the random number generator at a time, the outcome of the simulation will depend on the access order.

Ideally, all random numbers in a simulation should come from a single source. In a serial simulation this is trivial to implement, but in parallel simulations this would require shipping a large number of random numbers from a central random number generator (RNG) to all processes. This is impractical. Therefore, NEST uses one independent random number generator on each virtual process. Not all random number generators can be used in parallel simulations, because many cannot reliably produce uncorrelated parallel streams. Fortunately, recent years have seen great progress in RNG research and there is a range of random number generators that can be used with great fidelity in parallel applications.

Based on this knowledge, each virtual process (VP) in NEST has its own RNG. Numbers from these RNGs are used to

- Choose random convergent connections
- Create random spike trains (e.g. poisson_generator) or random currents (e.g. noise_generator).

In order to randomize model parameters in a PyNEST script, it is convenient to use the random number generators provided by NumPy. To ensure consistent results for a given number of virtual processes, each virtual process should use a separate Python RNG. Thus, in a simulation running on N_{vp} virtual processes, there should be $2N_{\mathrm{vp}} + 1$ RNGs in total:

- The global NEST RNG;
- One RNG per VP in NEST;
- One RNG per VP in Python.

We need to provide separate seed values for each of these generators. Modern random number generators work equally well for all seed values. We thus suggest the following approach to choosing seeds: For each simulation run, choose a master seed msd and seed the RNGs with seeds msd, $msd + 1$, $\dots msd + 2N_{\mathrm{vp}}$. Any two master seeds must differ by at least $2N_{\mathrm{vp}} + 1$ to avoid correlations between simulations.

By default, NEST uses Knuth's lagged Fibonacci RNG, which has the nice property that each seed value provides a different sequence of some 2^{70} random numbers (Knuth, 1998, Ch. 3.6). Python uses the Mersenne Twister MT19937 generator (Matsumoto and Nishimura, 1998), which provides no explicit guarantees, but given the enormous state space of this generator it appears astronomically unlikely that neighboring integer seeds would yield overlapping number sequences. For a recent overview of RNGs, see L'Ecuyer and Simard (2007). For general introductions to random number generation, see Gentle (2003), Knuth (1998, Ch. 3), or Plesser (2010).

18.6 Example 2: Randomizing Neurons and Synapses

Let us now consider how to randomize some neuron and synapse parameters in the sparsely connected network model introduced in Sect. 18.3. We shall

- Explicitly seed the random number generators;
- Randomize the initial membrane potential of all neurons;
- Randomize the weights of the recurrent excitatory connections.

We discuss here only those parts of the model script that differ from the script discussed in Sect. 18.3.1; the complete script brunel2000–rand.py is part of the NEST examples.

We begin by importing the NumPy package to get access to its random generator functions:

```
import numpy
```

After line 1 of the original script (cf. p. 545), we insert code to seed the random number generators:

```
r1  msd = 1000    # master seed
r2  msdrange1 = range(msd, msd+n_vp)
r3  n_vp = nest.GetKernelStatus('total_num_virtual_procs')
r4  pyrngs = [numpy.random.RandomState(s) for s in msdrange1]
r5  msdrange2 = range(msd+n_vp+1, msd+1+2*n_vp)
r6  nest.SetKernelStatus({'grng_seed': msd+n_vp,
r7                        'rng_seeds': msdrange2})
```

We first define the master seed msd and then obtain the number of virtual processes n_vp. On line r4 we then create a list of n_vp NumPy random number generators with seeds msd, msd+1, ... msd+n_vp−1. The next two lines set new seeds for the built-in NEST RNGs: the global RNG is seeded with msd+n_vp, the per-virtual-process RNGs with msd+n_vp+1, ..., msd+2∗n_vp. Note that the seeds for the per-virtual-process RNGs must always be passed as a list, even in a serial simulation.

After creating the neurons as before, we insert the following code after line 28 to randomize the membrane potential of all neurons:

```
r8  node_info   = nest.GetStatus(nodes, ['global_id','vp','local'])
r9  local_nodes = [(gid,vp) for gid,vp,islocal \
r10                    in node_info if islocal]
r11 for gid,vp in local_nodes:
r12   nest.SetStatus([gid], {'V_m': pyrngs[vp].uniform(-V_th,V_th)})
```

In this code, we meet the concept of *local* nodes for the first time (Plesser et al., 2007). In serial and multi-threaded simulations, all nodes are local. In an MPI-based simulation with m MPI processes, each MPI process represents and is responsible for updating (approximately) $1/m$-th of all nodes – these are the local nodes for each process. In line r8 we obtain an information triplet for each node: the global id, the id of the virtual process updating the neuron and a boolean value indicating whether the node is local. We then use a list comprehension to create a list of gid and vp tuples for all local nodes. The **for**-loop then iterates over this list and draws for each node a membrane potential value uniformly distributed in $[-V_{\text{th}}, V_{\text{th}})$, i.e., $[-20\text{mV}, 20\text{mV})$. We draw the initial membrane potential for each node from the NumPy RNG assigned to the virtual process vp responsible for updating that node.

As the next step, we create excitatory recurrent connections with the same connection rule as in the original script, but with randomized weights. To this end, we replace the code on lines 29–33 of the original script with

```
r13 nest.CopyModel('static_synapse', 'excitatory')
r14 for tgt_gid, tgt_vp in local_nodes:
r15   weights = pyrngs[tgt_vp].uniform(0.5*J_E, 1.5*J_E, C_E)
r16   nest.RandomConvergentConnect(nodes_E, [tgt_gid], C_E,
r17                                weight=list(weights),
r18                                delay=delay,
r19                                model='excitatory')
```

The first difference to the original is that we base the excitatory synapse model on the built-in static_synapse model instead of static_synapse_hom_wd, as the latter requires equal weights (and delays) for all synapses. The second difference to the original script is the way we connect the nodes.

For each local target, we draw an array of C_E random weights (line r15) uniformly distributed on $[0.5 \times J_E, 1.5 \times J_E)$. Using these weights and delays, we then call **RandomConvergentConnect** to create connections from C_E randomly chosen nodes in nodes_E. We need this loop over the set of local targets, because in parallel simulations, connection information is managed by the virtual process updating the target node (Morrison et al., 2005; Plesser et al., 2007). Thus, weights, delays and possibly other connection parameters can only be set by the MPI process to which the target node belongs. Drawing the weights from the RNG for the virtual process

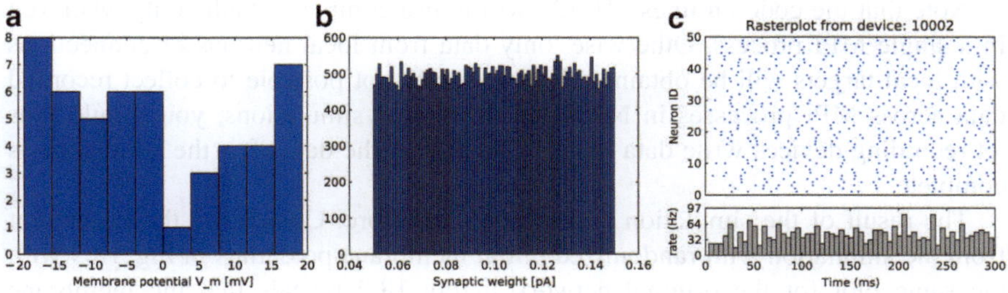

Fig. 18.3 (**a**) Distribution of membrane potentials V_m of 50 excitatory neurons after random initialization. (**b**) Distribution of weights of randomized weights of approximately 50,000 recurrent connections originating from 50 excitatory neurons. (**c**) Spiking activity of 50 excitatory neurons during the first 300 ms of network simulation; compare with the corresponding diagram for the same network without randomization of initial membrane potentials and weights in Fig. 18.2

updating the target node ensures that we will set the same weights independent of how many MPI processes underly a given number of virtual processes. Note that the corresponding randomization of weights for random divergent connections is more complicated; please see the online documentation on NEST's website for details.

Two remarks about the parameters on line r18: first, NEST functions mostly only accept Python lists as arguments, not NumPy arrays. We thus need to convert the array returned by uniform() into a list before passing. Second, when passing weight to **RandomConvergentConnect**, we also have to pass delay explicitly.

Before starting our simulation, we want to visualize the randomized initial membrane potentials and weights. To this end, we insert the following code just before we start the simulation:

```
r20 pylab.figure()
r21 V_E = nest.GetStatus(nodes_E[:N_rec], 'V_m')
r22 pylab.hist(V_E, bins=10)
r23 pylab.figure()
r24 ex_conns = nest.FindConnections(nodes_E[:N_rec],
r25                                 synapse_type='excitatory')
r26 w = nest.GetStatus(ex_conns, 'weight')
r27 pylab.hist(w, bins=100)
```

Line r21 retrieves the membrane potentials of all 50 recorded neurons. The data is then displayed as a histogram with 10 bins, see Fig. 18.3. Line r24 finds all connections that

- Have one of the 50 recorded excitatory neurons as source;
- Have any local node as target;
- And are of type excitatory .

In line r26, we then use GetStatus() to obtain the weights of these connections. Running the script in a single MPI process, we record approximately 50,000 weights, which we display in a histogram with 100 bins as shown in Fig. 18.3.

Note that the code on lines r21–r26 will return complete results only when run in a single MPI process. Otherwise, only data from local neurons or connections with local targets will be obtained. It is currently not possible to collect recorded data across MPI processes in NEST. In distributed simulations, you should thus let recording devices write data to files and collect the data after the simulation is complete.

The result of the simulation is displayed as before. Comparing the raster plot from the simulation with randomized initial membrane potentials in Fig. 18.3 with the same plot for the original network in Fig. 18.2 reveals that the membrane potential randomization has prevented the synchronous onset of activity in the network.

As a final point, we make a slight improvement to the rate computation on lines 54–57 of the original script. Spike detectors count only spikes from neurons on the local MPI process. Thus, the original computation is correct only for a single MPI process. To obtain meaningful results when simulating on several MPI processes, we count how many of the N_rec recorded nodes are local and use that number to compute the rates:

```
r28 N_rec_local_E = sum(nest.GetStatus(nodes_E[:N_rec], 'local'))
r29 rate_ex= events[0]/simtime*1000.0/N_rec_local_E
```

Each MPI process then reports the rate of activity of its locally recorded nodes.

18.7 Example 3: Plastic Networks

NEST provides synapse models with a variety of short-term and long-term dynamics. To illustrate this, we extend the sparsely connected network introduced in Sect. 18.3 with randomized synaptic weights as described in Sect. 18.5 to incorporate spike-timing dependent plasticity (Bi and Poo, 1998) at its recurrent excitatory-excitatory synapses.

```
node_E_info = nest.GetStatus(nodes_E, ['local','global_id','vp'])
node_I_info = nest.GetStatus(nodes_I, ['local','global_id','vp'])
local_E_nodes = [(gid,vp) for islocal,gid,vp
                                       in node_E_info if islocal]
local_I_nodes = [(gid,vp) for islocal,gid,vp
                                       in node_I_info if islocal]
for gid,vp in local_E_nodes + local_I_nodes:
  nest.SetStatus([gid],
                   {'V_m': pyrngs[vp].uniform(-V_th,0.95*V_th)})
```

As in the previous section, we first acquire information about the locality of each node. Here, we gather this information separately for the excitatory and inhibitory populations, as we will be making different types of connection to them. The initial membrane potentials are randomized as before.

We then generate a plastic synapse model for the excitatory-excitatory connections and a static synapse model for the excitatory-inhibitory connections:

```
nest.CopyModel('stdp_synapse_hom',
               'excitatory-plastic',
               {'alpha':STDP_alpha,
                'Wmax':STDP_Wmax})
nest.CopyModel('static_synapse', 'excitatory-static')
```

Here, we set the parameters alpha and Wmax of the synapse model but use the default settings for all its other parameters. Finally, we use these synapse models to create plastic and static excitatory connections with randomized initial weights:

```
for tgt_gid, tgt_vp in local_E_nodes:
  weights = list(pyrngs[tgt_vp].uniform(0.5*J_E, 1.5*J_E, C_E))
  nest.RandomConvergentConnect(nodes_E, [tgt_gid], C_E,
                        weight = weights, delay = delay
                        model='excitatory-plastic')
for tgt_gid, tgt_vp in local_I_nodes:
  weights = list(pyrngs[tgt_vp].uniform(0.5*J_E, 1.5*J_E, C_E))
  nest.RandomConvergentConnect(nodes_E, [tgt_gid], C_E,
                        weight = weights, delay = delay,
                        model='excitatory-static')
```

After a period of simulation, we can access the plastic synaptic weights for analysis:

```
w = nest.GetStatus(nest.FindConnections(nodes_E[:N_rec],
                            synapse_type='excitatory-plastic'),
                            'weight')
```

Plotting a histogram of the synaptic weights reveals that the initial uniform distribution has begun to soften (see Fig. 18.4). Simulation for a longer period results in an approximately Gaussian distribution of weights.

18.8 Example 4: Classes and Automatization Techniques

So far, we have encouraged you to try our examples line-by line. This is possible in interactive sessions, but if you want to run a simulation several times, possibly with different parameters, it is more practical to write a script that can be loaded from Python.

Python offers a number of mechanisms to structure and organize not only your simulations, but also your simulation data. The first step is to re-write a model as a *class*. In Python, and other object-oriented languages, a class is a data structure which groups data and functions into a single entity. In our case, data are the different parameters of a model and functions are what you can do with a model. Classes allow you to solve various common problems in simulations:

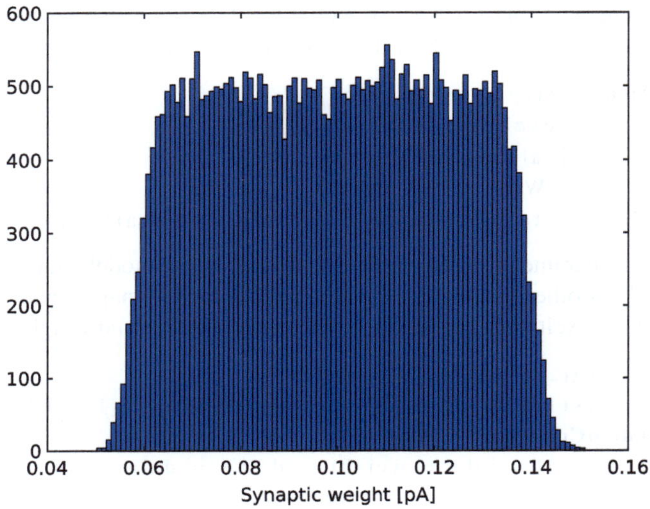

Fig. 18.4 Distribution of synaptic weights in the plastic network simulation after 300 ms

Parameter sets Classes are data structures and so are ideally suited to hold the parameter set for a model. Class inheritance allows you to modify one, few, or all parameters while maintaining the relation to the original model.

Model variations Often, we want to change minor aspects of a model. For example, in one version we have homogeneous connections and in another we want randomized weights. Again, we can use class inheritance to express both cases while maintaining the conceptual relation between the models.

Data management Often, we run simulations with different parameters, or other variations and forget to record which data file belonged to which simulation. Python's class mechanisms provide a simple solution.

We organize the model from Sect. 18.3 into a class, by realizing that each simulation has five steps which can be factored into separate functions:

1. Define all independent parameters of the model. Independent parameters are those that have concrete values which do not depend on any other parameter. For example, in the Brunel model, the parameter g is an independent parameter.
2. Compute all dependent parameters of the model. These are all parameters or variables that have to be computed from other quantities (e.g. the total number of neurons).
3. Create all nodes (neurons, devices, etc.)
4. Connect the nodes.
5. Simulate the model.

We translate these steps into a simple class layout that will fit most models:

```
class Model(object):
    """Model description."""
    # Define all independent variables.

    def __init__(self):
        """Initialize the simulation, setup data directory"""
    def calibrate(self):
        """Compute all dependent variables"""
    def build(self):
        """Create all nodes"""
    def connect(self):
        """Connect all nodes"""
    def run(self,simtime):
        """Build, connect and simulate the model"""
```

In the following, we illustrate how to fit the model from Sect. 18.3 into this scaffold. The complete and commented listing can be found in your NEST distribution.

```
class Brunel2000(object):
    """
    Implementation of the sparsely connected random network,
    described by Brunel (2000) J. Comp. Neurosci.
    Parameters are chosen for the asynchronous irregular
    state (AI).
    """
    g      = 5.0
    eta    = 2.0
    delay  = 1.5
    tau_m  = 20.0
    V_th   = 20.0
    N_E    = 8000
    N_I    = 2000
    J_E    = 0.1
    N_rec   = 50
    threads=2          # Number of threads for parallel simulation
    built=False        # True, if build() was called
    connected=False# True, if connect() was called
    # more definitions follow...
```

A Python class is defined by the keyword **class** followed by the class name, Brunel2000 in this example. The parameter object indicates that our class is a subclass of a general Python Object. After the colon, we can supply a documentation string, encased in triple quotes, which will be printed if we type help(Brunel2000). After the documentation string, we define all independent parameters of the model as well as some global variables for our simulation. We also introduce two Boolean variables built and connected to ensure that the functions build() and connect() are executed exactly once.

Next, we define the class functions. Each function has at least the parameter self, which is a reference to the current class object. It is used to access the functions and variables of the object.

The first function we look at is also the first one that is called for every class object. It has the somewhat cryptic name __init__():

```
c21     def __init__(self):
c22         """
c23         Initialize an object of this class.
c24         """
c25         self.name=self.__class__.__name__
c26         self.data_path=self.name+'/'
c27         nest.ResetKernel()
c28         if not os.path.exists(self.data_path):
c29             os.makedirs(self.data_path)
c30         print "Writing data to: "+self.data_path
c31         nest.SetKernelStatus({'data_path': self.data_path})
```

__init__() is automatically called by Python whenever a new object of a class is created and before any other class function is called. We use it to initialize the NEST simulation kernel and to set up a directory where the simulation data will be stored.

The general idea is this: each simulation with a specific parameter set gets its own Python class. We can then use the class name to define the name of a data directory where all simulation data are stored.

In Python it is possible to read out the name of a class from an object. This is done in line c25. Don't worry about the many underscores, they tell us that these names are provided by Python. In the next line, we assign the class name plus a trailing slash to the new object variable data_path. Note how all class variables are prefixed with self.

Next we reset the NEST simulation kernel to remove any leftovers from previous simulations.

The following two lines use functions from the Python library os which provides functions related to the operating system. In line c28 we check whether a directory with the same name as the class already exists. If not, we create a new directory with this name. Finally, we set the data path property of the simulation kernel. All recording devices use this location to store their data. This does not mean that this directory is automatically used for any other Python output functions. However, since we have stored the data path in an object variable, we can use it whenever we want to write data to file.

The other class functions are quite straightforward. Brunel2000.build() accumulates all commands that relate to creating nodes. The only addition is a piece of code that checks whether the nodes were already created:

```
    def build(self):
        """
        Create all nodes, used in the model.
        """
        if self.built: return

        # remaining code to create nodes
        self.built=True
```

The last line in this function sets the variable self.built to True so that other functions know that all nodes were created.

In function Brunel2000.connect() we first ensure that all nodes are created before we attempt to draw any connection:

```
    def connect(self):
        """
        Connect all nodes in the model.
        """
        if self.connected: return
        if not self.built:
            self.build()
        # remaining connection code
        self.connected=True
```

Again, the last line sets a variable, telling other functions that the connections were drawn successfully.

Brunel2000.built and Brunel2000.connected are state variables that help you to make dependencies between functions explicit and to enforce an order in which certain functions are called. The main function Brunel2000.run() relies on both state variables to build and connect the network:

```
    def run(self, simtime=300):
        """
        Simulate the model for simtime milliseconds and print the
        firing rates of the network during this period.
        """
        if not self.connected:
            self.connect()
        nest.Simulate(simtime)
        # more code, e.g. to compute and print rates
```

In order to use the class, we have to load the file with the class definition and then create an object of the class:

```
net=Brunel2000()
net.run(500)
```

Finally, we demonstrate how to use Python's class inheritance to express different parameter configurations and versions of a model. In the following listing, we derive a new class that simulates a network where excitation and inhibition are exactly balanced, i.e. $g = 4$:

```
class Brunel_balanced(Brunel2000):
    """
    Exact balance of excitation and inhibition
    """
    g=4
```

Class Brunel_balanced is defined with class Brunel2000 as parameter. This means the new class inherits all parameters and functions from class Brunel2000. Then, we redefine the value of the parameter g. When we create an object of this class, it will create its new data directory.

We can use the same mechanism to implement alternative version of the model. For example, instead of re-implementing the model with randomized connection weights, we can use inheritance to change just the way nodes are connected:

```
class Brunel_randomized(Brunel2000):
    """
    Like Brunel2000, but with randomized connection weights.
    """
    def connect(self):
        """
        Connect nodes with randomized weights.
        """
        # Code for randomized connections follows
```

Thus, using inheritance, we can easily keep track of different parameter sets and model versions and their associated simulation data. Moreover, since we keep all alternative versions, we also have a simple versioning system that only depends on Python features, rather than on third party tools or libraries. The full implementation of the model using classes can be found in the examples directory of your NEST distribution.

18.9 How to Continue from Here

In this chapter we have presented a step-by-step introduction to NEST, using concrete examples. The simulation scripts and more examples are part of the examples included in the NEST distribution. Information about individual PyNEST functions can be obtained with Python's help() function. For example:

```
>>>help(nest.Connect)

Connect(pre, post, params=None, delay=None, model='static_synapse')
    Make one-to-one connections of type model between the nodes
    in pre and the nodes in post. pre and post have to be lists
    of the same length. If params is given (as dictionary or
    list of dictionaries), they are used as parameters for the
    connections. If params is given as a single float or as
    list of floats, it is used as weight(s), in which case delay
    also has to be given as float or as list of floats.
```

To learn more about NEST's node and synapse types, you can access NEST's help system, using the PyNEST command. NEST's online help still uses a lot of SLI syntax, NEST's native simulation language. However the general information is also valid for PyNEST.

Help and advice can also be found on NEST's user mailing list where developers and users exchange their experience, problems and ideas. And finally, we encourage you to visit the web site of the NEST Initiative at www.nest-initiative.org.

Acknowledgements AM partially funded by BMBF grant 01GQ0420 to BCCN Freiburg, Helmholtz Alliance on Systems Biology (Germany), Neurex, and the Junior Professor Program of Baden-Württemberg. HEP partially supported by RCN grant 178892/V30 eNeuro.

Version Information

The examples in this chapter were tested with the following versions.
NEST: 1.9.8914, Python: 2.6.6, Matplotlib: 1.0.0, NumPy: 1.4.1.

References

Ananthanarayanan R, Esser SK, Simon HD, Modha DS (2009) The cat is out of the bag: cortical simulations with 10^9 neurons and 10^{13} synapses. In: Supercomputing 09: proceedings of the ACM/IEEE SC2009 conference on high performance networking and computing, Portland

Bi Gq, Poo Mm (1998) Synaptic modifications in cultured hippocampal neurons: dependence on spike timing, synaptic strength, and postsynaptic cell type. J Neurosci 18:10464–10472

Bower JM, Beeman D (1995) The book of GENESIS: exploring realistic neural models with the GEneral NEural SImulation system. TELOS, Santa Clara, CA

Brette R, Rudolph M, Carnevale T, Hines M, Beeman D, Bower J, Diesmann M, Morrison A, Goodman P, Harris F, Others (2007) Simulation of networks of spiking neurons: a review of tools and strategies. J Comput Neurosci 23(3):349–398. http://www.springerlink.com/index/C2J0350168Q03671.pdf

Brunel N (2000) Dynamics of sparsely connected networks of excitatory and inhibitory spiking neurons. J Comput Neurosci 8(3):183–208

Diesmann M, Gewaltig MO, Aertsen A (1995) SYNOD: an environment for neural systems simulations. Language interface and tutorial. Tech. Rep. GC-AA-/95-3, Weizmann Institute of Science, The Grodetsky Center for Research of Higher Brain Functions, Israel

Eppler JM, Helias M, Muller E, Diesmann M, Gewaltig M (2009) PyNEST: a convenient interface to the NEST simulator. Front Neuroinform 2:12. doi:10.3389/neuro.11.012.2008

Gentle JE (2003) Random number generation and Monte Carlo methods, 2nd edn. Springer, New York

Gewaltig MO, Diesmann M (2007) NEST (Neural Simulation Tool). In: Izhikevich E (ed) Scholarpedia encyclopedia of computational neuroscience, p 11204. Eugene Izhikevich. http://www.scholarpedia.org/article/NEST_(Neural_Simulation_Tool)

Hines ML, Carnevale NT (1997) The NEURON simulation environment. Neural Comput 9: 1179–1209

Hunter JD (2007) Matplotlib: a 2D graphics environment. Comput Sci Eng 9(3):90–95

Knuth DE (1998) The art of computer programming, vol 2, 3rd edn. Addison-Wesley, Reading

L'Ecuyer P, Simard R (2007) TestU01: a C library for empirical testing of random number generators. ACM Trans Math Softw 33:22, doi 10.1145/1268776.1268777

Matsumoto M, Nishimura T (1998) Mersenne twister: a 623-dimensonally equidistributed uniform pseudorandom number generator. ACM Trans Model Comput Simul 8:3–30

Migliore M, Cannia C, Lytton WW, Markram H, Hines M (2006) Parallel network simulations with NEURON. J Comput Neurosci 21(2):119–223

Morrison A, Mehring C, Geisel T, Aertsen A, Diesmann M (2005) Advancing the boundaries of high connectivity network simulation with distributed computing. Neural Comput 17: 1776–1801

MPI Forum (2009) MPI: a message-passing interface standard. Tech. rep., University of Tennessee, Knoxville. http://www.mpi-forum.org/docs/mpi-2.2/mpi22-report.pdf

Oliphant TE (2006) Guide to NumPy. Trelgol publishing (Online). http://www.tramy.us/numpybook.pdf

Pérez F, Granger BE (2007) Ipython: a system for interactive scientific computing. Comput Sci Eng 9:21–29. doi:http://doi.ieeecomputersociety.org/10.1109/MCSE.2007.53

Plesser HE (2010) Generating random numbers. In: Grün S, Rotter S (eds) Analysis of parallel spike trains. Springer series in computational neuroscience. Springer, New York, Chap 19, pp 399–411

Plesser HE, Eppler JM, Morrison A, Diesmann M, Gewaltig MO (2007) Efficient parallel simulation of large-scale neuronal networks on clusters of multiprocessor computers. In: Kermarrec AM, Bougé L, Priol T (eds) Euro-Par 2007: parallel processing. Lecture notes in computer science, vol 4641. Springer, Berlin, pp 672–681. doi 10.1007/978-3-540-74466-5

Index

N. Le Novère (ed.), *Computational Systems Neurobiology*,
DOI 10.1007/978-94-007-3858-4,

N

T

FSC
www.fsc.org
MIX
Papier aus verantwortungsvollen Quellen
Paper from responsible sources
FSC® C105338

If you have any concerns about our products,
you can contact us on
ProductSafety@springernature.com

In case Publisher is established outside the EU,
the EU authorized representative is:
Springer Nature Customer Service Center GmbH
Europaplatz 3, 69115 Heidelberg, Germany

Printed by Libri Plureos GmbH
in Hamburg, Germany